Nonlinear Optimization of Vehicle Safety Structures

Modeling of Structures Subjected to Large Deformations

To my wonderful wife Rachael for being the love of my life
—Jesper Christensen

I would like to dedicate this book to my wife Athina and my children Lucas and Lydia for their unconditional love, patience, kindness and support throughtout my life
—Christophe Bastien

Nonlinear Optimization of Vehicle Safety Structures

Modeling of Structures Subjected to Large Deformations

Jesper Christensen

Christophe Bastien

ELSEVIER

AMSTERDAM • BOSTON • HEIDELBERG • LONDON
NEW YORK • OXFORD • PARIS • SAN DIEGO
SAN FRANCISCO • SINGAPORE • SYDNEY • TOKYO
Butterworth-Heinemann is an Imprint of Elsevier

Butterworth-Heinemann is an imprint of Elsevier
The Boulevard, Langford Lane, Kidlington, Oxford OX5 1GB, UK
225 Wyman Street, Waltham, MA 02451, USA

ISBN: 978-0-12-804424-7

British Library Cataloguing-in-Publication Data

A catalogue record for this book is available from the British Library

Library of Congress Cataloging-in-Publication Data

A catalog record for this book is available from the Library of Congress

For information on all Butterworth-Heinemann publications
visit our website at http://store.elsevier.com/

Typeset by Thomson Digital

Table of Contents

Preface

Automotive safety engineering has moved a long way since the 1960s, when the airbag system was invented to reduce occupant injuries due to the lack of seatbelt usage. The introduction of legislative standards and crash test dummies in vehicle safety assessment has initiated a transformation of the safety landscape forever, creating a revolution and evolution of vehicles that are safer from one generation to the next. The introduction of consumer tests and insurance test ratings presented methods for directly assessing the chances of survival in the case of a vehicle accident. This was done by introducing a "star rating" to rank the relative safety of each vehicle, which has a direct link to the insurance premium of a specific vehicle, thereby making safety a direct and measurable variable to be considered when purchasing a new vehicle.

Despite these significant advances in vehicle safety, almost 2 million people die as a result of vehicle-related accidents each year, clearly indicating that a momentous task related to vehicle design, road infrastructures, and law enforcement still remains.

Vehicle architectures have evolved from ladder frame structures to monocoques. The material "journey" has ranged from mild-grade steels to high-strength steel and boron steels, with the "latest trend" being more "exotic," materials such as aluminum, magnesium, or even carbon fiber. Each material generally has one or several unique selling points in the areas of volumetric mass density, stiffness, or strength. Tall trees can, however, not grow into the sky, and a unique advantage often comes with one or several disadvantage(s) such as expensive raw material and the cost of manufacturing, all of which are passed onto the end user.

Apart from a brief period in the 1970s, vehicle mass and fuel economy was of little concern to the everyday motorist throughout the 20th century. Only fanatics craving the ultimate in performance, handling, and agility considered the mass of a vehicle. As the demand for increased levels of safety, comfort, and "gadgets" such as electric seats and infotainment systems has risen, so has the mass of the vehicles. In the meantime, the demands for emission reductions and increased fuel efficiency have also significantly increased, forcing vehicle manufacturers to resolve very complex and generally contradictive problems of enhanced safety and comfort levels with reduced mass.

Initially, the general response from the automotive manufacturers was to replace the "dense" steels used for vehicle manufacturing with "lighter" alternatives such as aluminum or to utilize high-strength steels, thereby enabling a reduction in the volume of material used. Traditionally, metals with lower volumetric mass densities are less stiff than their steel counterparts. Producing vehicle safety structures using these "non-steels" is, however, complex; they are, for example, difficult to deep draw to regain a stiffness equivalent to their steel counterparts. Although the "non-steels" are widely used, the difficulty associated with the manufacturing process has led vehicle manufacturers and material suppliers to explore other avenues such as twinning-induced plasticity (TWIP) steels. TWIP steels are very strong as well as very ductile, making them suitable for pressing of deep and complex geometry. All these developments have, in some respect, been a consequence of the evolution of vehicle safety, aiming to maximize the vehicle cross section shapes for energy absorption while reducing mass. High-end luxury vehicle manufacturers can "afford" non-steel materials, as their vehicles are sold at a premium providing larger profit margins. On the other hand, steel materials are mostly used for less expensive vehicles that are generally produced in significantly larger numbers, as steels are more affordable and mechanically suitable for all automotive applications. The search for a lightweight, safe, and easy-to-manufacture vehicle structure has, therefore, never been greater. This inherently leads to an increased interest and a need for optimization techniques from mechanical and automotive engineers (as well as countless other disciplines). This desire has, of course, also increased in the context of the exponential rise in computing capacity.

Despite the introduction of optimization techniques, it should be remembered that the "fundamental" vehicle layout using a front-mounted engine has largely remained unchanged for the past 50–60 years. This inherently means that "traditional" modern-day vehicle structures have been continuously refined (and optimized) for over half a century. This does not, however, eliminate the usage of optimization techniques for "traditional" vehicle layouts; this merely means that careful considerations should be made as to where they may be efficiently applied.

Since the turn of the millennium, there has been an escalation in the interest in electrical and hybrid vehicles, as well as very lightweight "L" class vehicles aimed at transporting one to two passengers. Standard equipment on almost all new vehicles includes autonomous emergency braking (AEB), which will reduce and, in some cases, mitigate accidents; furthermore, fully autonomous vehicles have also started to emerge. There is certainly no denying that the automotive landscape is dramatically changing and is at present at a turning point; future vehicles do not have to follow current vehicle architectures. Despite the great technological leaps that have been made in the last 50–60 years, it can be noted that the vehicle architecture has effectively stopped evolving. The first electrical vehicles produced at the turn of the 21st century are still based on a "standard" monocoque chassis. The question is whether or not the conventional vehicle layout and structure really is the optimum way forward?

Surely with enhanced computing capabilities, optimization methods, vehicle segments, and opportunities for redistribution of, for example, powertrain, the opportunities to redesign and re-invent the next generation of vehicles have never been greater. Although learning from past experience is important, the opportunities for creating increasingly efficient vehicle structures in terms of lightweighting and safety performance in addition to improving handling and performance, for example, through mass distribution must be explored; the vehicle structures of the future need to start from afresh – a clean sheet.

This book, "Nonlinear Optimization of Vehicle Safety Structures," will present the basis of loadpath generation inside the bounding vehicle architecture with an emphasis on safety and will provide salient optimization methods that are aimed at re-thinking the design of future vehicle architectures. These techniques are in no way exhaustive, and the methods proposed in this book aim to provide the reader with an understanding of what can be computed with confidence and what can be estimated due to algorithmic limitations.

The work presented in this book is partially inspired by two research projects funded by the Engineering and Physical Sciences Research Council (EPSRC), the UK's main agency for funding research in engineering and the physical sciences; Advantage West Midlands; and the European Regional Development Fund (ERDF). The "Low Carbon Vehicle Technology Project" (LCVTP) focused on the lightweighting of vehicle architectures in general. Part of the "Towards Affordable, Closed-Loop Recyclable Future Low Carbon Vehicle Structures" (TARF-LCV) project was to investigate more detailed structural optimization opportunities and techniques in the context of recycled materials for automotive manufacturing. Aspects of this work were performed with the collaboration of researchers and students from Coventry University; special recognition and acknowledgements go to Mr. Amit Prem and Mr. Oliver Grimes, who have supported some of the analytical work presented in this book.

Throughout the chapters, the reader will observe that errors may originate from the computation itself and in some cases in the interpretation of the results as a consequence of the type of finite element (FE) solver used. Understanding the limitations is fundamental to future vehicle design, as the engineer should ultimately decide the way forward based on the outputs from the optimization – not the other way a round!

This book suggests that the basis of future vehicle design is an optimization method denoted called topology optimization. This method is theoretically capable of determining optimum load path locations, which is perfect when a new design needs to be created from an outer design envelope, as the proposed solutions are often organic and minimalistic. The subject of localized structure optimization techniques is also discussed, including methodologies aimed at reducing variables and runtime is also presented.

Welcome to the future of vehicle architecture design – the journey starts on the next page.

Chapter | one

Vehicle Architectures, Structures, and Safety Requirements

CHAPTER OUTLINE

1.1 INTRODUCTION

Motor vehicle safety on the roads has greatly improved in the past 40 years. Statistics have shown that the number of fatalities per billion kilometers traveled has reduced by around 80% since 1970 (International Transport Forum, 2013), as illustrated in Figure 1.1.

These safety static improvements have been achieved thanks to concurrent and complementary engineering approaches:

1. The enhancement of the vehicle structure strength to reduce intrusions within the occupant safety space.
2. The management of decelerations and forces exerted on the occupant.
3. The enhancement of the vehicle structure stiffness for ride and comfort handling.

These improvements have been influenced by the introduction of better passive safety measures, for example the vehicle structure coupled with the restraint system limit structural intrusions in the cabin area. Coupling the occupant to the seat that has the effect of enabling a better engagement with the airbag system (Stubbs, 2013), (NHTSA, 1998) as well as mitigating occupant ejection (Crandall, 2013; Neal-Sturgess, 2013; NHTSA, 2013a). It is also suggested that speed management and effective drinking and driving policies have reduced

Nonlinear Optimization of Vehicle Safety Structures. http://dx.doi.org/10.1016/B978-0-12-804424-7.00001-4

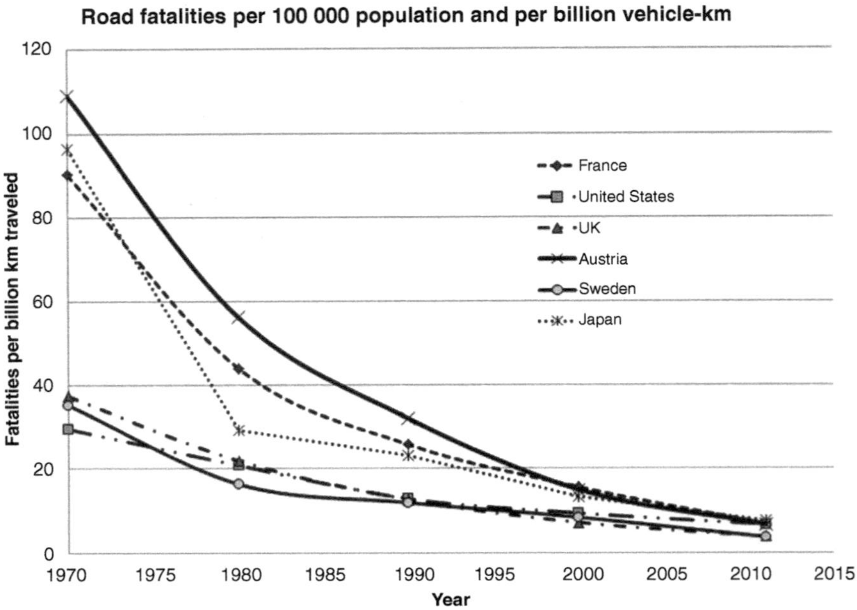

FIGURE 1.1 Fatalities per billion km traveled (1970–2010) (Bastien, 2014a).

fatalities by nearly half between 2000 and 2010 (International Transport Forum, 2013).

The first level of protection was implemented through passive safety; this is how vehicles are able to reduce accident severity without any driver interventions. Three distinct periods in the development history of automotive safety (American Iron and Steel Institute, 2004) have been observed.

The first period started from the turn of the 1900 until 1935, when developments to understand the extremely complex process of vehicle collisions, including the forces involved and the concept of energy absorbing capability of the vehicle structure, were researched. This led to the first crash test ever conducted in the early 1930s (American Iron and Steel Institute, 2004). Nevertheless, only basic automotive vehicle improvements were implemented. They included the reduction of tyre blowouts to avoid loss of vehicle control, the self-starter to eliminate injuries caused with engine cranking, the incorporation of headlamps to provide visibility at night, the installation of laminated glass to reduce facial lacerations, and the adoption of an all-steel body structure for better crash protection (American Iron and Steel Institute, 2004).

The second period spread from 1936 to 1965, when manufacturers introduced crash avoidance devices including turn signals, dual windshield wipers, and better headlamps. The safety of the vehicle interior was also improved by performing tests to simulate head impact into the instrument panel and engineering high penetration-resistant windshield glass. Means of restraining the occupants in the vehicles were believed to be important for occupant safety. In 1956,

Swedish inventor Nils Bohlin, who worked for Swedish manufacturer Volvo, designed an effective three-point belt (Happian-Smith, 2002), having demonstrated from accident statistics that unbelted occupants sustained fatal injuries throughout the whole range of speed scale (up to 60 m/h). His invention was granted US Patent 3,043, and in 1959 seatbelts were fitted as standard equipment. Additional research by General Motors was performed on car-to-barrier frontal crash test (launching a vehicle into a retaining wall), leading to improved vehicle structural performance based on observations of the crushed vehicle. The implementation of airbags started at the beginning of the 1960s to address the low seatbelt usage by implementing a safety device that needed no occupant intervention (Happian-Smith, 2002).

The third period spreads from 1966 until present day. It started with the creation of the National Highway Traffic Safety Administration (NHTSA) in 1966. The Federal Motor Vehicle Safety Standards (FMVSS) were introduced to regulate several aspects of vehicle crashworthiness and crash avoidance performance. As such, the collective vehicle safety technologies, together with improvements to highways and better driver education, have contributed to a large drop in fatalities (Figure 1.1). To support the development of vehicle safety various computer codes became available for numerical analysis; however they were still at their infancy. In 1990, a realistic adequate set of tools to address predictive vehicle safety design improvements became available and included contact algorithms, strain rate dependent material models (Cowper-Symonds), improved components self-contacts (Figure 1.2) (NCAC, 2013), as well as the first uniform pressure airbag model (LSTC, 2012).

In 1994, Arbitrary Lagrangian and Eulerian (ALE) were made available; this was the dawn of multiphysics applied to the field of vehicle safety. ALE codes allow the coupling fluid to structural element, for example the modeling of gas pressure expelled from the pyrotechnic ignition against the airbag cloth, allowing a more realistic modeling airbag deployment. In 1995, airbag venting

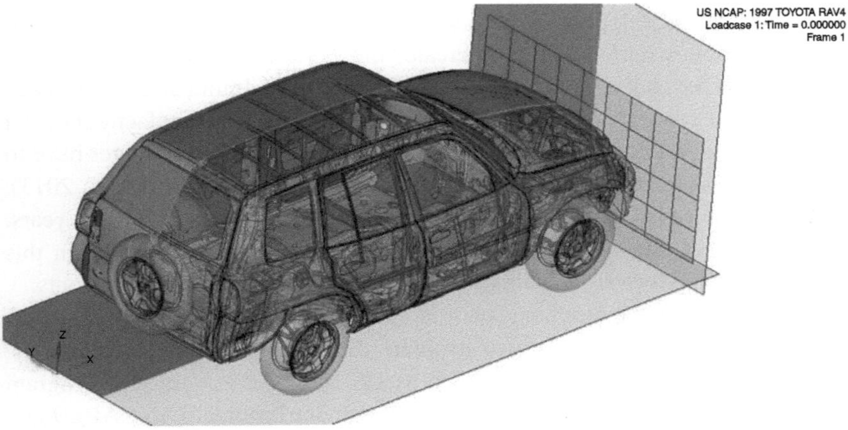

FIGURE 1.2 Example of standard CAE crash analysis. Rav4.0 NCAC crash model (US NCAP).

capabilities and improved airbag fabric materials were now available. The coupling capabilities between fluid and structure were, however, not implemented until 2000 by TASS with the "Gas Flow 1" product, as part of the MADYMO software (Mahangare et al., 2007; Mahangare, 2007).

Automotive vehicles have vastly changed in the last 40 years thanks to legislative requirements and consumer testing that have led to the reduction of fatalities on the roads. This achievement would have been impossible without the extensive use of computer modeling which is able to include a multitude of safety requirements expected by the Original Equipment Manufacturers' (OEM).

Many OEMs used these computer tools to improve the safety of their vehicles. In 1988 at Mercedes-Benz, the maximum vehicle model size was 10,000 elements, 80,000 in 1994, and 500,000 in 2000 (Du Bois, 2010). Nowadays (2015), OEMs run models of over 6,000,000 elements.

Due to all these engineering measures, since 1970 the fatality rate per 100 million miles traveled has dropped by 80% (International Transport Forum, 2013).

1.2 LEGISLATIVE REQUIREMENTS

All vehicles in Europe must pass legal safety requirements and obtain a Vehicle Certification Agency (VCA) 'approval certificate' prior to the vehicles being mass produced (VCA, 2013). In the United States, vehicles can be traded on the trust that they meet legal requirements. NHTSA (NHTSA, 2013b), part of "the Office of Defects Investigation, or ODI and as the US Department of Transport National Highway Traffic Safety Administration," can "conduct defect investigations and administer safety recalls on everything from vehicles and equipment to tires and child safety seats" (NHTSA, 2013b). It has to be noted that legal requirements between Europe and the USA are different and that some vehicle tests can be more stringent than others, for example interior head impact (FMVSS201, 2013) and vehicle frontal impact with unbelted occupant (FMVSS208, 2013). In 2010, discussions between the European Union and the United States started aiming to create "the world's largest free trade zone" and alleviate the lack of general standards (BBC, 2013). Initial discussions have started on the requirements for vehicle brake lights, as car manufacturers from Ford to BMW want standardized safety regulations so they "no longer have to manufacture separate parts to adhere to different safety standards" (BBC, 2013). A lot more is needed to standardize the legislation and this will take many years. The concept of Global Technical Regulations (GTR), discussed later in this chapter, is a step forward towards the integration of legislative requirements.

Several consumer programs such as European New Car Assessment Programme (EuroNCAP, 2013) also evaluate the safety level of new cars using laboratory crash tests. Others such as US New Car Assessment Program (US-NCAP), Australianasian New Car Assessment Program (ANCAP), Japanese New Car Assessment Program (JNCAP), and the Insurance Institute for Highway Safety (IIHS, 2013) perform similar tests, but their safety target limits usually exceed the legal safety requirements. Since 1996, EuroNCAP has become an important player in Europe and in the rest of the world in the improvement of vehicle safety. EuroNCAP has, since its introduction and up to 2009, tested in excess of 300 of the best-selling car models in Europe (EuroNCAP, 2013). It has to be noted at this point that a vehicle just meeting legal requirements may have a very poor consumer safety rating, as consumer tests' safety targets are in all cases much more stringent than the legislative ones.

This vehicle safety improvement trend has also been observed in a Swedish Folksam report (Folksam, 2013) based on data from both real-life accidents and crash tests, including a collision sample based on 105,000 car accidents that occurred between 1995 and 2008. This study involved 29,000 injured persons and considered data from two-car collisions; hence, the outcome of the collision is determined by the vehicle crashworthiness features and masses. Folksham's philosophy is that crash tests do not always correspond 100% to reality, they also do not take into account cross-category accidents, such as sport utility vehicles (SUV) accidents against super-minis. Consequently, it is advocated that a vehicle "should be chosen on the basis of the results of real-life accidents, and secondly on the basis of crash test results" (Folksam, 2013). It has been observed that "more modern cars usually have considerably higher safety standards and consequently fewer injured persons. A new small car may therefore be as safe as or safer than an old large car, but various makes and models differ considerably" (Folksam, 2013).

These recommendations are important for the customer who would wish to purchase a vehicle, but not very useful for the vehicle manufacturer, as they need a metric to design their vehicles.

Some further research has suggested that there was a positive link between the safety rating of vehicles and their performance in real-life accidents. It was observed that "5-star rated Euro NCAP cars were found to have a lower risk of injury and fatalities compared to 2-star rated vehicles" (Kullgren et al., 2010). This therefore confirms that OEMs are focusing their safety performance on the serious crash outcomes, as there is a "good concordance between Euro NCAP and Folksam real-world crash and injury ratings" (Kullgren et al., 2010).

To illustrate the comparison between legislative test severity and real-life accidents, a Dodge Neon, part of the NCAC database (NCAC, 2013), was chosen as bullet vehicle against Ford Fiesta of comparable mass (Bastien et al., 2013) and illustrated in Figure 1.3.

Numerical accident simulations scenarios were performed to compare a standard rigid wall impact test at 35 mph (56 km/h) compared with standard high-speed rear-impact scenarios (Bastien et al., 2013).These scenarios were based on the Assessment of Vehicle Safety Systems (ASSESS) FP7 European project (ASSESS, 2012a; ASSESS, 2012b), which categorized three categories of accident findings separated in accidents (Table 1.1):

- Rear collisions at constant speeds (A1A and A1C accident types)
- Rear collisions decelerating lead vehicle (A2A and A2D accident types)
- Rear collisions stopped lead vehicle (A3A and A3C accident types)

FIGURE 1.3 A1C impact between Neon (bullet) and Fiesta (target) vehicles (Bastien et al., 2013).

In order to record the deceleration levels, accelerometers are fitted into the computer model in nondeformable areas of the structure. The position selected is at the base of the "B" pillar, which is a standard location area used in physical crash tests.

From the assessment of the accidents recorded in Figure 1.4, it can be noted that most of the accident acceleration crash pulses recorded on the B pillar do not exceed 25 g in all cases, and hence are less severe than a 56 km/h rigid wall impact. For the accident cases A1C and A3C (Figures 1.3 and 1.4), the ASSESS accident deceleration values are the highest, which is expect as the relative velocity ΔV between both vehicles is the highest, which was also confirmed in other research (Berg and Rucker, 2012).

It can also be noted that for ΔV values of 40 km/h, which is the impact speed stipulated by FMVSS208 unbelted tests (impacts A1A), crash pulses recorded on the B pillar do not exceed 10 g.

Looking at the NCAC report for the Dodge Neon (NCAC, 2013), it can noted that the seat cross member average acceleration against a rigid wall at 56 km/h is around 34 g, at 50 ms after initial impact (Figure 1.4). One can therefore conclude that the accident patterns suggested by standard rear-end accident impact postulated by ASSESS are less severe than a standard full rigid barrier test conducted at 56 km/h, even in case A3C when the relative impact velocity ΔV is 80 km/h. This is to be expected, as the bullet vehicle is impacting a deformable structure. From these analyses, designing vehicle structures using a 56 km/h impact speed against a rigid wall seems a reasonable and safe approach.

Considering the road fatality trend from 1970 to 2010 (Figure 1.1) and the latest results published in the Road Safety Annual Report 2013 (Figure 1.5) (International Transport Forum, 2013), it can be observed that the number of fatalities overall is now steady with an average of 5.6 road fatalities per billion vehicle-kilometer (ignoring Czech Republic and Korea).

The passive technology has reached full maturity and structural loadpaths are very well understood (Grimes et al., 2013), especially in the safety cell area

TABLE 1.1 Real-Life Accident Assessment (ASSESS FP7)

Analysis Type	ASSESS Impact Code	Overlap Between Bullet and Target Vehicles (%)	Target Vehicle Speed (km/h)	Bullet Vehicle Deceleration (m/s^2)	Bullet Vehicle Speed (km/h)	Gap Between Bullet and Target Vehicles Prior to Braking (m)	Velocity of Target Vehicle After Deceleration and Just Before Accident (km/h)	Relative Speed Between Bullet and Target Vehicles (km/h)
Rear collision constant speed	A1A	100	10		50			40.0
	A1C	100	20		100			80.0
Rear collision decelerating lead vehicle	A2A	100	50	4	50	14	39.4	10.6
	A2D	100	80	4	80	45	61.0	19.0
Rear collision stopped lead vehicle	A3A	100	0		50			50.0
	A3C	100	0		80			80.0

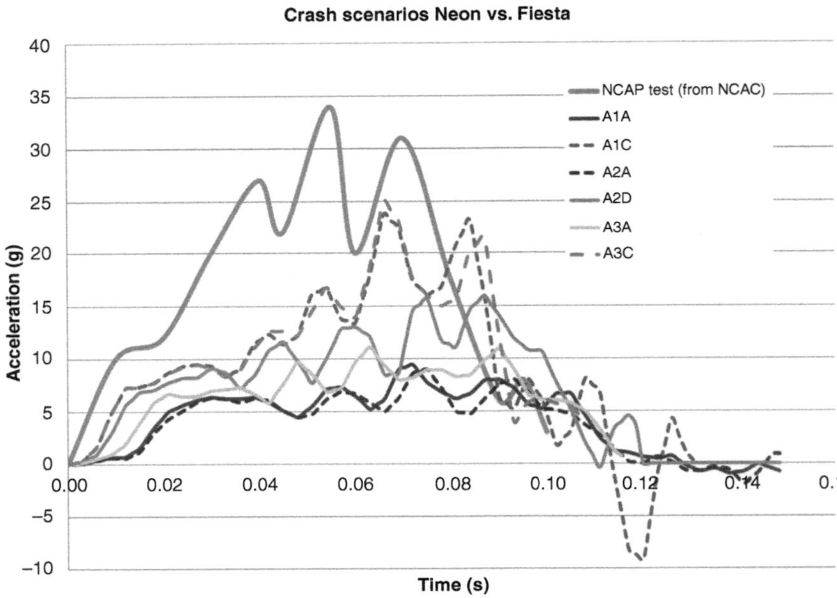

FIGURE 1.4 Assessment of vehicle collision between a Dodge Neon (bullet vehicle) and a Ford Fiesta (Bastien et al., 2013).

(Christensen et al., 2012a; Christensen et al., 2012b; Christensen et al., 2012c). The challenge for the vehicle manufacturers is to balance the impact energy absorption and the vehicle mass. Consider, for example, an Audi A2, EuroNCAP safety rating is only a 4 Star (EuroNCAP, 2013). The body-in-white (BIW) of an Audi A2 accounts for 20% of the total vehicle mass, hence increasing structural rigidity can only be achieved as part of the main structure, which will have a direct repercussion on the vehicle mass, with no obvious opportunity to remove vehicle mass elsewhere. The mass increase would lead to an augmentation of Greenhouse Gas (GHG) emissions, hence increasing the challenge of passing the EU vehicle emission standard requirements Euro5 (Delphi, 2012).

One of the biggest challenges to be met for lightweight electrical vehicles is safety crash compatibility (Grimes et al., 2013) when lighter vehicle collide against heavier ones. In such cases, cabin space intrusion can be a major concern (Figure 1.6), as the occupant survival space can be seriously compromised.

Some earlier work has shown that the safety cell stability is currently difficult to achieve (Grimes et al., 2013), due to un-optimized loadpath and, in the case of super lightweight vehicles, the excessive use of glass fiber for the door-opening panels (Figure 1.6), chosen for its lightweight properties and ease of manufacturing for low production volumes. Some changes in the regulations would be needed to address the safety of future lightweight vehicle during their introduction in the current road network.

In order to sell a vehicle, the first thing to understand is which market this vehicle is going to be sold in, as legislation requirements vary depending on each market.

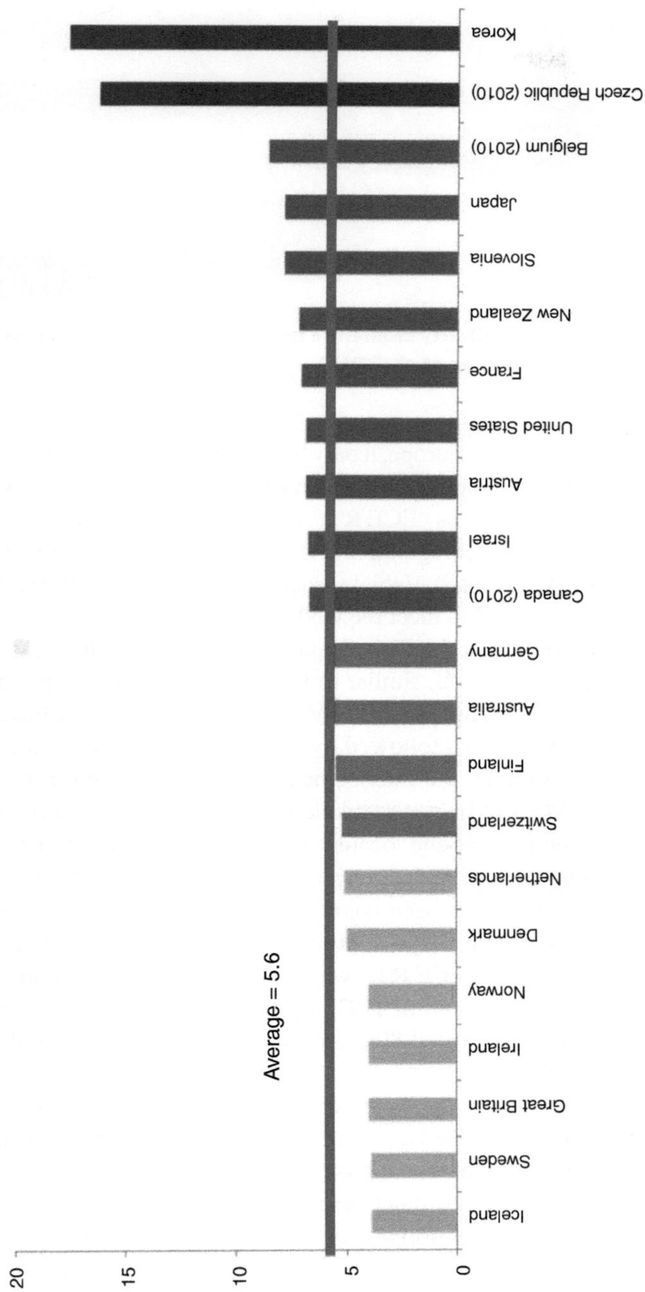

FIGURE 1.5 Road fatalities per billion vehicles-kilometers in 2011. (Bastien, 2014a).

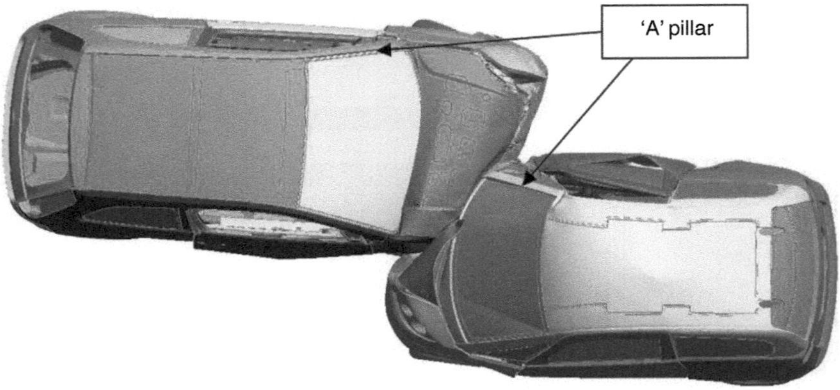

FIGURE 1.6 Crash compatibility challenge between a standard 'C' class vehicle and a lightweight EV (Grimes et al., 2013).

When a vehicle is sold in Europe, it only has to satisfy the United Nations Economic commission for Europe (UNECE), in short ECE (Crash Network, 2013). All the regulations are listed as "ECE R" ("R" stands for Regulation) and then a number which relates to a set regulation. No vehicle can be sold until the "ECE" stamp from the VCA is granted. Vehicles such as Renault, Peugeot, and Fiat, not sold in the USA, only need to meet the UNECE requirements.

A vehicle that is sold in the United States or Canada needs to meet safety regulations that are, in general, similar to UNECE in philosophy, but different in practice. US regulations are labeled "FMVSS," which stands for Federal Motor Vehicle Safety Standards, followed by a number that relates to the specific regulation. As an example, the interior head impact legislation in the United States is labeled FMVSS201; this regulation aims to protect the occupant from interior injuries. The number and location of head impact points to be analyzed is set; however some other impact areas may be chosen by the testing body to ensure that due diligence has been considered. This test assesses against head injury criteria (HIC), which will be discussed later in this chapter. In the EU, the closest regulation is labeled ECE R21, which uses a set number of impact points with an 80 g/3 ms clip instead of HIC. This criterion stipulates that no head impact deceleration must exceed 80 g for a duration of 3 ms during the whole duration of the impact.

This can be problematic as there is no straight mathematical relationship between these two injury criteria, and hence could be challenging to engineer should a vehicle be sold in both markets (US and Europe).

As previously mentioned, other levels of compliance can be envisaged, which the Global Technical Regulations (GTR) developed under the 1998 Global Agreement, aim to harmonize. GTR are mainly focusing on performance-oriented test procedures designed to quantify product behaviors as objectively as possible, and in some cases setting a defined level of performance or defining a set limit value for the results of the test procedures. This is the

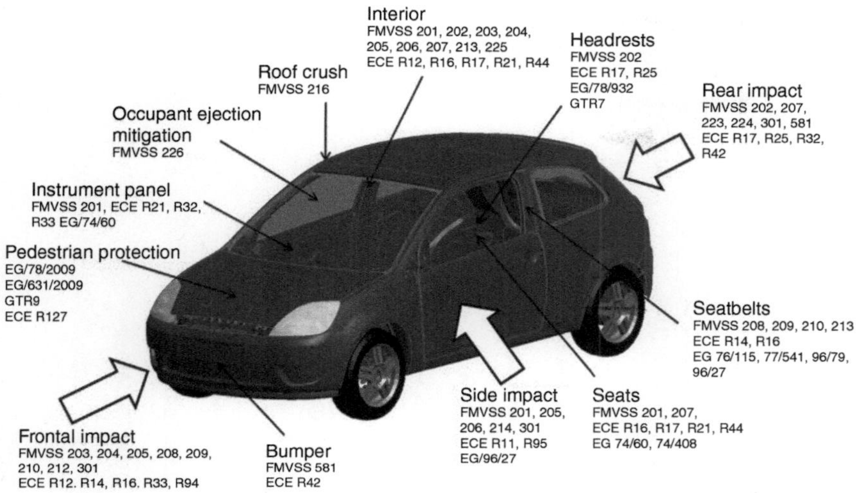

FIGURE 1.7 Location of EU and US legislative requirements.

case for the pedestrian protection regulations, which has a huge influence on the vehicle design and styling. The use of Global Technical Regulations is entirely voluntary, although Contracting Parties to the 1998 Global Agreement are nominally expected to transpose GTR specifications into their local legislation.

A summary of some of the US and EU regulations is given in Figure 1.7.

It is interesting to note that most FMVSS legal requirements have their ECE counterparts, except for Occupant Ejection Mitigation (FMVSS226) and Roof Crush (FMVSS216). This is due to the fact that, in the United States, 60% of fatalities in SUVs are caused in rollover conditions, as recorded by the National Automotive Sampling System (NASS) (NHTSA, 2013c). This could be caused by the fact the center of gravity of SUV vehicles is higher than sedans and super-minis, making them more prone to rollover. Another parameter may be that in Europe, the guardrails on main roads and motorways are very close to the verge. In the United States these guardrails are often further away, and do not steer the vehicle back into the road.

In the event of vehicle rollover and subsequent airbag curtain deployment, it must be ensured that the occupant remains within the vehicle. To certify that the occupant is not ejected, a dummy headform is fired towards the deployed airbag from the inside of the vehicle. A successful outcome is that the headform is contained within the vehicle (FMVSS226).

To assess the vehicle structural strength in the case of a rollover scenario, the roof is crushed using a plenum (FMVSS216) on the top of the "A" pillar (Figure 1.7), at dictated angles by legislation, with a force 3.5 times the empty weight of the vehicle. During the test, the roof must not deflect more than 250 mm (10 in), as measured by the movement of the test device.

The regulations usually specify occupant injuries (Table 1.2) as a means to assess the safety soundness of a vehicle. Occupants' injuries will be discussed in the next section of this chapter.

TABLE 1.2 Frontal Crash EU and US Legal Legislative Targets

Regulation/Test Modes	Market	Occupant Size	Tested Item	Requirements		Unit
UN-ECE Regulation 94.01	ECE	50th		Head performance criteria (36 ms)	1000	HIC
	ECE	50th		Head Acceleration (3 ms)	80	G's
	ECE	50th		Neck Extension (Upper)	57	Nm
	ECE	50th		Neck Tension Index (Mertz Table)	1	Index
	ECE	50th		Neck Shear Index (Mertz Table)	1	Index
	ECE	50th		V*C (Thorax Viscous - criterion)	1	m/s
• Full vehicle 34.8 m/h (56 km/h) (40% frontal offset belted)	ECE	50th	Hybrid III 50th % male driver and passenger	Chest Deflection	50	mm
	ECE	50th		Femur Load Index	1	Index
	ECE	50th		Tibia/Knee Displacement	15	mm
	ECE	50th		Tibia Axial Compression	8	kN
	ECE	50th		Upper Tibia Index	1.3	Index
	ECE	50th		Lower Tibia Index	1.3	Index
	ECE	50th		Lower Tibia Moment	None	Nm

FMVSS 208 (New Rule 2014) • Full vehicle 0–30/35 m/h (0–48 km/h) (90° frontal belted)	USA	50th	Driver and Passenger	Head Injury Criteria (15 ms)	700	HIC
	USA	50th		Chest Acceleration (3 ms)	60	G's
	USA	50th		Chest Deflection	63	mm
	USA	50th		Femur Load	10,000	N
• Full vehicle 20–25 m/h (32–40 k/h) (90° frontal unbelted) (30° angular unbelted)	USA	50th		Neck Injury Criteria (Upper)	1.0	Nij
	USA	50th		Neck Tension (Upper)	4170	N
	USA	50th		Neck Compression (Upper)	4000	N
FMVSS 208 (New Rule 2014) • Full vehicle 0–30/35 m/h (0–48 km/h) (90° frontal belted)	USA	5th	Driver and Passenger	Head Injury Criteria (15 ms)	700	HIC
	USA	5th		Chest Acceleration (3 ms)	60	G's
	USA	5th		Chest Deflection	52	mm
	USA	5th		Femur Load	6805	N
• Full vehicle 20–25 m/h (32–40 km/h) (90° frontal unbelted)	USA	5th		Neck Injury Criteria (Upper)	1.0	Nij
	USA	5th		Neck Tension (Upper)	2620	N
	USA	5th		Neck Compression (Upper)	2520	N

Courtesy of Steve Gough, Jaguar Land-Rover.

Recently, the rollover test (FMVSS216) is being challenged by a group of researchers proposing the Jordan Rollover System (JRS), which simulates a rotating vehicle landing on a moving platen representing the road. It was proposed that "it (the JRS system) provides reliable roof crush and roof crush speed comparisons between vehicles; and it (the JRS system) measures cumulative roof crush data believed by many experts in rollover to be a function of head–neck system injury severity" (Shirwa, 2010). Only time will tell if JRS will replace the platen quasi-static test.

When designing a vehicle structure for safety applications, the initial approach is to reduce the intrusions to a minimum, as the occupant is very vulnerable. Any contact with stiff components, like the toe-board, dashboard, steering column, "A" pillar, and the like, could be injurious or fatal, even if the airbags are deployed. Avoiding all "hard" contacts is a first step in preventing fatal injuries. Once intrusions are mitigated, the next challenge is to manage occupant's deceleration levels, as a low compliance structure will inherently transmit excessive acceleration loads.

The optimum vehicle would maximize the structural intrusions, as long as no contact with the occupant are observed; this would lead to lower transmitted forces on the occupant and the design of a less aggressive restraint system.

1.3 OCCUPANT INJURIES

In general, human tolerance to accelerations is a function of five factors that are related to the crash pulse and the design of the seat and restraint system. These tolerance levels include five key components in the acceleration characteristics (Shanahan, 2013):

1. The magnitude of the acceleration:
   The higher the acceleration or deceleration, the higher the probability of injury.
2. The direction of the acceleration:
   It has been observed that human tolerance in the sagittal plane, that is, in the frontal direction, is more tolerable than vertical and lateral.
3. The duration of the acceleration:
   In general, human tolerance is better for a shorter duration pulse of the same magnitude. Acceleration tolerance is usually considered to comprise two distinct types, abrupt acceleration and sustained acceleration. Most crash impacts have a duration of less than 250 ms, which is considered to be in the realm of abrupt acceleration. Human tissue and the vascular system respond considerably different to these very short duration pulses than they do to more sustained pulses.
4. The rate of acceleration/deceleration:
   For a given magnitude and duration of acceleration, the greater the rate of onset, the less tolerable the acceleration.
5. Position/restraint/support:
   This is one of the most critical factors determining human tolerance to a crash pulse. It refers to how well the occupant is restrained and supported by the

seat and restraint system and the degree to which the loads experienced in the crash are distributed over the body surface. It is this factor that is the primary determinant of fatality even when the crash event is survivable, if postcrash fire is excluded.

Legislative pass and fail "criteria" focus on occupant injury levels that are based on injury criteria. The term criteria is very important, as this relates to a computed injury, based on accelerometer, force, moment and displacement measurements. These are therefore only indicators which do not necessary relate to a specific trauma.

1.3.1 The crash test dummy families (or the tools to capture injury criteria)

The injury criteria targets of crash test dummies, or anthropometric crash test dummies (ATD), aim to replicate human tolerance values that have been evaluated on injury corridors defined on human cadaver tests.

Crash test dummies (Figure 1.8) are engineering measuring devices used to assess occupant kinematics in acceleration field and predict the severity of potential real-world injuries. They are very useful to develop engineering solutions, optimize restraint systems, tune vehicle trim components, and so on. The crash test dummy family has evolved over the years and represents three percentiles of the human population: 5th percentile (roughly equivalent to a 12-year-old European female), a 50th percentile European male, and 95th percentile European male. A percentile represents a statistical measure indicating a value below which a given percentage of observations in a group are witnessed. Consequently, a 5th percentile female ATD represents, statically, the shape of the female from the first 5% from the whole female population. The 5th and 95th

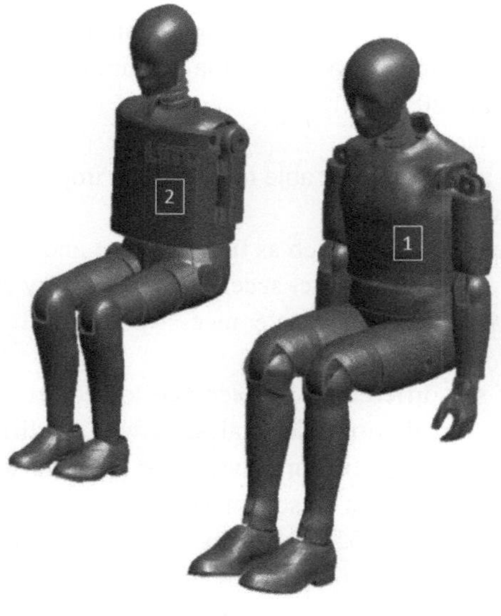

FIGURE 1.8 LSTC crash test dummies (side impact dummy - left; Hybrid III, right).

percentiles are also equal to two standard deviations on either side of the mean. The 50th percentile represents an average male (based on 1949 American anthropometry), while a 95th percentile ATD represents the first 95% of the whole male population (or two standard deviations), neglecting the last 5% who are the tallest people.

What is important to remember is that ATD are machines which provide repeatable results, meaning that, for an exact test configuration, a consistent response would be output. As these are machines, there provide no ethical issues compared to performing crash tests with cadavers or low deceleration volunteer sled tests, for example.

In general, these crash test dummies are nonbreakable; hence, they would not be able to, for example, represent a femur breakage in a vehicle crash. It would, nevertheless, record whether the load through the femur is high enough to warrant concern.

Some of the biofidelity is however questionable, especially in the neck area. In the case of the Hybrid III, compressive neck loads are not representative, as the current Hybrid III neck construction assumes the neck to be a beam with a compression load slightly off-center its center of gravity. The human neck is completely different and is "S" shaped.

Some new dummies are being designed and tested, such as the WorldSID or the THOR dummy, striving to replicate more human features.

It is important to recognize that these crash test dummies are not human surrogates, but engineering "machines." There is a specific crash test dummy for each loadcase, as they are not omni-directional. Each of them favors an impact acceleration direction. A Hybrid III ATD will be used for a frontal crash scenario, as depicted as crash test dummy family "1" in Figure 1.8. A side impact dummy (SID), as depicted as crash test dummy family "2" in Figure 1.8, will only be suited for a lateral loadcase. It can be noted in Figure 1.8 that this dummy does not have any arms, while a frontal dummy has.

These crash test dummies have some major limitations, as they do not take into account (Crandall, 2013):

- Age of the subject:
  Young, healthy adults are best able to withstand impact accelerations.
- Health of subject:
  Chronic medical conditions such as heart disease and osteoporosis degrade one's ability to withstand impact accelerations. History of previous injuries may also adversely affect one's tolerance.
- Sex of subject:
  There are clear sex differences in tolerance to acceleration. Females have a different mass distribution than males, as well as differences in muscle mass. This has been of particular concern for the neck where females have approximately one third less muscle mass than males of comparable stature. UK statistics suggests that 35% of males involved in vehicle crashes will be killed or seriously injured compared with 53% of females (Department for Transports, 2011a; Department for Transports, 2011b).

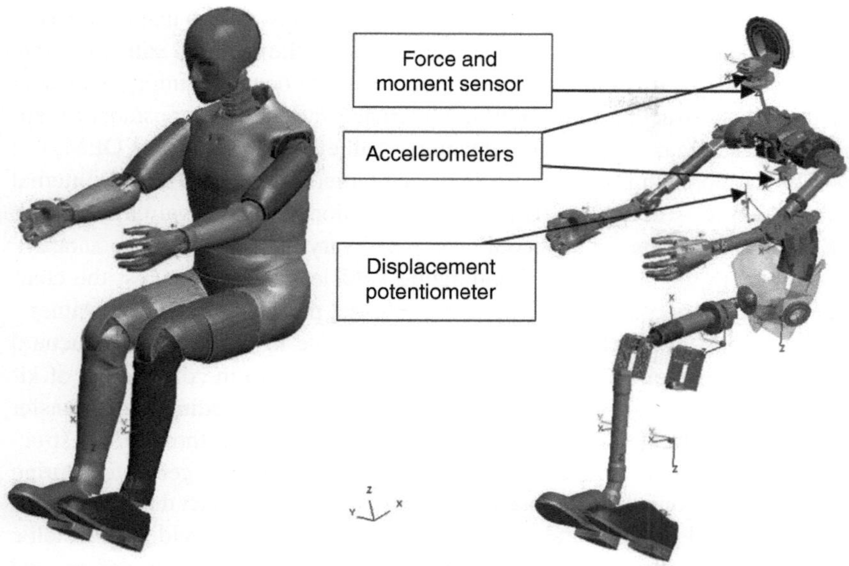

FIGURE 1.9 Hybrid III 50th percentile ATD (left) and accelerometer positions (right).

- Physical conditioning:
  Physical conditioning appears to increase one's tolerance to both abrupt and sustained acceleration, probably due to increases in muscle mass and strength.

In order to read injuries, ATD models are fitted with accelerometers, displacement, force, and torque sensors (AGARD, 1996). Accelerations can also be inherently transformed into forces, provided that the acceleration values follow a known reference system.

A typical seated ATD and its respective sensors are displayed in Figure 1.9. It can be observed that the arms, legs, and pelvis are rigid steel structures, unlikely going to break during a crash event.

In spite of these limitations, ATD are useful engineering tools that have had a great influence in the reductions of occupant fatalities (Bastien, 2014b).

1.3.2 Typical injury criteria

According to the market segment in which the vehicle is sold, a specific set of injury targets are set by the legislation (Table 1.2).

It can be observed that in the EU market, the 50th percentile ATD is mostly used, while in the United States tests will also include the 5th percentile female. Furthermore, as discussed in this chapter, injury criteria between markets differ, which can cause a serious engineering challenge should a vehicle be sold both in the United States and Europe. It can also be observed that the European legislation requires more injury output than the American market.

When the vehicle is engineered, manufacturers must ensure that any vehicle variant selected by the testing body is able to meet the required safety criteria. Consequently, as it is impossible to compute all the occupant injury responses for all vehicle structure manufacturing tolerance variants, a more statistical approach is usually used in order to encompass all eventualities. Some OEMs use a 6-Sigma approach for their vehicle design targets and ensure that all internal certification tests lie within 3σ (standard deviations) of the legal target. These OEM internal targets are kept confidential and vary from one OEM to another.

To add to the challenge of optimizing vehicle safety structures, the computation does not often exactly predict the 'exact physical' crash test "values". A vehicle impact usually involves high strains due to the speed of structural deformation as well as the large deformations related to the dissipation of kinetic energy. Impact events vary during the time domain, leading to the transfer of forces, structural displacements, and therefore energies through the structure. Some complexity is added by the fact that damage is generated during the event; components' response is changing during the impact due to atomical changes caused by structural collapse. A typical example would be a metallic vehicle crush can that during the impact would start to plastically deform and catastrophically buckle leading to an uneven transfer of force inside the vehicle structure. This damage may also mean that some masses can detach having the effect of un-coupling some vehicle structural and ancillary masses during the accident, causing complex and unidentified outcomes. This example is common in vehicle impact scenarios; as such the problem that the engineers are faced with is a time-dependent nonlinear problem.

Consequently, other design and therefore optimization targets need to be set which should include the prediction "mismatch" between computer-aided engineering (CAE) and the corrected mean to the legal tests adjusted with the 6-sigma method. A hypothetical example is presented in Table 1.3 to illustrate the difficulty in design optimization which starts by setting an adequate CAE target. This example is based on occupant head injury.

It is therefore evident from Table 1.3 that designing a vehicle for occupant safety is challenging, as the CAE design targets injury values have to be set much lower than the legal requirement to ensure a consistent and guaranteed legal compliance.

TABLE 1.3 A HIC Legal Target That Has Been Offset for Design and CAE (Hypothetical Example)

	Legal Requirement	6-Sigma Method for Robust Vehicle Legal Delivery (Test)	CAE Design Target
Head injury criteria (HIC)	1000	800	750

All the requirements are mostly read directly from the crash test dummy outputs. Some legal injury requirements, such as the head injury criteria (HIC), the neck injury criteria (NIJ), and the tibia index (TI), require multiple signal inputs. The computation of these injury levels would differ according to the type of dummy used in the analysis. For example a 95th percentile male would be more "robust" than a 5th percentile female, hence will withstand higher injuries. Some of these compound injury computations are listed later in this chapter.

Different software using a multibody architecture would have joints to connect limbs to each other. Limbs forces are then extracted directly.

1.3.2.1 HEAD INJURY CRITERIA (HIC)

The HIC value is computed from the scalar values of the head accelerations located on the center gravity in X, Y, and Z. The HIC value is then computed using HIC Equation (1.1) and is not occupant percentile dependent

$$HIC = \max\left[\frac{1}{t_2 - t_1}\int_{t1}^{t2} a(t)\,dt\right]^{2.5} (t_2 - t_1) \tag{1.1}$$

1.3.2.2 HEAD INJURY CRITERIA FOR FREE MOTION HEAD FORM (HIC(d))

In the case of a surrogate head impactor, as in the case of the Interior Head Impact legislation (FMVSS201), HIC(d) is used and computed as Equation (1.2)

$$HIC(d) = 0.75446 \cdot HIC + 166.4 \tag{1.2}$$

1.3.2.3 NECK INJURY CRITERIA (N_{ij})

The neck injuries are based on a compound set of loads, which are neck forces and neck moments, and depend on the dummy percentile.

- The neck forces, on their own, will have the effect of either "pulling" the head away from the neck (tension) or "pushing" it closer (compression).
- The neck moments, on their own, will have the effect of either rotating the head "forward" (flexion) or rotating it "rearward" (extension).

The N_{ij} values are computed from the scalar values of the neck forces and moments located at the occipital neck location, which is a location in the neck below the neck force and moment transducer (Equation (1.3). As the distance of this point varies between occupant anthropometry, a bending moment correction factor is usually integrated.

There are four N_{ij} values (tension/ flexion, tension/extension, compression/ flexion, and compression/extension) that are based on the signs of neck vertical force and the neck moment. F_{int} and M_{int} are occupant percentile-specific reference values (Table 1.3).

TABLE 1.4 Critical Neck Forces and Neck Moments per Occupant Percentile

Dummy Type	Tension (N)	Compression (N)	Flexion (Nm)	Extension (Nm)
Hybrid III 5th percentile female	3370	3370	155	62
Hybrid III 50th percentile male	4500	4500	310	125
Hybrid III 95th percentile male	5440	5440	415	166

$$N_{ij} = \frac{F_z}{F_{\text{int}}} + \frac{M_y}{M_{\text{int}}} \tag{1.3}$$

N_{ij} is occupant percentile dependent, meaning that according to the ATD tested, the injury levels will vary, as listed in Table 1.4.

It can be observed that a 95th percentile occupant can withstand a higher level injury compared with a 50th percentile male and a 5th percentile female.

The N_{ij} injury value is very useful, as usually one of the four modes will be clearly higher than the other. By considering the N_{ij} injury values, it is possible to interpret the relationship between the occupant's head and a deploying airbag.

1.3.2.4 TI (TIBIA INDEX)

As the N_{ij}, TI is anthropometric dependent (Table 1.5) and is calculated in the tibia index Equation (1.4), where M_x and M_y are the bending moments at the tibia ends.

$$TI = \left| \frac{\sqrt{(M_x)^2 + (M_y)^2}}{(Mc)_r} \right| + \left| \frac{F_z}{(Fc)_z} \right| \tag{1.4}$$

The occupant's critical bending moment (Mc) and critical compression force (Fc) are listed in Table 1.5, affirming again that a 95th percentile occupant can withstand more loads than a smaller occupant before being injured

TABLE 1.5 Anthropometric Tibia Tolerances

Dummy Type	*Mc*: Critical Bending Moment (N.m)	*Fc*: Critical Compression Force (N)
Hybrid III 5th percentile female	307	44.2
Hybrid III 50th percentile male	225	35.9
Hybrid III 95th percentile male	115	22.9

When tests are conducted, accelerometers, displacement potentiometers, force, and moment transducers are utilized to record the crash test dummy injuries and produce raw output signals. In order to record a signal with a consistent and adequate level of noise, filters are used, and are based on the SAE J211 (SAE J211, 1995; SAE International, 1995) filter classes listed in Table 1.6.

It can be noticed that not all transducers in the dummy have the same levels of filtering. The higher the filter class, the less modified the signal. Consequently, using a CFC1000 will have less influence than a CFC600 (SAE J211, 1995). During a crash event, the vehicule structure has the tendency to "ring" because of high velocity and high stiffness components contacting each other. Consequently, these filters must be included in the analysis/optimization of the dynamic crash phase and the occupant injury assessment.

TABLE 1.6 Filter Classes Used in a Dummy Instrumentation for Crash Testing

Standard SAE J211 Filter Classes and Their Applications			
Structural Instrumentation	Accelerometers	Total vehicle comparison/simulations/ sleds	CFC60
		component analysis	CFC600
		integrations (velocity/displacement)	CFC180
	Load cells	Steering column	CFC600
		barrier force/seat belt load cells	CFC60
Dummy Instrumentation	Head Accelerations		CFC 1000
	Neck	Forces	CFC 1000
		Moments	CFC 600
	Thorax	Spine Accelerations	CFC 180
		Rib Accelerations	CFC 1000
		Sternum Acceleration	CFC 1000
		Deflections	CFC 600
	Lumbar	Forces/Moments	CFC 1000
	Pelvis	Accelerations/Forces/Moments	CFC 1000
	Femur/Lower Leg	Forces/Moments	CFC 600
		Displacements	CFC 600

Stubbs, 2013.

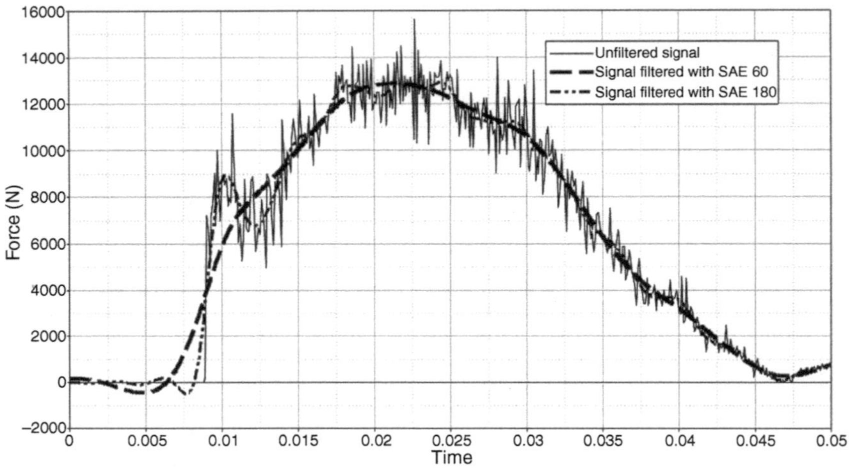

FIGURE 1.10 An example of filtering raw data. SAE 180 keeps initial signal shape, while SAE60 smoothes it excessively.

These filters remove the noise, but can have a tendency of over smoothing the signal (SAE J211, 2007). If the filter is too strong, the physical test response from the original raw data can be lost. This is well-depicted in Figure 1.10, where a signal is lightly filtered using SAE 180 and is strongly filtered using SAE 60 filters.

It can be observed that the signal filtered at SAE 180 is smooth and follows the original signal very well, even capturing the change of curvature at time 0.01 s. On the other hand, the signal computed with a SAE 60 filter is over-smoothed. Should the initial signal's curve inflection at time 0.01 s have an important physical meaning in the test, then using an SAE 180 would have been adequate, as this even can be visualized and interpreted, while using an SAE 60 filter would have been a bad choice.

1.3.3 Surrogate impactors

Some other crash test dummies are based on subsystems, that is, using subcomponents of the crash test dummy. This is the case for example of the FMVSS201 interior head impact legislation where the test relates to a Hybrid III 50th percentile head form impacting the vehicle interior upper environment. The test aims to evaluate the hard points of the cabin during a rollover scenario. Simulating this test with a full crash test dummy would unlikely be repeatable, hence the simulation of the head impact on its own against the vehicle interior. Some typical markers are presented in Figure 1.11 located at the top A pillar joint of a vehicle (AP1 and AP2), as well as the front header (FH1).

The injury criteria is modified, as no neck and no body mass are connected to this head form, leading to the HIC(d) injury criteria, which is a linear function of the standard head injury criterion (HIC), discussed previously in this chapter.

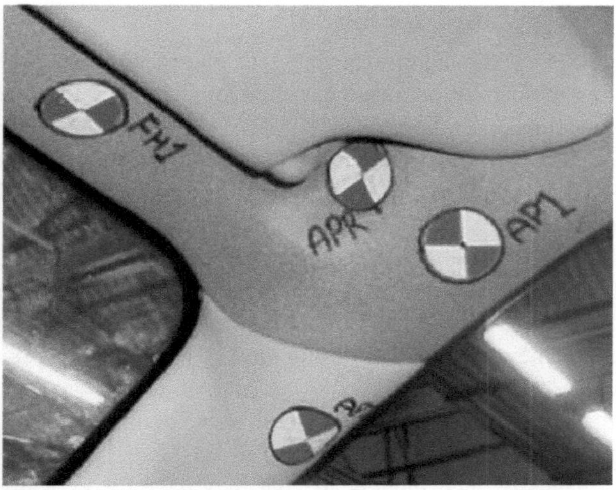

FIGURE 1.11 Top of "A" pillar FMVSS201 mark-up (Bastien and Williams, 2002).

Other methods of assessment are based on subsystems; this is the case with the pedestrian protection legislation where impacting a whole Hybrid III crash test dummy sideways would not be repeatable. As such, the pedestrian's protection is assessed via three separate "body parts," which are the lower leg, upper leg (or pelvis), and the pedestrian heads, which are illustrated in Figure 1.12 (UNECE, 2014). Whilst the head maximum HIC value set for a child and an adult are the same, that is, 1700, the head impact locations are different, as a child's head is likely to impact the front of the bonnet while an adult would impact the rear of the bonnet and even the windscreen.

Test protocols are very important. Even if an impactor is nonbreakable, like for example a pedestrian or FMVSS201 headform, as a general rule, at least 30 min must elapse between tests to allow the rubber skin to recover from the previous impact, due to creep and visco-elasticity.

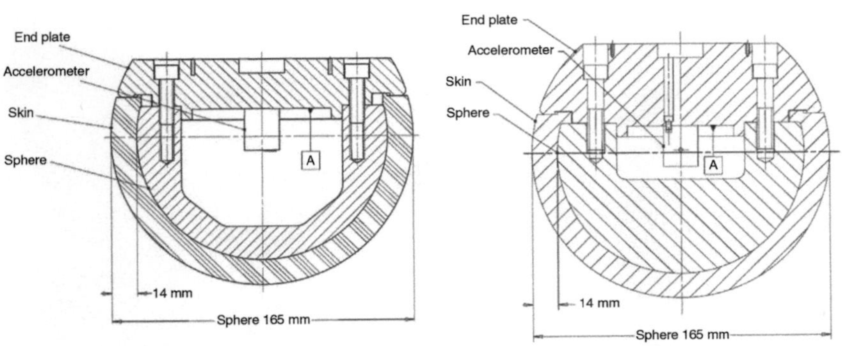

FIGURE 1.12 Pedestrian heads. Child headform (left) and adult headform (right) (UNECE, 2014).

1.3.4 Human computer models

Crash events can be separated into three distinct phases (Crandall, 2012): the vehicle impacting the barrier, the occupant impacting the restraint system, and finally the organs impacting the occupant internal body cavity (Campbell, 2013).

As such, more intricate occupant models are now being considered to assess the safety of vehicles (Fig. 1.13) to address organ injuries. Research is now considering assessing vehicles using an omni-directional human computer model (Toyota, 2010) which has already been used in injury trauma assessments (Evans, 2013). Human pedestrian models have already been used by Daimler (Mayer, 2013) to confirm the head impact contact time of a pedestrian struck by a vehicle. This new methodology using the THUMS human computer model has been accepted by EuroNCAP in a few crash test scenarios as a valid process to enhance the assessment of pedestrian deployable bonnets and rear seat child protection in frontal collisions (Mayer, 2013).

The advantages of such models are that they have theoretically more relevance to the overall human body compared to a crash test dummy. Even if the trauma parameters are still being debated by biomechanics experts (Crandall, 2012), it is overall agreed that this is the future direction research should take to further improve safety. As such models could include, in the future, parameters relevant to the entire population, such as gender, age, percentile, and other cases such as stage of pregnancy (Bastien, 2014b).

At this point in time, 2015, these human models are still under development and it would be impossible to optimize, let alone engineer, for all crash scenario eventualities, because of their excessive computer runtimes. ATD are used and will be used for the foreseeable future as they are convenient, repeatable, and produce output injury responses that can be interpreted using Newton's second law: when a mass is accelerated or decelerated, a force is generated upon it.

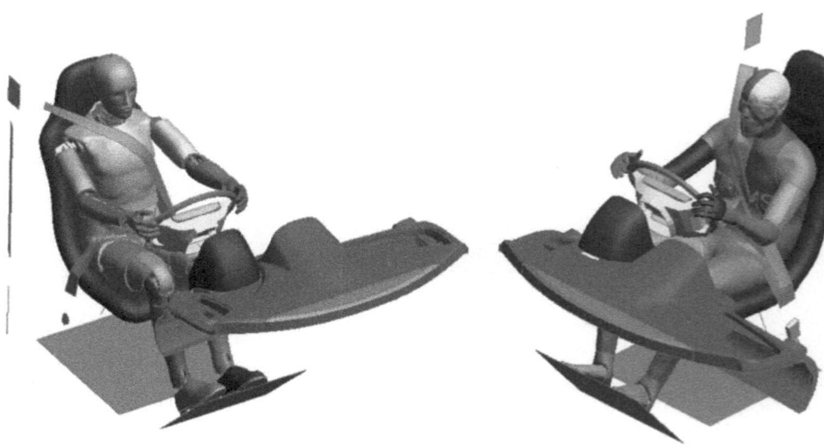

FIGURE 1.13 50th percentile occupants. ATD (left), THUMS4.0 (right).

1.4 TYPICAL VEHICLE ARCHITECTURES AND SCOPE FOR OPTIMIZATION

Current vehicle structures can be categorized into three main types of architectures:

- Ladder frame
- Tubular frame
- Integral

1.4.1 Ladder frame

Ladder frame is usually found in low-production niche vehicles that have a strong heritage from the beginning of the 1900, vehicles such as Ford model "T" vehicles. Due to their low volumes, they are often not tested by EuroNCAP but need mainly to meet ECE R12 (uniform provisions concerning the approval of vehicles with regard to the protection of the driver against the steering mechanism in the event of impact), ECE R14 (Seatbelt anchorages), and ECE R16 (Restraints and Safety Systems) regulations.

The ladder frame chassis construction mounts the cabin, a separate body, to a rigid frame that, which also supports the drivetrain. This method was the original method of building vehicles and continues to this day in niche automotive manufacturers (e.g. Morgan vehicles, pictured in Figure 1.14) as well as heavy goods vehicles. Ladder frames were initially made of wood (commonly ash) and moved to steel construction in the 1930s and are currently made of aluminum and magnesium (2015).

In general, ladder frames are:

- Heavier than an integral structure, causing lower performance and/or higher fuel consumption;
- Weaker in torsion which can compromise the handling characteristics and the road grip;

FIGURE 1.14 A typical ladder frame body (courtesy of Morgan Cars).

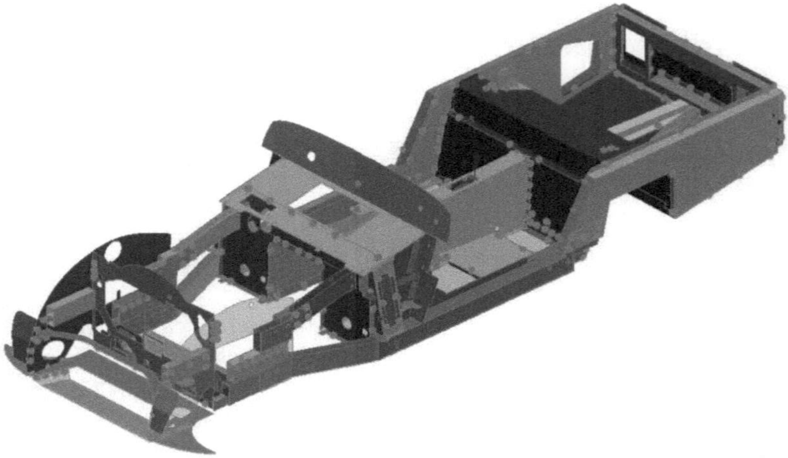

FIGURE 1.15 ISO view of Morgan Chassis (CAE Model).

- Not built with an efficient crumple zone, which generally may lead to tend to two different problems during a crash scenario:
    - Excessive deceleration on the occupants;
    - The cabin could detach from the frame and cause injuries due to vehicle intrusions into occupant space.

In order to reduce the weight of such a structure, current ladder frame chassis panels are often made of aluminum and magnesium and are riveted, illustrated in Figure 1.15 as dots. Generally, when torsion occurs, the rotation point is usually at the vicinity of the rear of the occupant cabin, which has been illustrated in Figure 1.16.

The opportunities for optimizing such structures are limited, as most sections are extrusions that have limited scope of being tuned overall as well as locally in the architecture. Nevertheless, optimizing ladder frame architectures can be achieved by investigating gauge and materials changes, by aiming to

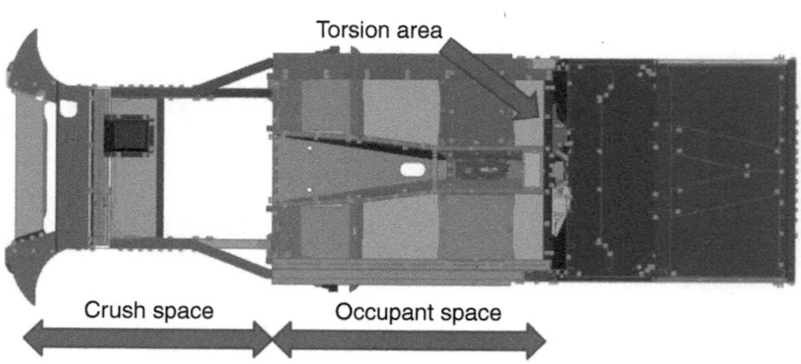

FIGURE 1.16 Bottom view of Morgan Chassis (CAE Model).

increase structural performance while minimizing the weight penalty increase caused by these changes.

1.4.2 Tubular structures

Tubular chassis structures are also produced for small volume vehicles that are usually niche vehicles. These chassis perform better than ladder frame chassis, as there are more opportunities to position beams or trusses in strategic areas of the vehicle architecture. Consequently, some of these chassis can be used for racing. The key advantage of such chassis is that they are affordable to build and require a minimum amount of assembly capabilities and space. A tabular structure is a perfect solution for low-budget enterprises, such as universities (Figure 1.17) or small-volume manufacturers.

Tubular chassis are quite difficult to assemble however as they are prone to asymmetry due to the rigging, the length tolerance due to the trimming of each tube, as well as the thermal expansion due to the welding process. Steel tubular structures are relatively easy to weld and usually require oxygen-acetylene welder, while aluminum tubes require expert welders and the use of tungsten inert gas (TIG).

When a tubular chassis is designed, the tubes are either connected together using sleeve brackets, or butt against one another and then are connected by either welding or adhesive bonding. The butt joint condition can be challenging for circular tubular sections, as some tubes will need to be trimmed like a "fish mouth" so that adjoining tubes can physically contact with a small tolerance before attachment is possible. Assembling noncircular sections (squares) can be easier to weld or glue than circular sections. Circular sections have, on the other hand, a constant second moment of area. Consequently, their orientation is not important during the assembly process, while a noncircular tube will need to be carefully oriented as it will have a critical role in the bending and buckling resistance of the chassis. This leads to a major disadvantage of the tubular chassis using nonaxisymmetric cross sections, as unexpected loads can have a detrimental outcome on the structural integrity.

FIGURE 1.17 Coventry University Formula Student 2010 Chassis Concept (left) and computer model (right).

Tubular chassis inherently carry the structural loads along the length on the welded beam structure. Beams work well in tension and it is very easy to ensure that the stress levels are acceptable.

As some members are in tension, others will be in compression to balance the structure. A beam in compression may have a stress level that can be acceptable; however the compression load may initiate buckling. Standard vehicles contain panels and other structures to counter any unexpected loads, which is not the case for a tubular structure. In the case of a tubular structure, any load that reverses its state from tension to compression may lead to catastrophic failure. In any design and optimization of a tubular structure, buckling must be a constraint parameter. Consequently, the critical force to initiate a beam linear buckling (F_{crit}) is a function of the second moment of area (I), the material Young's modulus (E) (assuming that the material is isotropic), its length (L), and slenderness ratio (k), as shown in general linear buckling Equation (1.5). This equation relates to the first-order Eulerian buckling mode.

$$F_{crit} = \frac{\pi^2 EI}{kL^2} \tag{1.5}$$

An application of this criterion is displayed in Figure 1.18, where the main longitudinal tubes are in compression, including the under-driver seat strengthener. Compression forces in Figure 1.18 are low (in the order of 757 N); however buckling verification is always needed when compression is present.

In order to assess the structural integrity of tubular structures, it is possible to use beam elements which are perfectly adequate for linear analysis and provide guidance on loads not exceeding the material's yield stress. For a crash analysis, the formulation of such beams is, however, not adequate, as will be discussed in Chapter 2. Consequently, beam elements subjected to compressive loads in linear computer models must be replaced by shell elements when the structure is validated using explicit crash analysis computation.

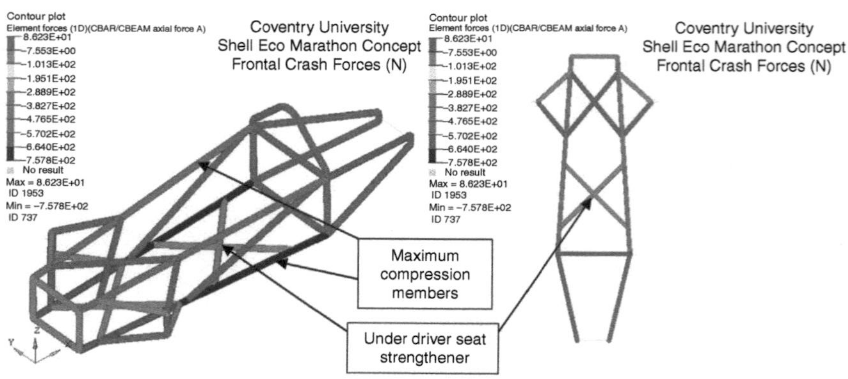

FIGURE 1.18 Coventry University Shell Eco Marathon Crash test. Axial forces (N).

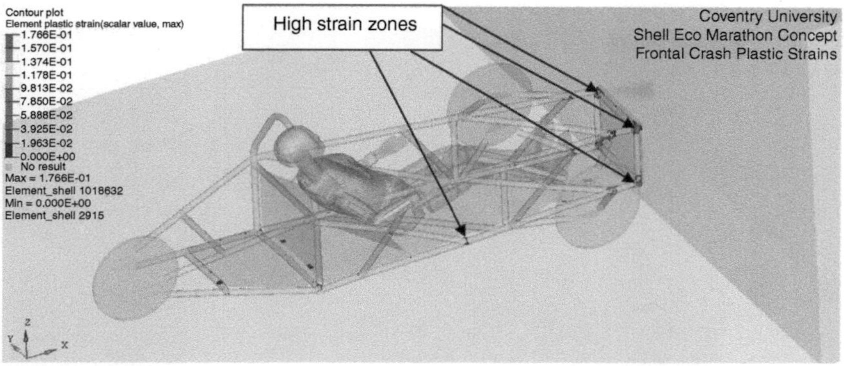

FIGURE 1.19 Coventry University Shell Eco Marathon Crash Test Model (plastic strains).

To illustrate this, a simple beam model of the Coventry University Shell Eco Marathon vehicle, which has successfully met the first Eulerian buckling mode using linear analysis, is subjected to a dynamic crash validation. For this to be performed, all beams have been changed to shell elements.

The results are now showing plastic strains in the order of 17% in the front and welded areas of the structure (Figure 1.19), which is a different outcome from the linear static analysis. This level of plastic strain is considerable for an aluminum structure and may now lead to catastrophic collapse of the frame.

This example confirms that simple stress computation on tubular beams may not be sufficient and that a simple linear buckling calculation must not be the end solution. However it is a good step in the right direction prior to a more appropriate full crash model development.

1.4.3 Integral structures

Most modern-day vehicles structures utilize integral structures, which are totally different from the ladder and tubular structures previously discussed in this chapter.

Integral structures are based on skin panels that are connected to a skeleton which provides the stiffness and strength of the structure, as illustrated in Figure 1.20.

These skin panels add shear resistance to the underlying structure, as well as some level of bending (doors on side impact) and buckling stiffness. The skins panels on a bonnet illustrated in Figure 1.21, located in the central hexagon area, fulfil important design requirements, as they are designed to withstand aerodynamic loads, palm printing abuse loads (or oil canning), pedestrian head impact loads, as well as bonnet slam loads. At the same time, this bonnet must be fitted with crash initiators which are needed to induce a critical bending to relieve frontal impact loads to the rear bonnet latches. This is a desirable feature, as a bonnet delatching would separate from the vehicle, contact the windscreen, and could potentially injure the occupant. These crash initiators illustrated in

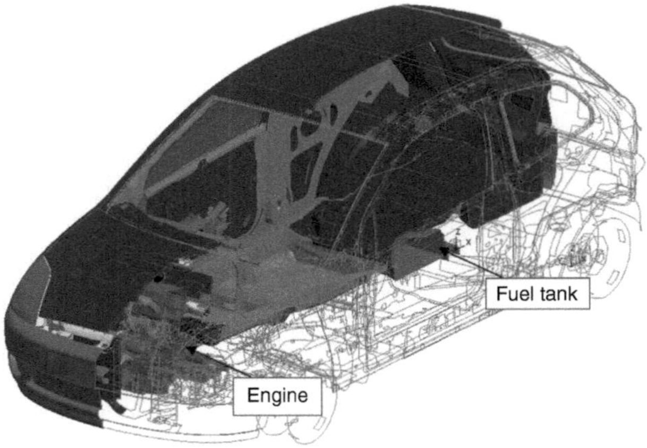

FIGURE 1.20 Crash model of Ford Fiesta. Coventry University.

Figure 1.21 are engineered by reducing the bonnet local underlying skeleton cross section (Bastien and Harvey, 2004).

In general, an integral vehicle structure is made of the same type of material, that is, metallic and is ideally of the same class of metal, for example all steel or all aluminum. Some vehicles also use composites as well as a multi-materials.

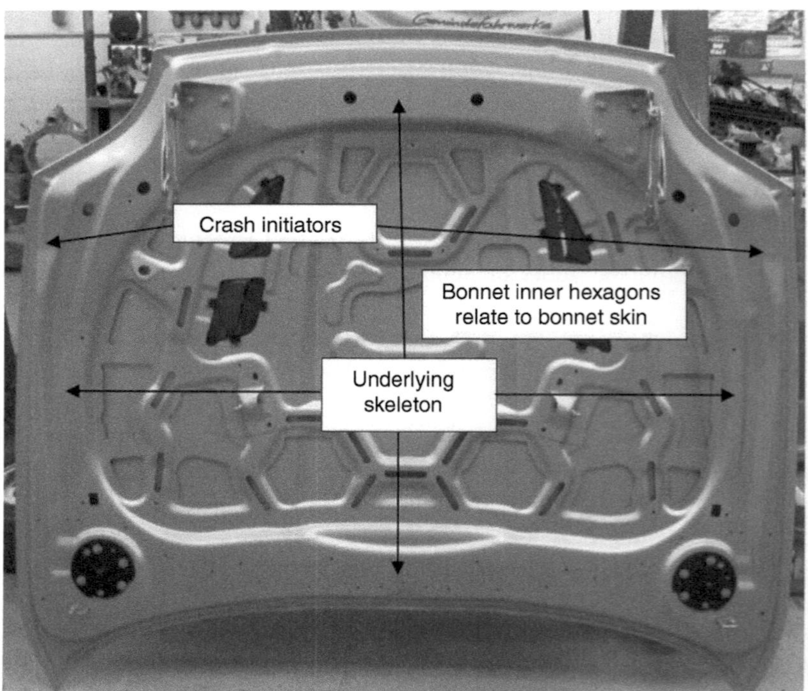

FIGURE 1.21 Jaguar XK aluminum bonnet (active pyrotechnic activation).

Since 2006, Bastien has noticed that vehicle steel structures adopted aluminum closures in order to save further mass.

One of the key closure panels is the bonnet, as it tends to be physically a large component. It has been observed by many OEMs that aluminum was the preferred material of choice for this panel (Figure 1.21), especially for pedestrian head impact protection, because of its reduced inertia, reduced stiffness, and low strain rate response to high-speed impact, relative to steel.

In other instances, aluminum architectures use high-strength steel bumper beams to absorb more energy in case of frontal impact, such as in the Jaguar XK.

Integral structures are usually made of large floor pressing, on which the side structure, or door-opening panels (DOP), are primary framed, spot-welded, and then fitted with the engine and powertrain followed by the closures (Figure 1.20).

This process is commonly used for the high-volume production of vehicles, such as the Ford Fiesta. It can be noticed that the fuel tank is positioned at the rear of the Ford Fiesta, under the passengers' seat, and that the engine is located at the front of the vehicle (Figure 1.20). In case of frontal collisions, the management of the vehicle kinetic energy will depending on the vehicle crash space, mass as well as the engine position within the vehicle architecture.

To illustrate this, a comparison between the vehicle acceleration pulses (NHTSA, 2014) of four chosen vehicles listed in Table 1.7 was made and displayed in Figure 1.22.

The vehicle configurations were as follows:

- 2011 Ford Fiesta: front transversely mounted engine, front wheel drive, five seats;
- 2008 Mini Cooper: front transversely mounted engine, front wheel drive, four seats, short front overhang;
- 2004 Smart FourTwo: rear transversely mounted engine, rear wheel drive, two seats, and very short front overhang;
- 2004 Honda S2000: front longitudinal mounted engine, rear wheel drive, two seats, long front overhang, no roof load path;

TABLE 1.7 Vehicle Crash Pulse Comparison (NHTSA, 2014)

Vehicle	Model Year	Test Mass (kg)	Impact Speed (km/h)	NHTSA Test Number	NHTSA Channel (Acceleration Global X)
Ford Fiesta	2011	1359	56.5	6996	84
Mini Cooper	2008	1371	56.3	6291	89
Smart FourTwo	2008	1057	55.9	6332	86
Honda S2000	2003	1465	57.0	4462	89

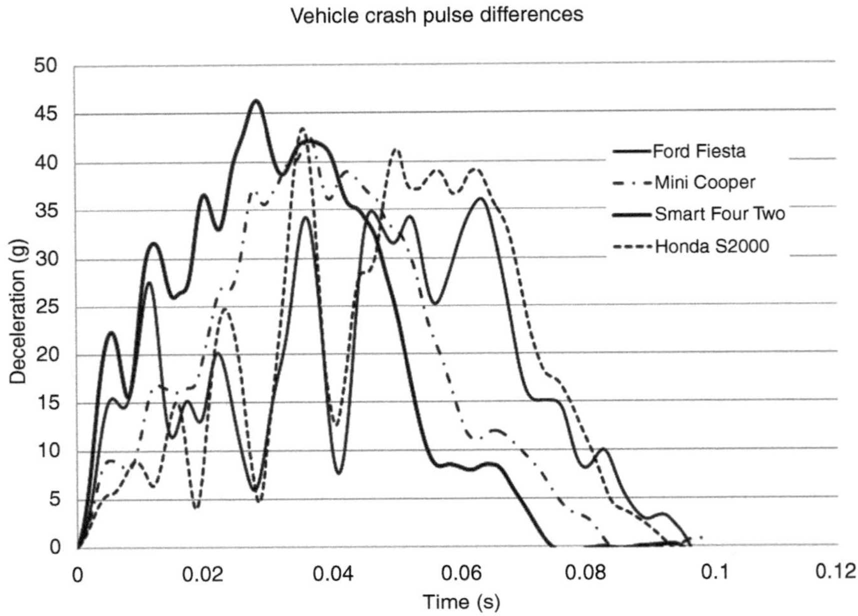

FIGURE 1.22 Vehicle deceleration response with different architectures.

It can clearly be observed from Figure 1.22 that engine type and transmissions have a major influence on the load the vehicle experiences during a crash event. Also, it can be noted that the Smart FourTwo is a small vehicle size compared with the other ones selected; hence, the energy must be managed rapidly to avoid excessive intrusions. Consequently, the load is highest compared with the vehicles presented (Figure 1.22) and has to occur in a shorter time duration, here 46 g in 0.03 s compared with 43 g in 0.04 s (in the case of the Honda S2000 and Mini Cooper).

In the case of lower volume production, like the Microcab (Microcab, 2015). the floor is made of a composite material compound called the "tub" on which the upper structure, made mainly of glass fiber, is fastened by means of adhesive bonding. The engines located at the front of the vehicle (Figure 1.23) are much smaller than those of a Fiesta and are powered by a fuel cell, located behind the rear bumper (Figure 1.24), which draws hydrogen from the tank located underneath the rear passengers' seat. As the heaviest components of the Microcab are located more rearwards than the Fiesta, it is suggested that the crash pattern will also differ as already presented in Figure 1.22.

Comparing these two vehicles suggests that, should vehicle architecture encompass all the means of propulsion, that is, standard internal combustion engine, electric, hybrid, or other means, then the behavior of the crash structure is likely going to be contradictory for each engine variant. Research has shown that vehicles encompassing all propulsion variants would be, in general, 50% heavier than the one with a single variant (Prem and Bastien, 2013).

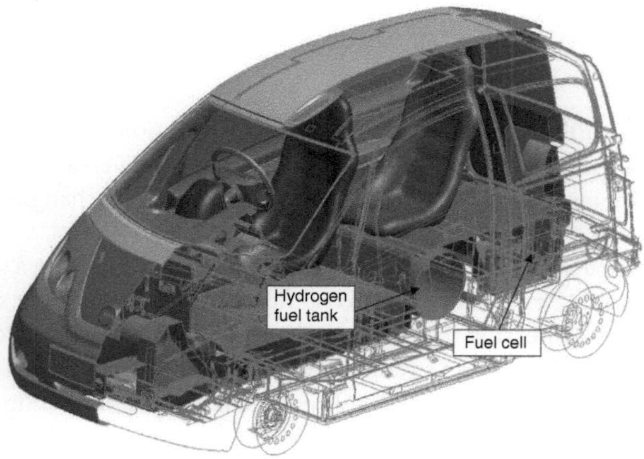

FIGURE 1.23 Crash model of The Microcab Hydrogen Fuel cell. Coventry University.

FIGURE 1.24 Crash model of The Microcab Hydrogen Fuel cell. View underneath Vehicle. Coventry University.

Integral vehicle architectures provide a bigger scope of optimization compared with the ladder and tubular frame options due to the more complex manufacturing process, as well as the tool development. These main optimization opportunities are in loadpath optimization, shape sizing and material selection.

1.4.4 Shape and size

In vehicle structures, a lot of the stiffness, or the ability to resist displacements, depends on two main parameters:

1. The shape of the structure;
2. The material the structure is made of.

Vehicle stiffness is inversely proportional to the Young's modulus (atomic property of the material) and the second moment of area (I). Equation (1.6), which is a property of the structural shape and thickness.

$$stiffness \propto \frac{1}{EI} \tag{1.6}$$

As the second moment of area considers the position of the cross section relative to the center of gravity of the section, it is a more versatile and efficient parameter to obtain the required structural stiffness than just changing the material gauge.

Should shape and size optimization be performed separately on a sheet metal structure, it would be recommended to optimize the structure using shape variable first before addressing the component gauge, which should be considered at the finalization stage of the structural optimization.

During the crash phase, small details are needed to control the structural and the structure bend and collapse; hence, local shape changes can have a huge effect of the structural performance. This is true in the case of the frontal and rear crash when the vehicle structure is subjected to nonlinear buckling. Buckling as mode of energy dissipation is chosen here because of the large crush zone available between the barrier and the occupant. In frontal and rear impacts, the available crush zone needed to absorb the impact is usually in the order of 500 mm, while in side impact it is around 150 mm. A controlled buckling mode is very efficient (frontal and rear impacts), as more energy is taken compared with bending (main mode of energy dissipation for side impact). If the buckling is not controlled, then the structure becomes plastically deformed and bent, then it cannot take any more energy. In this eventuality, the loadpath does not follow the main line of the structure, i.e., the crush can and main vehicle longitudinal members.

In the case of the side impact, the distance between the impacting barrier and the occupant is shortest; the aim of the structure is therefore to keep intrusions to a minimum and as such to limit the B pillar flex and collapse. On the occupant injury levels, should a contact occur with the vehicle interior, it is recommended the impact velocity to be at a minimum. Consequently, the B pillar has a bending behavior that is primary controlled by the bending stiffness, hence the second moment of area.

Considering the frontal crash requirements, local details can cause a different response to the vehicle structure and to the occupants. Such local design change parameters may be considered during the optimization stage. The following geometrical shapes concerning the crush initiators are illustrated in Figure 1.25:

a. Corner notches
b. Corner holes
c. Surface beads

The crush initiators stabilize the profile during the buckling deformations with the ultimate aim to crush the structure in an axisymmetric deformation while avoiding bending collapse.

The energy profile without any crush initiator is not the one that absorbs most of energy, when considering Figure 1.26 based on previous research (Kumar, 2008). This means that the initiators do not always decrease the profile

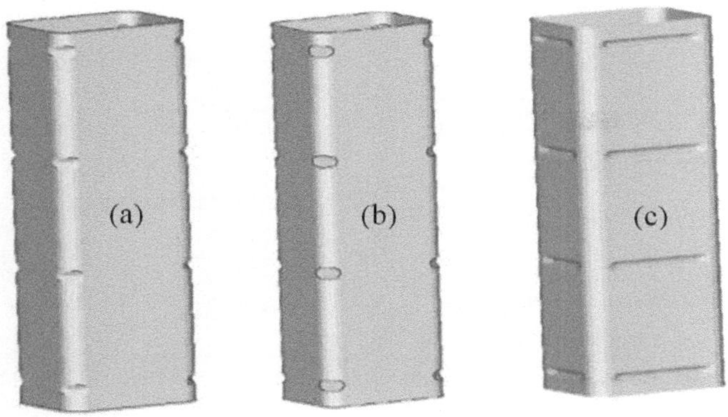

FIGURE 1.25 Three typical crush initiators. a) corner notches, b) corner holes, and c) then surface beads.) (Pirard, 2009).

of energy absorption. It can also be observed that because of the severity and complexity of the impact, the initiator densities do not have a linear influence with the energy absorption.

Considering Figure 1.26, it can be noticed that, compared with a structure with no crash initiators, the corner holes configuration consistently absorbs less energy. On the other hand, structures with up to six surface beads regularly absorb more energy than a structure with no crash initiators. After six beads, the crash absorption performance is reduced; after 10 beads, it becomes detrimental. For the corner notches configuration, the energy capabilities are initially reduced, but are best after six notches. It can be concluded from the study

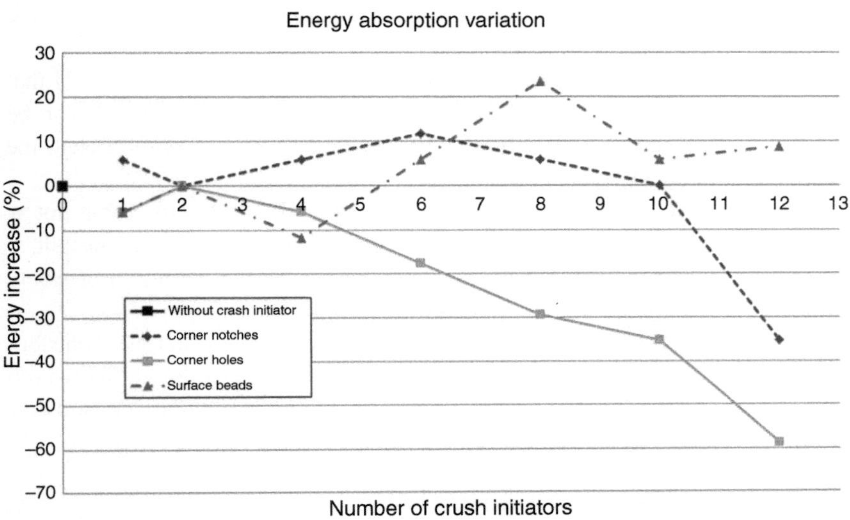

FIGURE 1.26 Energy absorption variation at a function of number of crush initiators (based on Kumar, 2008).

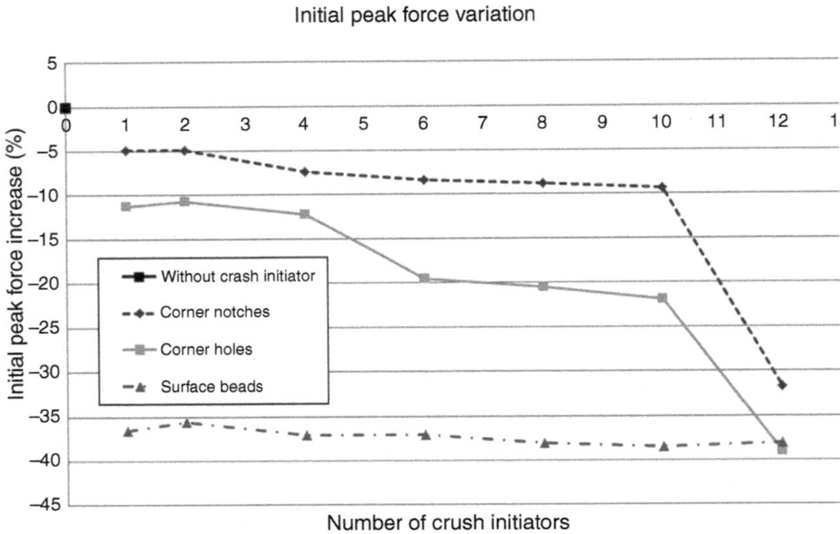

FIGURE 1.27 Initial peak force variation at a function of number of crush initiators (based on Kumar, 2008).

(Kumar, 2008) that holes are detrimental to energy absorption and that notches and beads can improve energy absorption performance of a structure in the order of 10 to 20%.

It is also desirable to decrease the initial peak force as much as possible as high peak forces can cause serious injuries to the vehicle occupants (Shanahan, 2013).

Reducing the initial peak force can be achieved by creating a shape in the structure profile which will easily fold.

Considering the initial peak forces, it can be observed from Figure 1.27 that the maximum peak force relates to the profile without any initiator. It can be consistently noticed that the more the design feature density is increased, the lower the initial peak force.

Corner notches and corner holes decrease the initial peak force but not as much as the surface beads configuration, which remains lowest and constant.

Overall, the shape parameter plays a major role in the definition of safety structures as an overall response, as well as a detailed response.

When looking at vehicle overall structural response, the material's contribution is also critical; Young's modulus plays a major contribution in the overall isotropic stiffness, while the strength is controlled by the nonlinear/plastic area of the material characteristic, as the next paragraph will demonstrate.

1.4.5 Materials and manufacture

In the early 2000s, vehicle structures were mainly built using mild steel grades, leading to vehicles with a good level of torsional rigidity capability; however, they were under-optimized in crash performance. The current trend in the automotive industry (2015) is either producing:

- Very high-volume affordable and safety performant vehicles (usually the likes of Renault, Ford, Mazda, etc.);
- High volumes of high market and safety performant vehicles (usually the likes of Jaguar, Audi, Mercedes, etc.);
- Very low volumes of luxury and safety performant vehicles (usually the likes of Ferrari, Bugatti, etc.).

Some OEMs may aim at one or two market segments for their vehicle model; consequently, the choice of material is critical (Davies, 2003). As an example, entry level vehicles would aim to achieve good EuroNCAP requirements, while at the same time being competitively priced. Steel constructions are still the most cost efficient for high-volume vehicles (Table 1.8); unfortunately, there is a mass penalty to be paid for choosing this material.

Selecting aluminum, on the other hand, looks like an attractive solution for lightweight design; however its stiffness is a third that of steel. Interestingly, the stiffness-to-mass ratio of steel and aluminum is the same (as the stiffness of aluminum is also a third of steel). Surprisingly, changing the material of an existing steel structure (with a set design envelope) to aluminum does not yield to a vehicle weighing a third of the original mass.

The reason behind comes from the fact that pressing aluminum is very challenging and that it tends to split if:

- The radius when bent is too small (usually minimum radius is 12.5 mm), or
- The draw depth is excessive.

As a consequence, the second moment of area of aluminum structures is often inferior to their steel counterparts, hence a loss of stiffness, leading to the necessity of adding new panels to compensate for this deficiency.

New research suggests that new techniques in aluminum hot forming deep drawing are being investigated by heating the die and the punch. These techniques are still at their infancy; nevertheless, they can potentially yield to improved vehicle structural performances.

As can be observed from Table 1.8 that cost is a driving factor in the choice of materials. In the case of aluminum structures, the main architecture is usually built using a 5000 series material, while the outer panels are manufactured of a 6000 series that is more resistant to stone chipping impacts. To implement such materials, the relative cost compared with steel panels will be in the order of four to five times more expensive (Table 1.8).

While observing the stress–strain material responses (Figure 1.28), it can be noticed that:

- For the same type of material, for example, steel, Young's modulus remains a constant, regardless of the material grade. This is an important statement, which leads to the conclusion that material grade has an influence on the structural strength under impact conditions, but not on static structural stiffness.
- Aluminum and magnesium will absorb less energy than a steel structure (for the same shape and thickness). It has been documented that in buckling,

TABLE 1.8 Cost of Material Choice in Vehicle Design (Davies, 2003)

	Design Parameters			Ease of Manufacturing (1= Difficult to Process; 10 = Few Production Problems)		Relative Cost
Material	Yield Stress (MPa)	Young's Modulus (GPa)	Density (g/cm3)	Forming	Joining	Forming Steel = 1.0
Mild steel	140	210	7.9	8	9	1.0
High strength steel	300	210	7.9	6	8	1.1
Ultra high strength steel	1050	210	7.9	4	7	1.5
Aluminum 5000 series	110	69	2.7	4	7	4.0
Aluminum 6000 series	120	69	2.7	6	5	5.0
Magnesium sheets	160	45	1.8	4	4	4
Titanium sheets	880	110	4.5	6	5	60
Glass reinforced plastics	950	40	2.0	8	7	8.0
Carbon fiber composite	1100	120–250	1.6–1.9	8	7	50+

aluminum panels would need to be thickened in the order of 44% to 50% to match the performance of their steel counterpart (Prem and Bastien, 2013).

Magnesium's mechanical properties suggest a lower specific density for a cost comparable to aluminum (Table 1.8). Nevertheless, for structural stiffness, aluminum looks to be a better option than magnesium because of its higher Young's modulus and easier joining techniques (steel rivets vs. aluminum rivets and bonding) (Table 1.8).

It has to be noted that all material curves used in computer models in crashworthiness analyses use a true stress–strain and not a simple engineering

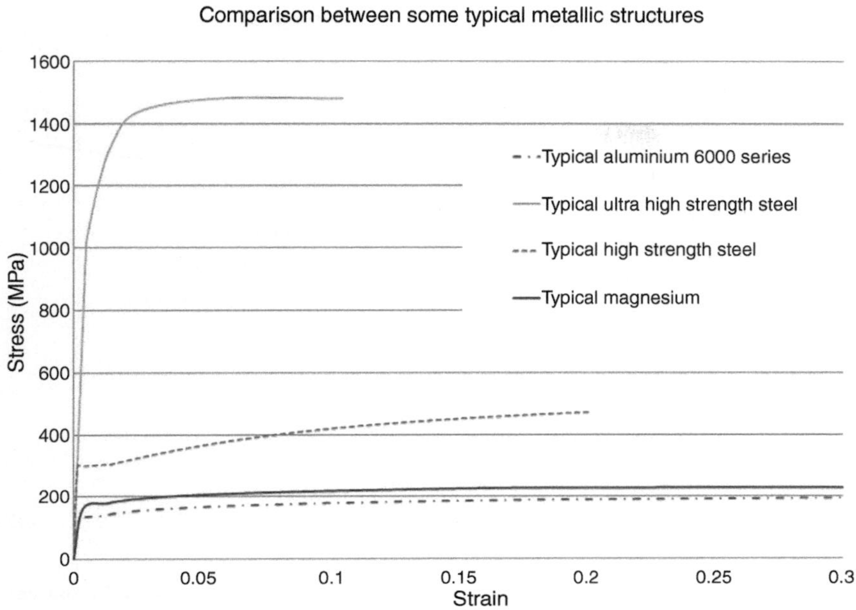

FIGURE 1.28 Stress–strain comparison between metallic structures.

stress–strain characterization. This is due to the fact that during an impact, or a dynamic stamping simulation, for example, the material thickness changes.

Another parameter in the material properties which needs consideration is the strain-rate effect. The strain-rate effect affects the structure performance, as the material is stiffer the faster it is deformed, as illustrated in Figure 1.29. This can be modeled using a tabulated set of curves, which characterize the material response as the strain rate increases. Considering Figure 1.29, it can be noticed that the curvature of the plastic region follows a power law. When the material has reached the yield stress, it follows a nonlinear behavior.

Two methods can therefore be used to capture all the strain-rate responses, based on the initial quasistatic test response: the Johnson–Cook and the Cowper–Symonds mathematical strain-rate amplification formulations.

The Johnson–Cook uses a natural logarithmic factor, while the Cowper–Symonds uses a power law. These mathematical models are commonly available and have an advantage over the tabulated formulation, as the strain rate is used directly to calculate the stress–strain response, while the tabulated formulation needs to extrapolate the strain rate between tabulated curves to compute a stiffness response.

All of these material computer models inherently have another problem, which is a disconnect between the mathematical formulation and the physical phenomenon of material failure. When metals are strained with increasing speed, they tend to be more brittle.

To date, no known computer material model has the ability to adjust the failure strain along with strain rate computed in the model.

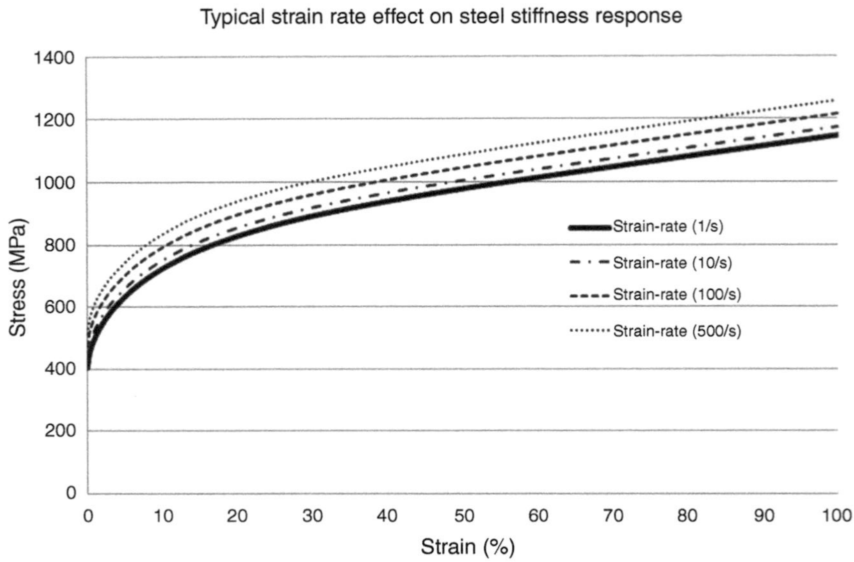

FIGURE 1.29 Typical strain rate effect on high strength steel stiffness response.

Consequently, all the curves proposed in Figure 1.29 contain the same strain level as the quasistatic curve to keep the computation stable at higher strain rates, which is not physically correct.

Another fundamental limitation of using elastic–plastic models is that they are only valid up to the necking point of the metal, at which little loads can cause very large and unrealistic deformations. Consequently, care must the taken when excessive plasticity and deformations occur.

Overall, it has been demonstrated that in the design and optimization of nonlinear safety structures, many materials are available, which provide a huge scope for lightweighting while ensuring the adequate vehicle structural integrity. Nevertheless, beyond the mathematical formulae, sound engineering judgments on the quality and accuracy of the computation is required to sanitize the optimized solution.

After considering commonly used vehicle structures previously presented, including potential optimization avenues, such as shape, size, and material, the next section will address a more generic approach to vehicle structural optimization in which the vehicle exterior styling is known but the underlying structure starting point is unknown and unbiased.

1.5 HOLISTIC APPROACH TO VEHICLE DESIGN

The fundamental purpose of vehicle safety structural design is to save the occupant from fatalities in case of an accident; hence, the first rule of vehicle safety is to limit, or avoid, vehicle intrusions inside the cabin. No intrusion means that the vehicle stiffness and strength must be adequate so that the maximum dynamic displacements do not contact the occupant. When a vehicle is to be designed, this should be the first axiom to implement, as it is the condition *sine qua non* to meet a legislative architecture. The second rule is that the deceleration forces generated by the accident must not injure the occupant through inertial forces. This can be achieved with the help of the restraint system, which primarily includes seatbelts and airbag systems.

All other requirements are numerous and complex, namely durability and noise vibration and harshness (NVH), which are necessary requirements but not legislative ones.

If NVH bears no legal requirements, it is still extremely relevant as it can make the vehicle more desirable by reducing cabin noise and limiting warranty costs on components exposed to cyclic and random vibration loads.

Consequently:

Vehicle Architecture (legislative)
=
Must limit intrusions in accident cases (legislative)
+
Must limit inertial forces on the occupants using the restraint system (legislative)
+
Should meet other durability, NVH and other requirements (important but optional)

In a generic structural loadpath definition, the inertial forces on the occupants are difficult to obtain with accuracy at the earliest stage of the vehicle architecture design, as has been shown from Figures 1.26 and 1.27, where very small details in the definition of the crush can, for example, have a huge influence in the initial peak force and impact energy capability of the front end structure.

Consequently, the legislative optimization of vehicle architecture would require a sequential optimization stage:

1. Overall architecture design for structural intrusion limitation
2. Local shape and sizing for:
   a. Occupant safety regarding inertial forces (primarily)
   b. Durability, NVH, and other requirements (as a matter of specific OEM comfort and durability targets)

1.5.1 Overall architecture design for structural instruction limitation

One of the major questions that needs to be asked when a vehicle structure is being evaluated is whether the new vehicle structure needs to follow standard vehicle architecture conventions or deviate from them. Overall, it can be observed that, if the chassis types have somehow changed during the years (ladder frames to integral structures), most upper structures have remained mostly unchanged. These upper structures can carry important loads, like the A and D pillars for frontal, rear impact, roof crush, and the B pillar for side impact protection.

Due to the development of computer optimization codes, which are going to be discussed in this book, it is now possible to ask this question and investigate

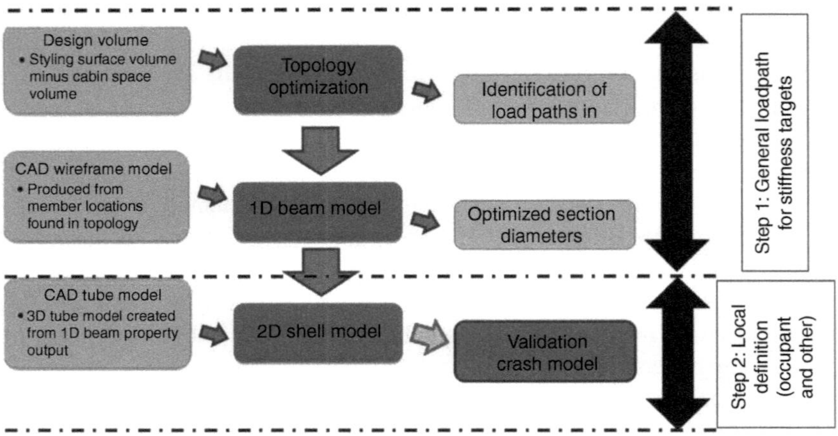

FIGURE 1.30 Structural optimization model for initial vehicle architecture design (Christensen et al., 2012a).

possibilities to find ways of improving the architecture and reducing weight while keeping acceptable levels of stiffness and strength (Christensen et al., 2012a). This first stage is based on structural optimization, which will compute a structural loadpath only when the necessary load-bearing structure is required. A vehicle architecture process is proposed in Figure 1.30 and in the earliest structural architecture would define a load path followed by a beam model sized adequately to meet stiffness targets. Consequently, at the early stage, only the Young's modulus (in the case of isotropic structures) is required as material properties, as plastic response will be assessed subsequently in a local computer model where a higher level of definition will be required.

Everything would start from the design space, which is the available volume to "route" structural components. This design space comprises, in general, a Boolean operation between the external vehicle shape and the desirable habitable occupant cabin space including all necessary ancillaries (engine, suspension, etc.).

As an illustration, in order to compute Step 1, "General loadpath for stiffness targets," from Figure 1.30, a vehicle design envelope is created and illustrated in Figure 1.31.

Relevant legislative loadcases are then applied on the design volume, as illustrated in Figure 1.32.

The challenge of this phase is to generate the values of these loads, as in reality these loads are transient dynamics, that is, they are forces that vary with time. Benchmarking vehicles of similar masses and dimensions against crash test data can be used to evaluate the average force and the crush space used to absorb the energy.

Two simple options are available:

1. Choose the peak force from the benchmark as input to the computer model. It is the easiest option, but the load applied will certainly be more than necessary, which will lead to an over-stiff and heavier structure than what it should be

FIGURE 1.31 Vehicle design envelope, TARF EPSRC (Prem and Bastien, 2013).

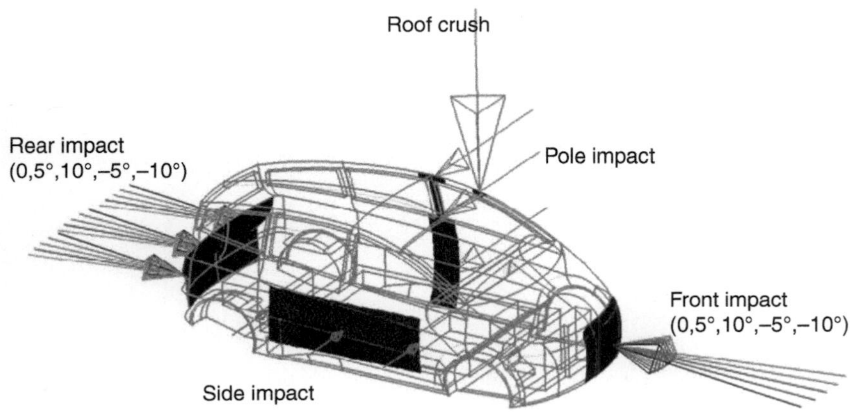

FIGURE 1.32 Vehicle legislative structural loadcases (Christensen et al., 2012b).

2. Assuming that:
   a. The impact energy is the same between the benchmark vehicle and the one being designed;
   b. A reasonable crush length is chosen for the designed vehicle;
   c. An average force on the model is evaluated from the benchmark vehicle.

It is possible to compute the average impact force which can be applied on the designed vehicle. This method is not perfect, as the load is transient, but it is a compromise.

A linear topology optimization can be run revealing the optimized structural loadpath, as illustrated in Figure 1.33.

Structural loadpaths can be interpreted and drawn as in Figure 1.34 and the architecture sized for overall stiffness compliance using beam elements (Figure 1.35).

The sizing of the beam model takes into account the second moment of area, as well as the buckling requirement that has been given in the tubular chassis section, as illustrated by Equation (1.5). This is very relevant for the definition of the A pillar of the vehicle, as this pillar has to withstand a large amount of the frontal impact transmitted to the occupant safety cell. Consequently, a buckling

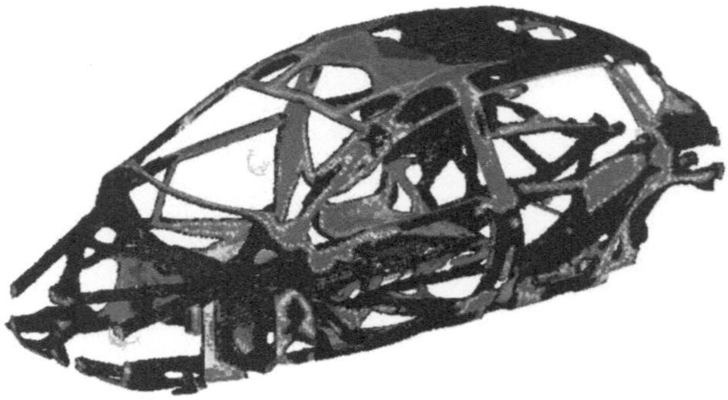

FIGURE 1.33 Vehicle topology optimization results, TARF EPSRC (Prem and Bastien, 2013).

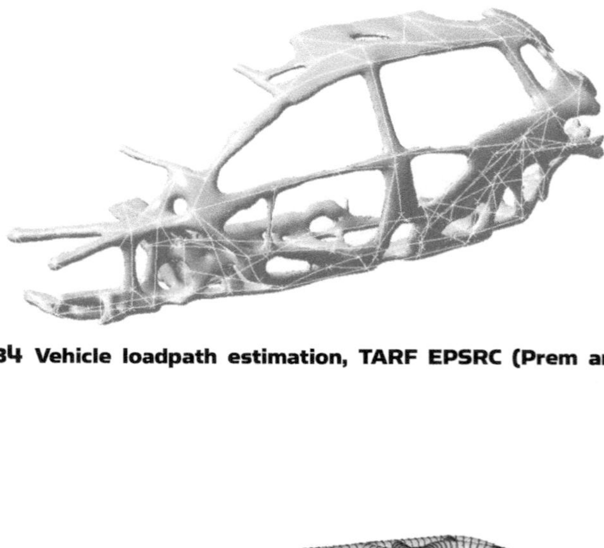

FIGURE 1.34 Vehicle loadpath estimation, TARF EPSRC (Prem and Bastien, 2013).

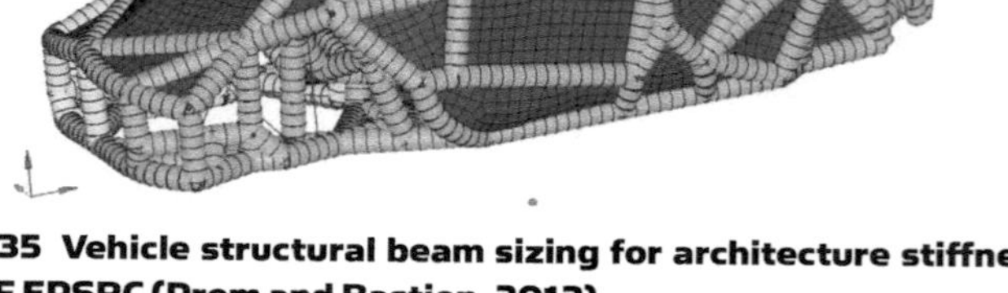

FIGURE 1.35 Vehicle structural beam sizing for architecture stiffness compliance, TARF EPSRC (Prem and Bastien, 2013).

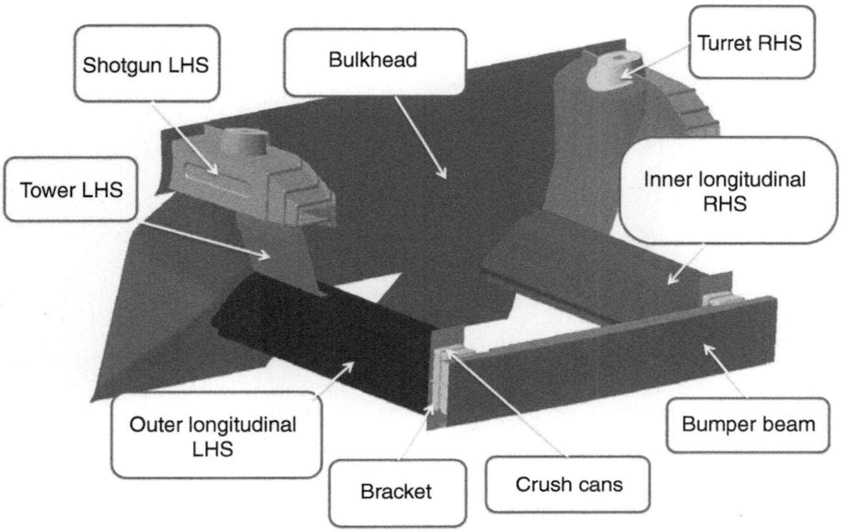

FIGURE 1.36 TARF vehicle front structure, with package and design targets based on beam model TARF EPSRC (Prem and Bastien, 2013).

limit force computed can be extracted at this stage and used in the detailed model as the critical load limit not to reach; this is called an optimization constraint. For the case of the B pillar, buckling check is less relevant, as this pillar is exposed to a combined bending mode comprising stretch and bending due to the lateral impact forces.

1.5.2 Local shape and sizing for legal and other desirable structural requirements

Details of the front crash structure, for example, can be defined locally with very detailed 2D shell meshes (Figure 1.36), which will be able to provide a platform for crash pulse optimization, intrusion validation, and restraint system development.

The setup of such a computer model will be explained in detail in chapter 2. The final deformed crush space is displayed in Figure 1.37, where the

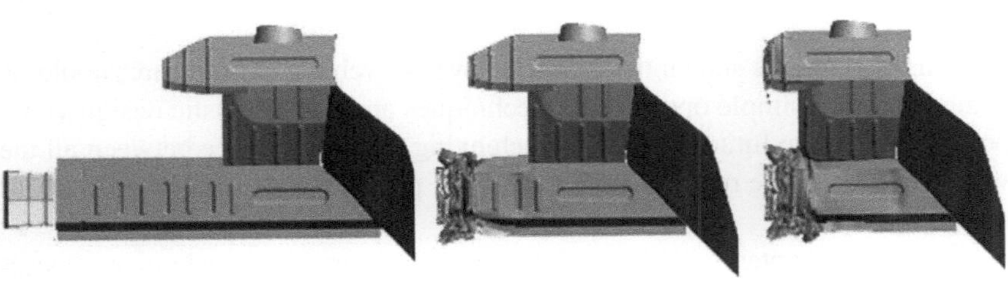

FIGURE 1.37 TARF vehicle front structure crash shape. Side view (Prem and Bastien, 2013).

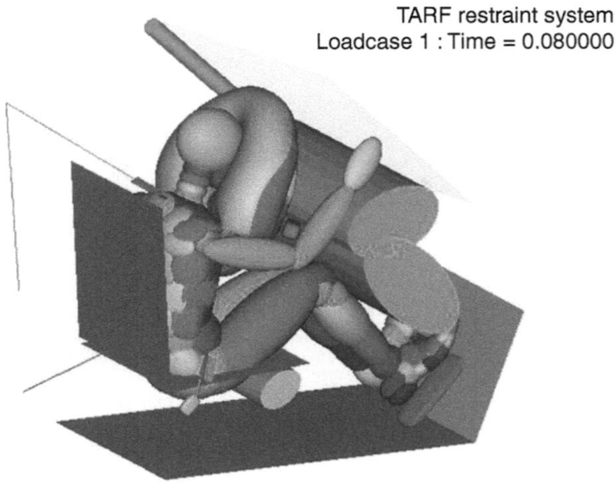

FIGURE 1.38 Optimization of the TARF restraint system.

maximum crush length has been achieved while controlling and preventing buckling of the A pillar.

The restraint system can therefore be optimized following the optimum crash structural performance, using the crash pulse signature from the occupant seat located at the base of the B pillar as an input to the restraint system optimization (Figure 1.38). All these key stages will be discussed in details in this book.

1.6 CONCLUSIONS AND OPPORTUNITIES

This chapter has outlined the general principles of vehicle safety structure design and engineering and has covered aspects such as accident data, legislative requirements and their relevance, crash test dummies, injuries, vehicle architectures, and materials.

As optimization techniques are used more widely in the automotive industry, there are some exciting opportunities to challenge the current vehicle architectures in view of the new implementation of fully electrical vehicles, as well as the development of lightweight 'L' category vehicles (for example, the Renault Twizzy).

This chapter has shown that some innovative vehicle architectures could be evaluated using simple optimization techniques and that a holistic design viewpoint could be the future method for achieving the best balance between all the vehicle performance requirements as early as possible at the early stages of the design cycle.

The next chapter will discuss in depth the numerical techniques necessary to set up computer model as a starting point of the optimization, while the subsequent chapters will address a range of optimization theories to obtain the desired results.

REFERENCES

AGARD. (1996). Anthropomorphic Dummies for Crash and Escape System Testing, AGARD Advisory Report 330, Advisory Group For Aerospace Research & Development. Technical Report Available from <https://www.cso.nato.int/Pubs/rdp.asp?RDP=AGARD-AR-330>, North Atlantic Treaty Organization.

American Iron and Steel Institute. (2004). Vehicle Crashworthiness and Occupant Protection. Technical report Available from <www.autosteel.org> Accessed 05.07.13.

ASSESS. (2012a). Assessment of Vehicle Safety Systems. FP7 project Available from <www.assess-project.eu> Accessed 01.07.13.

ASSESS. (2012b). ASSESS D4.1b – FINAL – Action plan pre_crash evalution-PUBLIC-2010-03.01. Technical Report Available from <www.assess-project.eu/downloadables/Public%20Deliverables/ASSESS%20D4.1b%20-%20FINAL%20-%20Action%20plan%20pre_crash%20evalution-PUBLIC-2010-03.01.pdf> Accessed 1.07.13.

Bastien, C. (2014a). The prediction of kinematics and injury criteria of unbelted occupants under autonomous emergency braking. PhD Thesis, Coventry University. Awarded 02/11/2014.

Bastien, C. (2014b). People move on. Article Crash Test International Magazine. Available from <http://viewer.zmags.com/publication/ecc7d46a#/ecc7d46a/1> Accessed 27.09.14.

Bastien, C., & Harvey, M. (2004). Jaguar Sports Car Deployable Bonnet Research - Conflicts Between Bonnet Functionality And Pedestrian Head Impact Protection, Conference Proceedings, Institute of Mechanical Engineering Conference, London, 2004.

Bastien, C., & Williams, H. (2002). Sub-structured Approach for Occupant Head Injury Simulation. Time Compression Technology (TCT) 2002. Held 16–17 Oct 2002 in Manchester, UK.

Bastien, C., Blundell, M., & Neal-Sturgess, C. (2013). Influence of vehicle secondary impact following an emergency braking on an unbelted occupant's neck, head and thorax injuries. *International Journal of Crashworthiness*, *18*(3), 215–224.

BBC. (2013). Transatlantic trade talks open amid tensions. Article available from <www.bbc.co.uk/news/business-23221503> Accessed 01.07.13.

Berg, A., & Rucker, P. (2012). Crash test using a car with automatic pre-crash braking. ICRASH2012, international crashworthiness conference proceedings. Held 14–15 July 2012 at Milano.

Campbell, D. (2013). Impact! A Horizon Guide to Car Crashes. BBC4. [24.10.13, 23:00].

Christensen, J., Bastien, C., Blundell, M. V., Grimes, O., Appella, A., O'Sullivan, K., & Bareham, G. (2012a). Generation of optimised hybrid electric vehicle body in white architecture from a styling envelope. *Global Journal of Researches in Engineering B: Automotive Engineering*, *12*(1), 1–7. Available from http://www.engineeringresearch.org/index.php/GJRE/article/view/426/375.

Christensen, J., Bastien, C., Blundell, M. V., & Ravenhall, N. (2012b). Development of front end crash structure for lightweight hybrid electric vehicle. *Global Journal of Researches in Engineering Automotive Engineering*, *12*, issue 3, Version 1.0, 2012.

Christensen, J., Bastien, C., & Blundell, M. (2012c). Effects of roof crush loading scenario upon body in white using topology optimisation. *International Journal of Crashworthiness*, *17*(1), 29–38. Available from http://dx.doi.org/10.1080/13588265.2011.625640.

Crandall, J. R. (2012). Simulating the road forward: the role of computational modelling in realizing future opportunities in traffic safety. In: *IRCOBI 2012 international conference*. Available from <www.ircobi.org>. Held in September at Trinity College Dublin, Ireland.

Crandall, J. R. (2013). International advanced course on injury biomechanics. Course notes. Held 11–15 March 2013 at Barcelona, Spain.

Crash Network. (2013). The Web Portal for the Crash Test Engineer. Technical website. Available from <www.crash-network.com/Regulations/ECE_Regulations/ece_regulations.html> Accessed 29.12.13.

Davies, G. (2003). *Materials for automobile bodies*. Oxford, UK: Butterworth Heinemann.

Delphi. (2012). Worldwide emission standards – passenger cars and light duty vehicles. Technical Report. Available from <delphi.com/pdf/emissions/Delphi-Passenger-Car-Light-Duty-Truck-Emissions-Brochure-2011-2012.pdf> Accessed 05.07.13.

Department for Transports. (2011a). Drivers in reported accidents by gender, number injured, road user type and age, Great Britain. Technical database. Available from <https://www.gov.uk/government/organisations/department-for-transport/series/road-accidents-and-safety-statistics> Accessed 18.01.14.

Department for Transports. (2011b). Road transport forecasts 2011, results from the Department for Transport's National Transport Model. Technical Report. Available from <assets.dft.gov.uk/publications/road-transport-forecasts-2011/road-transport-forecasts-2011-results.pdf> Accessed 25.07.13.

Du Bois, P. (2010). Crashworthiness and Impact Engineering with LS-Dyna. Engineering Course Notes, California, Livermore Software Technology Corporation.

EuroNCAP. (2013). www.euroncap.com Accessed 3.07.13.

Evans, R. J. (2013). Injury prevention within a car, focusing on the inclusion of seatbelt and how best to model the steering column. Unpublished MSc thesis. Coventry University.

Folksam. (2013). How Safe is your Car? Technical Report. Available from <www.folksam.se/polopoly_fs/1.11226!/webbversioneng_R6546.pdf> Accessed 3.07.13.

Grimes, O., Christensen, J., & Bastien, C. (2013). Lightweighting of a hydrogen fuel cell vehicle whilst meeting urban accident criteria. In: *EVS27 international symposium conference proceedings*. Held 17–20 November 2013 at Barcelona, Spain.

Happian-Smith, J. (2002). *An introduction to modern vehicle design*. Oxford, UK: Butterworth Heinemann.

IIHS. (2013a). Insurance Institute for Highway Safety. Insurance testing. Available from <www.iihs.org/> Accessed 3.07.13.

International Transport Forum. (2013). Road Safety Annual Report 2013. Technical Report Available from <www.internationaltransportforum.orr/Pub/pdg/13IrtadReport.pdf> Accessed 17.03.13.

Kullgren, A., Lie, A., & Tingvall, C. (2010). Comparison between Euro NCAP Test Results and Real-World Crash Data. *Journal of Traffic Injury Prevention, 11*(6), 587–593.

Kumar, S. (2008). A Numerical Study on the Axial Crush Characteristics of Thin Walled Rectangular Tubes Subjected to Dynamic Impact. SAE, 2008-01-0242, Society of Automotive Engineers.

LSTC. (2012), LS-Dyna. Keyword User's Manual. Volume 1. Technical Manual. Available from <ftp.lstc.com/anonymous/outgoing/trent001/manuals/ LS-DYNA_manual_Vol_I_R6.1.0.pdf> Accessed 12.03.12.

Mahangare, M. (2007). To model an airbag for out-of-position occupants in frontal impacts. PhD Thesis, Coventry University.

Mahangare, M., et al. (2007). Investigation into the effectiveness of advanced driver airbag modules designed for OOP injury mitigation. In: *ESV2007 international conference*. Available from <www-nrd.nhtsa.dot.gov/pdf/esv/esv20/07-0319-O.pdf>. Held 18–21 June 2007 at Lyon, France Accessed 12.01.12.

Mayer, C. (2013). Application of numerical human body models within automotive development aspects of model verification and validation. In: *9th ARUP European conference proceedings*. Held 2–4 June 2013 at Manchester, UK.

Microcab (2015) 'Microcab H2EV' available from <www.microcab.co.uk> Accessed 10.11.13.

MIRA. (2014). Motor Institute Research Association. Available from http://www.mira.co.uk/.

NCAC. (2013). Finite element model archive. Computer Models Database. Available from <www.ncac.gwu.edu/vml/models.html> Accessed 1.07.12.

Neal-Sturgess, C. (2013). Injury Causation & Rollover 2013. Unpublished Course Notes, Coventry University.

NHTSA. (1998). Development of improved injury criteria for the assessment of advanced automotive restraint systems. Technical report. Available from <ntl.bts.gov/lib/17000/17800/17880/PB2001104885.pdf> Accessed 28.12.13.

NHTSA. (2013a). Three-reasons-everyone-should-wear-a-seat-belt. Available from <www.nhtsa.org/2013/04/08/three-reasons-everyone-should-wear-a-seat-belt> Accessed 16.07.13.

NHTSA. (2013b). What we are and What We Do. Available from <www.nhtsa.gov/About(NHTSA/Who(We(Are(and(What(We(Do> Accessed 10.01.13.

NHTSA. (2013c). National Automotive Sampling System (NASS). Technical Database. Available from <www.nhtsa.gov/NASS> Accessed 8.07.13.

NHTSA. (2014), Crash test results. Available from <www.nhtsa.gov/Research/Databases(and(Software> Accessed 28.10.14.

Pirard, P. (2009). Design an impact test rig. Master's Degree dissertation. Unpublished, Coventry University, June 2009.

Prem, A., & Bastien, C. (2013). Optimisation of future lightweight multi-material vehicle crash structures, *Proceeding of IMECHE conference*, held on 3/12/2013, Gaydon, Warwickshire.

SAE International. (1995). INSTRUMENTATION FOR IMPACT TEST—PART 1—ELECTRONIC INSTRUMENTATION, Standard available from <https://law.resource.org/pub/us/cfr/ibr/005/sae.j211-1.1995.html> Accessed 12.10.14.

SAE J211. (2007). SAE J211-1 (1995): Instrumentation for Impact Test, Part 1, Electronic Instrumentation. Technical Document. Available from <https://law.resource.org/pub/us/cfr/ibr/005/sae.j211-1.1995.pdf> Accessed 25.10.14.

Shanahan, D. (2013). Human tolerance and crash survivability, NATO Technical Report. Available from <www.academia.edu/3636171/Human_Tolerance_and_Crash_Survivability> Accessed 26.09.14.

Shirwa, E. C., Stephenson, R., Batzer, S., & Grzbieeta, R. (2010). Review of the Jordan Rollover System (JRS) vis-à-vis other dynamic crash test devices. *International Journal of Crashworthiness*, *15*(5), 553–569.

Stubbs, D. (2013). Vehicle Signal Analysis. Unpublished Course Notes. Coventry University.

Toyota. (2010). THUMS (Total Human for Safety). Technical Information. Available from <www.toyota-global.com/innovation/safety_technology/safety_measurements/thums.html> Accessed 1.07.11.

UNECE. (2014). GTR 9: Proposed Draft Global Technical Regulation (Gtr) On Pedestrian Protection. Technical Report. Available from www.unece.org/fileadmin/DAM/trans/doc/2004/wp29grsp/ps-116.pdf Accessed 25.10.14.

VCA. (2013). Vehicle Certification Agency, Vehicle Legislation Approvers. Available from <www.dft.gov.uk/vca> Accessed 16.07.13.

Chapter | two

Numerical Techniques for Structural Assessment of Vehicle Architectures

CHAPTER OUTLINE

Many options exist for assessing the structural (safety) performance of vehicle architectures, including real-world testing, manual calculations, and computer simulations. There are three main parameters that dominate the decision process for the choice of analysis approach to the structural assessment of vehicle architectures: flexibility/versatility, accuracy, and precision.

On the basis of the variety of architectures and general geometry demonstrated in Chapter 1, it is clear that a flexible and versatile approach is required to systematically (and with relative ease) enable the analysis of a wide variety of components and systems. In addition to this a high level of accuracy and (generally) a high level of precision are both desirable to guide the design process with a significant level of confidence and clarity. Figure 2.1 illustrates the difference between precision and accuracy.

If asked to hit the bullseye (center) in Figure 2.1, then the series of dots on the left-hand side illustration represent a "precise" attempt because the relative

Nonlinear Optimization of Vehicle Safety Structures. http://dx.doi.org/10.1016/B978-0-12-804424-7.00002-6

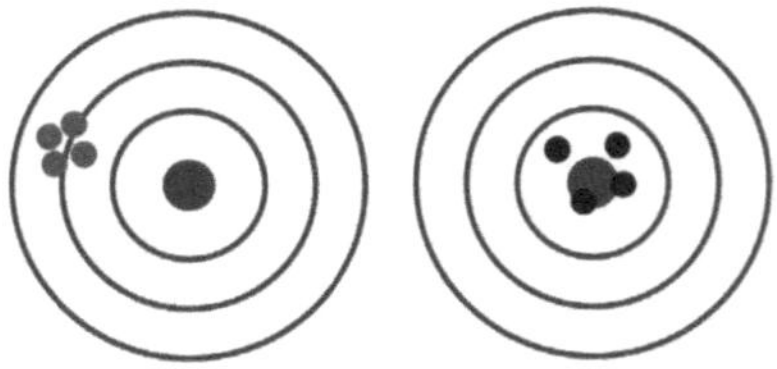

FIGURE 2.1 Precision versus accuracy.

change of location between these individual dots is small. These do, however, not represent an "accurate" attempt, as the dots are situated relatively far away from the target, that is, the bullseye. The dots on the right-hand side illustration in Figure 2.1 do, on the other hand, represent an "accurate" attempt, as all of these are relatively close to the target, that is, the bullseye. The dots on the right-hand side illustration are, however, not precise when compared with the left-hand side illustration, as the relative distances between the individual dots (right-hand side) are much greater than those of the left-hand side illustration.

The above explanation may seem trivial, but possessing the correct definitions of precision and accuracy plays a vital role in the overall understanding of the contents of this book.

The finite element method (FEM) is well suited for general and systematic formulation of many different types of engineering problems; however, many others, such as the finite difference method (FDM), the finite element limit analysis (FELA), or meshless methods, also exist. Always remember that:

FE is a tool, not the only tool.

Despite the existence of competing methods, FE is however (in general) the most widely used method, primarily due to the high level of flexibility/versatility combined with the potential of high levels of accuracy and/or precision.

2.1 INTRODUCTION TO FINITE ELEMENT ANALYSIS (FEA)

By the use of computer-aided design (CAD) software such as Catia or SolidWorks, it is possible to create 3D models displaying minute details of any component or system, for example, an entire vehicle, such as that illustrated in Figure 2.2.

With CAD models containing the level of detail such as the one displayed in Figure 2.2, is there still a need for finite element analysis (FEA)?

The answer to this question is simply YES, there is still need for FEA!

Theoretically, the CAD models can "replace" the FE models, for example by the use of meshless methods; however, in practice it is simply not feasible in the vast majority of cases, as will be further explained in Section 2.2. The primary reason for this is that the CAD models such as the one illustrated in Figure 2.2 generally utilize B-splines or nonuniform rational B-splines (NURBS) to represent

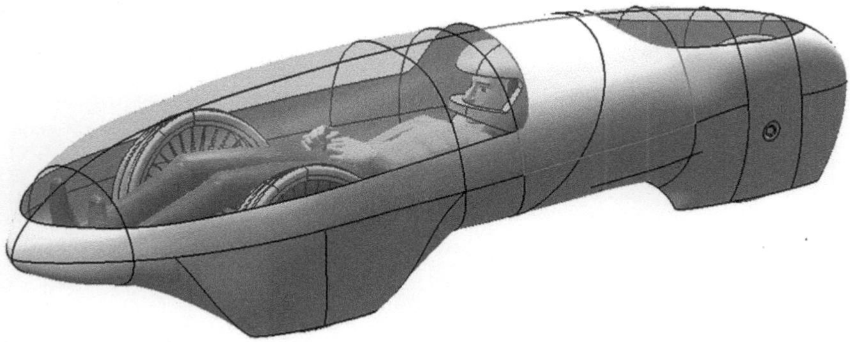

FIGURE 2.2 Example of CAD model.

lines and surfaces. These lines and surfaces represent continuous geometry, which can only be solved mathematically by the use of differential equations and/or integral expressions, as will be demonstrated in Section 2.2. These equations are very difficult and consequently extremely time consuming to solve. In almost all cases, these differential equations are impossible to solve analytically, and thus may only be solved numerically. In many cases, the fundamental differential equations may also be very cumbersome or even impossible to define, meaning that this method is neither very robust nor versatile in the context of solving complex solid mechanical (also known as mechanics of materials) problems. From a mathematical viewpoint, FE offers a "simpler," more versatile, and robust numerical approach to solving these types of problems; this is one of the main reasons why FE is utilized. Please note that even though CAD cannot replace FE, they are often used in combination to solve engineering problems by importing CAD-based geometry into FE software.

A general overall flowchart for solving an FE problem may be illustrated as done in Figure 2.3, which is an adaptation of Lund (2006).

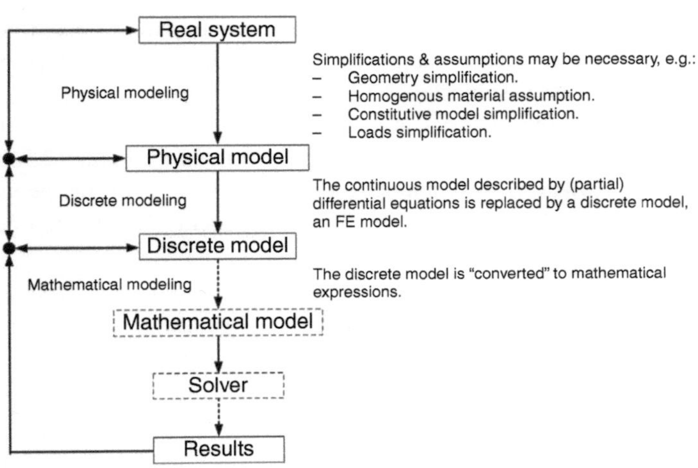

FIGURE 2.3 General flowchart for solving an FE problem.

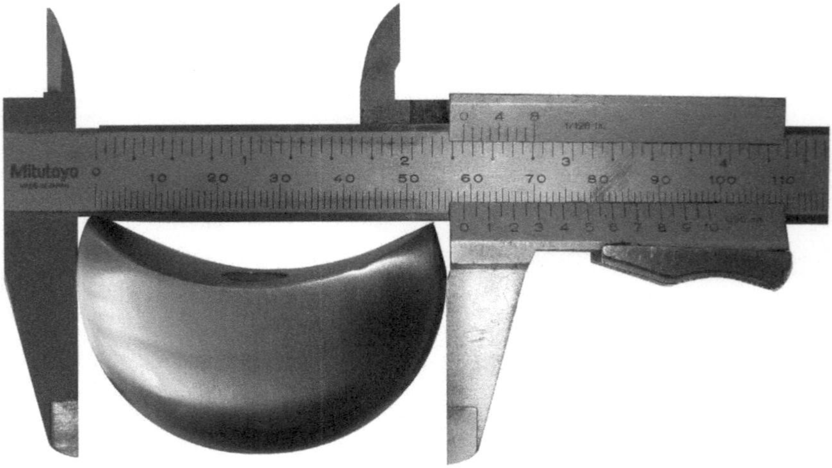

FIGURE 2.4 Measuring with a vernier caliper.

The "solid line" boxes and arrows, including associated text, in Figure 2.3 represent the "manual steps" of the procedure, i.e., the steps that the engineer must carry out manually.

The first step, that is, "physical modeling" is the step between the "real system" and "physical model" steps. This consists of measuring characteristics of the real (world) system; this could, for example, include conducting measurements of simple geometry by means of a vernier caliper (Figure 2.4), and utilize this to create a CAD model (physical model).

The second step in Figure 2.4 is the "discrete modeling" step. This step involves creating a "discrete model" (e.g. an FE model) from a "physical model" (e.g. a CAD model). This step involves "discretization" of the physical model.

To better understand the "discrete modeling" step, that is, converting a physical model into a discrete model, consider the bracket displayed in Figure 2.5. Imagine that the impending task of the engineer is to find the stresses throughout the structure, the displacement at point A, and the reaction forces at points B, C, and D when the truss is subjected to the force P (Figure 2.5).

Assuming that the material is isotropic and behaves in a linear manner, an obvious method of solving the problem illustrated by Figure 2.5 is to use beam theory. However, this cannot be directly solved using the geometry of Figure 2.5; instead, it is necessary to discretize the continuous CAD geometry

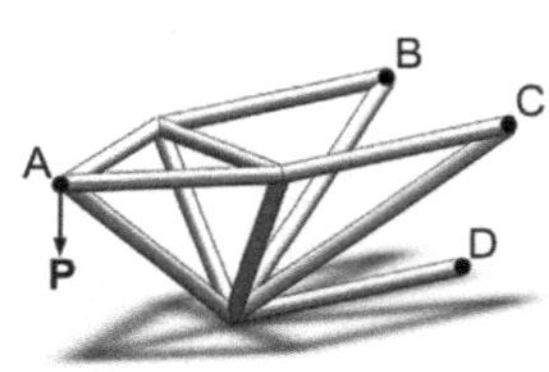

FIGURE 2.5 Continuous (CAD) model of truss structure.

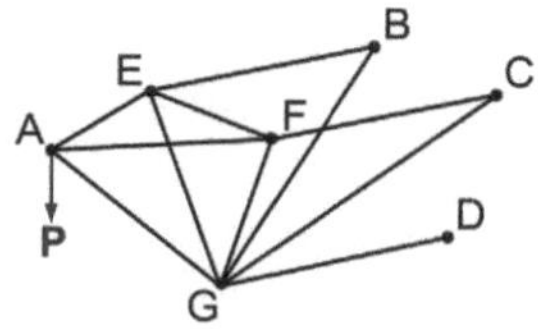

FIGURE 2.6 Discrete model of truss structure.

in Figure 2.5 is replaced by a "discontinuous," discrete geometry, such as the one illustrated in Figure 2.6.

In order to solve the task, it is imperative that information relating to the following aspects is available:

- Geometry.
- Applied loading and constraints/boundary conditions (BC).
- Constitutive/material characteristics.

The previous list should be apparent to anyone who possesses basic knowledge of solid mechanics. Thereby, the task set forth may be solved; the stresses, displacements, and reaction forces may be found, for example, via FEA using the discrete model illustrated in Figure 2.6. The system may be solved either analytically or numerically by:

Obtaining equilibrium between the applied loading and the (deformation) energy stored in the component/system!

This statement is central to both FE, solid mechanics, and many other aspects of engineering; always remember that the problem is solved by means of the equations of equilibrium. The energy stored in the system or component may assume many different forms; in terms of mechanical engineering it is most often stored as elastic- and/or plastic deformation and/or displacement.

The "dashed" lines and boxes in Figure 2.3 may, in general, be expected to be carried out internally in the commercial FE software/solver (if used). This means that the users will not normally be directly involved in these steps. However, it is also in connection with these steps that the subsequent optimization algorithm(s) generally will be implemented; hence, it is important that the reader possesses an insight into these steps.

Furthermore, the postprocessing and extraction of optimization results may be complex; hence, an "automated" method of extracting these may be very useful, as will subsequently be discussed.

In addition, extra attention should be paid to the "multidirectional" arrows on the left-hand side of Figure 2.3; these indicate that the actions of a "skilled and observant engineer" who will always relate and compare the results obtained via FEA to the preceding modeling steps in order to check, substantiate, and validate the accuracy of the obtained results. The steps illustrated by Figure 2.3 may, in general, be divided into three different categories:

1. Preprocessing, this involves the "physical modeling" and "discrete modeling" steps;
2. Solving, this includes the "dotted" lines and boxes;

3. Postprocessing, this is the final step, that is, extracting, viewing, interpreting, and concluding upon the results of the FEA and/or optimization analysis.

As mentioned, the preprocessing phase includes discretizing the structure to be analyzed; this is done by discretizing using an absolute (or finite) number of elements hence the name: FEM.

An element is defined by nodes (with geometric coordinates); the displacements in any FE analysis will only be calculated for the nodes, consequently forces, etc., will only be transferred from element to element via the nodes. This also means that constraints and external force may only be applied on nodes; an important aspect to keep in mind in connection with the implementation of optimization algorithms within FEA.

The following section will extract the fundamental equation of FEA using theory of elasticity. This derivation is important as it will be used as a starting point for implementation of optimization algorithms throughout this book.

2.2 THEORY OF ELASTICITY

The purpose of this section is to provide a brief mathematical introduction to the fundamental equation of FEA, which forms the basis of any FE analysis.

The derivations of this section are based on Cook, Malkus, Plesha, & Witt (2001), Lund (2006), and Christensen, Montes Schutte, Nesari, & Dedic (2007) and will be carried out using energy considerations from (linear) theory of elasticity. Therefore, it would be beneficial to possess prior knowledge of the concepts of this theory to enable a better understanding of the derivations. Additional information may also be found in Ugural and Fenster (2003).

The total elastic potential is a key principle in the theory of elasticity. It summarizes the internal work U and the external work Ω. The total elastic potential Π_p may therefore be defined as Equation (2.1)

$$\Pi_\mathrm{p} = U + \Omega \tag{2.1}$$

The external work may be perceived as the deformation and/or displacement of the component/system to which the external loading is applied. The internal work may be perceived as the (elastic) deformation energy stored in the component/system; that is, the potential energy that will return the component/system to its original state (shape) when the external loading is removed. As an example, imagine pulling a rubber band. The external work carried out by the external loading is used to elongate and deform the rubber band. Once the external load is removed, the rubber band will return to its original length and shape courtesy of the elastic (deformation) energy stored within the rubber band. Please note that energy conservation is assumed; this means that the amount of energy used to, for example, increase the temperature of the rubber band is ignored.

Analog to solid mechanics, work is equal to the product of force and displacement, the energy U_0 required to deform a unit volume of material is given by Equation (2.2)

$$U_0 = \int \{\sigma\}^T \{d\varepsilon\} = \frac{1}{2}\{\sigma\}^T \{\varepsilon\} = \frac{1}{2}\{\varepsilon\}^T \{\sigma\} \tag{2.2}$$

where U_0 is defined in Equation (2.3)

$$U = \int U_0 \; dV \tag{2.3}$$

Equation (2.3) thus states that the total internal energy may be obtained by integrating the expression of the unit volume over the total volume of the elastic body (component), that is, summarizing the internal energy in a continuum. Assuming that the material in question is indeed elastic, the energy stored during deformation may be recovered once the external loading is removed (as was the case in the rubber band example). In this instance, U is denoted as the elastic strain energy density.

Furthermore, by assuming that the material behavior is linear and isotropic, Hooke's generalized law is valid, Equation (2.4)

$$\{\sigma\} = \{E\}\{\varepsilon\} = \overbrace{\frac{E}{1-\upsilon^2}\begin{bmatrix} 1 & \upsilon & 0 \\ \upsilon & 1 & 0 \\ 0 & 0 & \dfrac{1-\upsilon}{2} \end{bmatrix}}^{[E]}\{\varepsilon\} \tag{2.4}$$

If the (linear elastic isotropic) body is exposed to initial stresses, $\{\sigma_0\}$, Equation (2.4) takes the form of Equation (2.5). An example of this could be that the rubber band is already suspended between two columns, before it is further elongated (pulled).

$$\{\sigma\} = [E](\{\varepsilon\} - \{\varepsilon_0\}) + \{\sigma_0\} \tag{2.5}$$

By inserting Equation (2.5) into Equation (2.2) and subsequently integrating, Equation (2.6) is obtained.

$$U_0 = \frac{1}{2}\{\varepsilon\}^T [E]\{\varepsilon\} - \{\varepsilon\}^T [E]\{\varepsilon_0\} + \{\varepsilon^T\}\{\sigma\} \tag{2.6}$$

Equation (2.6) thus describes the deformation energy per unit volume of a prestressed body. According to Equation (2.3), the total strain energy of the body may be obtained by integrating over the volume of the body, as in Equation (2.7)

$$U = \int_V U_0 \; dV = \int_V \frac{1}{2}\{\varepsilon\}^T [E]\{\varepsilon\} - \{\varepsilon\}^T \{\varepsilon_0\} + \{\varepsilon^T\}\{\sigma\} dV \tag{2.7}$$

As previously stated, the potential of the external work Ω is dependent upon the externally applied loading. This loading may be divided into volumetric loads (e.g. gravitational loads), surface loads (e.g. pressure), and point loads (e.g., single point forces). The volumetric loads and the associated displacements are defined in Expression (2.8)

$$\{f\}=\begin{Bmatrix} f_x \\ f_y \\ f_z \end{Bmatrix} \quad ; \quad \{d\}=\begin{Bmatrix} u \\ v \\ w \end{Bmatrix} \tag{2.8}$$

By utilizing the vectors defined in Expression (2.8), the potential for work (i.e. energy) lost by the force applied to an infinitesimal (i.e. very small) unit of volume dV can be expressed as Equation (2.9)

$$d\Omega_V = -\{u\}^T\{f\}dV \tag{2.9}$$

In Equation (2.9), V denotes that the loads in question are volumetric; the negative sign denotes a lost potential. By integrating Equation (2.9) over the total volume of the (elastic) body, the work carried out by all the volumetric loads; the total lost potential, may be obtained, as in Equation (2.10)

$$\Omega_V = -\int_V \{u\}^T\{f\}dV \tag{2.10}$$

By adopting the same approach as above, similar expressions may be obtained for the surface and point loads, as in Equations (2.11) and (2.12), respectively.

$$\Omega_S = -\int_S \{u\}^T\{\Phi\}\,dS \tag{2.11}$$

$$\Omega_P = -\{D\}^T\{P\}\,dP \tag{2.12}$$

In Equation (2.11), the subscript S denotes a surface load (or integration over the surface); Φ denotes the surface tractions.

For clarity, the P vector in Equation (2.12) denotes the global single point loads vector, while D represents the global displacement vector.

The total lost potential by the applied forces may thus be obtained by summation of Equations (2.9) through (2.12), leading to Equation (2.13)

$$\Omega = \Omega_V + \Omega_S + \Omega_P \tag{2.13}$$

The total elastic potential Π_p as defined in Equation (2.1) is thus the sum of Equations (2.7) and (2.13), yielding Equation (2.14)

$$\begin{aligned}\Pi_p = &\overbrace{\int_V\left(\frac{1}{2}\{\varepsilon\}^T[E]\{\varepsilon\}-\{\varepsilon\}^T[E]\{\varepsilon_0\}+\{\varepsilon\}^T\{\sigma\}\right)dV}^{U} \\ &\overbrace{-\int_V\left(\{u\}^T\{f\}\right)\,dV-\int_S\left\{\{u\}^T\{\Phi\}\right\}dS-\{D\}^T\{P\}}^{\Omega}\end{aligned} \tag{2.14}$$

Equation (2.14) thereby represents a "typical" elastic problem, with infinite degrees of freedom. In this case the functional in Equation (2.15) is a valid solution.

$$\Pi_{\mathrm{p}} = f\left(\frac{\partial\{u\}}{\partial x}, \frac{\partial\{u\}}{\partial y}, \frac{\partial\{u\}}{\partial z}, \{u\}\right) \tag{2.15}$$

A functional is in essence a function that takes a vector as an input argument and returns a scalar. This is very "handy" as the objective is to achieve static equilibrium of the total elastic potential, which is analog to, for example, a simple beam deflection analysis. The total elastic potential will achieve static equilibrium when Equation (2.16) is true.

$$\left\{\frac{\partial \Pi_p}{\partial\{D\}}\right\} = \{0\} \tag{2.16}$$

This means that a linear elastic system, such as the previously used example of the rubber band, with an infinite number of DOF (remember that this is a continuous system; it has not been discretized as such) may be solved by defining a functional (2.15), so that Equation (2.16) is true.

Please note that Expressions (2.14) through (2.16) consist of a series of differential equations and integral expressions. As previously stated, these expressions are very difficult if not impossible to define and/or solve for the majority of problems. Therefore, it is necessary to introduce an approximation of this continuum with a finite number of DOF. If it was indeed possible to define and solve these expressions in general, it would be extremely interesting from an optimization viewpoint, as an optimization algorithm could be defined and integrated within these. In practice, this would mean that a "continuous function" could be defined to, for example, describe the displacement at any given point within any structure – this would also imply that there is no need for FEA. Always remember that

> *"FE is not a science but a compromise; software vendors do the arithmetic, not the physics."* (Du Bois, 2007)

For the time being this hypothetical solution methodology is abandoned, and the focus returns to the solution utilizing a finite number of DOF. The discretization can in essence be conducted in many different ways, one of which is the Rayleigh–Ritz method; this provides a way of replacing all displacement modes of a system (component) with infinite DOF $\{u\}$ (i.e. a continuous system) with a finite subset $\{\tilde{u}\}$.

Thereby, the continuous problem (2.15) may be "approximated" by the discrete problem formulation of Equation (2.17)

$$\Pi_p = f\left(\frac{\partial\{\tilde{u}\}}{\partial x}, \frac{\partial\{\tilde{u}\}}{\partial y}, \frac{\partial\{\tilde{u}\}}{\partial z}, \{\tilde{u}\}\right) \tag{2.17}$$

By using an approximation (2.17) of the continuous problem (2.15), an exact solution will not be possible in the vast majority of cases. The obtained solution will (ideally) converge toward the exact value of the minimum total elastic potential with an error that is dependent upon the number of DOF of the discrete model. The original Rayleigh–Ritz method uses only one approximation for the displacements; this is also known as the displacement hypothesis for the entire system.

As FE discretizes the structure by a finite number of elements, subsequently a displacement hypothesis is made for the individual element. This is a very important aspect, which can be summarized as stated below:

> *In FE, the geometry is approximated by a finite number of elements discretization, the displacement relating to these are also approximated, using the displacement hypothesis!*

In FEA, the elastic potential is therefore formulated for the individual element, and not (initially) for the structure as a continuum. Instead, the total elastic potential is obtained via summation, thereby Equation (2.14) becomes Equation (2.18)

$$\begin{aligned}\Pi_p = &\sum_{i=1}^{n_{els}}\left(\int_V\left(\frac{1}{2}\{\varepsilon\}_i^T[E]\{\varepsilon\}_i-\{\varepsilon\}_i^T[E]\{\varepsilon_0\}+\{\varepsilon\}_i^T\{\sigma\}_i\right)dV\right)\\&-\sum_{i=1}^{n_{els}}\left(\int_V\left(\{u\}_i^T\{f\}_i\right)dV\right)-\sum_{i=1}^{n_{els}}\left(\int_S\left\{\{u\}_i^T\{\Phi\}_i\right\}dS\right)-\{D\}^T\{P\}\end{aligned} \tag{2.18}$$

Equation (2.18) thereby represents the "discrete" approximation of the total elastic potential, where i represents the current element number and n_{els} represent the total number of elements in the model. The D and P vectors represent the global displacement and load vectors, respectively. Consequently, these are not functions of the individual element i, which is also distinguishable in Equation (2.18).

Equation (2.18) is of high interest in an optimization context as it can be used to evaluate the "structural performance" of the system.

At this point, it is necessary to introduce shape functions. The purpose of a shape function, N_j, is to (mathematically) define the possible displacements of any node, j, as the problem has been discretized and is not continuous. The shape function will help bridge the discontinuities generated by the discretization process. The shape functions are thus associated with the nodes of the individual element. For example, the shape function N_1 belongs to node 1 of a specific element. A four-noded element, such as a "standard" shell element, will thus have four shape functions associated with it, N_1 to N_4. These may be "compiled" into the shape function matrix $\lfloor N_j \rfloor_i$ belonging to element i. Shape functions may be linear, bilinear, quadratic, or even cubic; this will have a significant impact on the central processing unit (CPU) time and often also upon the accuracy and/or precision of the results.

Equation (2.19) may be used to define the approximate displacements associated with element i.

$$\left\{\tilde{u}_j\right\}_i = \left[N_j\right]_i \left\{d_j\right\}_i \tag{2.19}$$

In Equation (2.19), $\{d_j\}$ represents the displacement vector of node j, which may be combined to obtain the element displacement vector $\left\{d_j\right\}_i$. For simplicity, the node reference j will be omitted in the following derivations.

Analog to solid mechanics/beam theory, the strains of a component/element can be found by differentiation of displacements, as given in Equation (2.20)

$$\{\varepsilon\}_i = [\partial]\{\tilde{u}\}_i \tag{2.20}$$

By utilizing Equation (2.19), Equation (2.20) becomes Equation (2.21)

$$\{\varepsilon\}_i = [B]_i \{d\}_i \tag{2.21}$$

$[B]_i$ in Equation (2.21) is often referred to as the strain–displacement matrix, and is defined in Equation (2.22)

$$[B]_i = [\partial][N]_i \tag{2.22}$$

In Equation (2.22), $[\partial]$ contains the derivatives of the shape functions (with respect to x, y, and z). The size of this matrix will depend upon the element type (e.g. 1D, 2D, or 3D) and of course the number of nodes within the element itself.

At this point, it is convenient to "shorten" the notation of the equations. This is done by utilizing Equations (2.20) and (2.21). Thereby, the element stiffness matrix $[k]_i$ may be introduced in Equation (2.23)

$$[k]_i = \int_V \left([B]_i^T [E][B]_i\right) \, dV \tag{2.23}$$

The (local) element stiffness matrix defined in Equation (2.23) includes the matrix $[E]$ containing material data and is of particular interest in connection with optimization, it play a pivotal role in connection with the VDM-SIMP and ESLM optimization approaches as shall be subsequently discussed. Equation (2.23) also contains the strain–displacement matrix $[B]$, Equation (2.22), which contains information regarding the allowable displacements of the individual nodes.

In addition to the local stiffness matrix it is also convenient to introduce the consistent load node vector $\left\{r_e\right\}_i$, which contains all the local nodal loads, of node j belonging to element i. This may be obtained by utilizing Expressions (2.18) through (2.22), leading to $\left\{r_e\right\}_i$ as defined in Equation (2.24)

$$\begin{aligned}\left\{r_e\right\}_i &= \int_V \left([B]_i^T [E]\{\varepsilon_0\}\right) \, dV - \int_V \left([B]_i^T \{\sigma_0\}\right) \, dV \\ &\quad + \int_V \left([N]_i^T \{f\}_i\right) dV + \int_S \left([N]_i^T \{\Phi\}_i\right) \, dS\end{aligned} \tag{2.24}$$

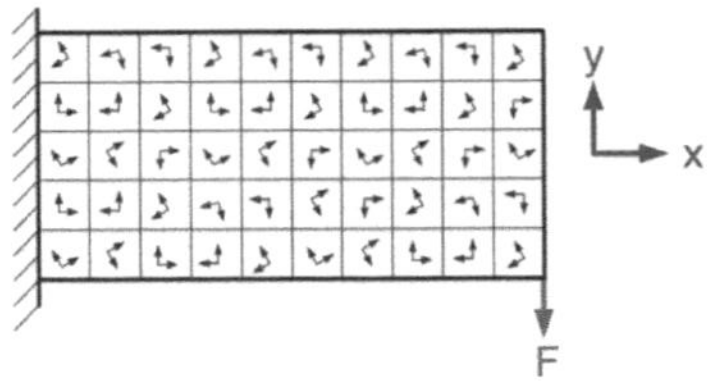

FIGURE 2.7 Local (element) coordinate systems.

The consistent loadvector for node j of element i, Equation (2.24), is listed in local coordinates. It is however convenient to solve the general problem, Equation (2.18), in global coordinates so that all displacements and rotations are relative to the global coordinate system, and not the local ones. To appreciate this, consider Figure 2.7. If the system (2.24) was solved using the local coordinate systems, then the displacements of the individual nodes would be relative to the local coordinate systems such as those illustrated in Figure 2.7. This would be very confusing and difficult to interpret from both an analysis and an optimization viewpoint. Instead, it is generally more convenient if all displacements and rotations are relative to a predefined global coordinate system, such as that illustrated in Figure 2.7. Please note that the "random" orientations of the local coordinate systems in Figure 2.7 are not a gross exaggeration; FE meshes of complex geometries often lead to a wide variety in the orientation of local (element) coordinate systems.

In order to solve the system (2.24) in global coordinates, the $\{d\}$ vector containing the displacements and rotations of element i (in local coordinates) must be related to the global displacement vector $\{D\}$, containing all DOF of the system. This is done by the transformation matrix $[L]_i$, which rotates the local axis orientations into the global ones using basic trigonometry. At this point, it is sufficient to note that the transformation matrix $[L]_i$ contains only the values 0 and 1. $[L]_i$ is therefore an m_i by m matrix, in which i represents the DOF of the current node. Thereby, Equation (2.25) can be defined.

$$\{d\}_i = [L]_i \cdot \{D\} \tag{2.25}$$

By using Equations (2.23) and (2.24), Equation (2.24) may be rewritten as Equation (2.26)

$$\Pi_\mathrm{p} = \frac{1}{2}\sum_{i=1}^{n_\mathrm{els}}\left(\{d\}_i^T [k]_i \{d\}_i\right) - \sum_{i=1}^{n_\mathrm{els}}\left(\{d\}_i^T \{r_e\}_i\right) - \{D\}^T \{P\} \tag{2.26}$$

By inserting Equation (2.25) into Equation (2.26), Equation (2.27) appears.

$$\Pi_p = \frac{1}{2}\{D\}^T\left(\sum_{i=1}^{n_{els}}[L]_i^T [k]_i [L]_i\right)\{D\} - \{D\}^T\left(\sum_{i=1}^{n_{els}}[L]_i^T [r_e]_i\right) - \{D\}^T \{P\} \tag{2.27}$$

The local stiffness matrices $[k]_i$ may be used to construct the global stiffness matrix $[K]$; as in Equation (2.28)

$$[K]=\sum_{i=1}^{n_{els}}[L]_i^T[k]_i[L]_i \tag{2.28}$$

All loads in Equation 2.27, that is, the volumetric-, surface-, and single-point loads as listed in Equation (2.13), may be combined to a single load vector, denoted $\{F\}$, seen in Equation (2.29)

$$\{F\}=\sum_{i=1}^{n_{\mathrm{els}}}\left([L]_i^T\{r_e\}_i\right)+\{P\} \tag{2.29}$$

By inserting Equations (2.28) and (2.29) into Equation (2.27), the total elastic potential may be rewritten as Equation (2.30)

$$\Pi_{\mathrm{p}}=\frac{1}{2}\{D\}^T[K]\{D\}-\{D\}^T\{F\} \tag{2.30}$$

Equation (2.16) can now be used to determine when the total elastic potential, that is, Equation (2.30), obtains static equilibrium, as in Equation (2.31)

$$\frac{\partial\Pi_{\mathrm{p}}}{\partial\{D\}}=\frac{1}{2}[K]\{D\}-\{F\}=0 \tag{2.31}$$

It can be shown that Equation (2.31) will be true if Equation (2.32) is true.

$$\{F\}=[K]\{D\} \tag{2.32}$$

Equation (2.32) is often referred to as the fundamental equation of FEA. This is the primary equation governed with FEA. It is an equation of equilibrium attempting to "balance" the external loading, $\{F\}$ with the response of the system, represented by the global stiffness matrix $[K]$, and the unknown nodal displacements and rotations in $\{D\}$. Please note that this vector is also often represented as $\{U\}$.

Thereby, the reader should have obtained a "scientific" and "mathematical" appreciation of how the fundamental equation of FE has been derived, including pinpointing particular aspects which are interesting from a structural optimization viewpoint. The next section will be dedicated to elements that are pivotal to FE analysis and optimization.

2.3 ELEMENTS

The purpose of this section is to provide a basic overview of some of the different types of elements that can be used in connection with finite element analysis. The section will commence by presenting different types of elements. In general, FE elements may be divided into four different types:

- Zero-dimensional elements (0D)
- One-dimensional elements (1D)
- Two-dimensional elements (2D)
- Three-dimensional elements (3D)

It is convenient to commence the explanation of the four types listed above with 1D elements, because these should be familiar to the reader.

2.3.1 One-dimensional elements

The "dimension" of an element is defined in the exact same manner as used in solid mechanics. For example, solving the classic engineering problem of finding the displacement $\delta(x_i)$ of the cantilever beam in Figure 2.8 is a 1D problem. This is because the deflection is merely a function of a single coordinate, x_i, as the beam is perpendicular with the x-axis, in this case $x = x_i$, provided that the origins of the two are coincident.

Note that even though the beam in Figure 2.8 is defined in a 2D geometrical space, the xy-plane, the problem still remains 1D as the resulting deflection, $\delta(x_i)$, is solely a function of one coordinate, namely x_i.

Always remember that the dimension of the element does not necessarily equal the geometrical space (dimension) in which it is defined; that is, a 1D element may be defined using a 2D or even 3D geometrical space. In other words, the line element defined in a 3D (x, y, z) space with endpoints A(0,0,0) and B(1,1,1) is a 1D element, provided that the deflection, δ, is merely a function of the location, x_i, along the axis of the beam. The same approach may be taken to applying different load-cases to a beam as illustrated by Figure 2.9, which is also a 1D problem, as the deflections (and compression) caused by F_1 and F_2 are merely functions of the single coordinate x_i.

Examples of 1D elements are beams and rods; remember that the difference between a beam and a rod is the inability of the latter to react a bending moment. A good example of how a problem defined in a 3D geometrical space is modeled by 1D elements can be understood by returning to the truss example of Figure 2.5, which is a 3D (CAD) model in a 3D space, where Figure 2.6 represents the 1D (FE) model in a 3D space. Please note that components such as springs and dampers may also be used in FEA, these are also (in general) 1D elements. The mesh associated with 1D elements is normally denoted as a "line mesh" and is often illustrated by line(s) such as indicated in Figures 2.8 and 2.9.

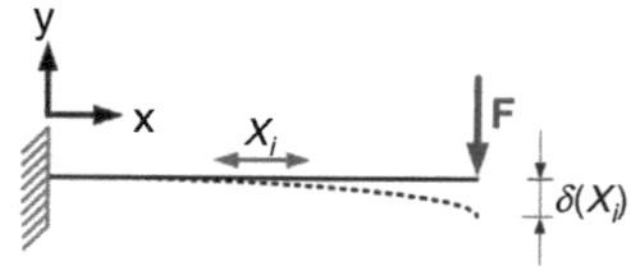

FIGURE 2.8 Example of a 1D problem.

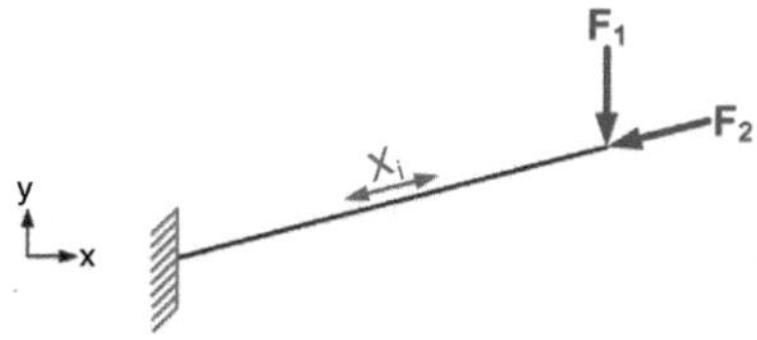

FIGURE 2.9 Another example of a 1D problem.

2.3.2 Two-dimensional elements

Two-dimensional elements are often referred to as "shell elements." In line with the definitions above, the usage of 2D elements indicates that, for example, displacements are functions of two parameters. Remembering that one of the basic assumptions of beam theory is the height (or width) to length ratio λ must be "low", λ should typically be less than 1/10 for conventional beam theory to be valid. As opposed to the "long" and "slender" beams, an example of a "short" and "fat" beam may be seen in Figure 2.10.

For example by setting L to 1.0 (Figure 2.8), h to 0.6 and w to 0.1 means that the beam violates the slenderness ratio, λ assumption of beam theory, and 1D elements should not be used. Instead 2D elements may be used; in this case the displacement becomes a function of two coordinates $\delta(x_i, y_i)$, where x_i and y_i are parallel to the x- and y-axis, respectively.

2D elements may either be triangular or square elements (also sometimes referred to as "quad" elements), as illustrated in Figure 2.11; no other 2D shapes are generally used in FEA.

It is worth noting that the number of nodes of the respective elements illustrated in Figure 2.11 is the minimum number of nodes the elements may have, but not the maximum. For instance, a 2D square element may have eight (or more) nodes; this is directly related to the shape functions.

Also please note that the side lengths L in Figure 2.11 of the ideal element are identical for all three sides (triangular element) or four sides (square

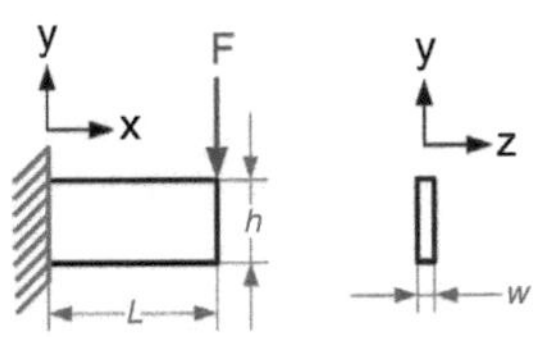

FIGURE 2.10 Example of a "short" and "fat" beam.

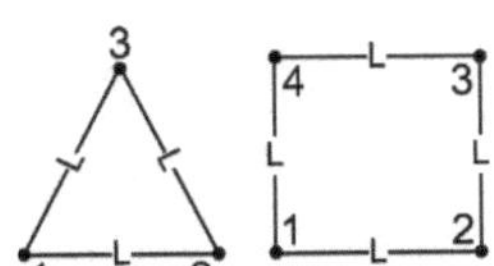

FIGURE 2.11 Triangular and square 2D (shell) elements.

element; this is often referred to as the mesh size. For example, if an FE model is said to contain a mesh size of 1.0 mm it means that the side length L of the ideal element is 1.0 mm. The majority of models will, however, contain elements deviating from this size (and indeed shape); the "severity" of this "deviation" is indicated by the Jacobian matrix/determinant. The determinant of the Jacobian matrix is also often referred to as the "Jacobian." The Jacobian is in essence a scale factor between the "perfect" element and the actual element. For 1D elements this is a length ratio, for 2D elements it is an area ratio, and for 3D elements it is a volume ratio. An undistorted 1D (actual) element will be exactly the same length as the "perfect" element; hence, the Jacobian (for the element) will be 1.0, which is the maximum value. If the element is distorted, the Jacobian will decrease; generally, Jacobian values of 0.7 or above are acceptable. If the Jacobian value becomes negative, an FE analysis will not run.

2.3.3 Three-dimensional elements

Three-dimensional elements are often referred to as "solid elements." These are elements where, for example, the (nodal) displacements of the element are functions of three parameters. 3D elements may either consist of triangular or square sides. These are denoted tetrahedron (plural: tetrahedral) and hexahedron (plural: hexahedral) elements, or simply "tetra" and "hex" or "brick" elements. The two types, that is, tetra and hex elements, are illustrated in Figure 2.12. Please note that the tetra elements may be either five- or six-noded (as a minimum); the shapes of the two are significantly different as illustrated in Figure 2.11.

In line with the 2D elements in Figure 2.11, it should be noted that the number of nodes of the respective elements illustrated in Figure 2.11 are the minimum number of nodes the elements may have, but not the maximum. Furthermore, the mesh size refers to the average side lengths L of the individual elements.

An example of when it is appropriate to use 3D elements can be appreciated by considering the model illustrated in Figure 2.13.

Based on solid mechanics the "engineering intuition" indicates that the displacements and stresses at any point within the structure in Figure 2.13 are dependent upon three variables, namely $\delta(x_i, y_i, z_i)$, where x_i, y_i, z_i are parallel to the x-, y-, and z-axis, respectively.

Initially, it might seem that 3D elements would be the ideal element choice for discretizing the majority of geometries, as the "real system" (Figure 2.5) is in

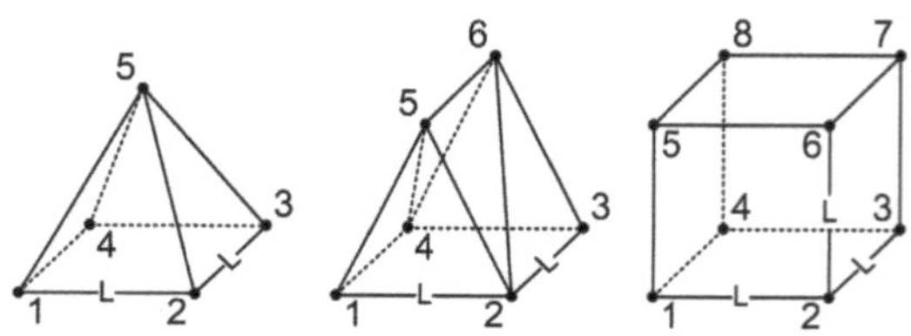

FIGURE 2.12 Five-node tetrahedron, six-node tetrahedron, and eight-node hexahedron 3D (solid) elements.

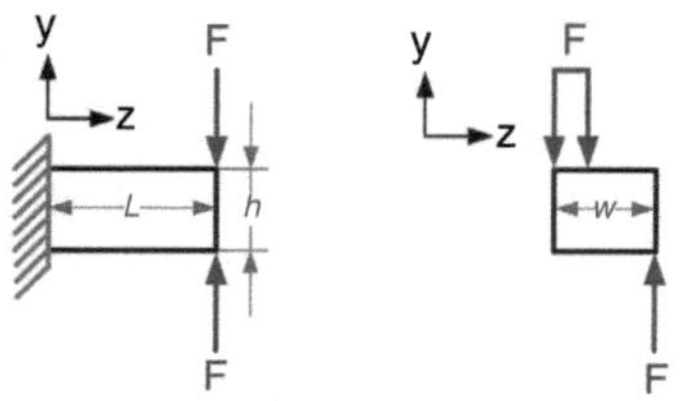

FIGURE 2.13 Example of a model where 3D elements are appropriate to use for discretization.

essence always in a 3D geometrical space. However, discretizing using solid elements does not automatically ensure an increased level of accuracy (Figure 2.1). In addition to this, solving a 3D problem is significantly more complex and time consuming than solving the "corresponding" problem in 1D. For instance, consider the complexity of manually solving the deflection of a simple cantilever beam such as that in Figure 2.13 without simplifying it to a 1D problem. The same principle applies to the usage of FEA in this context. Solving the problem in 3D will lead to a considerable increase in CPU time. For a relatively small number of elements (such as the cantilever beam), the actual time will most likely be insignificant for the user. As an example consider a CPU time for the 1D model of 30 s, and the CPU time of the 3D model as 45 s. In real time the difference is insignificant. However, the 3D model represents a 50% increase in CPU time. This is, of course, directly linked to the number of DOF within the system or model. Generally, the increase in CPU time versus number of DOF (nodes) is not a linear relationship but an exponential one, as illustrated by Figure 2.14.

The significance of the graph in Figure 2.14 is very important, both from an FEA and also an optimization point of view, and can be appreciated by considering a typical example of industrial application of FEA. In the automotive industry FE models are widely used to predict crash performance of new vehicles, such as passenger cars. A large (but not atypical) FE crash model of this kind may contain in excess of two million elements. If this was to be solved using solid (3D elements) the real time (not CPU time) would most likely be measured in weeks. However, using 2D elements an "identical" model may be solved within 2 to 3 days or less. The above is not an exaggerated example and shows

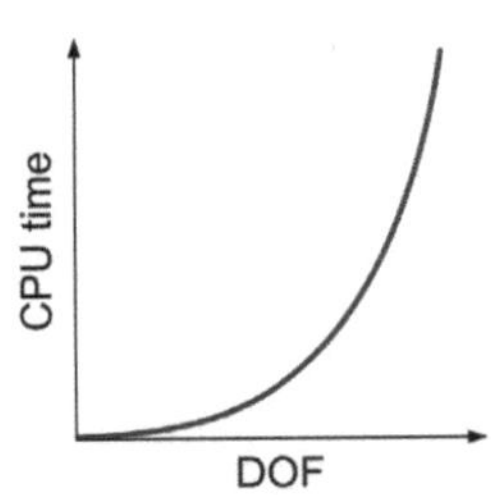

FIGURE 2.14 CPU time versus DOF.

the significance of Figure 2.14, thus underlining the significance of choosing an appropriate element for the subsequent FE analysis.

> *The task of selecting appropriate element type(s) for any FE analysis is a task that any engineer will face on a regular basis. An essential part of ensuring a sensible selection process is heavily dependent upon the engineer possessing a reasonable level of understanding and appreciation of the fundamentals of FE.*

2.3.4 Zero-dimensional elements

Following the definition of 1D, 2D, and 3D elements in the preceding subsections, the intuition of what constitutes a zero-dimensional (0D) element may seem somewhat cumbersome. However, in FE terms, the 0D element is very simple to understand. From the previous definitions, it has been established that the "dimension" of the element is directly linked to the number of parameters needed to define for example, the relative displacement of a single node. The same methodology is applicable to 0D elements where 0 parameters (or variables) are needed to define the relative displacement of any node within the element. In other words, this means that the relative displacement of any node within the element is identical for all nodes. In solid mechanical terms, this is the equivalent of rigid body displacement, that is, the element in question will not deform during FEA; this type of element is often referred to as "rigid" elements. It should be noted that different FE solvers may implement this in different ways; for example the "rigid" attribute may also come from the material definition; not the element itself. In general 0D elements may also be used to represent aspects such as single point masses. In terms of graphics, the 0D (rigid) element is most often represented as a line with one node either end, similar to a 1D element.

Figure 2.15 can be used to appreciate the difference between a 0D and a 1D element.

As Figure 2.15 reveals, the 0D element does not compress at node 2, simply because node 1 is rigidly clamped and no relative displacement or rotation is allowed between the two nodes, which is not the case for the 1D element. This means that the CPU cost for including a 0D element in an FE analysis is extremely low. Because of the behavior of the 0D element it may seem trivial and have no usage in FEA. However, the great advantage of using 0D (rigid) elements is linked to the trend illustrated by Figure 2.12, which describes the relationship between DOF and CPU time. This may be appreciated by considering

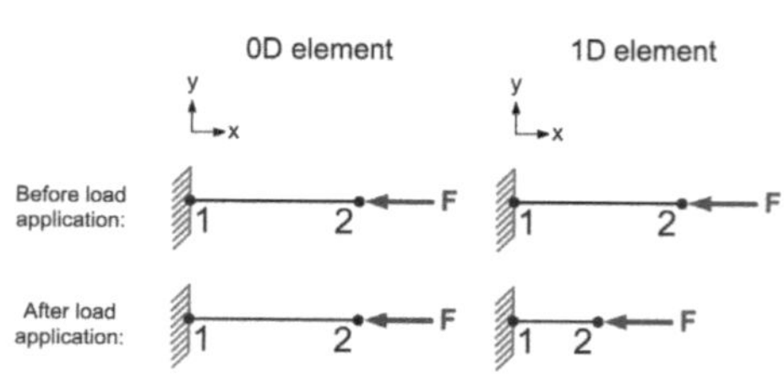

FIGURE 2.15 Difference between 0D and 1D elements.

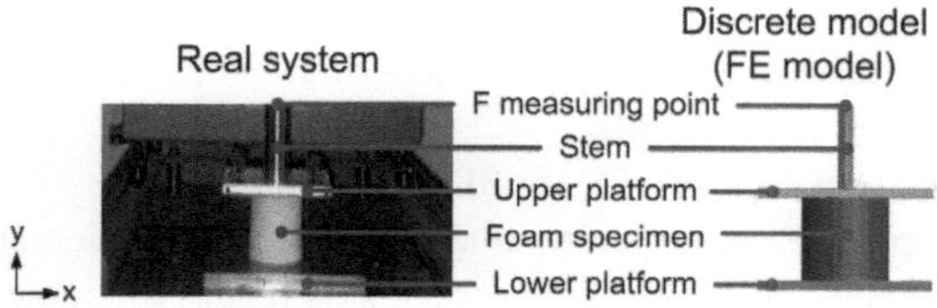

FIGURE 2.16 Example of convenient usage of 0D (rigid elements).

the example illustrated by Figure 2.16, which represents the material testing of a foam sample.

The real system is illustrated on the left-hand side of Figure 2.16. The intention is to perform a uni-axial pressure test of the foam specimen, that is, a force is applied in the negative y-direction. The force and displacement of the sample is measured at the top point of the stem, thus approximating the force absorbed by the foam specimen. This is remodeled in the FE model (i.e. the discrete model), where the force is also measured at the top of the stem (Figure 2.16). The stem, the upper, and the lower platforms are all made from a mild grade steel, which may be perceived as infinitely stiff when compared with the foam sample. For example, Young's modulus of the steel parts is 210 GPa, whereas Young's modulus of the foam specimen is approximately 10 GPa. Thereby, the deformations of the steel parts during the test (force application) are infinitesimally small compared with that of the foam specimen. This means that the steel parts may be modeled using rigids, as no significant deformation is anticipated. By doing so, accurate results of the force and displacement values at the top of the stem may be obtained by using rigid elements as opposed to, for example, 3D elements, however at a much lower CPU cost.

As always, it is very important to consider the potential consequences of simplifying any given model in this case by using rigid elements.

Other examples where rigid elements are often used are to represent assemblies, such as nuts, bolts or welds, where the stress levels in these regions are of low or no interest to the specific task at hand.

The following section will, as previously mentioned, highlight some of the aspects the engineer should consider during the element selection process.

2.3.5 Meshing strategy

The task of discretizing a system or a structure using FE is often referred to simply as "meshing," which is not a simple task. In most cases, the meshing process is generally based upon personal experience, or a "standardized" method of meshing within a given company or industry. As the mesh needs to be adapted for each individual model, it is impossible to write a "complete guide" of how to discretize a model. The discretization of any structure can be completed in many different ways; this means that in general there is strictly speaking no right or wrong way of meshing. There are, however, poor or good ways of meshing. It is extremely important that the consequences and implications of the meshing strategy, which may have a significant influence upon the results of the FEA, and thereby also the optimization are understood. When meshing, four key aspects generally need to be defined:

- Element type (e.g. 0D, 1D, 2D, or 3D)
- Element shape (i.e. triangular or square for 2D and 3D elements)
- Element size (i.e. mesh size; the average side length of element)
- Element shape function

The selection of shape function may usually be defined or changed subsequent to the actual meshing, that is, discretizing the CAD model with finite elements, and is therefore not of immediate concern here. Before commencing to mesh a model, it is always (as a minimum) important that the following questions are answered:

- What is the purpose of the FE analysis?
- What is the desired accuracy of the FE analysis?

By answering the above questions, the first steps toward selecting the four meshing parameters (as stated above) have already been taken. For instance, is the purpose of the analysis to conduct a final validation of an engine component or is it simply to test the feasibility of a basic principle or idea at the conceptual stages of designing a bracket? In the case of the engine component high levels of (geometrical) detail are most likely available and high accuracy of the model is required. Determining the initial feasibility of the bracket does, however, most likely not contain high levels of detail, nor are high levels of accuracy needed, only "ball-park figures." For any model it should always be considered if all details of the geometry are important and necessary to include in the FE model. As an example, consider the cantilever beam illustrated in Figure 2.17.

If asked to conduct an FE analysis of the beam in Figure 2.17, should the hole in the plate (at point B) be included? The answer to this question depends on the purpose of the analysis. If the purpose of the task is to obtain the stresses throughout the structure, then the hole should most certainly be included in the FE model. On the other hand, if the purpose of the FE analysis is to obtain the reactive forces at point A, then it is not necessary to include the hole, as this will have no effect on the reactive forces, subject to certain assumptions. The major advantage of not including the hole is obviously related to the amount of time saved, firstly in connection with the discretization and secondly the CPU time spent to solve the model. It is also important to consider the effects of this in connection with optimization; as a significant number of iterations, i.e. FEA is required, hence the increase in CPU time might be further scaled by a substantial factor. However, depending on the purpose of the analysis, the hole may be

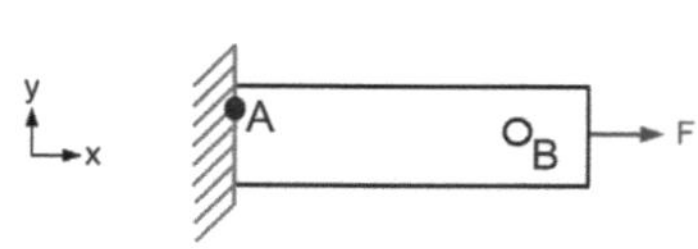

FIGURE 2.17 Cantilever beam.

essential to the FEA. Similarly, the choice of element type is related to the above questions. Always keep the following in mind when defining an FE model:

> *"An FE model should be as simple as possible, but as complex as necessary."* (Sorokin, nd)

The selection of element type is most often associated with the question of CPU time versus desired accuracy. The previous sections listed examples of when 1D, 2D, 3D, and even 0D elements are appropriate to use. When making this choice, it is important to consider the limitations of the individual element type. Again, comparisons can be drawn between solid mechanics and FEA. For instance, consider the limitations of using 1D beam theory to solve a 3D problem. In solid mechanics assumptions and simplifications are most often introduced in order to reduce the difficulty of solving the problem, which just as in FEA is a question of time spent versus accuracy obtained. It is also very important to remember that any choices made at this stage will have significant influence on the impending optimization.

2.3.6 Element type

The beam examples from the previous section, shown in Figure 2.10, were used to discuss appropriate scenarios of when to employ 1D, 2D, or 3D elements. If the beam is "long and slender," the magnitude of L, Figure 2.10, thereby far exceeding those of h and w then 1D elements are appropriate to use. This is, for example, because the stress values are not expected to vary greatly as a function of neither the y nor z coordinate (Figure 2.10). On the other hand, if the beam is "short and fat," that is, the magnitude of h is comparative to L but w is not, then 2D elements are appropriate to employ, because the stress values are expected to vary significantly as a function of both the x and y coordinates. Finally, if L, h, and w are all of comparable magnitude then 3D elements may be appropriate to use, as it is expected that the stress values vary significantly as a function of both the x, y, and z coordinates. It should be noted that the above examples are also dependent upon other factors such as material properties, loading scenarios, and constraints. It should also be noted that 0D, 1D, 2D, and 3D elements can all be used in the same FE model.

The above may introduce complexities with respect to structural optimization, as will be discussed later.

2.3.7 Element shape

For the discussion in this section, the "element shape" simply refers to the shape of the sides of the element, i.e. triangular or square. This means that the elements in question are either 2D or 3D elements; in other words, shell (Figure 2.11) or solid elements (Figure 2.12).

In general, good engineering practice dictates using as many quad (2D) or hex (3D) square elements as possible. The reason for this is simply that the quad elements are generally more accurate than the triangular elements, which is directly related to the shape functions. There are, however, certain exceptions to this. Consider, for example, a cantilever beam exposed to pure bending; one "side" of the beam will be exposed to tensile stresses while the other will be exposed to compressive stresses. As a consequence of using linear (first-order) shape functions, the stress in the individual element becomes constant. In order to capture the tensile and compressive stresses on either side of the neutral axis, it is necessary to increase the number of elements (when using first-order shape functions). Alternatively, a higher order shape function may be applied; for example, quadratic (second order). Thereby, there is generally no need to increase the number of elements to capture the tensile and compressive stresses in the FEA. The methodology of increasing the number of elements can also be referred to as "*h* refinement" (where *h* refers to the element size). The methodology of increasing the order of the shape function (or polynomial degree, *p*) is often referred to as "*p* refinement." Finally, it is also possible to rearrange the nodes of an FE model (without increasing the number of elements); this is referred to as "*r* refinement."

Despite the fact that quad elements are preferable from an FEA accuracy viewpoint, they are not always "practical" to use; this is primarily in connection with the discretization or meshing itself. Keeping in mind that elements are used to discretize continuous lines or surfaces that are dependent upon the geometry, it is not always feasible to only use square elements. To better appreciate this, consider the task of meshing a circular plate in a 2D geometrical space. The left-hand side of Figure 2.18 illustrates the plate discretized using only quad elements of relative size 1.0; the right-hand side of Figure 2.18 illustrates the plate discretized using triangular elements also of size 1.0.

Figure 2.18 clearly shows the disadvantages of using quad elements in this connection, as many of these elements are severely distorted, that is, are no longer square. This "level of distortion" is, as previously stated, "checked" using the Jacobian matrix. The triangular elements in Figure 2.18 do however seem to retain their "original" shape. Thereby, it is evident that triangular elements are sometimes required simply due to the geometry of the component to be discretized. It could be argued that reducing the mesh size of the quad elements may lead to "less distorted" quad elements; however, this means that

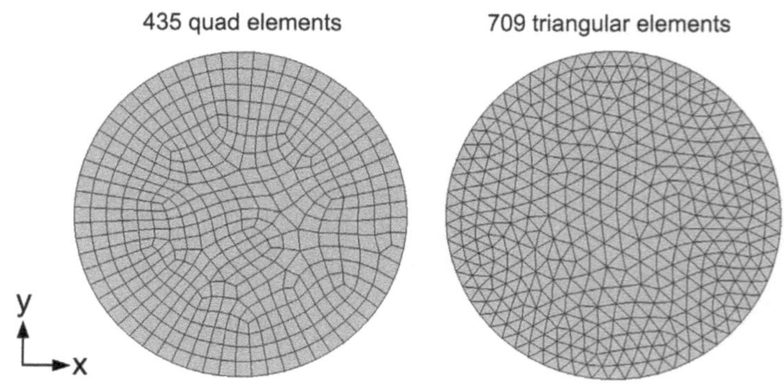

FIGURE 2.18 Circular plate discretized using quad- and triangular elements, *L* = 1.0.

the number of nodes will increase, consequently so will the CPU time. Furthermore, the general problem will not be eliminated, simply due to the fact that a rectangle is not a very good representation of a circle.

The quad mesh in Figure 2.18 contains 467 nodes, whereas the triangular mesh only contains 387 nodes; thereby, the CPU time of the latter will be less, as discussed in Section 2.3.3. Reducing the relative mesh size to 0.5 means that the quad mesh now consists of 1658 elements and 1721 nodes, representing a significant increase including CPU time. The size reduced quad mesh is illustrated on the left-hand side of Figure 2.19.

An alternative way of meshing is simply to use quad and triangular elements in combination (or tetra and hex in 3D); this type of mesh is often referred to as a "mixed mesh," often representing a "best" compromise. An example of a mixed mesh can be seen on the right-hand side of Figure 2.19. The mixed mesh primarily consists of quad elements; however, the addition of triangular elements helps avoid significant distortion of the quad elements, thus leading to a better "element quality." The mixed mesh contains 400 nodes (and 375 elements), which is only marginally more than the triangular mesh of Figure 2.18, which contains 387 nodes (and 709 elements).

If it is necessary to include triangular or tetra elements in the mesh, as a rule of thumb always aim to keep the number of triangular (or tetra) elements below 2% of the total number of elements. This value is commonly used in industry as it limits the risk of the structure being excessively stiff and in some cases locking when exposed to extreme buckling. Membranes, like airbags, on the other hand, can be fully meshed using first-order triangular triangles, as they have no bending stiffness. In such cases, the advantage of utilizing triangular elements is to prevent element warping (distortion).

When selecting the element shape, the following general statements may be useful:

- Strive to use quad or hexahedral elements over triangular or tetrahedral elements.
- If necessary to include triangular or tetrahedral elements, aim to use less than 2% of these in any FE model.

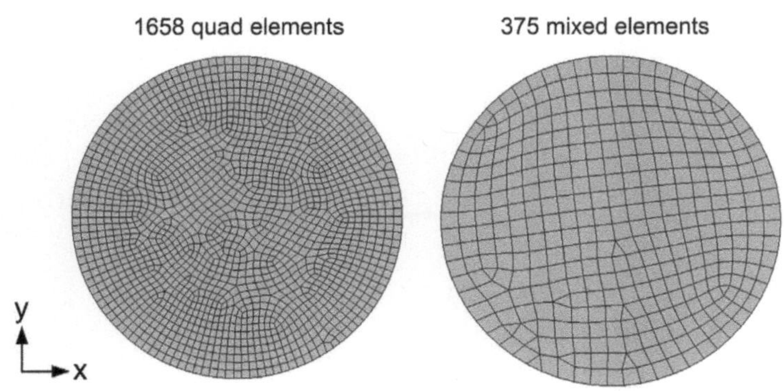

FIGURE 2.19 Circular plate discretized using quad elements *L* = 0.5, and mixed elements *L* = 1.0.

2.3.8 Element size

The task of selecting an appropriate element size also relates to the "purpose and accuracy questions" previously listed. The task of selecting an appropriate element size is important for both 1D, 2D, and 3D elements; for 0D elements the task is fairly straightforward, as the elements do not deform. The "anticipated" degree of deformation is an essential part of selecting an appropriate element size, as the FE analysis should "capture all modes of deformation." To better understand this, consider the cantilever beam in the upper left quadrant of Figure 2.20.

Imagine that the upper right quadrant of Figure 2.20 represents the results from a physical experiment, showing the "true" deformation modes of the beam. A first attempt at correlating the physical experiment to an FE model is illustrated in the lower left quadrant of Figure 2.20, using a single element. This model captures the compression of the beam (along the x-axis); however, it does not capture the displacement of the beam in the y-direction. In other words, the first model has not captured all modes of deformation. The results of a remeshed model (using a smaller mesh size) can be seen in the lower right quadrant of Figure 2.20, displaying both modes of deformation; that is, the second FE model captures all modes of deformation.

The above example is, of course, highly idealized; nevertheless, it provides a simple illustration of the important term "capturing all modes of deformation."

In order to take full advantage of the above, it is necessary to know the modes of deformation beforehand, which is seldom the case. However, most often a "guesstimation" may suffice. Furthermore, the necessity to capture all modes of deformation is of course also directly related to the purpose of the FE analysis, including optimization, and the desired accuracy as previously discussed.

From the above example, it may be deduced that a "simple" way of ensuring all modes of deformation are captured is to make the mesh size very small. Even though this may initially solve the problem, it will inherently lead to other severe issues. The CPU time will dramatically increase (Figure 2.14), which may ultimately lead to error termination of the FE software, for instance, due to lack of memory. Always keep in mind that:

The ideal solution to FEA discretization problems is seldom to continuously reduce the mesh size!

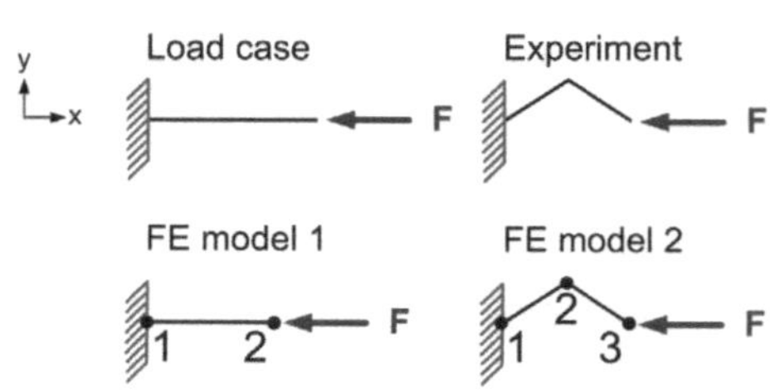

FIGURE 2.20 Capturing all modes of deformation.

Furthermore, issues such as singularities may also become a problem, and may in turn "distort" the outcome of any optimization. The problem of singularities is relatively easily understood by the following analogy from solid mechanics. The direct stress S is a relative measure of force intensity, and is defined as force F over area A, as in Equation (2.33)

$$S = \frac{F}{A} \tag{2.33}$$

In FE loads/forces may only be applied onto nodes. The nodes are of course related to an element, the area A of which is determined by the side lengths L, representative of the mesh size. If force F is applied to node N belonging to element E, then the stress magnitude S can be straightforwardly calculated using Equation (2.33). It is clear that if the area A of element E is halved, the stress magnitude of S will be doubled. If the mesh size is continuously reduced, Equation (2.34) is valid.

$$A \to 0 : S = \frac{F}{A} \Rightarrow S \to \infty \tag{2.34}$$

Thus, the term "singularity" in connection with FEA can simply be interpreted as the tendency that stress value(s) are rapidly approaching infinity, which may cause "havoc" in terms of optimization, wherein the maximum stress may be used as a parameter.

A more straightforward approach to initially determine an appropriate mesh size is simply to compare the created mesh to the original "geometry," for example, a CAD file.

An example of this can be seen in Figure 2.21, where the (original) CAD geometry is illustrated in the upper left quadrant. The remaining three quadrants in Figure 2.21 represent three different relative mesh sizes of 1.0, 1/2, and 1/4.

Even to the untrained eye, the discretization in the upper right quadrant (relative element size of 1.0) of Figure 2.21 does not "smoothly" define the hole in the plate. Smoothness is also an important phrase in relation to meshing:

Always strive to make the discretized model (i.e. mesh) "smoothly" and approximate the original geometry!

The above statement is obviously dependent upon the purpose and desired accuracy of the FE analysis.

Clearly, the lower two meshes in Figure 2.21 represent increasingly smoothed discretization relative to the one in the top right quadrant.

However, which (if any) of the lower two represents the "best mesh" to conduct an FE analysis? Clearly, the lower right quadrant contains a smoother mesh; however, it also contains a considerably larger number of nodes, leading to increased CPU time, and possible singularity issues.

A systematic approach to answering these questions is to conduct a so-called "mesh convergence" study. The purpose of a mesh convergence study is simply to investigate the effects of mesh size (or shape) upon the FEA results.

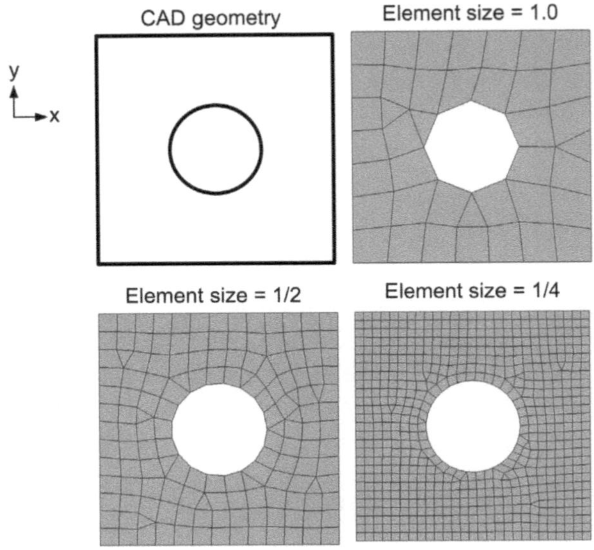

FIGURE 2.21 Original geometry versus element size.

If optimization is to be conducted it is pivotal that the FE model (if applicable) is truly representative of the problem in question. A mesh convergency study is generally carried out by conducting a series of FE analyses and comparing the results to each other. The results to be compared may be chosen from task to task; a "classical" scenario for linear static analysis is to compare maximum stress levels and general stress distributions. As a minimum, three FE analyses should be used for a mesh convergence study.

Figure 2.22 illustrates the resulting Von Mises' (VM) stress plots from three FE analyses of the models in Figure 2.21. The models are all fully constrained along the y-axis on the left-hand side of the plate and exposed to a tensile force in the positive x-direction on the right-hand side of the plate.

From Figure 2.22 it can thus be seen that the stress distribution varies considerably between the three models. Furthermore, the maximum VM stress also varies significantly between models, strongly indicating that the mesh has not

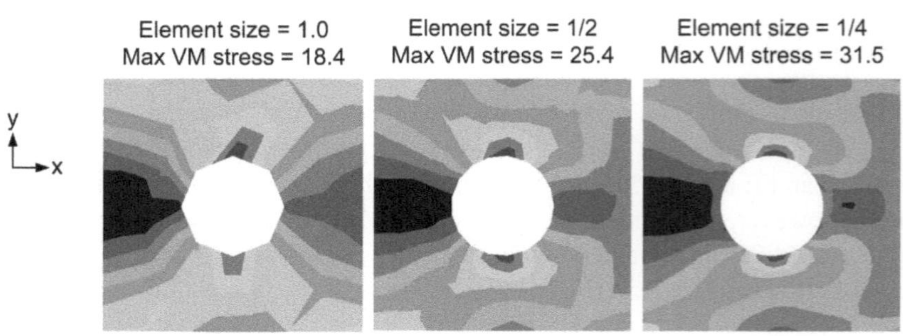

FIGURE 2.22 Von Mises' (VM) stress plots from three FE models in Figure 2.21.

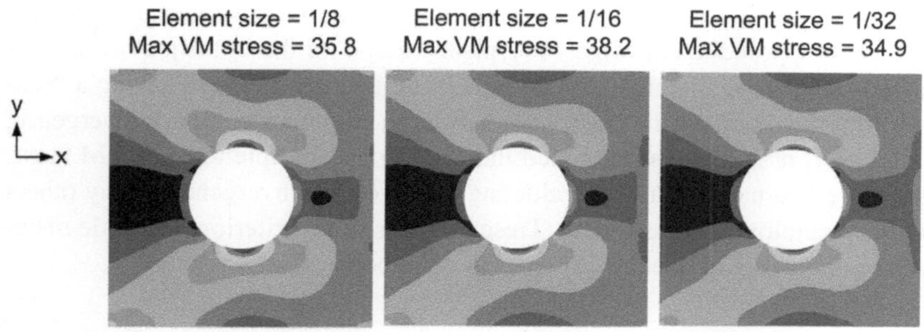

FIGURE 2.23 Von Mises' (VM) stress plots of additional FE models for convergency study.

yet converged. Therefore, it is necessary to conduct additional FE analyses to determine when the mesh has converged. This is done by conducting a further three FE analyses with relative mesh sizes of 1/8, 1/16, and 1/32; the resulting stress plots of these can be seen in Figure 2.23.

Observing the stress distributions of the six FE models in Figures 2.22 and 2.23, it can be seen that the stress distribution (in general) remains constant from a relative mesh size of 1/8 and less. The same is true for the maximum VM stress that changes less than 10% between 1/8 and 1/16; between 1/8 and 1/32 the difference is less than 2.5%. This means that the mesh has converged at a relative element size of 1/8. A good way of illustrating this is to plot a graph of the number of elements (inverse of element size) versus maximum VM stress, i.e. a mesh convergency graph (Figure 2.24).

The graph in Figure 2.24 substantiates what was concluded above, namely that the mesh has converged at a relative element size of 1 / 8, point A (Figure 2.24). This is because the slope of the graph for a number of elements less than point A is very large, whereas it is moderately low after point A. The shape of the graph in Figure 2.24 can also be used to substantiate the previous claim that a minimum of three FEA should be included in any mesh convergence study; this is simply because the graph in Figure 2.24 is not linear, even though the actual FE analysis is.

Note that 10% was selected as the acceptable convergence value of the maximum VM stress in the above example; this could also be 2%, 25%, or any other number; it all depends upon the intended purpose and desired accuracy of the FE analysis.

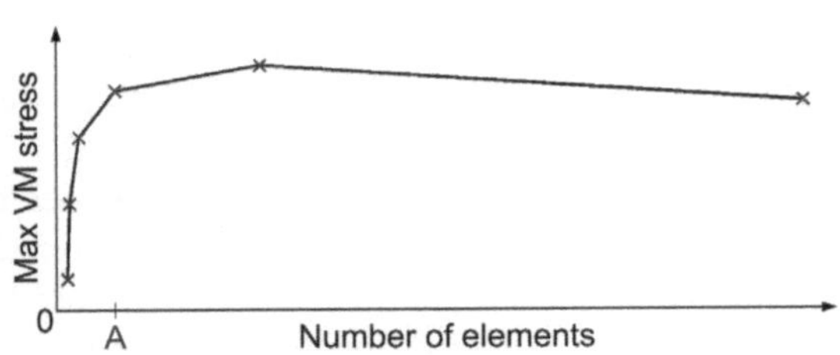

FIGURE 2.24 Mesh convergence graph.

At this point, it should also be noted that the mesh convergence study may diverge as opposed to converge. A "typical" reason for this is the previously discussed problem of singularity. This underlines the importance of having a "sensible" starting point for the mesh size, in order to minimize the risk of divergence.

Finally, it should also be noted that the above example utilized VM stress as the predominant means of evaluating the level of convergence. Many others could be employed; for example, Tresca's shear stress criterion for brittle materials or observing extrapolated values of neighboring Gauss points.

2.4 FUNDAMENTAL EXPLICIT AND IMPLICIT FINITE ELEMENT ANALYSIS

This section will briefly outline the fundamental solution methodology for implicit (linear static) FEA. This solution methodology is generally not applicable to large deformation and nonlinear analysis as typically found in connection with vehicle safety and crashworthiness analysis. It is, however, necessary to have some insight into this method as it will be used in subsequent discussions of current optimization algorithms. Furthermore, it will be useful in connection with the presentations and discussions of explicit FEA in the following section. The starting point for the implicit method is the fundamental equation of FEA Equation (2.32) that, for convenience, is repeated below

$$\{F\} = [K]\{D\}$$

Remember that $\{F\}$ and $[K]$ are known quantities, whereas the displacements, $\{D\}$, also often referred to as $\{U\}$, are the unknown quantities. From a mathematical viewpoint the vector matrix system in Equation (2.32) can be solved in two different ways. The first is to "guess" the entries in $\{D\}$ until Equation (2.32) is true (or has very small residuals); this approach is referred to as explicit FEA and will be the focus of attention in Section 2.5. The second option of solving Equation (2.32) is simply to invert $[K]$ and multiply by $[K]^{-1}$ to obtain the unknown displacements; this approach is denoted as implicit FEA, the methodology of which is demonstrated in Expression (2.35)

$$\{F\} = [K]\{D\} \Leftrightarrow [K]^{-1}\{F\} = \{D\} \tag{2.35}$$

Noting that one of the general characteristics of the global stiffness matrix is that it is symmetrical across the leading diagonal and will therefore always have a positive determinant and consequently be invertible, hence $[K]^{-1}$ will always exist. The methodology of the implicit solver is thus relatively easy to appreciate, as this "simply" involves the inversion of a single matrix.

Consider the beam example of Figure 2.8, which is assumed only to have a single DOF, $\delta(x_i)$, and imagine that the beam model (in Figure 2.8) represents an actual strain controlled physical experiment that is to be correlated by means of FEA. The stress–strain curve obtained from the physical experiment is represented by the fully drawn line in Figure 2.25.

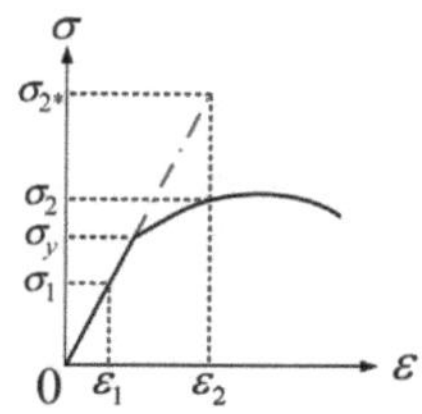

FIGURE 2.25 Stress versus strain.

Firstly, the beam is deflected to the strain value, ε_1, and the corresponding F value is obtained. The F value is subsequently entered into the load vector $\{F\}$ and the FE model is solved using implicit FEA, Expression (2.35). The solution, i.e. the displacement vector $\{D\}$, is obtained in a single step, as $[K]$ is inverted only once. Thereby, the stress value rises from 0 to σ_1 according to the slope of the material curve at the origin (0,0), which is represented by Young's modulus, E. This FE solution methodology produces accurate results as long at the stress levels in the beam do not exceed the yield-stress value σ_y, thereby remaining in the linear (elastic) region of the material curve.

Next the physical test deflects the beam to the strain value ε_2 (Figure 2.23). Analog to before the force value is found and entered into the FE model. The implicit solver will still utilize a single step to obtain equilibrium of the system by means of Expression (2.35). The implicit solver will once again do this utilizing the stiffness at (0,0). Thereby, the implicit solver will obtain equilibrium with the stress value σ_{2*}, which is highly inaccurate, as the actual stress value is σ_2 (Figure 2.25). The reason for this inaccuracy is simply the fact that the implicit solver (generally) obtains a solution with only one single step. This makes the implicit solver very useful for linear (static) FEA, because it provides an unconditionally stable solution procedure.

However, as illustrated in the above example the implicit solver may have problems with highly nonlinear FE models. Another potential issue with using an implicit solver is the fact that inverting extremely large (stiffness) matrices may take very long time and require a large amount of memory.

The implicit solver is therefore most often used in connection with linear analysis, static, or low strain rate analysis; it can however also be used for nonlinear FEA, by introducing an iteration cycle, that is, "applying the load (in the FE model) in small increments;" the size of these increments is often referred to as "timesteps," Δt, and can be used in connection with both implicit and explicit FEA. The timesteps will be further explained in connection with the explanations in the following section. Further information on implicit FEA can, for example, be found in (Cook et al., 2001).

2.5 NONLINEAR EXPLICIT FINITE ELEMENT ANALYSIS

The aim of this section is to provide a pragmatic approach to nonlinear explicit computation applied to the field of vehicle safety. The main equations of explicit FEA are derived and interpreted to present the reader with an overview of the computational method, list of necessary inputs, troubleshooting techniques, and tools for initial assessment of model validity. As highlighted above, explicit FEA generally utilizes an iterative process to "assess" the "state" of the FE model in relatively small increments. This can be appreciated by reconsidering Figure 2.25, but imagine that the first (horizontal) axis now represents time as opposed to strain. Initially, at $t = 0$ no loading is applied; the loading is now gradually applied in small "increments" (time-steps) until the loading reaches the value σ_1. The equation of equilibrium $\{F\}=[K]\{D\}$ is solved and the results stored for each time-step Δt. Observing Figure 2.25, it is clear that little additional information would be obtained via this "iterative" (explicit) approach when compared with the "single step" approach of implicit FEA presented above. The advantage can, however, be appreciated if the load is now increased from σ_1 to σ_2. This will be further explained in the following section.

2.5.1 Understanding the need for explicit FEA in connection with vehicle safety assessment

In the case of implicit FEA, a global stiffness matrix is computed, inverted, and applied to the nodal out-of-balance loads in order to obtain a displacement increment, assuming that an iterative process is used. The advantage of this approach is that the timestep Δt size may be selected by the user; hence, the solution may be computed using a small number of iterations. Implicit simulations therefore typically involve a relatively small number of central processing unit (CPU) expensive timesteps, as the stiffness matrix $[K]$ must be defined, stored, and inverted for every single iteration (Jensen, 2007). As previously explained, the basic problem being solved in any FE analysis is to determine the unknown displacements, d, of any node of the model, that is, determine $\{D\}$ in Equation (2.32). Remembering that the state of the system is time-dependent; when, for example, analyzing a vehicle crash, the increments in nodal displacements must be calculated for the "next step" at time $t_{n+1,}$ that is, d_{n+1} must be determined. If the incremental nodal displacement is denoted as x the first derivate of the location, the velocity is denoted as $\dot{x}$, and the second derivative, the acceleration, is denoted as $\ddot{x}$, the governing equation of equilibrium for the current state, t_n, can be defined as Equation (2.36)

$$[M]\{\ddot{x}\}+[C]\{\dot{x}\}+[K]\{x\}=\{F\} \tag{2.36}$$

$[M]$, in Equation (2.36), is the lumped mass matrix that represents the mass of the system, while the damping matrix $[C]$ represents the natural damping of the system, which, for example, will depend on the material characteristics; while $\{F\}$ is the external load vector initially presented in Section 2.2. It should be noted that Equation (2.36) is in essence an "expanded" version of Equation (2.32) as $\{x\} = \{D\}$; Equation (2.36) does, however, also incorporate the dynamic aspects of velocity and acceleration, which were previously ignored.

The incremental changes of displacement and velocity for a single node from state (time) t_n to t_{n+1} can now be determined via integration. If the Newmark β-method is used for this integration step, the updated displacement x_{n+1} and velocity $\ddot{x}_{n+1}$ can be obtained using Equations (2.37) and (2.38), respectively (Jensen, 2007).

$$x_{n+1} = x_n + \Delta t \dot{x}_n + \frac{\Delta t^2}{2}(1-2\beta)\ddot{x}_n + \beta \Delta t^2 \ddot{x}_{n+1} \quad (2.37)$$

$$\ddot{x}_{n+1} = \dot{x}_n + (1-\gamma)\Delta t \ddot{x}_n + \gamma \Delta t \ddot{x}_{n+1} \quad (2.38)$$

In Equations (2.37) and (2.38) β and γ are both integration constants while Δt is the previously presented timestep; the "size" of the increment. By combining Equations (2.37) and (2.38) it can be proven that (for $\beta > 0$) Equation (2.39) is true.

$$\ddot{x}_{n+1} = \frac{x_{n+1} - x_n - \Delta t \dot{x}_n - \frac{\Delta t^2}{2}(1-2\beta)\ddot{x}_n}{\beta \Delta t^2} \quad (2.39)$$

Equation (2.39) shows that the solution to (2.36), the equation of equilibrium at state (time) t_n, depends on the nodal displacement at state t_{n+1}, which is presently unknown. In addition, the individual nodal displacement x_{n+1} at state t_{n+1} are both functions of the current state t_{n+1} and the previous state, t_n, Equation (2.40)

$$x_{n+1} = f(\ddot{x}_{n-1}, \ddot{x}_{n+1}, x_n, \dot{x}_n, ...) \quad (2.40)$$

Equation (2.40) therefore implies that an iterative procedure is needed to compute the equilibrium at time t_{n+1}.

One of the primarily used methods to solve a system of equations simultaneously is the Newton–Raphson method (Jensen, 2007), which will be fully introduced in section 3.6. In addition to the integration steps outlined above, it is also necessary to determine the mass matrix [*M*] and the global stiffness matrix [*K*] before the equation of equilibrium, i.e. Equation (2.36), can be solved.

Solving the above using implicit FEA has some clear advantages if equilibrium is indeed obtained; as Δt can be automatically adjusted for subsequent steps enabling a rapid convergence toward a numerically feasible solution. It can therefore once again be highlighted that implicit FEA is good for problems with high degrees of linearity, slowly varying loads such as quasistatic and little or no dynamic effects.

Nevertheless, should the equilibrium not be achieved, then Δt needs to be recalculated as well as the matrix coefficients; [*M*] and [*K*] which in terms of computational requirements is a very expensive operation, remembering the potential sizes of these matrices.

Relating to the aspects of vehicle structural safety assessment that inherently involves dynamic and nonlinear behavior, it becomes evident that implicit FEA is not an ideal solution method as it has considerable and persistent stability issues. The stability issues are primarily contributed to the following aspects of FEA modeling (Jacob & Goulding, 2002):

- Short duration transient loading <100 ms)
- Discontinuous loadings exhibiting large force variations and significant inertial effects
- Large displacements
- Contacts (see Section 2.7 for more details and definitions)
- Rigid body motions (unconstrained bodies with very large rotations)
- Extreme plasticity of metals

Due to the issues listed above, almost all FE-based analyses relating to vehicle structural safety assessment utilize explicit FEA. Even quasistatic events with relatively low inertial loading such as a roof crush scenarios where a load is applied at the top of the A-pillar, are computed using explicit FEA. This is due to the numerous contacts (self-touching panels as will be explained later), the relatively large deformation of the roof structure as well as the plasticity of the "conventional" materials used.

For these reasons an alternative method to implicit FEA is required to accurately solve highly nonlinear problems. This method should be more forgiving and should not depend on a total criterion of stability; the generally used method is therefore explicit FEA.

2.6 EXPLICIT FEA APPLIED TO VEHICLE SAFETY ASSESSMENT

The purpose of this section is to present selected aspects of explicit FEA in the context of vehicle safety assessment. The following individual subsections will focus on separate aspects in this context, commencing with the governing fundamental mathematical expressions and methodologies.

2.6.1 Standard explicit equations and convergence criteria

The starting point in this subsection is Equation (2.36) from the previous section. Acknowledging that the equation is based upon Newton's second law, it can be perceived that the internal forces $\{F_{\text{int}}\}$ within the structure (or system) are represented by the $[K]\{x\}$ term. $\{F_{\text{int}}\}$ does in essence ensure the "stability" of the structure while being exposed to the external loading $\{F_{\text{ext}}\}$. Utilizing the above terminology and ignoring the damping parameters (for simplicity), Equation (2.36) may be rewritten as Expression (2.41)

$$[M]\{\ddot{x}\}+[K]\{x\}=\{F_{\text{ext}}\}\Rightarrow[M]\{\ddot{x}\}+\{F_{\text{int}}\}=\{F_{\text{ext}}\} \tag{2.41}$$

Observing the right-hand side equation of (2.42) it can be appreciated that it is possible to calculate an increment in acceleration for state t_{n+1} without the need to generate and invert large matrices. This can be done by reorganizing Expression (2.41). For clarity, this has been done for a single node, Equation (2.42)

$$\ddot{x}_{n+1} = \frac{\left(f_{ext} - f_{\text{int}}\right)_{n+1}}{m} \tag{2.42}$$

Based on Equation (2.42) the velocity and displacement can be obtained by integration over a certain interval Δt, which will increment for each or state. The increment will usually be of identical magnitude but may, however, vary between iterations. The definition of the interval, or timestep Δt, is defined in Equation (2.43)

$$\Delta t = \frac{t_{n+1} - t_n}{2} \tag{2.43}$$

By combining Equations (2.42) and (2.43) expressions for the velocity and the displacement, (2.44) and (2.45) can be obtained, respectively.

$$\dot{x}_{n+\frac{1}{2}} = \Delta t \cdot \ddot{x}_{n+1} + \dot{x}_{n-\frac{1}{2}} \tag{2.44}$$

$$x_{n+1} = \Delta t_{n+1} \cdot \dot{x}_{n+\frac{1}{2}} + x_n \tag{2.45}$$

At this point it is important to note that the solving technique is very much an iterative process that will continue until a user predefined end time (state) is reached, that is, t_{end}; this is also often referred to as the termination time. It is critical that t_{end} is set "correctly," as a "premature" termination might mean that important effects are not being considered in the analysis; that is, the computation may terminate before the full crash energy is absorbed by the crash structure. Setting a prolonged termination time means that the CPU time is excessively and unnecessarily long; this is often referred to as "overshoot." In the case of premature termination it may be possible to "restart" the computation; there are however some limitations to this and it is indeed not always possible to do so.

As previously explained, the computation process is iterative in itself and it can be observed that there is no "actual" requirement for equilibrium to be obtained at each Δt for the computation to proceed, as was the case with the implicit method. This makes the explicit method well suited for problems in which equilibrium is not usually present, for example rigid body motions including very large rotations.

Subsequent to solving (2.41), strain rates, strain, stress, and force magnitudes are computed in the order defined, using the expressions listed in (2.46).

$$\begin{aligned}
&\textit{Strain rates}: && \dot{\varepsilon}_{i,\Delta t} = \frac{\dot{u}_{i+1,\frac{\Delta t}{2}} - \dot{u}_{1,\frac{\Delta t}{2}}}{L} \\
&\Downarrow && \\
&\textit{Strain magnitudes}: && \varepsilon_i = \dot{\varepsilon}_{i,\Delta t} \cdot \Delta t \\
&\Downarrow && \\
&\textit{Stress magnitudes}: && \sigma_{i,\Delta t} = E \cdot \varepsilon_{i,\Delta t} \\
&\Downarrow && \\
&\textit{Force magnitudes}: && F_{i,\Delta t} = \sigma_{i,\Delta t} \cdot A
\end{aligned} \tag{2.46}$$

In Equation (2.46) " · " denotes the first derivative, "," denotes differentiation, "*A*" is the element surface area while "*E*" is Young's modulus. At this point it is convenient to highlight particular fundamental aspects of FEA apparent from Equation (2.46): strain rates, strains, stresses, and force levels traveling through the structure are all functions of the element size, *L*, Equation (2.46). As such, variations in force magnitudes will be obtained between different mesh sizes. Hence, it is important to note that the problem solution accuracy will be defined during the meshing process, as previously presented.

The next state (time) of the computation sequence t_{n+1} can then be obtained by adding the current timestep Δt_n. The internal and external forces are updated and the iteration loops continue until the termination time is reached (Jacob & Goulding, 2002). This sequence can be mathematically formulated as Equation (2.47)

$$\Delta t_{n+1} = t_n + \Delta t_n \big| t_n < t_{\text{termination}} \tag{2.47}$$

However, as the process is incremental it is possible that the obtained values diverge from the actual solution, as opposed to converge. The divergence tendency may be enhanced by the use of relatively large Δt values. It may however be desirable to increase Δt, as this will (significantly) reduce the overall CPU time. These facts further highlight the importance of the Δt parameter solidifying it as a key factor for the stability of an explicit FEA, as will be further discussed in the following sections.

As part of this discussion the following subsection will discuss the link between the time step Δt and vehicle safety simulations.

2.6.2 Stress wave propagation and timestep

The purpose of this subsection is to present how a stress wave propagates through a continuum and the relationship to the time step; this is pivotal not only for explicit FEA but also for vehicle crashworthiness/optimization in general.

Consider the scenario of a frontal vehicle collision, for example, a vehicle colliding head on with a (rigid) concrete barrier. The actual duration of the impact is very short in real time, most often measured in milliseconds. Nevertheless, there is a slight delay between the vehicle impacting the barrier and the impact "being felt" by the drive/occupants. This delay is caused by a stress wave that propagates from the point(s) of impact (at the front end of the vehicle) toward the occupants (and beyond).

To better understand this phenomenon, consider the simple (1D) rod element illustrated in Figure 2.26.

The rod is initially traveling with a velocity v immediately before impacting the rigid wall. On contact with the wall, a stress wave of amplitude σ propagates along the rod at a wave velocity C.

The length, L, of the (now) stationary and compressed section of the rod is a function of C and the impact duration t, as defined in Equation (2.48)

$$L = Ct \tag{2.48}$$

The mass, m, of the compressed section generated by the impact for a rod of constant cross sectional area, A and volumetric mass density, ρ, is defined in Equation (2.49)

$$m = \rho AL = \rho ACt \tag{2.49}$$

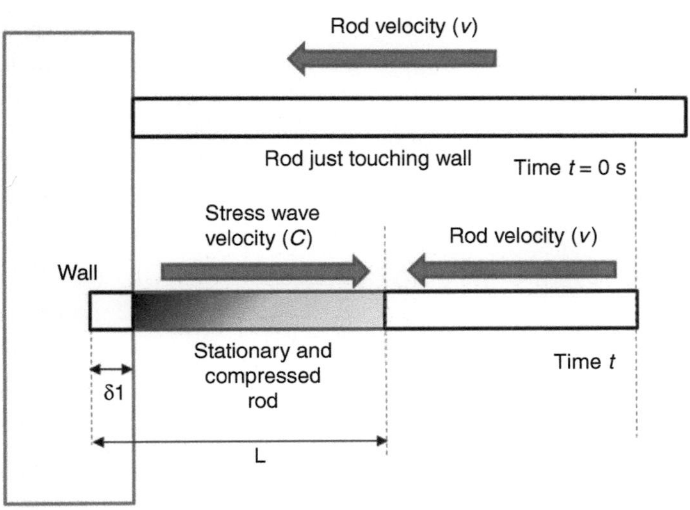

FIGURE 2.26 Stress wave traveling through a rod impacting a rigid wall.

The magnitude of the compressive strain, ε, in the stationary rod can be calculated according to Equation (2.50), where v is the impact velocity.

$$\varepsilon = \frac{-v}{C} \tag{2.50}$$

By balancing the impulse generated by the impact velocity against the momentum of the compressed section of the rod (at an instance in time), Equation (2.51) can be deduced.

$$-\sigma At = \rho ACtv \tag{2.51}$$

Equation (2.51) can be simplified to Equation (2.52)

$$\sigma = \rho Cv \tag{2.52}$$

By combining Equations (2.50) and (2.52) whilst utilizing Hooke's law, it can be concluded that the stress wave propagation speed C is proportional to the material specific properties E and ρ, that is, Young's modulus and the mass density, Equation (2.53)

$$C = \pm\sqrt{\frac{E}{\rho}} \tag{2.53}$$

Some significant considerations can be deducted from Equation (2.53) (Macaulay, 1987):

- The wave velocity is constant and independent of stress magnitudes (until the stress value reaches the yield stress);
- The positive and negative solutions prove that the stress wave can travel in more than one direction, suggesting that the stress in the continuum is proportional to the impact velocity and that there is a cut-off velocity for which the stress will be greater than the yield stress of the structure.

Analog conclusions on the level of proportionality can be drawn from the stress wave propagation through shell (2D) elements and solid elements, Equations (2.54) and (2.55), respectively (LSTC, 2012).

$$C = \pm\sqrt{\frac{E}{\rho(1-v^2)}} \tag{2.54}$$

$$C = \pm\sqrt{\frac{K + \frac{4}{3}\cdot G}{\rho}} = \pm\sqrt{\frac{E(1-v)}{\rho(1+v)(1-2v)}} \tag{2.55}$$

The stress wave propagation speed in any continuum is thereby a function of the material microstructure, as C is a function of Young's modulus and the

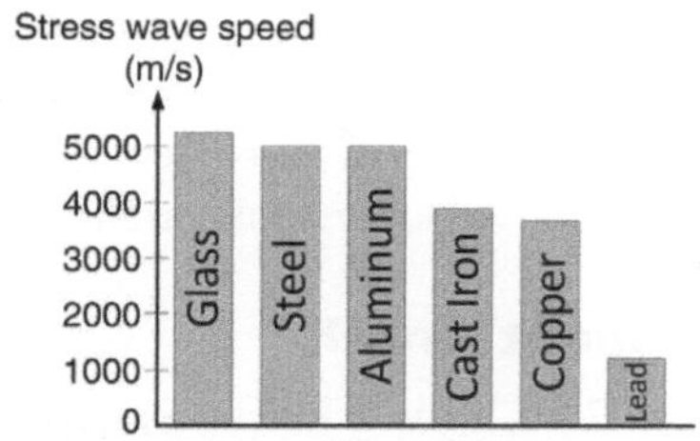

FIGURE 2.27 Stress wave speed of selected materials.

mass density. To appreciate the significant influence of the material parameters, Figure 2.27 demonstrates the variation of C value for different materials.

Figure 2.27 illustrates the diversity of the stress wave velocities for different materials; it can also be noted that due to the identical specific stiffness of steel and aluminum the stress wave speed will also be identical for both metals.

2.6.3 Relating the timestep to explicit FEA for vehicle safety assessment

In Section 2.6.1, it was demonstrated that Δt is utilized as the integrator to extract nodal velocities and displacements from accelerations in connection with explicit FEA.

In order to appreciate how this methodology is applied to "real-world engineering" in connection with vehicle safety analysis, consider the 1D rod with a uniform mesh density L (Figure 2.28).

A force of magnitude F is now applied as illustrated in Figure 2.28, giving rise to a stress wave that will travel sequentially from one element to another. In order to ensure that the "information" travels through every single element correctly, the timestep Δt has to be equal to or less than the time it takes for the stress wave C to reach the next node of the next element. This is pivotal to ensure the information travels chronologically through the structure and does not "skip" parts of it leading to calculation instabilities. This is referred to as the Courant–Friedrichs–Lewy condition.

As such a key criterion of ensuring stability and accuracy using explicit FEA is to abide to Expression (2.56)

$$\Delta t \leq \frac{L}{C} \tag{2.56}$$

In order to meet the requirement of Equation (2.56) it is recommended that at least 90% of the "original" value of Δt is retained in order to ensure a stable

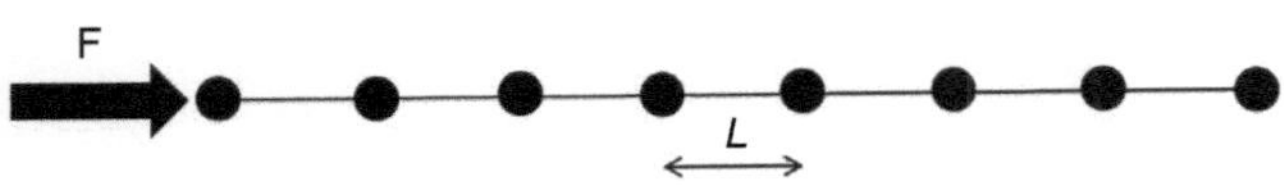

FIGURE 2.28 Rod discretized by elements of length *L*.

structural analysis computation (LSTC, 2013; Du Bois, 2010; Jacob & Goulding, 2002). Should stability problems still persist, this ratio can be further reduced. As a rule of thumb, a timestep of 1e-6 s (or 1 μs) is a good starting point for initial analysis for automotive safety applications.

2.6.4 Critical element length

As defined in Expression 2.56, Δt is a function of the element length and the stress wave speed; it is therefore important to understand how the critical length of each individual element is calculated. As defined throughout Section 2.3, and illustrated in Figures 2.11 and 2.12, the extremities of any element are defined by nodes. For 1D elements such as beam, truss, and rod elements, the critical length cannot be misconstrued, as the critical length the stress wave travels through the element is simply the length of the 1D element.

For all other elements, that is, 2D and 3D elements, the minimum interior dimension of the element is the critical element length, $L_{\text{critical}} \cdot L_{\text{critical}}$ will define the element timestep alongside the material dependent C; hence, Expression (2.56) can be reworked to define the timestep, Equation (2.57)

$$\Delta t = \frac{L_{\text{critical}}}{C} \tag{2.57}$$

Defining the minimum interior dimension of a 2D or a 3D element is not always a straightforward task, particularly when the element shape is "irregular," as illustrated in Figure 2.29.

The expressions in (2.58) can be used for calculating the critical element length, L_{critical}, in connection with the timestep computation for 2D elements.

$$\begin{aligned} L_{\text{critical}} &= \frac{(1+\beta)A}{L_{\max}} = \frac{(1+\beta)A}{\text{Longest side}} \\ L_{\text{critical}} &= \frac{(1+\beta)A}{D_{\max}} = \frac{(1+\beta)A}{\text{Longest diagonal}} \end{aligned} \tag{2.58}$$

Either of the two expressions of (2.58) can be used to calculate L_{critical}; generally $\beta = 0$ is used for quadrilateral elements while $\beta = 1$ is used for triangular shell elements.

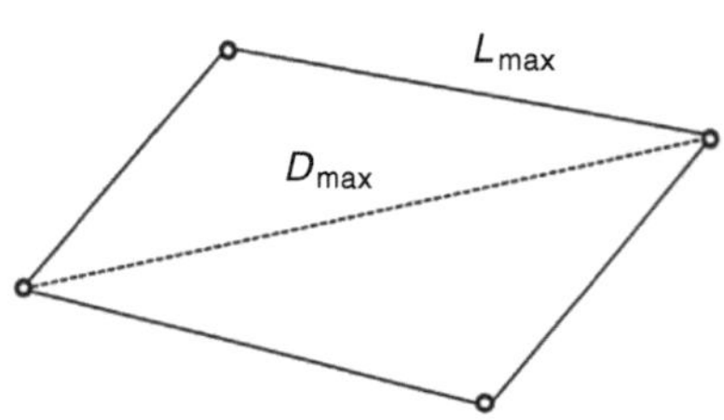

FIGURE 2.29 Length of side and diagonal of a quadrilateral element.

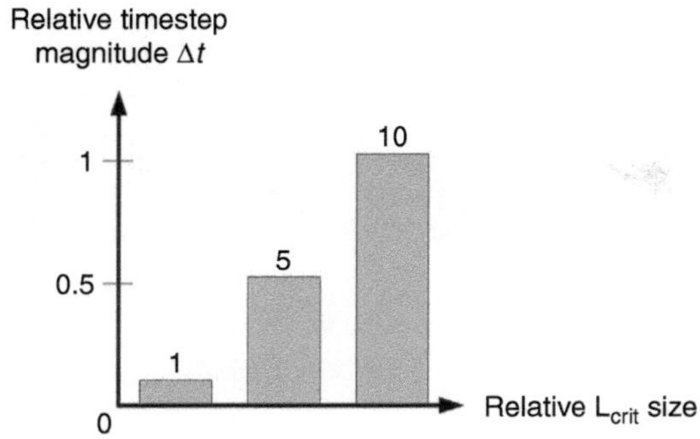

FIGURE 2.30 Relative timestep magnitude as function of mesh size.

For 3D, solid elements, the ratio between the internal volume and the largest side surface area is utilized to define $L_{critical}$, Equation (2.59)

$$L_{critical} = \frac{V}{A_{max}} \tag{2.59}$$

Considering Expression (2.57), Figure 2.30 illustrates the fact that the element critical length has a considerable effect on the timestep; the smaller the element critical length the smaller the timestep.

1D spring elements are an exception to the above rules, as their length does not control the timestep; hence, they have no critical length, $L_{critical}$ in this context. Instead, the timestep calculation for spring elements is based on the spring stiffness, k, and the mass values, m_1 and m_2, associated with the two nodes defining the individual element. The equation for calculating the timestep for a spring element is defined in Equation (2.60)

$$\Delta t = \sqrt{\frac{4(m_1 \cdot m_2)}{k(m_1 + m_2)}} \tag{2.60}$$

In general, it is good engineering practice to ensure that each of the spring end-nodes is connected to a node with mass assigned to it, that is, to another element. This helps ensure that the solver computes an unambiguous timestep value.

2.6.5 Summation of factors influencing the timestep magnitude

The purpose of this subsection is to highlight the primary parameters influencing the timestep value, as presented in previous subsections.

Firstly, Section 2.6.4 highlighted the influence of the element size/length. For otherwise "identical" elements in terms of element formulation and material

TABLE 2.1 Effects of Material Property and Element Size on Timestep Magnitude

Material	Young's Modulus (MPa)	Poisson's Ratio	Density (t/mm^3)	Stress Wave Speed (Shell Element) (mm/s)	Element Size (mm)	Timestep (s)
Generic plastic	3,000	0.4	1.2e-9	1,725,164	5	2.9e-6
Generic plastic	3,000	0.4	1.2e-9	1,725,164	1.6	9.3e-7
Generic steel	210,000	0.3	7.8e-9	5,439,283	5	9.2e-7

properties the significant differences in stress wave speed for a shell element are demonstrated in Figures 2.27 and 2.30.

As demonstrated in Table 2.1, it will take longer for the stress wave to travel through a longer element than a shorter one. Keeping this in mind while considering the overall CPU time requirements, an appropriate mesh size for an initial analysis of a full metallic (steel or aluminum) vehicle structure is generally approximately 5 mm.

The second highlighted factor that influences the timestep magnitude is the material properties. Table 2.1 illustrates the variation of the stress wave speed as a function of the material parameters; hence, the trend and influence displayed in Table 2.1 will be identical for the timestep calculation, provided the element length remains constant. It is, however, possible to obtain "equal magnitude timesteps" using different material models by adjusting the element size.

As demonstrated in Table 2.1 and initially presented above, a plastic component meshed with a shell mesh size of 1.6 mm will require the same timestep as a 5 mm "steel" shell element in order to meet the stability criterion, Expression (2.56).

The third highlighted factor that influences the timestep magnitude is the element choice. For any given problem, for example, a rod in compression, the timestep will depend on the type of element chosen during the discretization process, the "simpler" the element the larger the Δt. Examples of the variation of timestep magnitude for identical element sizes as a function of different element types are listed in Table 2.2.

This subsection has thus summarized the main parameters influencing the timestep magnitude. At this point, it is important to note that the magnitude of the stress wave speed will be greater for a 5 mm steel element than a 1.6 mm plastic element using an identical timestep value and element type. This is because the stress wave propagation per timestep will be 5 mm in the "steel elements" compared with 1.6 mm in the "plastic elements." This demonstrates that in this specific case, the computation of the stress wave would be compromised

TABLE 2.2 Timestep Magnitude as a Function of Element Type

Element Type	Stress Wave Speed	Timestep Reduction	Relative Element Timestep Magnitude
1D elements	$\sqrt{\frac{E}{\rho}}$	N/A	1
2D elements	$\sqrt{\frac{E}{\rho(1-\nu^2)}}$	$\sqrt{(1-\nu^2)}$	0.954
3D elements	$\sqrt{\frac{E(1-\nu)}{\rho(1+\nu)(1-2\nu)}}$	$\sqrt{\frac{(1+\nu)(1-2\nu)}{(1-\nu)}}$	0.862

by the discretization of the structure, as will be further discussed in the following subsection.

2.6.6 Importance of consistent mesh size

As previously presented, the stress wave travels through the structure one element at a time, and is "sampled" at intervals of Δt, during the numerical computation sequence.

If the mesh is of a consistent size, for example 5 mm, the stress wave will propagate at a speed of approximately 1 μs per element, utilizing a shell element and mild grade steel material properties.

If the mesh is nonuniform, that is, utilizes elements of varying size, the stress wave speed will "accelerate" for larger elements, as the timestep magnitude of an FE analysis is defined based on the size of the smallest element in the model, as outlined in Section 2.6.4. This may have negative effects on the accuracy of the analysis, as the timing of the stress wave in the structure is incorrect, which may lead to incorrect propagation of the stress wave(s).

This potential issue is illustrated in Figure 2.31. When a mesh of a consistent size is used, the distance traveled by the stress wave is linear. For the nonuniform mesh, at time Δt_3, the distance traveled is δ_3 and not δ_4, due to the larger element in the nonuniform mesh (Figure 2.31).

Figure 2.31 thus highlights the importance of obtaining a consistent mesh size for each component, but not necessarily the entire model. It is imperative that this aspect is considered during the discretization stage, and may not only have an effect on the accuracy of the FE analysis but also upon the optimization results as will be discussed in the subsequent chapters.

2.6.7 Manipulating timestep magnitude

Based on the meshing requirements, considerations, and potential pitfalls discussed throughout this chapter, it is evident that for complex structures such as a full vehicle model, the definition of an appropriate timestep magnitude leading

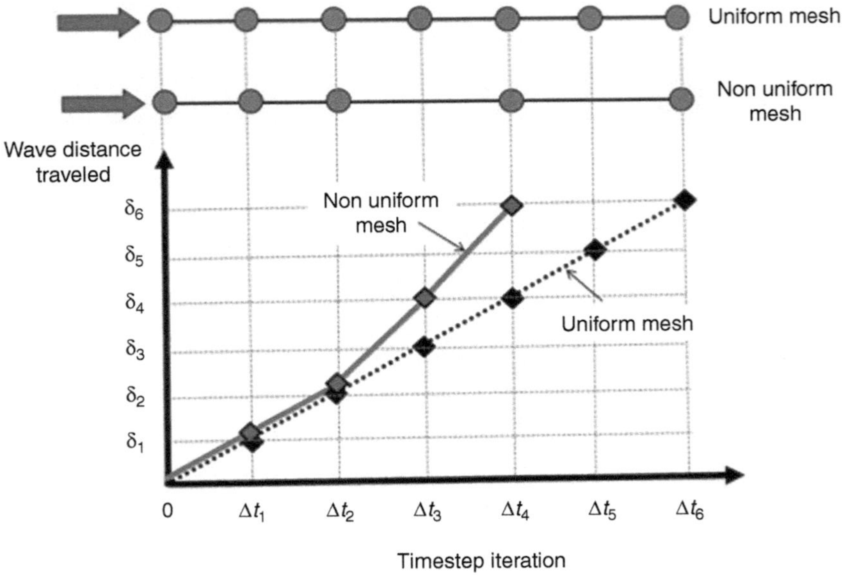

FIGURE 2.31 Stress wave distance traveled as function of mesh size.

to a "sensible" computing run-time providing and at the same time yielding an accurate/precise solution is both a challenging and time-consuming task, as illustrated in Figure 2.32.

As the Courant–Friedrichs–Lewy condition must be fulfilled, it should be noted that it only takes one "very small" element to reduce the timestep magnitude and consequently increase the full computation runtime, most often by a significant amount.

Revisiting Equation (2.57), which defined the timestep magnitude, it can be observed that two general options are available to overcome this undesirable

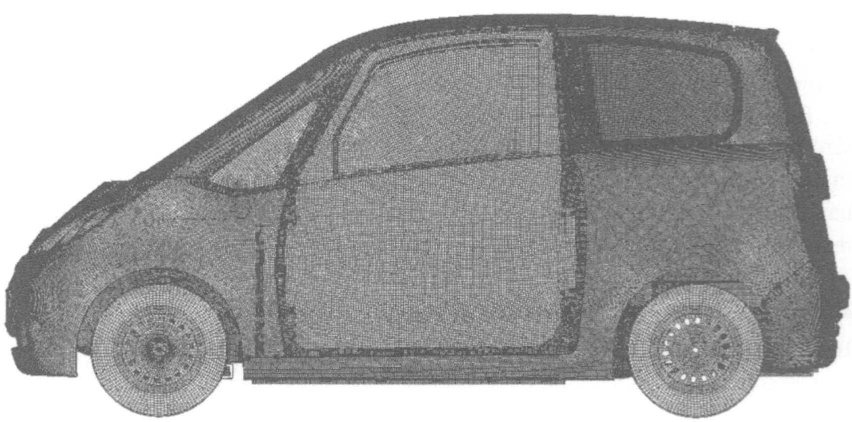

FIGURE 2.32 Microcab crash test model.

increase in CPU time. The first one is to remesh the offending zones in order to increase the overall problem timestep by increasing the element size, that is, $L_{critical}$, Equation (2.57); this may of course lead to issues with accurately representing the original geometry in the FE model. The second method is to manipulate the timestep of the offending element without remeshing, that is, "modify" the stress wave propagation speed, C, Equation (2.57); this approach is often referred to as "mass scaling." The ideology is to locally add mass in order to increase the density of the structure, thus increasing the timestep, Δt, Equation (2.57), via C, Equations (2.53), (2.54), and (2.55). The latter equations consist of two dominant parameters, Young's modulus, E, and the element density, ρ, both of which could potentially be customized. Nevertheless, it has to be noted that E is fundamental to structural contact stiffness, as will be discussed in Section 2.7. Hence, it is not recommended to modify E, as it could have grave repercussions for, for example, model validation. Consequently, only the density ρ should be altered in order to increase Δt; this is the reason why the approach is referred to as mass scaling. Assuming a predefined "ideal" timestep magnitude $\Delta t_{specified}$, the local element density, $\rho_{offending\ element}$, of any "offending element" with a timestep Δt less than $\Delta t_{specified}$ can be scaled according to Equation (2.61)

$$\rho_{offending\ element} = \frac{E \cdot \left(\Delta t_{specified}\right)^2}{\left(l_{offending\ element}\right)^2} \tag{2.61}$$

Most modern day FE computations utilize the mass scaling approach, as this method has its clear advantages. Mass scaling brings stability, especially when "low stiffness" structures, such as compliant foams are present, as these will generally deform significantly when subjected to external loading. Consequently, the element sizes will change significantly and greatly reduce the timestep during the iterative explicit FE computation sequence. Furthermore, the use of mass scaling also enables a more predictable CPU runtime to solve the individual FE model, as a predefined timestep, problem end time (termination time), and known computer hardware specifications will be known.

On the other hand, mass scaling will change the mass of the structure to align all the elements to a set timestep. As such, this could generate inertial effects, as additional mass could be present in the model than wished. As a rule, mass scaling exceeding 2% of the overall "original" structural mass is not recommended. Although mass scaling may be used to solve FEA specific problems as listed above; it does have a significant drawback in terms of using the variable density method for optimization, as shall be discussed in Section 4.1.

2.7 CONTACTS

In general, contacts are used and applied in a wide variety of ways in connection with FEA. As a result, many different types of connections are available within commercial FE software, including panel-to-panel contacts. The two main types of contacts generally used for modeling vehicle safety structures are

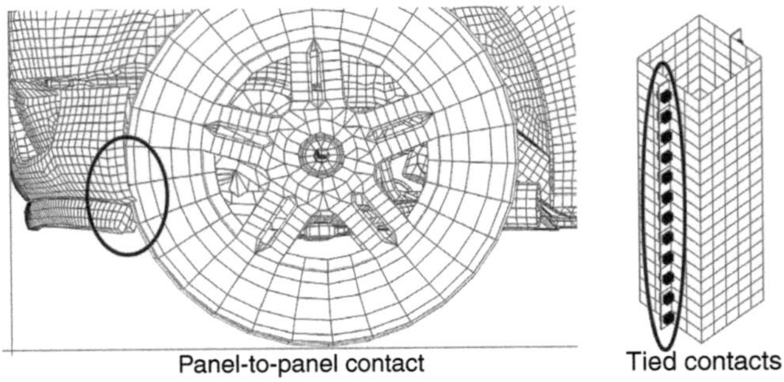

FIGURE 2.33 Typical contacts: panel-to-panel contacts and tied spot-welds.

component contacts, such as panel-to-panel contacts as well as tied contacts. Panel-to-panel contacts are, for example, used in connection with crash analysis to ensure that Newton's third law of motion is not violated. During a front crash scenario it must, for instance, be ensured that the load is transferred from the main body structure to the wheel when the former is pushed back to "contact" the latter. Tied contacts are generally used to represent mechanical connections such as spot welds. Illustrative examples of both types of contacts can be seen in Figure 2.33.

The following subsections will discuss the two types of contacts illustrated in Figure 2.33 in greater detail.

2.7.1 Panel-to-panel contacts

Panel-to-panel contacts used in connection with explicit FEA analysis of vehicle safety structures generally utilize a penalty formulation. This means that the contact reaction force is proportional to the degree of encroachment into the element space, in agreement with Newton's third law of motion. The principle of a contact is to prevent "slave" nodes from penetrating "master" segments (Du Bois, 2010), as depicted in Figure 2.34.

The contact is activated when a slave node penetrates the master element volume and is initiated when the node "penetrates" the "external" surface of the element (Figure 2.34); this is referred to as penetration. During the contact search, the norm of projection of a slave node is computed against the value of half the shell thickness, and is calculated as a scalar product, Expression (2.62).

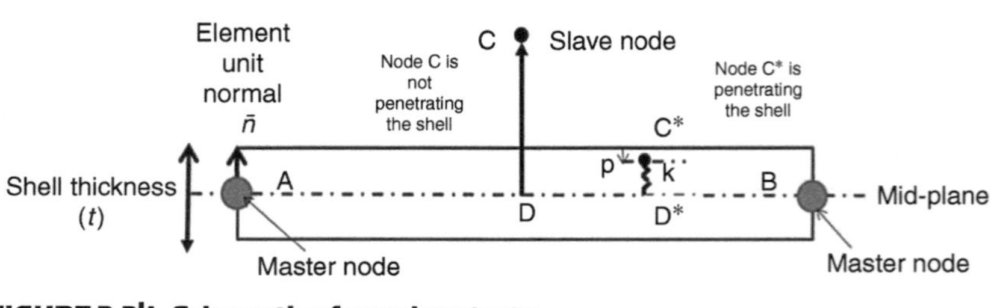

FIGURE 2.34 Schematic of panel contacts.

$$\begin{aligned} \left\|\bar{n}\cdot\overline{CD}\right\| &\geq \frac{t}{2} = \text{No penetration} \\ \left\|\bar{n}\cdot\overline{CD}\right\| &< \frac{t}{2} = \text{Penetration} \end{aligned} \tag{2.62}$$

In most instances, the element normal is straightforwardly computed, as it is based on 2.62. Contact problems can occur when shell elements are excessively warped as the shell normal becomes difficult to uniquely define. If a penetration is detected a linear tension-only spring of stiffness, k, is automatically added to the model. The recoil force of the slave node violating the master segment is proportional to the linear spring stiffness and the penetration, p, as illustrated in Figure 2.34. Provided the slave node does not cross the mid-plane of the element, the explicit computation will remain stable. The contact stiffness is element type dependent, as indicated in Expression (2.63)

$$\begin{aligned} &\text{2D (shell) elements:} \quad k = \frac{f_s \cdot A \cdot K}{\text{Max diagonal}} \\ &\text{3D (solid) elements} \quad k = \frac{f_s \cdot A^2 \cdot K}{\text{Volume}} \end{aligned} \tag{2.63}$$

In Equation (2.63), A is the surface area of the element; the diagonal and the volume also refer to the individual elements.

It can be noted that k depends on the mesh size, as it is linked with the contact area between the slave nodes and the master segments, as well as the material stiffness, K (bulk modulus). As the bulk modulus is linked with Young's modulus (E), Equation (2.63) reinforces the previous statement that changing the value of E to stabilize a computational run or perform mass scaling is not a preferred option. The value of the contact penalty factor, f_s, is per default set to 0.1 in order to keep contact force magnitudes relatively small in an effort to maintain numerical stability (LSTC, 2013). Great care should therefore be exerted before the value of this parameter is modified. The default value of the penalty factor has been set to optimize the stiffness of the contact spring for impact of two components with identical material characteristics and equal sized elements (Du Bois, 2010).

Contacts generally work better when elements in contact are of the identical size, as this ensures that the recoil force is spread evenly across the slave segments, hence avoiding any local force gradients and potential risks of element hourglassing.

Element hourglassing is also often referred to as a spurious mode, zero energy mode, or singular mode among others. If element hourglassing occurs in the individual element, it "simply" means that the individual element has no resistance to the specific applied loading, and thus the element does not contain any strain energy, hence zero energy mode. Hourglassing is in essence caused by limitations of the discretization and numeric analysis. Additional information on hourglassing is available from many different FEA books, for example, Cook et al. (2001).

Prior to solving any model containing contacts using explicit FEA, it is pivotal that the contact stability criterion, Expression (2.64), is checked for all individual contact springs.

$$\Delta t_{\text{Contact}} > \Delta t \tag{2.64}$$

In (2.64) $\Delta t_{\text{Contact}}$ is the timestep associated with the individual contact spring, while Δt is the timestep determined by all other elements in the specific model.

As Δt is calculated from the element geometry and material properties, it can be computed directly in the preprocessor environment, while the specific contact timestep, $\Delta t_{\text{Contact}}$, will fluctuate throughout the explicit computation iterations and will subsequently be computed by the solver. This is because $\Delta t_{\text{Contact}}$ depends upon the nodal masses of penetrating nodes/elements, which are unknown prior to solving the problem, as the deformation levels and magnitudes are unknown. Furthermore, also unknown which elements will come into "contact" with each other.

If the default contact values are used, then the contact stability criterion is usually met. Any unrealistic stiffness changes can cause contact instability; if needed, Δt can be reduced to meet the requirements of Expression (2.64).

2.7.2 Tied contacts

For FEA of automotive structures, one of the most commonly used applications for tied contacts is the spot-weld, but this also includes other connectors such as adhesive strips. Prior to the year 2000, spot-welds were generally modeled as short beams directly connected from one node of a flange panel to a second node on another flange panel. The primary reason for this was that the implementation of tied contacts into commercial FE software was not sufficiently robust, and the "automation" (of connection multiple nodes on different panels) was in its infancy.

Utilizing the short beams nonetheless produced acceptable FE crash models that were a great step forward toward accurately modeling and improving vehicle crash scenarios. Nevertheless, this method had its significant drawbacks, and was extremely cumbersome to fully implement. The issues included that:

- The method was highly manual; therefore, a considerable amount of man-hours were required to complete a single model.
- It was not consistently possible to match the (exact) spot-weld locations from CAD models to the FE models.
- It was difficult to "automate" the modeling process, as the nodes on both flanges had to be aligned in order to generate a spot-weld perpendicular to flanges.
- This method inherently bridged two nodes using a very stiff beam; hence, the risk of structural instability due to hourglassing was significantly increased.

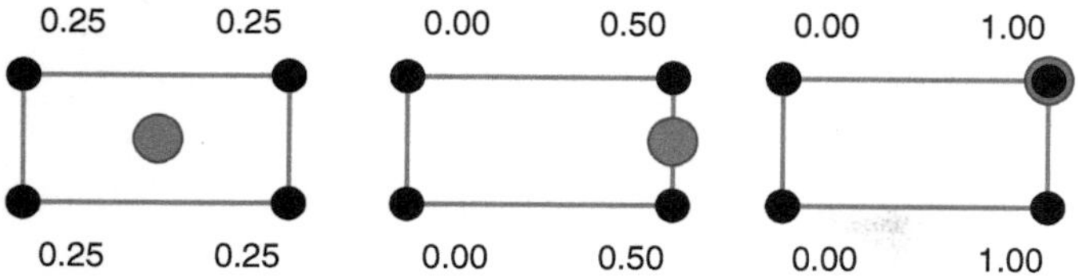

FIGURE 2.35 Load transfer ratio between a spot-weld connect to a panel via a tied contact.

The successful implementation of tied contacts resolved all of the above-listed issues: any spot-weld or adhesive can now be automatically connected to panels with a minimum amount of effort. It is now possible in some user interfaces (ARUP, 2013) to automatically generate these connections from a text file containing the spot-weld/adhesive locations and their respective panels to assemble.

When a weld is created in FE software, the tied contact creates a connection with the elements which it is attached to. The tied contacts are generally beam elements, but may also be 3D solid elements depending on the application. The load transferred from the weld to the panel (modeled with shell elements) follows the scheme displayed in Figure 2.35, where the grey dots represent spot-welds and the black dots represent element nodes. The decimal points indicate the relative fraction of the force transmitted through the individual node as a function of its vicinity to the spot-weld.

Consequently, a spot-weld located at the center of an element and tied to this element will "spread" its force evenly to all nodes of the element. For a four-noded element, 0.25 of the load is transmitted per node. The same logic applies for a spot-weld aligned with an element's edge or aligned with an element's corner: the transmitted load is then 0.5 per edge node and 1 for the latter, as illustrated in Figure 2.35. The positive attributes of the tied contact method mean that the chance of transmitting the full load through a single weld is remote, therefore vastly reducing the likelihood of instability caused by hourglassing.

Despite the fact that the introduction of the tied-contact resolved the majority of the issues associated with modeling spot-welds in FE, the inherent small timestep of the (steel) spot-weld does however still remain. In general, the lengths of the "beams" created are in the region of 1 mm, significantly less than the "proposed" 5 mm mesh size previously suggested for FEA of (steel-based) vehicle structures, thereby decreasing the timestep magnitude in the order of 80%. As the timestep is proportional to the element length, material density, and Young's modulus, any change in these parameters will alter the definition of the spot-welds' overall stiffness, which in general is highly undesirable.

As the spot-weld is directly connected to a panel and does not use a penalty formulation for the contact, as is the case with the panel-to-panel contact, it is possible to modify Young's modulus (E) and the spot-weld diameter to obtain a sensible timestep magnitude while maintaining the flexural rigidity (overall stiffness). This is because the spot-weld diameter dictates the second moment of area, I, of the spot-weld. Therefore, Expression (2.65) can be used to calculate the "new" E_{after} after updating the spot-weld diameter, while maintaining the flexural rigidity.

$$
\begin{aligned}
& E_{before} \cdot I_{before} = E_{after} \cdot I_{after} \\
\Downarrow \\
& E_{before} \cdot \frac{\pi d_{before}{}^4}{64} = E_{after} \cdot \frac{\pi d_{after}{}^4}{64} \\
\Downarrow \\
& E_{after} = E_{before} \left(\frac{d_{before}}{d_{after}} \right)^4
\end{aligned}
\tag{2.65}
$$

Expression (2.65) can now be implemented into Equation (2.57) to form Equation (2.66). Equation (2.66) can now be used to calculate the updated timestep Δt_{after}, which can be compared to the desired timestep Δt.

$$
\Delta t_{after} = \frac{l}{\sqrt{\dfrac{E_{before} \left(\dfrac{d_{before}}{d_{after}} \right)^4}{\rho}}}
\tag{2.66}
$$

As an example, for a set spot-weld length, the timestep could be increased from 1.93E-7 s to 7.71E-7 s, just by changing the diameter from 5 to 10 mm, without modifying the spot-weld stiffness characteristics, as listed in Table 2.3.

Assuming a predefined timestep Δt of 1E-6 s equal to 1 μs, the global model timestep could be slightly lowered by increasing the spot-weld diameter to 10 mm, or Δt could be retained at 1 μs by mass scaling the spot-welds. In general, the stiffness-to-density ratios of adhesives are larger than those of metals; hence, the stress wave propagation will be greater. As a consequence, there are usually no timestep issues related to adhesive (solid) elements.

TABLE 2.3 Scaling of Spot-Weld Stiffness for Timestep Improvements

	Original (Before)	**Updated (After)**
Spot-weld length (mm)	1	1
Spot-weld diameter (mm)	5	10
Young's modulus (MPa)	210,000	13,125
Spot-weld timestep, Δt_{SW} (s)	0.193E-06	0.771E-06

2.8 EXAMPLE CONVERGENCE STUDY OF EXPLICIT FEA

This section will cover a worked example of a convergence study focusing on the identification of an adequate mesh size for structural assessment of an automotive crush can using explicit FEA.

A good approach to evaluating the accuracy of mesh size is to consider the following:

- Firstly, take a pragmatic approach and focus on the smoothness of the deformed mesh and compare with expected outcomes (Du Bois, 2007);
- Subsequently, if additional analyses are completed with different mesh densities observe the energy plots as well as force levels and investigate the mesh size(s) for which changes are within what is acceptable (Du Bois, 2010).

As an example, a typical crash test on a top hat section would have the deformed shape response displayed in Figure 2.36 (American Iron Institute, 2004). Any comparable crush test that does not display a comparable smoothness in the folds would be too coarsely meshed.

It is important that the convergence study should only reflect the mesh refinement, as the computer model's response also depends on element formulation, hourglass formulation, the number of integration points (IP) through the thickness of a shell element, contact formulations (including tied contacts), mass scaling, element quality, and mesh orientation (Du Bois, 2010).

If various mesh computation runs are available based on a stable FE model, then a more scientific approach can be undertaken by comparing the following responses:

- Energy plots, including total energy, kinetic energy, energy of deformation, and hourglass energy;
- Section force plots in relevant cross sectional areas of the structure and/or of contact interactions.

FIGURE 2.36 Crush can displaying a "smooth" buckling behavior.

TABLE 2.4 Generic Crush Can Material Properties

Density (t/mm³)	Young's Modulus (MPa)	Poisson's Ratio	Yield Stress (MPa)	Etan (MPa)
7.8E-9	210,000	0.3	300	1000

Disregarding the previously presented issues of singularities, it is evident that a finer (smaller) mesh size will generally yield more accurate results. Nevertheless, in explicit FEA there is a price to pay as the runtime will most likely increase significantly primarily due to two complementary effects: namely the increase in DOF of the model and the reduction in timestep magnitude.

As presented in Section 2.3.8, culminating with Figure 2.24, the purpose of a mesh convergence study is to assess the effects of the mesh density (size) upon the results of the FEA. The ultimate purpose is to determine an adequate discretization level to accurately assess the engineering problem.

While Section 2.3.8 focused on a linear static model, this section will focus on a large deformation nonlinear and dynamic analysis using explicit FEA. The component under investigation is the generic crush can composed of two top hat sections, as illustrated on the right hand side of Figure 2.33.

The two top hat sections are connected using tied-contacts (spot-welds), and will be impacted by a rigid plate, with a mass of 1 ton, traveling at a constant velocity of 11 m/s (40 km/h).

The crush can is made of mild grade steel and will utilize an elastic–plastic material behavior model, for which the properties are listed in Table 2.4.

Four FE models, denoted 1 through 4 were created, each incorporating a relative mesh density of 1, 1/2, 1/4, and 1/8, respectively. No other model parameters were modified between the four different models.

By observing Figure 2.37, the final shape(s) at the end of the analysis, it is clear that these are not identical for all four models. It can be noted that in the

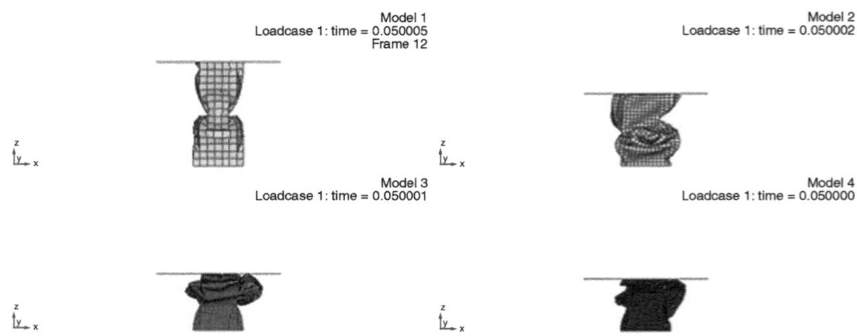

FIGURE 2.37 Comparison of crush can collapse shape as a function of mesh density.

case of Model 1 the crush can is significantly less deformed compared with the remaining models, where the collapse due to buckling is much clearer.

If Model 2 has buckled, its shape is very different from that of Model 3 and Model 4; the reason is that the deformed elements of Model 2 do not create a "smooth" surface. For Model 3 and Model 4, it can be noted that the deformations are "smoothly" represented and tend to converge/concur with each other. These facts are good indicators to use as a first assessment in connection with selecting an appropriate (and converged) mesh density or element size.

2.8.1 Contact forces

In order to further verify the deformed shape results, it is possible to extract the contact forces between the plate and the actual crush cans as a function of time; the resulting graphs are illustrated in Figure 2.38.

The largest contact forces of all four models should be found in model 1, as the structure has "locked" and thus prevented additional buckling deformation, as can be confirmed by observing Figure 2.38. The first peak on the "Model 1 graph" is the plate initially contacting the crush cans (6.5E5 N) and the second peak, at time 0.004 s, is the force generated by the structure locking (7.6E5 N).

For Model 2, Model 3, and Model 4, the initial contact force peak reaches a consistent value of 5.0E5 N (Figure 2.38). Differences are subsequently seen; this is caused by the poor smoothness of the deformed mesh of Model 2, which generates a higher force resistance, 4.0E5 N at time 0.004 s (Figure 2.38). Furthermore, it can also be observed that for Model 3 and Model 4 the overall force pattern is comparable; however, Model 3 still dictates a marginally higher peak force compared with Model 4.

2.8.2 Kinetic energy

It is always good practice to verify that the energy level postulated in the problem is input accordingly. As such, a simple energy check can be performed to ensure correct boundary conditions are present by plotting the initial energy input values from the FEA (Figure 2.39).

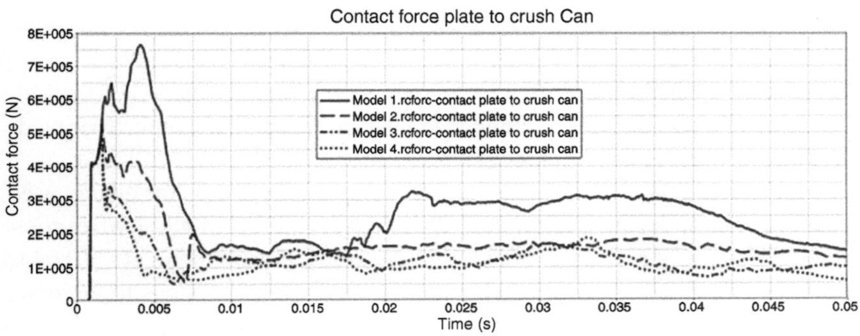

FIGURE 2.38 Comparison of crush can collapse force as a function of mesh density.

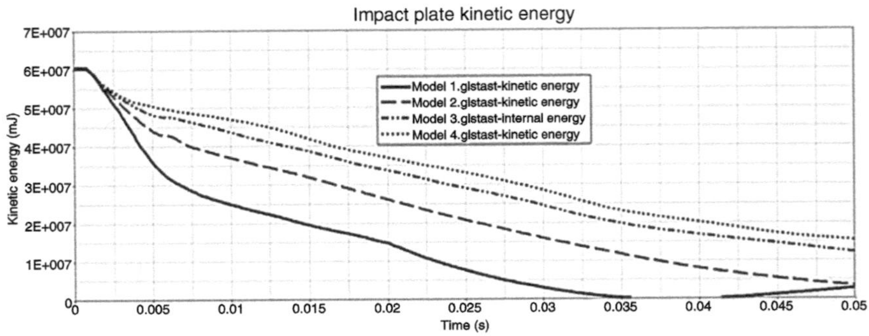

FIGURE 2.39 Comparison of crush can collapse kinetic energy as a function of mesh density.

Figure 2.39 confirms that the initial energy is consistent for all four models. In this particular case, the input energy at $t = 0$ equals the kinetic energy (KE) of the 1 ton plate traveling at 11 m/s. Hence, the magnitude of 6E7 mJ listed in Figure 2.39 can be straightforwardly compared with a manual calculation for kinetic energy, Expression (2.67)

$$\text{KE} = \frac{1}{2}\cdot M\cdot V^2 = \frac{1}{2}\cdot 1{,}000\left[\text{kg}\right]\cdot\left(11\left[\frac{m}{s}\right]\right)^2 = 6.05E7\ \text{mJ} \qquad (2.67)$$

Expression (2.67) thus confirms that the initial energy level of the explicit FE model is correct. By observing the KE levels throughout the duration of the FEA, it can be noted that the coarsely meshed structure (Model 1) "locks" as the KE of the impact plate becomes 0 mJ at $t = 0.038$ s and subsequently exhibits a state of rebound, represented by a change of curvature in the energy curve. The KE of the impact plates in the remaining three models continues to reduce throughout the duration of the analyses, meaning that these structures continue to crumple gradually. This suggests that all of the models are converging toward each other, and that the only difference in the overall structural compliance occurs within the first 0.005 s of the analysis run time.

2.8.3 Internal energy

The focus of attention is now turned to the internal energy (IE) of the four models which is illustrated in Figure 2.40.

Figure 2.40 reveals that the IE, which in essence represents the energy used to deform the crush can, is higher for (the relatively stiffer) Model 1 when compared with the remaining three models of the study. Furthermore, at $t = 0.04$ s, when the impact plate starts to rebound, the IE of Model 1 stops to increase, confirming that the internal energy "is used" to deform the structure.

Figure 2.40 reinforces the trend identified in Figure 2.39; namely that the only difference in the overall structural compliance between Models 2, 3, and 4

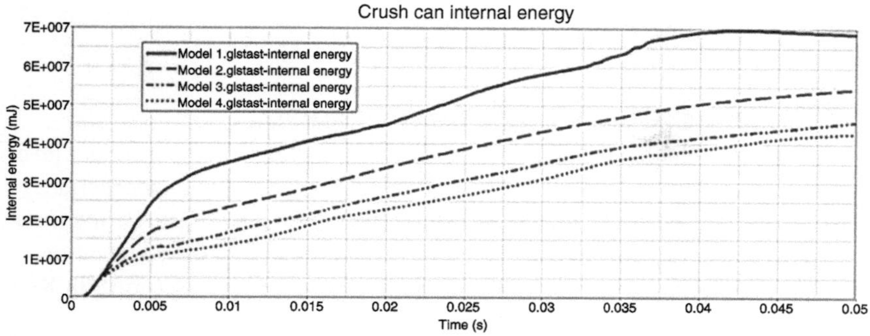

FIGURE 2.40 Comparison of crush can collapse internal energy as a function of mesh density.

occurs within the first 0.005 s of the analysis run time. This is because the IE of these three models is consistent in curvature and gradient after $t = 0.05$ s.

2.8.4 Total energy

Another check to perform in connection with explicit FEA is the energy balance (EB). For clarity, only two models will be discussed in this context; namely Model 1 and Model 3, the EB of which are illustrated in Figure 2.41.

The impact scenario of the rigid plate to the crush can considered in this particular case represents a conservative system as no external work is provided to the system; hence Equation (2.68) must be true.

$$\text{Total Energy (TE)} = \text{Kinetic Energy(KE)} + \text{Internal Energy(IE)}. \quad (2.68)$$

By plotting the three energy variants of Expression (2.68) for the crush can models, it is possible to check if any of the models are numerically (un)stable. Observing Figure 2.41 while considering Expression (2.68), it can be noted that Model 1 is numerically unstable as the IE exceeds the TE. This means that Model 1 is not suitable for engineering investigations including structural optimization.

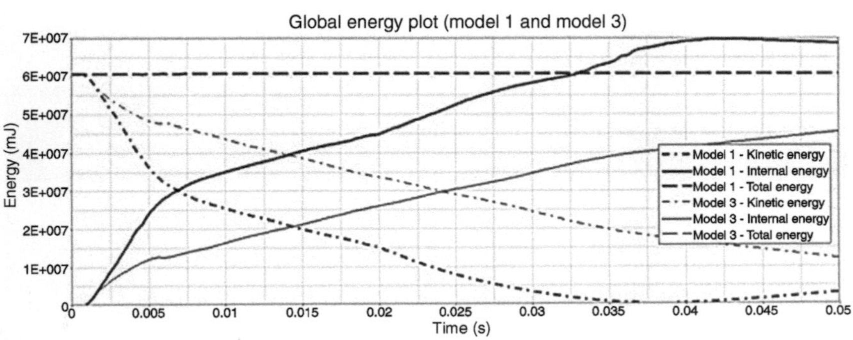

FIGURE 2.41 Comparison of crush can collapse total energy as a function of mesh density.

Model 3, on the other hand, is stable as the sum of IE and KE equates to the TE (Figure 2.41); hence, Equation (2.68) is true.

2.8.5 Summation of convergence study

One of the key challenges facing any FE user is to define an appropriate level of convergence; what is "good enough?" What mesh density is appropriate to simulate the engineering problem at hand? The importance of answering these two questions is further emphasized in the context of structural optimization.

This highlights the necessity to answer the two questions set forth in Section 2.3.5, repeated below for convenience:

- What is the purpose of the FE analysis?
- What is the desired accuracy of the FE analysis?

These are indeed pivotal questions, the answers to which may have critical outcomes, not only on stand-alone analysis but most certainly also on structural optimization. By assuming that the crush can model belongs to a vehicle model required to comply with a safety-related legislative requirement such as ECE-R95 for frontal impacts, the importance becomes very clear. Furthermore, the analysis is conducted as part of the final sign-off stage immediately before a prototype is built and tested and is used to inform an algorithm used to develop and fine tune the firing sequence and timing of the driver airbag. Choosing an inappropriate mesh density can lead to unrealistically high force outputs (Figure 2.38), which in turn will feed incorrect information into the development process of the restraint system control unit.

At this point, it is convenient to remember the quote from Section 2.3.5, repeated below:

> *"An FE model should be as simple as possible but as complex as necessary."* (Sorokin, nd)

Another aspect to take into account is the overall CPU consumption required to numerically solve the model, as discussed throughout this chapter.

The primary parameters to be used for deciding which mesh density to utilize for the crush can study are listed in Table 2.5.

TABLE 2.5 Primary Parameters

Model Number	CPU Time (s)	Model Converged?	Model Numerically Stable?	Scale Factor of CPU Time From First Converged Model
1	95	No	No	N/A
2	517	No	Yes	N/A
3	3715	Yes	Yes	1.0
4	27,120	Yes	Yes	7.3

Reviewing the data of Table 2.5, it immediately becomes clear that Model 1 and Model 2 cannot be used given the above-described scenario; hence Model 3 and Model 4 remain. Considering the significant increase in CPU time between Model 3 and Model 4, it is suggested that the mesh density of Model 3 would be appropriate to use in this particular example.

Evaluating the information in Table 2.5, it can be concluded that:

> *"If you only perform one calculation you do not know how wrong you are."* (Schwer, 2008)

With the above, the brief introduction of relevant FE theory, principles, practice, and suggestions is complete. These are important not only for the stand-alone analysis used for the development and validation of many mechanical systems including vehicle safety structures; but also play a pivotal role in connection with any FE based structural optimization. The next chapter will introduce basic concepts of structural optimization, including the identification of potential applications in connection with vehicle safety structures.

REFERENCES

American Iron Institute. (2004) *Crashworthiness and occupant protection*'. Technical document Available from <http://www.autosteel.org/~/media/Files/Autosteel/Research/Safety/safety_book.pdf> Accessed 01.12.13.

ARUP. (2013) *OASYS PRIMER*. User manual Available from < http://www.oasys-software.com/dyna/en/downloads/documentation.shtml> Accessed 17.01.14.

Christensen, J., Montes Schutte, H. U., Nesari, M., & Dedic, N. (2007). Stress and deformation analysis of link arm DMS7 semester project, Aalborg University, Aalborg, Denmark.

Cook, R. D., Malkus, D. S., Plesha, M. E., & Witt, R. J. (2001). *Concepts and applications of finite element analysis* (4th ed.). Wiley: Hoboken, New Jersey, USA.

Du Bois, P. (2007). Crashworthiness course. Engineering lecture. Held January 5, 2007 at Corus Automotive Engineering, Coventry, UK.

Du Bois, P. (2010). *Crashworthiness and impact engineering with LS-Dyna*. California: Livermore Software Technology Corporation, Engineering Course Notes. LSTC. Livermore, California, USA.

Jacob, P., & Goulding, L. (2002). *An explicit finite element primer*. Glasgow: NAFEMS.

Jensen, M. (2007). *Introduction to LS-Dyna implicit*. Lecture notes available from Livermore Software Technology Corporation, California.

LSTC. (2012), *LS-Dyna. Keyword user's manual. Volume 1. Technical manual* Available from <ftp.lstc.com/anonymous/outgoing/trent001/manuals/LS-DYNA_manual_Vol_I_R6.1.0.pdf> Accessed 12.03.12.

Lund, E. (2006). *Introduction to the finite element method*. Aalborg, Denmark: Aalborg University, Lecture notes from lectures held in the autumn semester 2006.

Macaulay, M. (1987). *Introduction to impact engineering*. London; New York: Chapman and Hall.

Schwer L. (2008). Blast modeling using LS-Dyna. In: *ARUP LS-Dyna User Conference Proceedings*. Held January 18, at Solihull, UK.

Sorokin, S. (nd) cite from Professor Sorokin, Aalborg University, Aalborg, Denmark.

Ugural, A. C., & Fenster, S. K. (2003). *Advanced strength and applied elasticity*. New Jersey: Pearson Education, Upper Saddle River, New Jersey, USA.

Chapter | three

Introduction to General Optimization Principles and Methods

CHAPTER OUTLINE

The first two chapters of this book were aimed at introducing the reader to fundamental aspects of vehicle structures as well as finite element (FE) theory and finite element analysis (FEA), with an emphasis on vehicle safety structures. The purpose of this chapter is to introduce the reader to some of the general mathematical optimization principles, techniques, and methods, giving the reader an overall understanding of what constitutes structural optimization from an engineering viewpoint. This is not meant to be a complete and in-depth overview, but merely an appetizer to give the reader an insight into the vast field of optimization, with an emphasis on structural optimization. Chapter 4 will be a direct continuation of Chapter 3, where the focus will significantly change from the general optimization principles to very specific aspects of vehicle structural optimization and associated techniques.

3.1 WHAT IS STRUCTURAL OPTIMIZATION?

The majority of people, not just engineers, will most likely have an inherent and possibly relatively vague understanding of the phrase "optimization." It is, therefore, very important to first define the term "structural optimization":

Nonlinear Optimization of Vehicle Safety Structures. http://dx.doi.org/10.1016/B978-0-12-804424-7.00003-8

An optimized structure is a structure that performs at its "best," that is, at its most "efficient" – the process through which a structure is optimized is denoted optimization.

Whenever the term "optimization" is used throughout the remainder of this book, it will relate to the above definition, unless otherwise stated. It is worth noting that the term "structure" in the above definition can straightforwardly be replaced with, for example, "system" or "component." It should also be noted that the term "most efficient" does not necessarily mean the absolute highest efficiency of the structure, as will be continuously discussed.

The definition of a fundamental optimization problem will, as a minimum, generally require the characterization of the two following parameters:

1. An objective: It is important to determine which "asset" of the structure is to be optimized. For structural optimization this is generally a measurable characteristic. This could, for example, be the strain energy obtained from an FE analysis, as discussed in Chapter 2, or the mass of the structure.
2. Design variable(s): These are parameters of the structure which may be altered during the optimization process to affect the objective, and via these obtain the optimized structure.

Further refinements, such as constraints, are very often added to the optimization problem to reduce the potential number of solutions to the problem and "guide" the optimization toward a solution, as will be further discussed throughout Chapters 4 through 8.

The n dimensional space created by inserting one axis per design variable is referred to as the design space; however, it may also be referred to as the feasible region, feasible set, search space, and (feasible) solution space, pending the introduction of constraints. In most structural optimization applications, it is not possible to graphically illustrate the design space; however, for problems containing up to three design variables, it is possible to graphically illustrate the design space as will be explained later.

3.2 HOW ARE OPTIMIZATION PROBLEMS GENERALLY SOLVED?

This section could very easily be expanded to an entire book or even a series of books, simply due to the vast number of potential answers. In general, the method in which an optimization problem is solved is heavily dependent upon the actual application. Broadly speaking, optimization solvers may be divided into three main categories:

1. Optimization Algorithms can be defined as a number of self-contained step-by-step operations to be performed (often sequentially) which terminate after a pre-specified number of cycles or steps.
2. Iterative Methods also consist of a number of self-contained step-by-step operations; these do, however, terminate according to a convergence criterion.
3. Heuristics and Meta-Heuristics Methods may be considered to be a type of "shortcut" to an optimized solution, favoring a rapid solution generally with a tradeoff in accuracy and precision.

To the end-user, for example, the user of a commercial FE program with optimization capabilities built-in, there is most likely not a noticeable difference between the above methods. In fact, in terms of structural optimization, many of the actual methods contain aspects of two or more of the above methods, as will be further explained later. To better understand the difference between the three "pure" methods above, a flowchart of the fundamental optimization process, that is, an optimization defined via an objective function and design variables as described in Section 3.1 has been created (Figure 3.1).

As can be seen in Figure 3.1, the general difference between an optimization algorithm and an iterative method may not be significant. In fact, many FE-based optimization methods available in commercial software utilize a hybrid version between an optimization algorithm and an iterative method as a convergence criterion is specified, but a maximum number of iterations is also specified. The optimization will, therefore, terminate once the convergence criterion (there may be more than one) is met before the maximum number of iterations; otherwise, the optimization will terminate once the maximum number of iterations is reached. The overall flowchart for a heuristic or meta-heuristic

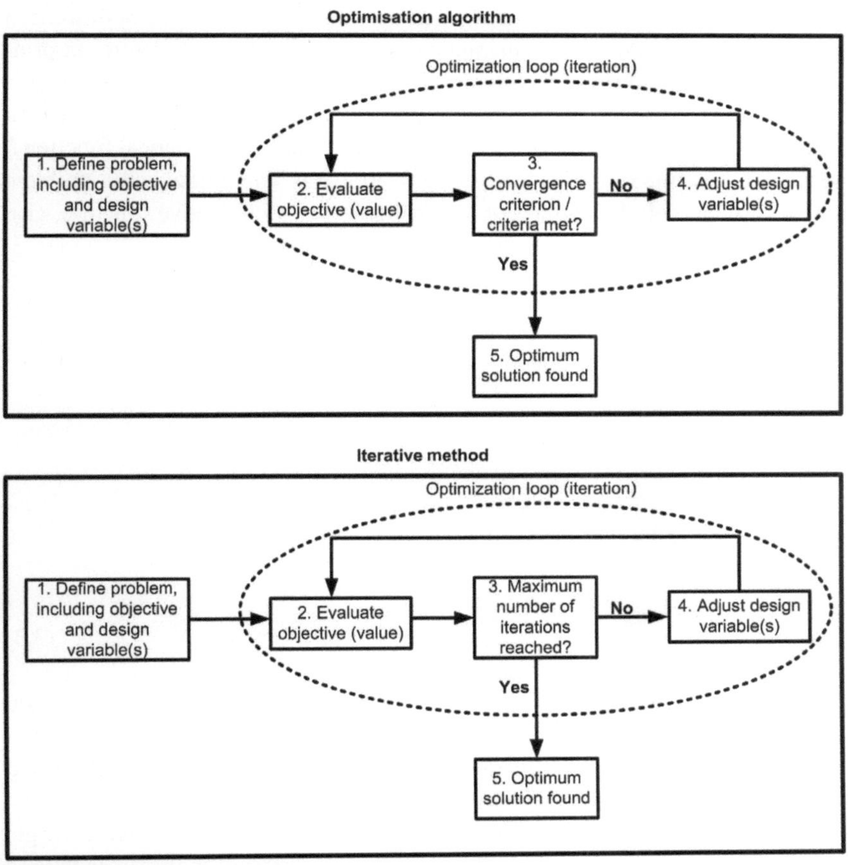

FIGURE 3.1 Optimization methods.

optimization process may be identical to either of the two illustrated in Figure 3.1, or may indeed also utilize a hybrid version between an algorithm and an iterative method. The difference lies in Step 4 (Figure 3.1), as will be discussed throughout Chapters 5 through 7.

For the above reasons, there will be no particular distinction between an optimization algorithm and an iterative method throughout the remainder of this book.

Figure 3.1 denotes a single optimization objective; however, in some cases, it is desirable to have multiple objectives, although this may significantly complicate the optimization process as will be discussed later. Figure 3.1 also demonstrates the definition of an optimization loop, which is the iterative process linking step 4, adjustment of design variable(s), to step 2, evaluation of objective, and ultimately determination of whether or not the "convergence criteria" has been met. At this stage, it is convenient to introduce mathematics to enhance the readers understanding of a "fundamental" optimization problem. Although all optimization approaches/algorithms discussed throughout this book do contain mathematical expressions, they are not all "classed" as being mathematical-based optimization. It may, therefore, be difficult to distinguish between mathematical-based and non-mathematical-based algorithms; in this book, a mathematical-based algorithm is defined as one where the entire optimization process is purely based on mathematical principles and can therefore, in principle, be defined and subsequently directly solved analytically, not numerically. To enhance the readers' understanding, assume that the optimization objective is to minimize the system characteristic described via a mathematical function f, which is a function of the (single) design variable x. The optimization problem may thereby simply be defined (mathematically) as an objective function, also sometimes referred to as a cost function, Expression (3.1)

$$\min(f(x)) \tag{3.1}$$

As an example of the objective function/cost function assume that:

$$f(x) = x^2 - 1 \tag{3.2}$$

For the time being, the solution x equals zero will be ignored. Now, if Equation (3.2) was to be optimized, that is, minimized in this example, a possible solution would be to rearrange Equation (3.2) as Equation (3.3)

$$x^2 = 1 \tag{3.3}$$

An obvious solution to Equation (3.3) is that x equals 1; however, -1 would also be a viable solution. This ambiguity is often referred to as "uniqueness of solution" or "multimodal optimization" and may be demonstrated by observing Figure 3.2, which plots the objective function defined in Equation (3.2).

Many structural optimization problems do not necessarily have a single unique solution and if they do it may be extremely difficult to prove, as will be

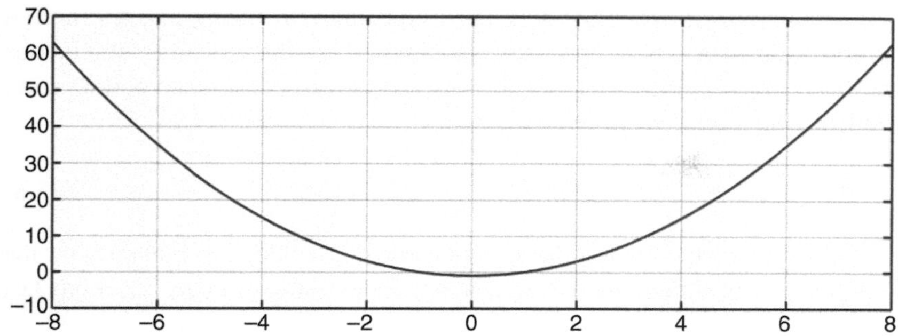

FIGURE 3.2 Example objective function.

demonstrated continuously throughout this book. It is of the utmost importance that the uniqueness of solution is proven or at least addressed for any optimization algorithm. The primary reason for this is that it needs to be verified or at the very least justified that the solution obtained is indeed an optimum solution, as opposed to an arbitrary one. For example, if x equals 0 is now re-introduced as a feasible solution to the objective function of Equation (3.2), x equals 1 or -1, is no longer an optimum solution, it is an arbitrary solution.

3.3 GENERAL OPTIMIZATION METHODS AND PRINCIPLES

Optimization problems can generally be split into two different categories: constrained and unconstrained. The above example where x equals zero was ignored as a viable solution is an example of a constrained optimization problem, whereas the inclusion of x equals 0 is an example of an unconstrained optimization problem. In this particular case, the constraint was imposed directly onto the design variable; however, the constraint(s) may also be imposed indirectly, for example by defining a maximum cross-sectional area of the beam which inherently is a function of the radius, x, but the constraint only "indirectly" discards specific x values. To demonstrate this, assume that Equation (3.2) describes the mass of a bar with a circular cross-section and x describes the radius of that cross-section. The objective is to minimize the mass of the bar, but now subject to a maximum stress limit. The latter is implemented indirectly as the stress is a direct function of the radius; but the mass (which is the objective function) is not a direct function of the stress; in fact there is no link between them at all. It is "obvious" that the radius x must be non-zero and positive. Equation (3.2) can therefore be mathematically formulated as Equation (3.4)

$$\min\left(f(x)\right)|x > 0 \tag{3.4}$$

Equation (3.4) is thereby a constrained version of the unconstrained optimization problem in Equation (3.2), although it does not include the stress constraint. Additional constraints may also be applied either directly or indirectly. The radius of the aforementioned bar may, only assume integer values, due to manufacturing constraints. If such a constraint was imposed, the design variable(s) is denoted as discrete, as opposed to the previous continuous design variable(s). Equation (3.5), therefore, denotes a constrained optimization problem with a discrete design variable.

$$\min\left(f(x)\right)\middle| x>0 \wedge x \in \mathbb{Z} \tag{3.5}$$

An example of a constraint applied indirectly could, for example, be the definition of a maximum stress (σ_{max}), or displacement (d_{max}) value, Equation (3.6)

$$\begin{aligned} & \min\left(f(x)\right)\middle| x>0 \wedge x \in \mathbb{Z} \\ \text{subject to:} \quad & \sigma \le \sigma_{max} \wedge d \le d_{max} \end{aligned} \tag{3.6}$$

Equation (3.6) thereby represents an optimization problem with a discrete design variable and indirectly, as well as directly, applied constraints.

As demonstrated above, it may be very beneficial to apply constraints to an optimization problem; in fact, most structural optimization problems will be constrained optimization problems. The application of constraints can also help improve the "efficiency" of the optimization algorithm by reducing the potential number of solutions. For example, the number of potential solutions to Equation (3.2) was halved by the introduction of the single constraint in Equation (3.4). Caution must, however, be applied to ensure that the problem does not become over-constrained, which will "inhibit" the optimization algorithm from obtaining a truly optimum solution, as was previously demonstrated by the inclusion or omission of x equals zero in the above example.

Equation (3.2) in the previous section listed an example objective function; it is apparent that this particular equation is of the second order, that is, a nonlinear function. Equation (3.7) is an example of a linear objective function; the difference between a linear and a nonlinear objective function can, therefore, be understood by comparing Equations (3.2) to (3.7)

$$f(x) = x - 1 \tag{3.7}$$

It is important to note that it is the design variable x that is subjected to the power; this is significant as Equation (3.8), where c is a predefined constant, is also an example of a linear optimization problem, not a nonlinear optimization problem.

$$f(x) = c^2 \cdot x - 1 \tag{3.8}$$

In general, nonlinear optimization problems (from a mathematical viewpoint) can be defined as having nonlinear objective functions, although problems with a linear objective function and, for example, a nonlinear constraint function may also be defined as nonlinear optimization problems.

Despite the fact that Equations (3.2) and (3.7) are examples of relatively simplistic objective functions, it should be noted that solving nonlinear optimization problems is generally significantly more difficult than solving linear optimization problems, as will be demonstrated throughout this book.

In Section 3.2, it was stated that optimization solvers can generally be divided into three main categories: optimization algorithms, iterative methods and (meta-) heuristics. Previously in this section, it was dictated that optimization problems may, in general, be divided into two different types of problems: constrained and unconstrained problems. In fact, there are many different ways in which optimization problems and indeed solution strategies may be divided; this often is dependent upon the context in which the optimization is to be applied. For example, structural optimization may be very different from optimization of computing networks or economic profit. Regardless of the specific application area, optimization problems are most often solved using computer programming. Some of the primary programming subfields include, but are not limited to:

- convex programming
- linear programming
- quadratic programming
- integer programming
- nonlinear programming
- stochastic programming
- robust programming

In the majority of cases above, the phrase "programming" is directly interchangeable with "optimization." Although the above list is not complete and describes programming subfields/optimization areas in general, the ones listed are very relevant to structural optimization, as will be demonstrated. Before the individual areas listed above are presented, it is worth exploring the reasons why different programming and optimization techniques are required as the most "simplistic" method of obtaining a truly optimum solution is simply to calculate all possible permutations?

3.4 THE CURSE OF DIMENSIONALITY

The majority of engineering optimization problems contain more than one design variable; the sheer number of design variables is very often a cause for concern. The most obvious example of this is topology optimization, as will be discussed in Chapter 4, where every single element can be considered to be a design variable. As it is not uncommon for "medium sized" FE models to contain in excess of 500,000 elements, the number of elements is a central topic to most engineering optimization problems; but just how significant is this problem?

In the subsequent sections, a number of different methods to solve optimization problems will be introduced including linear programming techniques and gradient-based methods. However, perhaps the most obvious way to definitely obtain a truly optimum solution is to calculate all possible permutations and simply choose the optimum solution. As an example, consider the task of minimizing the deflection of a cantilever beam with a circular cross-section of radius

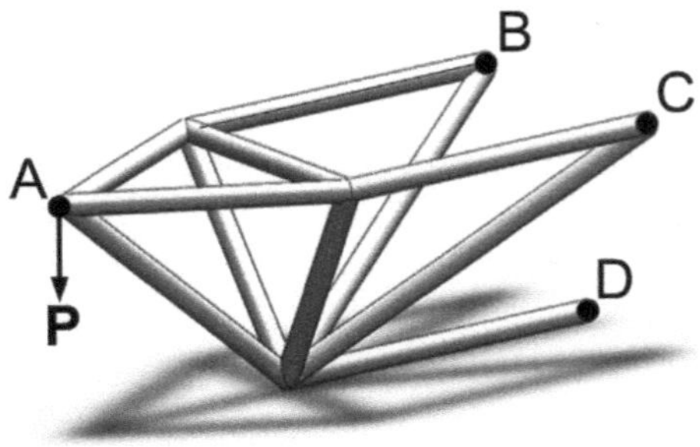

FIGURE 3.3 Truss structure.

r subject to maximum stress and mass constraints. Due to manufacturing constraints, the radius r can only assume five discrete values, for example, 1, 2, 3, 4 and 5. The simplest method for determining the optimum solution is to calculate the deflection, stress, and mass values for all five potential solutions and then to select the optimum solution.

In practice, the number of possible solutions does however rapidly become very large. To better understand the impact of this, consider the truss structure in Figure 3.3.

The truss in Figure 3.3 may be perceived to consist of 11 individual "beams" and is exposed to a single load P at point A while being constrained at points B, C, and D. The optimization task of minimizing the mass and maximizing the stiffness of the truss structure can simply be treated as a discrete optimization problem where 0 means that a specific beam should be removed and 1 means that the beam in question should remain. This means that there are a total of 11 design variables each with two possible "states," that is 0 or 1. As mentioned above, one potential solution to this is to calculate all possible permutations and simply choose the optimum solution. The simple optimization problem represented by Figure 3.3 does, however, have a total of 2048 possible solutions, seen in Equation (3.9)

$$\text{Number of potential states}^{\text{Number of design variables}} = 2^{11} = 2{,}048 \tag{3.9}$$

In general, if one beam in Figure 3.3 is perceived to represent one element in an FE model, it is obvious that the number of possible solutions rapidly becomes infinite; for example, 25 elements lead to approximately 33.5 million possible solutions, while 30 elements lead to in excess of 1 billion possible solutions. When the number of elements exceeds 1,000, the number of possible solutions effectively becomes infinite. This phenomenon is generally referred to as the curse of dimensionality and eradicates the method of calculating all potential permutations for most practical applications. At this stage, aspects such as geometrical symmetry can be taken into consideration. Nevertheless, if three-plane symmetry is utilized for a 3D component, the total number of elements before an infinite number of solutions can be found is a mere 8,000. The curse of dimensionality is a central topic to optimization and will be continuously discussed throughout the remainder of this book.

3.5 CONVEX PROGRAMMING AND OPTIMIZATION

Convex optimization is the study of a specific type of function, namely, a continuous (and generally differentiable) mathematical function. For the time being, only continuous, differentiable, and convex functions will be considered; the consequences will be thoroughly discussed in subsequent sections in relation to real-world structural optimization applications. Utilizing a convex function indicates that only a single minimum (extreme) point exists, that is, only one single optimum solution exists for a convex optimization problem. The question of whether or not only a single truly optimum solution exists is a pivotal aspect of any optimization problem as will be continuously discussed throughout the remainder of this book. In the interim only continuous mathematical (and differentiable) functions will be considered, such as the second-order function illustrated in Figure 3.4.

Figure 3.4 represents a continuous convex function with the global minimum point indicated; solving this type of problem is often referred to as convex optimization. If an optimization problem is defined as convex, it is implied that the task is to minimize the objective function. The polar opposite of convex optimization is termed concave optimization, where the objective function is maximized. It should be noted that for structural optimization "identical attributes" may be obtained via both maximizing and minimizing a specific measure. For example, maximizing the stiffness of a structure may be obtained by minimizing the compliance that in turn minimizes the displacement values.

Please note that Figure 3.4 essentially represents a 1D optimization problem containing a single variable. For multivariable optimization (not to be confused with multicriteria optimization, see Section 3.8.4) Figure 3.4 would have as many axes as parameters or elements.

The fact that the function in Figure 3.4 is continuous simply means that $f(x)$ is defined for every value of x. From observing the graph in Figure 3.4, it is straightforwardly clear where the global minimum of the function is. Determining the extremum points of a continuous convex or concave function $f(x)$ is

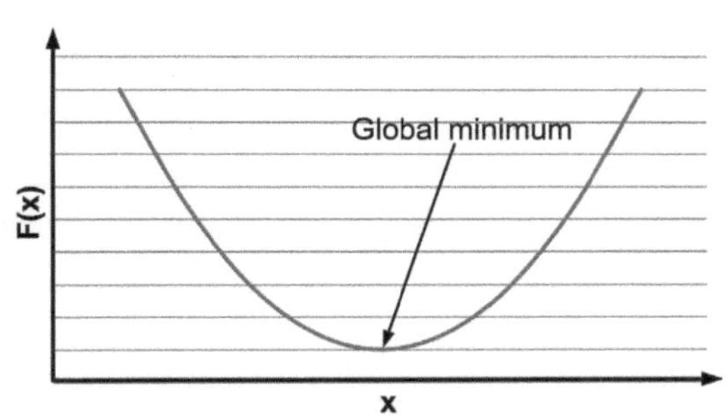

FIGURE 3.4 Example of a continuous convex function.

mathematically straightforward as this can be done by means of differentiation, the first term in Expression (3.10)

$$\begin{aligned} &f'(x) = 0 \\ &f'(x) \geq 0\,(\text{minimum}) \\ &f'(x) \leq 0\,(\text{maximum}) \end{aligned} \tag{3.10}$$

As stipulated by the first term in Expression (3.10), the extremum points are located at the only positions where the slope of the tangent is equal to zero. Using differentiation will thus reveal the two global maximum points of the function in Figure 3.4, as well as the single global minimum; as the single differentiation of 3.10 merely identifies extremum points. As single differentiation is used Equation 3.10 is often referred to as the first derivative test. At this point, it should be noted that if the objective was to maximize the convex function in Figure 3.4, there would be two equally valid solutions as there are two distinct points where $f(x)$ reaches its maximum value for different x values. This is the reason why it is assumed that a convex function is minimized, not maximized. If the task indeed was to maximize the convex function, there would be no preference to choose one optimum solution over another; the complication of this general dilemma will be discussed in greater detail in the subsequent chapters. Remembering that using Equation (3.10) will return three extremum points in the above example, how can it be determined if a specific extremum point is maximum or minimum? One method would be to "observe" the $f(x)$ values in the vicinity of the extremum points and determine if they are reducing or increasing. There is, however, a much more straightforward way to determine the "type" of extremum; namely to use the second derivative test, as seen in the last two terms in (3.10).

The principle of using differentiation in Expression (3.10) is equivalent to observing the gradient; hence, this type of approach is generally referred to as gradient-based optimization and represents a very efficient way of obtaining an optimum solution, even for a very large number of values of a single design variable. As an example of this, consider the task of minimizing the deflection of a cantilever beam with a fixed width w and a variable height h. In this example, the displacement is a function of a second-order function, namely the second moment of area I, a continuous and convex function of h. Even though h is assumed to be a continuous variable, the maximum value (that in turn will minimize the deflection) can straightforwardly (and very rapidly) be found using Expression (3.10). Gradient-based optimization methods will be further discussed in Section 3.6.

3.5.1 Linear programming

One of the most widespread methods of solving linear optimization problems is through the use of linear programming (LP). In order to use LP, it is pivotal that the objective function, as well as the constraint functions are linear. Although this is not the case for convex or concave optimization problems, LP does provide a basis for solving nonlinear optimization problems through sequential linear programming (SLP), which will be the focus of attention in Section 3.5.2. LP represents a very efficient way of solving linear optimization problems. Consider the optimization problem of minimizing the objective function $f(x_i)$ defined in Expression (3.11)

$$f(x_i) = 6x_1 + 5x_2 \tag{3.11}$$

The linear objective function $f(x_i)$ in Expression (3.11) could, for example, represent the bending stress levels in a beam subjected to multiple loads, as a function of the cross sectional geometry parameter x_i. The "beam" of Expression (3.11) is also subjected to three constraints, for example, relating to mass, cross-sectional height-to-width ratio and stress levels. Assume that these constraints are also linear functions of the two design variables x_1 and x_2 as specified in Expression (3.12)

$$\begin{aligned} &x_1 + 2x_2 \le 7 \\ &3x_1 - 2x_2 \ge 0 \\ &x_1 - x_2 \le 1 \end{aligned} \tag{3.12}$$

The optimization problem defined by Expression (3.11) and (3.12) can, of course, be solved analytically, and can also be conveniently solved using LP. The first step is to isolate one of the variables in all the three constraint functions of Expression (3.12); this is done in Expression (3.13)

$$\begin{aligned} &x_1 + 2x_2 \le 7 \Leftrightarrow x_1 \le 7 - 2x_2 \\ &3x_1 - 2x_2 \ge 0 \Leftrightarrow x_1 \ge \frac{2}{3}x_2 \\ &x_1 - x_2 \le 1 \Leftrightarrow x_1 \le x_2 + 1 \end{aligned} \tag{3.13}$$

Based on the right-hand side expressions in (3.13), it is possible to plot the inequalities (constraints) in the $x_1 - x_2$ plane (Figure 3.5).

As the lines in Figure 3.5 are defined using the constraints of the optimization problem, it therefore follows that the solution to the optimization problem must reside within the boundaries of the "triangle" defined by points A, B, and

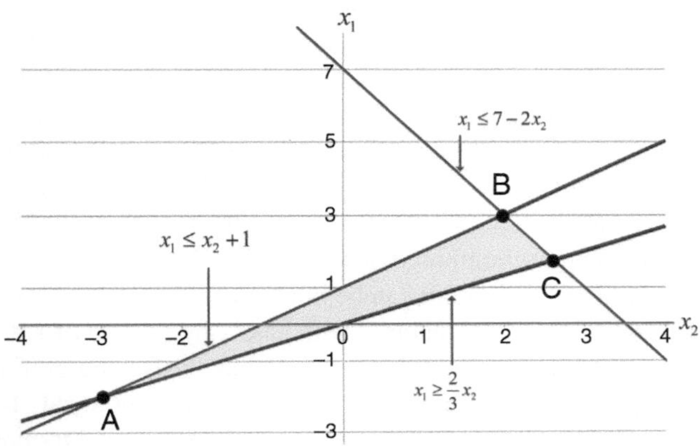

FIGURE 3.5 Constraints plot.

C in Figure 3.5. Now imagine that Figure 3.5 is expanded to a 3D plot by adding the objective function in Equation (3.11) to Figure 3.5 (which in effect would be plotted perpendicular to the x_1–x_2 plane). The resulting space is generally referred to as the permissible solution space and is in this case 3D. In most cases, the solution space is much larger than 3D and can in principle be infinite. It can be mathematically proven that the extremum points, that is, maximum and minimum values of an inequality problem as stipulated by (3.11) and (3.13) are located at the intersection points between the different constraint functions, that is, at points A, B, and/or C (Figure 3.5). The next step of the LP solution methodology is, therefore, to determine these points. In this particular example, these can straightforwardly be found using Figure 3.5. It should, however, be noted that Figure 3.5 is not part of the general LP process, but is merely included as a visual aid in this example. Determining the intersection points between the three constraint functions can be achieved by equating the three expressions of (3.13) individually to obtain the x_2 values, as in Expression (3.14)

$$\begin{array}{c|c|c} 7-2x_2=\frac{2}{3}x_2 & 7-2x_2=x_2+1 & \frac{2}{3}x_2=x_2+1 \\ \Downarrow & \Downarrow & \Downarrow \\ x_2=\frac{21}{8}\approx 2.63 & x_2=2 & x_2=-3 \end{array} \tag{3.14}$$

With the x_2 values of the extremum points known, the associated x_1 values can be straightforwardly obtained, using Expression (3.13), leading to Expression (3.15)

$$\begin{array}{c|c|c} x_2\approx 2.63: x_1=7-2x_2 & x_2=2: x_1=x_2+1 & x_2=-3: x_1=\frac{2}{3}x_2 \\ \Downarrow & \Downarrow & \Downarrow \\ x_1\approx 1.74 & x_1=3 & x_1=-2 \end{array} \tag{3.15}$$

As the design variable values x_1 and x_2 are now fully known at the extremum points of the constrained optimization problem, i.e. at A, B, and C in Figure 3.5, the objective function values can now be straightforwardly evaluated at these points by inserting the relevant values into Equation (3.11). It, therefore, follows that the optimum solution to the optimization problem of minimizing (3.11) subjected to the constraints of (3.12) is that x_1 equals −2 and x_2 equals −3. Even for those with limited programming knowledge, it is relatively easy to see how the derivations in (3.14) and (3.15) can be implemented into a software code to solve any problem of the above nature "with ease", by the use of LP. Expanding the solution to encompass additional constraints is also straightforward, in which case matrix and vector principles are most often employed. The above methodology coupled with a systematic approach to "test" the extremum points in order to find the optimum solutions is often referred to as the Simplex method.

3.5.2 The Simplex method

In general, the Simplex method utilizes the canonical form of the matrix system coupled with the introduction of slack variables. Although the example of Section 3.5.1 demonstrated the use of the Simplex method, it is worth explaining the detailed workings of the method in greater detail due to its widespread use and versatility.

The Simplex method is an iterative method that consists of six main steps, as illustrated in Figure 3.6.

To enhance the readers' understanding, the Simplex method will now be demonstrated through the use of another example. In this case, a linear optimization problem containing three linearly independent variables x_1, x_2, and x_3 will be used as defined by the objective function in Expression (3.16)

$$f(x_i) = 2x_1 - 3x_2 + 4x_3 \tag{3.16}$$

For this particular example the objective will be to minimize the above function subject to the constraints of Expression (3.17)

$$\begin{aligned} 4x_1 + 3x_2 + x_3 &\leq 3 \\ x_1 + x_2 + x_3 &\leq 10 \\ 2x_1 + x_2 - x_3 &\leq 10 \end{aligned} \tag{3.17}$$

The optimization problem defined in (3.16) and (3.17) is an LP problem; step 1 in Figure 3.6 now needs to be introduced to convert the LP problem into a set of linear equations. As previously stipulated, this is done by introducing slack variables to literally take up the slack by the introduction of equalities as opposed to the inequalities of (3.17). This leads to the updated system of equations in Expression (3.18).

$$\begin{aligned} 4x_1 + 3x_2 + x_3 + s &= 3 \\ x_1 + x_2 + x_3 + t &= 10 \\ 2x_1 + x_2 - x_3 + u &= 10 \\ 2x_1 - 3x_2 + 4x_3 + v &= 0 \end{aligned} \tag{3.18}$$

Note that the objective function of (3.16) is inserted last in (3.18). A total of four slack variables have been introduced in (3.18); these are s, t, u, and v. Step 1 in

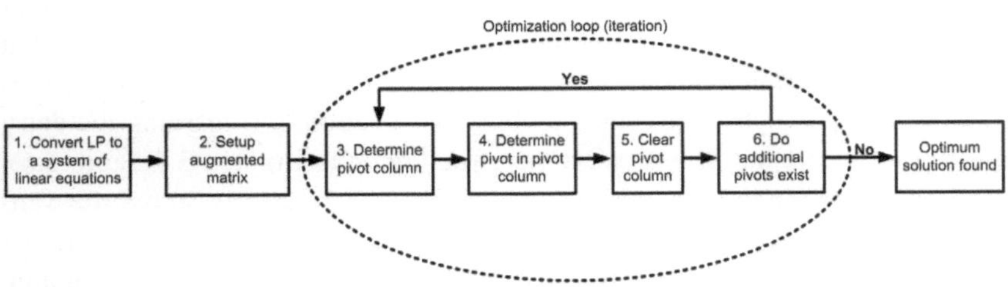

FIGURE 3.6 Flowchart of Simplex method.

Figure 3.6 is now complete as the LP problem has been converted into a series of linear equations. The next steps are simply to set up the augmented matrix, ensuring that the objective function occupies the last row of the augmented matrix and that the right-hand column represents the "results" column, that is, the right-hand side of the equilibrium signs in (3.18). This leads to the augmented matrix in Expression (3.19)

$$\begin{array}{cccccccc} x_1 & x_2 & x_3 & s & t & u & v & \\ \end{array}$$
$$\begin{bmatrix} 4 & 3 & 1 & 1 & 0 & 0 & 0 & 3 \\ 1 & 1 & 1 & 0 & 1 & 0 & 0 & 10 \\ 2 & 1 & -1 & 0 & 0 & 1 & 0 & 10 \\ 2 & -3 & 4 & 0 & 0 & 0 & 1 & 0 \end{bmatrix} \tag{3.19}$$

The augmented matrix in (3.19) thus represents the outcome of step 2 (Figure 3.6). Associated with each augmented matrix (which is changed once per iteration) is a so-called basic solution. This solution is obtained by identifying active and nonactive variables. The active variables are determined by identifying columns that are cleared, that is, only contain a single nonzero entry. In (3.19), it can be seen that the s, t, u, and v columns only contain a single nonzero entry, meaning that they are active variables. The specific value of an active variable is simply found by dividing it into the entry on the results column (corresponding to the nonzero entry row). The values of the remaining inactive variables are simply set to 0. In the case of (3.19), this leads to the solution listed in Expression (3.20)

$$\begin{aligned} x_0 &= x_2 = x_3 = 0 \\ s &= \frac{3}{1} = 3 \\ t &= \frac{10}{1} = 10 \\ u &= \frac{10}{1} = 10 \\ v &= \frac{0}{1} = 0 \end{aligned} \tag{3.20}$$

By inserting the basic solution of (3.20) into (3.18), it can be verified that the basic solution listed in (3.20) is an actual solution to the optimization problem. It is, however, unlikely to be an optimum solution, as $f(x_i)$ is 0 and the task at hand was to minimize $f(x_i)$; furthermore, it can be mathematically proven that the optimum solution has not been obtained as long as there are negative values in the last row, which represents the objective function, disregarding the entry in the results column. The process therefore continues to step 3 (Figure 3.6), which is to determine the pivot column. This task is relatively straightforward as it merely consists of choosing the largest (numeric) negative value in the last row. In the example of (3.19), this means the second column, i.e. the x_2 column. If there are multiple identical values, either can be chosen. Step 4 (Figure 3.6) is to determine the pivot entry of the pivot column, thereby determining the pivot row. This is done by means of the following three rules/steps:

1. The pivot must always be a positive nonzero number.
2. For each entry b in the pivot column, a test ratio a/b must be calculated, where a is the entry in the results column corresponding to the same row number as b is found in.
3. The pivot is finally chosen based on the smallest test ratios.

To continue the example, the test ratios for the second column in (3.19) must, therefore, be calculated as done in (3.21)

$$\begin{array}{cccccccc} x_1 & x_2 & x_3 & s & t & u & v & \\ \end{array}$$
$$\begin{bmatrix} 4 & 3 & 1 & 1 & 0 & 0 & 0 & 3 \\ 1 & 1 & 1 & 0 & 1 & 0 & 0 & 10 \\ 2 & 1 & -1 & 0 & 0 & 1 & 0 & 10 \\ 2 & -3 & 4 & 0 & 0 & 0 & 1 & 0 \end{bmatrix} \begin{array}{l} \text{test ratio: } 3/3=1 \\ \text{test ratio: } 10/1=10 \\ \text{test ratio: } 10/1=10 \\ \\ \end{array} \tag{3.21}$$

As the smallest test ratio is 1, the first row is selected; the pivot entry of the matrix therefore becomes the second entry in the first row. If there are multiple identical test ratios, for example, if the smallest test ratio found had been 10, either of these entries may be chosen. Step 5 (Figure 3.6) is now initiated; this consists of clearing the pivot column by means of pivoting. Pivoting is done for each row, and merely consists of scaling rows as well as summation of individual entries to clear the pivot column. For (3.21), the first row to be pivoted is the second row; in the interim this will be referred to as R2, and R1 will refer to the pivot row. The first step of pivoting is to determine the scaling of each row to obtain an identical pivot column entry. For R2 the entry in the pivot column is 1, and for R1 it is 3. An identical entry can, therefore, be obtained by multiplying R1 by 1 and R2 by 3. As both entries are of identical signs (positive in this case), the two rows must be deducted from each other; otherwise, they must be added. To update the second row of (3.21) through pivoting, Equation (3.22) must be followed.

$$\text{pivoted second row: } 3 \cdot \text{R2} - 1 \cdot \text{R1} \tag{3.22}$$

Completing this for (3.21) leads to the partially updated augmented matrix in (3.23)

$$\begin{array}{cccccccc} x_1 & x_2 & x_3 & s & t & u & v & \\ \end{array}$$
$$\begin{bmatrix} 4 & 3 & 1 & 1 & 0 & 0 & 0 & 3 \\ -1 & 0 & 2 & -1 & 3 & 0 & 0 & 27 \\ 2 & 1 & -1 & 0 & 0 & 1 & 0 & 10 \\ 2 & -3 & 4 & 0 & 0 & 0 & 1 & 0 \end{bmatrix} \tag{3.23}$$

The pivoting is now continued for the third row (R3) of (3.23) by the pivoting equation of (3.24)

$$\text{pivoted third row}: 3 \cdot \text{R3} - 1 \cdot \text{R1} \tag{3.24}$$

Equation (3.24) in turn leads to the updated matrix in (3.25)

$$\begin{matrix} x_1 & x_2 & x_3 & s & t & u & v & \\ \end{matrix}$$
$$\begin{bmatrix} 4 & 3 & 1 & 1 & 0 & 0 & 0 & 3 \\ -1 & 0 & 2 & -1 & 3 & 0 & 0 & 27 \\ 6 & 0 & -4 & -1 & 0 & 3 & 0 & 27 \\ 2 & -3 & 4 & 0 & 0 & 0 & 1 & 0 \end{bmatrix} \tag{3.25}$$

Step 5 (Figure 3.6) may now be completed by pivoting the last row using Equation (3.26)

$$\text{pivoted fourth row}: 1 \cdot \text{R4} + 1 \cdot \text{R1} \tag{3.26}$$

Equation (3.26) in turn leads to the completed pivoted matrix in (3.27)

$$\begin{matrix} x_1 & x_2 & x_3 & s & t & u & v & \\ \end{matrix}$$
$$\begin{bmatrix} 4 & 3 & 1 & 1 & 0 & 0 & 0 & 3 \\ -1 & 0 & 2 & -1 & 3 & 0 & 0 & 27 \\ 6 & 0 & -4 & -1 & 0 & 3 & 0 & 27 \\ 6 & 0 & 5 & 1 & 0 & 0 & 1 & 3 \end{bmatrix} \tag{3.27}$$

The matrix of (3.27) thereby represents the outcome of step 5 (Figure 3.6). As there are no additional negative numbers in the last row of (3.27), an optimum solution has been found, as there are no additional pivots in the matrix. The basic solution for (3.27) may now be determined as it was for (3.19) in Expression (3.20). It can be seen in (3.27) that x_1, x_3, and s are inactive variables, whereas x_2, t, u, and v are active variables. The basic solution of (3.27) may, therefore, be calculated as done in Expression (3.28)

$$\begin{aligned} x_1 &= x_3 = s = 0 \\ x_2 &= \frac{3}{3} = 1 \\ t &= \frac{27}{3} = 9 \\ u &= \frac{27}{3} = 9 \\ v &= \frac{3}{1} = 3 \end{aligned} \tag{3.28}$$

The validity of the solution in (3.28) may now be confirmed by inserting the values into the linear equations of (3.18), as done in Expression (3.29)

$$\begin{aligned}
&4x_1 + 3x_2 + x_3 + s = 4\cdot 0 + 3\cdot 1 + 0 + 0 = 3\\
&x_1 + x_2 + x_3 + t = 0 + 1 + 0 + 9 = 10\\
&2x_1 + x_2 - x_3 + u = 2\cdot 0 + 1 - 0 + 9 = 10\\
&2x_1 - 3x_2 + 4x_3 + v = 2\cdot 0 - 3\cdot 1 + 4\cdot 0 + 3 = 0
\end{aligned} \tag{3.29}$$

The objective function value can also be found by inserting the values of (3.28) into (3.16), thereby obtaining the value of -3 which is the minimum value the function can obtain without violating the constraints of (3.17).

Although this description of the Simplex method has been somewhat lengthy, most readers should be able to visualize the relative ease in which the Simplex method can be converted into computer programming and thus solve large multivariable problems with relative ease. In the examples of the current and preceding sections, the optimization problems have however been linear; it is easy to identify the shortcomings of the above examples in terms of many real-world engineering applications. The LP and Simplex method can, however, also be advantageously utilized for nonlinear optimization problems, as will be discussed in the subsequent sections.

3.5.3 Application to real-world engineering problems

In the previous sections LP and the Simplex method was employed to solve a linear optimization problem. This was, however, subjected to a number of caveats, predominately the fact that the objective and constraint functions must be linear. Most structural engineering optimization problems do, however, not conform to this, even at the most rudimentary levels of solid mechanics. In the previous section, it was stipulated that the constraints in (3.12) represented different attributes of the beam: mass, height, and stress state. In (3.12), the expressions were linear to apply LP; this stipulation does however rapidly unravel. For example, using conventional beam theory to obtain the bending stress of a cantilever beam involves determining the second moment of area which is an inherent nonlinear function of the cross-sectional geometry. At this point, it is convenient to introduce an example. Consider the optimization problem of minimizing the mass of a cantilever beam of length l exposed to a single static force F, as illustrated in Figure 3.7.

The beam is made from an isotropic material with a solid circular cross-section of radius r uniform along the length of the beam. Initially, only the shear load generated by F will be considered. The optimization parameters are:

- Objective: minimize the shear deformation (γ) of the beam.
- Constraints:
  - Maximum shear stress (τ): 50 MPa
  - Maximum mass of beam (M): 20 kg
- Design variable: radius r of beam

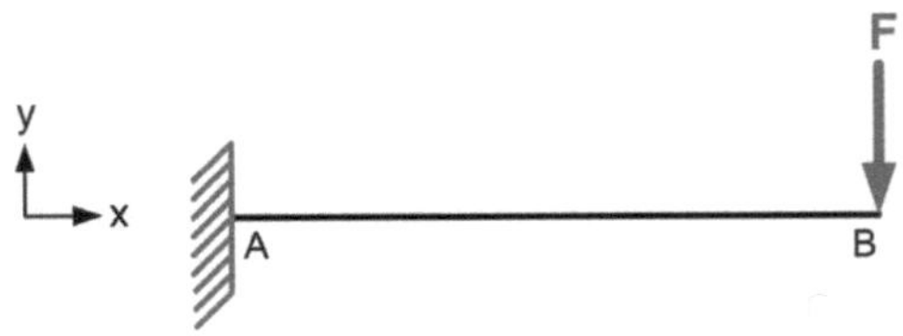

FIGURE 3.7 Cantilever beam example.

Using the basic principles of solid mechanics, the current deflection, bending stress at point B, and mass as a function of the radius can straightforwardly be obtained. Figure 3.8 plots the deflection, stress, and mass for the above example for 10 iterations, that is, different radius values.

The optimum solution to the above example problem can now be found using Figure 3.8. As anticipated, the deflection of the beam reduces and converges toward zero as the radius increases. "Interpreting" the deflection versus radius graph is in essence step 2 of the optimization loop (Figure 3.1). Remember that the prescribed optimization problem is a constrained one and so the constraints, that is, maximum stress and mass, would have to be evaluated alongside the objective function in step 3 (Figure 3.1). Please note that if the constraints were removed the design variable, that is, the radius, would simply move toward infinity and a feasible engineering solution would not be obtained. This would also be the case if only the maximum mass constraint was to be removed. The maximum mass constraint is included to further ensure a feasible solution from a structural viewpoint. Following step 3 the design variable would be adjusted and the optimization loop (iteration) would return to step 1 until the ideal solution is obtained. Steps 2, 3, 4, and 5 could, in essence, be obtained in a wide variety of ways, for example, using mathematical, evolutionary, entropy, or even a manual approach. It can, however, not be solved using LP as explained in the previous section, simply because one of the constraint functions is nonlinear. It is, however, possible to extend the principles of LP to sequential/successive linear programming (SLP) or sequential/successive quadratic programming (SQP), as will be discussed in Section 3.5.4.

For the time being, assume that the problem will be solved using a "manual" approach via Figure 3.8, which enables the link between steps 2 and 4 (Figure 3.1). But how is it known what the optimum solution is; how can it be known when to stop the iteration process? This question is very difficult if not impossible to answer, even for a relatively simple optimization problem such as the beam example above. Mathematically, the minimum (numeric) value of the deflection is zero; however, engineering intuition/theory of elasticity dictates that this is not possible as the beam is compliant; in other words, the deflection should converge toward zero, but will never actually become zero. Hence, the optimum and feasible solution cannot be obtained by observing the objective function in isolation. Observing the constraint functions separately, Figure 3.8 reveals that a mass of 20 kg is obtained for an approximate radius value of 23.5 mm, whereas

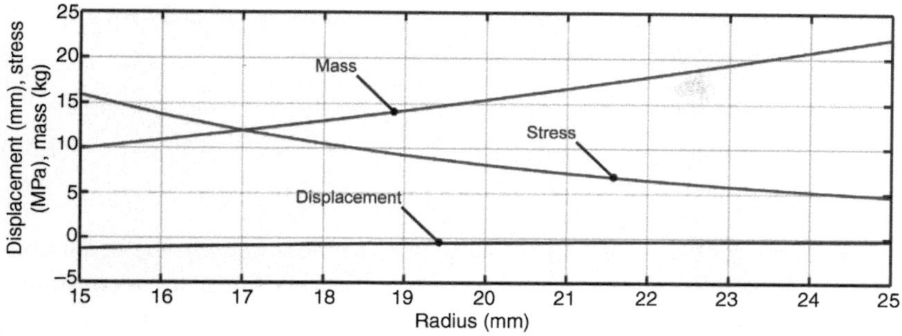

FIGURE 3.8 Radius versus displacement, mass, and bending stress.

a bending stress value of 10 MPa is obtained for an approximate radius of 18.5 mm. The constraints thereby dictate that the optimum solution to the constrained optimization problem lies within the radius interval of 18.5–23.5 mm; however, looking at the constraints in isolation does not provide a unique solution, either. If the objective function is now reconsidered, it becomes obvious that the ideal solution to the problem is a radius of 23.5 mm, as the deflection is at its lowest value without violating the two prescribed constraints. In this particular example, the "trump card" was the objective function, while the two constraints were not "weighted" against each other; in other words, no preference was given to an increase in mass versus a decrease in stress. If a preference to find the "ideal" tradeoff between mass and stress was now to be introduced, a radius of 17 mm would be of particular interest as the mass and stress curves intersect at this specific value, disregarding the fact that the original stress limit is exceeded for this radius value. The important thing to note here is that although the objective may be of primary concern "preferences" or weightings, for example, relating to the constraints may also be introduced. Alternatively, the constraints may be "upgraded" to objective functions, in which case the problem becomes one of multiobjective optimization (MOO).

3.5.4 Sequential linear programming

Many dedicated textbooks for sequential/successive linear programming (SLP) exist; this section is not intended to provide an in-depth detailed understanding of the technique, but merely to inform the reader of the main underlying principles. SLP was introduced by Griffith and Stewart (1961) who referred to the method as mathematical approximation programming (MAP) and was initially used and developed in the petrochemical industry. As suggested in the previous section, SLP is in essence an adaptation of LP. It generally works by linearizing a nonlinear problem based on a single starting point. In other words, SLP applies and solves a successive sequence of linear problems to solve a nonlinear optimization problem. Consider the single parameter optimization problem of (3.30) where $f(x)$ is the objective function, c_i represents the constraints with the individual boundary value b_i, and u_j and l_j represent the upper and lower boundaries of the permissible x_j solution values.

$$
\begin{aligned}
&\text{min/max}: f(x) \\
&\text{subject to:} \quad c_i(x) \le b_i \quad | \, i = 1,2,3,\ldots,m \\
&\qquad\qquad\quad\; u_j \ge x_j \ge l_j \quad | \, j = 1,2,3,\ldots,n
\end{aligned}
\tag{3.30}
$$

The above problem definition could, for example, represent the beam optimization problem presented in Section 3.4 that has a nonlinear stress constraint. A continuous function such as the bending stress of a beam can be linearized around a local point in a number of ways; often the most simple way is simply to use differentiation, but other methods such as first-order Taylor approximations can also be applied. The methodology of SLP is to determine (or simply guess) a starting point, linearize the function at that point, apply the Simplex method to solve the (now linear) problem, and then use the linear solution to determine the next point to linearize before the cycle is repeated. This cycle will repeat/iterate until a specified termination condition is met. The formulation of SLP proposed by Griffith and Stewart (1961) takes the form of (3.31)

$$
\begin{aligned}
&\text{min/max}: && \sum_{j=1}^{n} d_j \Delta x_j = f - f(x_k) \\
&\text{subject to}: && \sum_{j=1}^{n} a_{ij} \Delta x_j \le b_i - c_i(x_k) \quad | \, i = 1,2,3,\ldots,m \\
& && u_j - x_{jk} \ge x_j \ge l_j - x_{jk} \quad | \, j = 1,2,3,\ldots,n \\
&\text{where}: && \Delta x_j = x_j - x_{jk} \qquad \wedge \; d_j = \frac{\partial f(x_k)}{\partial x_j} \quad \wedge \; a_{ij} = \frac{\partial c_i(x_k)}{\partial x_j}
\end{aligned}
\tag{3.31}
$$

In (3.31), it is important to note that x_{jk} is the "current solution," which must comply with (3.32). In other words, x_{jk} must be a permissible solution.

$$
x_{jk} \in x_j \tag{3.32}
$$

Expression (3.31) represents the linearized version of the nonlinear optimization problem of (3.30). The difference may initially seem somewhat trivial; a subsequent example will, however, aid the readers' understanding. Before this is done, it is important to note that the Simplex method does not accept negative Δx_j values (Griffith & Stewart, 1961) and therefore made a simple change to the variable as stipulated by (3.33)

$$
\Delta x_j = \Delta x_j^{+} - \Delta x_j^{+} \;\left|\;
\begin{aligned}
\Delta x_j^{+} &= \begin{cases} \Delta x_j & |\Delta x_j \ge 0 \\ 0 & |\Delta x_j < 0 \end{cases} \\
\Delta x_j^{-} &= \begin{cases} -\Delta x_j & |\Delta x_j \le 0 \\ 0 & |\Delta x_j > 0 \end{cases}
\end{aligned}\right.
\tag{3.33}
$$

Inserting (3.33) into (3.31) provides a more "useable" form of the linear programming problem (3.34), which may now be solved using the Simplex method (Griffith & Stewart, 1961).

$$\begin{aligned}
\text{min/max:} \quad & \sum_{j=1}^{n} d_j \Delta x_j^+ - \sum_{j=1}^{n} d_j \Delta x_j^- = f - f(x_k) \\
\text{subject to:} \quad & \sum_{j=1}^{n} a_{ij} \Delta x_j^+ - \sum_{j=1}^{n} a_{ij} \Delta x_j^- \le b_i - c_i(x_k) \quad | \, i = 1,2,3,\ldots,m \\
& \Delta x_j^+ - \Delta x_j^- \le (u_j - x_{jk}) \quad | \, j = 1,2,3,\ldots,n \\
& \Delta x_j^+ - \Delta x_j^- \ge (l_j - x_{jk}) \quad | \, j = 1,2,3,\ldots,n
\end{aligned} \tag{3.34}$$

Once (3.34) has been solved for a given x_{jk} the iterative process may proceed by determining an updated x_{jk} value using Equation (3.35)

$$x_{jk+1} = x_{jk} + \Delta_j^+ - \Delta_j^- \tag{3.35}$$

It is important to note that (3.34) is a constrained optimization problem. Broadly speaking, the MAP method (or SLP) also works for unconstrained problems which are "mildly nonlinear" (Ravindran, Ragsdell, & Reklaitis, 2006). Constraints are, however, generally included to control the relative changes of the independent variables in order to ensure a more "robust" optimization algorithm, as will be further discussed in Section 3.8.3.

In (3.34) this is implemented by the addition of the $\Delta x_j^+ - \Delta x_j^- \le (u_j - x_{jk})$ and $\Delta x_j^+ - \Delta x_j^- \ge (l_j - x_{jk})$.

Inequalities directly influence the "rate of change" of the updated solution x_{jk} between iterations, as specified in (3.35). This is specifically controlled by the "predefined" lower l_j and upper u_j limits of the design variables. It is pivotal that these limits are set appropriately, as too low a l_j value will reduce the "effectiveness" of the optimization algorithm by increasing the number of iterations required to obtain an optimum solution, whereas too high a u_j will most likely lead to infeasible solutions violating (3.32). This phenomenon is a general concern for all optimization algorithms (regardless of the type of algorithm) and may, in general, be referred to as the "aggressiveness" of the optimization algorithm, as will be discussed throughout the remainder of this book.

In order to aid the readers' understanding of SLP, a simplified and classic example of SLP and the Simplex method will now be introduced. For the purpose of this example, consider a simple cantilever I-beam exposed to two separate loadcases; for example, a bending moment and a single point force. Assuming that the objective is to minimize the deflection by maximizing the cross-sectional area (CSA) of the flange x_1 and the web x_2, the objective function of the optimization problem can therefore be defined as (3.36)

$$\max: CSA(x) = 2x_1 + x_2 \tag{3.36}$$

The I-beam is, however, also exposed to certain geometrical constraints as defined by (3.37)

$$\begin{aligned}(x_1)^2+(x_2)^2&\le 25\\(x_1)^2-(x_2)^2&\le 7\end{aligned}\tag{3.37}$$

Before the optimization can commence the limiting boundaries (or the "aggressiveness") of the algorithm will be set, as seen in (3.38)

$$(u_j-x_{jk})=(l_j-x_{jk})=1\tag{3.38}$$

Finally, an initial guess or starting point for the optimization, at iteration 1 ($k = 1$) must be set, in this case unity; $x_{1,1} = 1$, $x_{2,1} = 1$, that is, $x = (1,1)$. Expression (3.34) may now be used to set up the first iteration of the SLP; the objective function in (3.36) becomes expression (3.39) before the starting point of $X_1 = (1,1)$ is inserted (3.39)

$$\begin{aligned}&\text{max:} && 2\Delta x_1^+ + \Delta x_2^+ - 2\Delta x_1^- - \Delta x_2^- = CSA(1,1)-(2x_{1k}+x_{2k})\\&\text{subject to:} && 2\Delta x_{1k}\Delta x_1^+ + 2\Delta x_{2k}\Delta x_2^+ - 2\Delta x_{1k}\Delta x_1^- - 2\Delta x_{2k}\Delta x_2^- \le 25-\left((x_{1k})^2+(x_{2k})^2\right)\\& && 2\Delta x_{1k}\Delta x_1^+ - 2\Delta x_{2k}\Delta x_2^+ - 2\Delta x_{1k}\Delta x_1^- + 2\Delta x_{2k}\Delta x_2^- \le 7-\left((x_{1k})^2-(x_{2k})^2\right)\\& && j=1\begin{cases}\Delta x_1^+ - \Delta x_1^- \le 1\\ -\Delta x_1^+ + \Delta x_1^- \le 1\end{cases}\\& && j=2\begin{cases}\Delta x_2^+ - \Delta x_2^- \le 1\\ -\Delta x_2^+ + \Delta x_2^- \le 1\end{cases}\end{aligned}\tag{3.39}$$

Inserting the starting point $x = (1,1)$ into (3.39) yields (3.40)

$$\begin{aligned}&\text{max:} && 2\Delta x_1^+ + \Delta x_2^+ - 2\Delta x_1^- - \Delta x_2^- = CSA-3\\&\text{subject to:} && 2\Delta x_1^+ - 2\Delta x_2^+ - 2\Delta x_1^- + 2\Delta x_2^- \le 23\\& && 2\Delta x_1^+ + 2\Delta x_2^+ - 2\Delta x_1^- - 2\Delta x_2^- \le 7\\& && j=1\begin{cases}\Delta x_1^+ - \Delta x_1^- \le 1\\ -\Delta x_1^+ + \Delta x_1^- \le 1\end{cases}\\& && j=2\begin{cases}\Delta x_2^+ - \Delta x_2^- \le 1\\ -\Delta x_2^+ + \Delta x_2^- \le 1\end{cases}\end{aligned}\tag{3.40}$$

There are four unknown (and linearly independent) variables in (3.39) and (3.40); namely $\Delta x_1^+, \Delta x_2^+, \Delta x_1^-$, and Δx_2^-. The problem may now be solved using the Simplex method that gives the solution to "iteration 1," that is, $k = 1$ (3.41). In addition, it is straightforward to calculate the current value of the objective function: $CSA(1,1) = 3$.

$$\Delta x_1^+ = 1 \quad \Delta x_2^+ = 1 \quad \Delta x_1^- = 0 \quad \Delta x_2^- = 0\tag{3.41}$$

Equation (3.35) may now be used to update x_1 and x_2 expression for the next iteration, $k = 2$, seen in Expression (3.42)

$$\begin{aligned} x_1 &: x_{12} = x_{11} + \Delta x_1^+ - \Delta x_1^- = 1 + 1 - 0 = 2 \\ x_2 &: x_{22} = x_{21} + \Delta x_2^+ - \Delta x_2^- = 1 + 1 - 0 = 2 \end{aligned} \tag{3.42}$$

The updated solution, $X_2 = (2,2)$, yields an updated objective value $CSA(2,2) = 6$. The iteration process may now be repeated by inserting X_2 into (3.39) leading to (3.43)

$$\begin{aligned} \text{max:} \quad & 2\Delta x_1^+ + \Delta x_2^+ - 2\Delta x_1^- - \Delta x_2^- = CSA - 6 \\ \text{subject to:} \quad & 4\Delta x_1^+ + 4\Delta x_2^+ - 4\Delta x_1^- - 4\Delta x_2^- \le 17 \\ & 4\Delta x_1^+ - 4\Delta x_2^+ - 4\Delta x_1^- + 4\Delta x_2^- \le 7 \\ & j = 1 \begin{cases} \Delta x_1^+ - \Delta x_1^- \le 1 \\ -\Delta x_1^+ + \Delta x_1^- \le 1 \end{cases} \\ & j = 2 \begin{cases} \Delta x_2^+ - \Delta x_2^- \le 1 \\ -\Delta x_2^+ + \Delta x_2^- \le 1 \end{cases} \end{aligned} \tag{3.43}$$

Solving (3.43) leads to the values listed in (3.44)

$$\Delta x_1^+ = 1 \quad \Delta x_2^+ = 1 \quad \Delta x_1^- = 0 \quad \Delta x_2^- = 0 \tag{3.44}$$

The updated solution X_3 may now be found (3.45)

$$\begin{aligned} x_1 &: x_{13} = x_{12} + \Delta x_1^+ - \Delta x_1^- = 2 + 1 - 0 = 3 \\ x_2 &: x_{23} = x_{22} + \Delta x_2^+ - \Delta x_2^- = 2 + 1 - 0 = 3 \end{aligned} \tag{3.45}$$

The updated objective function value is now $CSA(X_3) = CSA(3,3) = 9$, and the next iteration can now commence, seen in (3.46)

$$\begin{aligned} \text{max:} \quad & 2\Delta x_1^+ + \Delta x_2^+ - 2\Delta x_1^- - \Delta x_2^- = CSA - 9 \\ \text{subject to:} \quad & 6\Delta x_1^+ + 6\Delta x_2^+ - 6\Delta x_1^- - 6\Delta x_2^- \le 7 \\ & 6\Delta x_1^+ - 6\Delta x_2^+ - 6\Delta x_1^- + 6\Delta x_2^- \le 7 \\ & j = 1 \begin{cases} \Delta x_1^+ - \Delta x_1^- \le 1 \\ -\Delta x_1^+ + \Delta x_1^- \le 1 \end{cases} \\ & j = 2 \begin{cases} \Delta x_2^+ - \Delta x_2^- \le 1 \\ -\Delta x_2^+ + \Delta x_2^- \le 1 \end{cases} \end{aligned} \tag{3.46}$$

Solving (3.46) leads to the values in (3.47)

$$\Delta x_1^+ = 1 \quad \Delta x_2^+ = \frac{1}{6} \quad \Delta x_1^- = 0 \quad \Delta x_2^- = 0 \tag{3.47}$$

The updated solution X_4 may now be found, seen in (3.48)

$$\begin{aligned} x_1 &: x_{14} = x_{13} + \Delta x_1^+ - \Delta x_1^- = 3 + 1 - 0 = 4 \\ x_2 &: x_{24} = x_{23} + \Delta x_2^+ - \Delta x_2^- = 3 + \frac{1}{6} - 0 = \frac{19}{6} \end{aligned} \tag{3.48}$$

The updated objective function value is now $CSA(X_4) = CSA\left(4, \frac{19}{6}\right) = \frac{67}{6} \approx 11.17$ and iteration 5 can now commence, as in (3.49)

$$\begin{aligned} \text{max:} \quad & 2\Delta x_1^+ + \Delta x_2^+ - 2\Delta x_1^- - \Delta x_2^- = CSA - \frac{67}{6} \\ \text{subject to:} \quad & 8\Delta x_1^+ + \frac{19}{3}\Delta x_2^+ - 8\Delta x_1^- - \frac{19}{3}\Delta x_2^- \le -\frac{37}{36} \\ & 8\Delta x_1^+ - \frac{19}{3}\Delta x_2^+ - 8\Delta x_1^- + \frac{19}{3}\Delta x_2^- \le \frac{37}{36} \\ & j = 1\begin{cases} \Delta x_1^+ - \Delta x_1^- \le 1 \\ -\Delta x_1^+ + \Delta x_1^- \le 1 \end{cases} \\ & j = 2\begin{cases} \Delta x_2^+ - \Delta x_2^- \le 1 \\ -\Delta x_2^+ + \Delta x_2^- \le 1 \end{cases} \end{aligned} \tag{3.49}$$

Solving (3.49) leads to the values in (3.47)

$$\Delta x_1^+ = 0 \quad \Delta x_2^+ = 0 \quad \Delta x_1^- = 0 \quad \Delta x_2^- \approx 0.162 \tag{3.50}$$

The updated solution X_5 may now be found as in (3.51)

$$\begin{aligned} x_1 &: x_{15} = x_{14} + \Delta x_1^+ - \Delta x_1^- = 4 + 0 - 0 = 4 \\ x_2 &: x_{25} = x_{24} + \Delta x_2^+ - \Delta x_2^- = \frac{67}{6} + 0 - 0.162 \approx 3.004 \end{aligned} \tag{3.51}$$

The updated objective function value is now $CSA(X_5) = CSA(4, 3.004) \approx 11.011$, and iteration 6 can now commence, as seen in (3.52)

$$\begin{aligned} \text{max:} \quad & 2\Delta x_1^+ + \Delta x_2^+ - 2\Delta x_1^- - \Delta x_2^- = CSA - 11.011 \\ \text{subject to:} \quad & 8\Delta x_1^+ + 6.009\Delta x_2^+ - 8\Delta x_1^- - 6.009\Delta x_2^- \le -0.026 \\ & 8\Delta x_1^+ - 6.009\Delta x_2^+ - 8\Delta x_1^- + 6.009\Delta x_2^- \le 0.026 \\ & j = 1\begin{cases} \Delta x_1^+ - \Delta x_1^- \le 1 \\ -\Delta x_1^+ + \Delta x_1^- \le 1 \end{cases} \\ & j = 2\begin{cases} \Delta x_2^+ - \Delta x_2^- \le 1 \\ -\Delta x_2^+ + \Delta x_2^- \le 1 \end{cases} \end{aligned} \tag{3.52}$$

Solving (3.52) leads to the values in (3.53)

$$\Delta x_1^+ = 0 \quad \Delta x_2^+ = 0 \quad \Delta x_1^- = 0 \quad \Delta x_2^- \approx 0.004 \tag{3.53}$$

The updated solution X_6 may now be found, seen in (3.54)

$$\begin{aligned} x_1 : x_{16} &= x_{15} + \Delta x_1^+ - \Delta x_1^- = 4 + 0 - 0 = 4 \\ x_2 : x_{26} &= x_{25} + \Delta x_2^+ - \Delta x_2^- = 3.004 + 0 - 0.004 = 3 \end{aligned} \tag{3.54}$$

The updated objective function value is now $CSA(X_6) = CSA(4,3) = 11$. It can be seen that the change of objective function value, CSA, between iterations 5 and 6 is very small relative to the changes between all preceding iterations. Furthermore, the change of the design variable values, x_1 and x_2, is also very small relative to previous iterations; in fact, x_1 does not change at all between iterations 5 and 6. All of this information indicates that the optimization problem has indeed converged toward an (optimum) solution. Convergence is a critical aspect of any optimization algorithm, and is used for a number of purposes, including termination criteria, as previously discussed. From a mathematical viewpoint, convergence may straightforwardly be identified by reviewing the objective function value per iteration. Figure 3.9 illustrates the objective function value, CSA, as a function of iteration number for the above SLP example.

As illustrated in Figure 3.9, the slope of the objective function value (as a function of iteration number) is relatively large between iterations 1 and 4, while it decreases dramatically between iterations 4 and 6. In fact, the slope of the line is effectively zero between iterations 5 and 6. In addition to plotting the objective function per iteration, the design variables, in this example x_1 and x_2, may also be plotted as a function of iteration number (Figure 3.10).

Figure 3.10 reveals an identical trend to that of Figure 3.9, namely that there is a noticeable change (decrease) in the slope of the lines between iterations 4 and 6, with only marginal change (of x_2) between iterations 5 and 6. It is also worth noting that x_1 remains constant from iteration 4 onward. Figures 3.9 and 3.10

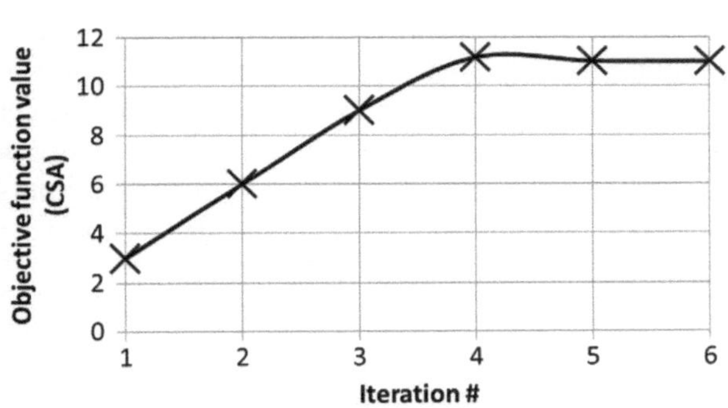

FIGURE 3.9 Objective function versus iteration #.

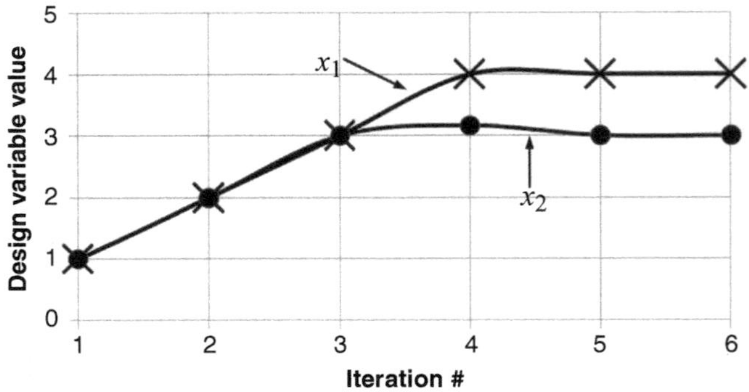

FIGURE 3.10 Design variables versus iteration #.

thus demonstrate that there is very little change of both objective function value and design variable value between iterations 4, 5, and 6. It may seem a trivial statement that both the objective function value CSA and the design variables x_1 and x_2 remain constant due to their direct link $CSA(x_1, x_2)$. It is, however, evident that a substantial number of different solutions may give the same objective function value. Inserting $(x_1,x_2)=(1,1)$ will lead to the same objective function value as, for example, $(x_1,x_2)=\left(\frac{1}{2},3\right)$ or $(x_1,x_2)=(5,-7)$ when inserted into Equation (3.36), discarding any optimization constraints. The point is that it is not always sufficient to simply observe the objective function value when evaluating whether or not an optimization problem has converged, as will be subsequently discussed.

In the next section, another type of optimization technique that is also very adept at solving convex and concave optimization problems will be introduced; this type is denoted as gradient-based optimization.

3.6 GRADIENT-BASED METHODS AND LINE SEARCH METHODS

As presented in Section 3.5, gradients may be used to solve optimization problems of continuous functions, the fundamental principles of some of the most relevant techniques in connection with structural optimization will be presented in this section. All the techniques presented in this section are iterative methods and generally utilize the gradient of the function to decide the search direction, that is, determine an updated solution, hence the term gradient-based method (GBM). Furthermore, the majority of methods presented utilize the generalized expression of (3.55) to determine the next iteration, that is, updated solution.

$$x_{i+1} = x_i + sf_i \cdot p_i \quad \Big| \quad i \geq 0 \tag{3.55}$$

In (3.55), i is the iteration number, x is the current "solution guess," which may either be a single number or indeed a vector containing multiple values, sf_i is a scale factor and p_i is the search direction. Iterative methods that use the generalized expression of (3.55) to determine the updated solution are generally referred to as line search methods (LSM). The general difference between different LSMs is the specific definition of p_i as will be subsequently demonstrated. There will also be dedicated sections that contain MatLab examples for each of the presented methods.

3.6.1 Gradient descent method

The gradient descent method (GDM) is also often referred to as "steepest descent" or the "method of steepest descent"; the latter is not to be confused with a mathematical method for approximating integrals of the same name. As the name suggests GDM utilizes the steepest gradient in order to search for an optimum, i.e. maximum or minimum, point for any given function. It should be noted that this optimum point may be local or global, as will be subsequently discussed. As the focus is currently on concave and convex functions, the single local and global minimum (convex) or maximum (concave) are identical. For this type of function the global optimum may be obtained in an iterative manner using differentiation, as initially stipulated in Section 3.5, that is, the gradient $\nabla f(x)$ of the function $f(x)$ must be determined, Equation (3.56)

$$\nabla f(x) = f'(x) \tag{3.56}$$

The optimum solution is obtained when the slope of the gradient is equal to zero, as previously stipulated in Expression (3.10). According to the principles of GDM, the minimum value of a function, and thereby the optimum solution (for a minimization problem), is obtained most rapidly if the search direction moves in the negative direction of the gradient; vice versa, the maximum is obtained if moving in the positive direction of the gradient. This means that to initiate the method a starting point x_0 must be chosen. To minimize a problem, the search direction must therefore be $-\nabla f(x_0)$. By using the GDM method an updated solution point, x_1, or more generally, x_{i+1} where i represents the iteration number, may now be found as dictated in Equation (3.57)

$$x_{i+1} = x_i - sf_i \nabla f(x_i) \quad | \quad i \geq 0 \tag{3.57}$$

In (3.57), sf_i is a scale factor that may be adjusted from iteration to iteration. sf_i is, however, not a trivial variable and should be chosen with caution, as it in essence controls the "aggressiveness" of each iteration. sf_i is analog to the timestep size of explicit FEA; if too large it may lead to instability and divergence; if chosen too small it will significantly increase the number of iterations, that is, time, before an optimum solution is found. The increase in time may be insignificant for smaller problems with a single variable; however, this increase will be exponential as the number of variables increase. There are a number of ways in which sf_i may be determined; one of the most common methods is to use a line search algorithm. Assuming for the time being that sf_i has been chosen appropriately, the ideology of GDM is that the function value for each iteration will sequentially decrease (minimization problem) or increase (maximization problem), as in Expression (3.58).

$$\begin{array}{l} \text{Minimization:} f(x_0) \geq f(x_1) \geq f(x_2) \geq ... f(x_i) \\ \text{Maximization:} f(x_0) \leq f(x_1) \leq f(x_2) \leq ... f(x_i) \end{array} \tag{3.58}$$

To aid the readers' understanding of the use of GDM, consider the simple optimization problem of minimizing the function $f(x) = x^2$; as a starting point 1 is selected, that is, $x_0 = 1$. The problem will be solved using three (constant) sf_i values: 0.1, 0.5, and 2.0. It should be noted that sf_i should not generally exceed 1.0; the sf_i value equal to 2.0 in this particular case is used for illustrative purposes only. The results from the example problem are illustrated in Figure 3.11.

As displayed in Figure 3.11, the GDM with $sf_i = 0.1$ smoothly follows the "true" $f(x) = x^2$ curve; after 20 iterations, the "solution" is that $x_{20} = 0.00922$ which leads to $f(x_{20}) = 0.00013$. Although the value is approaching zero (which is the true optimum solution in this case) the convergence toward this value is however relatively slow. After 100 iterations, i.e. at x_{100}, the "solution" is $x_{100} = 1.3 \cdot 1^{-10}$, in effect zero, but strictly speaking not actually zero. The second graph illustrated in Figure 3.11 where $sf_i = 0.5$ actually achieves the true optimum solution in the first iteration as $x_1 = 0$. For the final graph where $sf_i = 2.0$ the GDM actually diverges in the first instance as $x_1 = -3$ leading to $f(x_1) = 9$. At the 20th iteration $x_{20} = -0.04323$ and $f(x_{20}) = 0.00187$; at iteration 100 $x_{100} = -6.1 \cdot 1^{-10}$ and $f(x_{100}) = 3.7 \cdot 1^{-10}$ demonstrating that the GDM is now converging. This is most likely due to the relative smoothness and uniformity of the example function. This simple example thus demonstrates the potential influence of choosing an appropriate sf_i value to ensure convergence and to a minor extent stability of the GDM.

The gradient descent method can straightforwardly be extended to cover multiple variables by using a vector and matrix approach as defined in Equation (3.59).

$$\{x_{i+1}\} = \{x_i\} - sf_i [J] \quad | \quad i \geq 0 \tag{3.59}$$

In Equation (3.59) $[J]$ represents the Jacobian matrix that contains the first derivatives, that is, gradients of all $f(x_i)$ functions included in the problem; this should not be confused with the Jacobian matrix of FEA, although the strictly mathematical meaning of the two is identical.

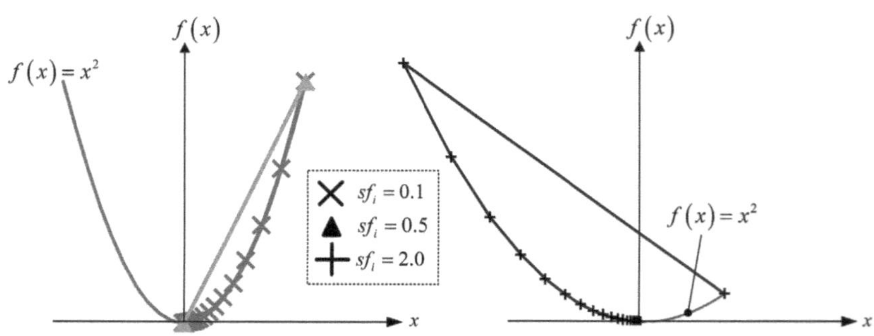

FIGURE 3.11 Gradient descent method example problem.

3.6.2 MatLab example of gradient descent method

The gradient descent method presented in the previous section is relatively simple, which also means it is relatively simple to generate a computer code for. Using MatLab, this section will introduce how to create a simple, yet effective, MatLab code to solve minimization problems using the gradient descent method. It is assumed that the reader possesses basic MatLab knowledge. As with all programming tasks, it is essential to create a flowchart before any programming is undertaken. The flowchart for the GDM algorithm is relatively simple, as illustrated in Figure 3.12.

It is generally good programming practice to clear all variables, memory, and workspace when a new program is started. Hence, the first two commands to be entered are:

Clear that will clear the workspace memory

Clc that will clear the command line window.

For the time being, it is assumed that the scale factor sf_i, Equation (3.57), will remain constant throughout the iterations; hence, a static value will be given to this variable. A single line is, therefore, added to define the *sf* variable. The program now looks like this:

```
clear
clc
sf = 0.1;
```

Note the ";" after the variable has been defined; this ensures that this command is not written to screen. For this example, a scale factor of 0.1 will be used. Next, the objective function and the associated first derivative will be defined; in this example, $f(x) = x^2$; $\nabla f(x) = x^2$ will be solved. This is entered as

```
fx = @(x) x^2;
dfdx = @(x) 2*x;
```

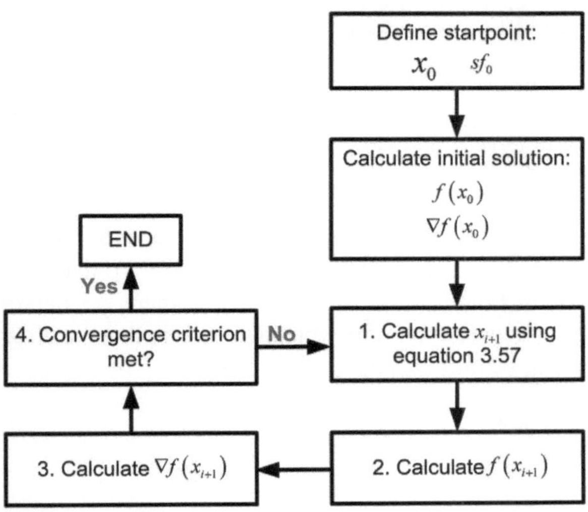

FIGURE 3.12 Flowchart of gradient descent method.

By using the above, it is only necessary for the user to change the right-hand side of the above fx and dfdx definitions to solve another problem, for example, $f(x)=3x^2-2x$. Next the start guess will be defined; this variable is simply referred to as "startguess":

```
startguess = 1;
```

For this example program we will introduce two simple measures for terminating the program if an optimum solution has been obtained, a maximum number of iterations, "maxiter," and a simple termination tolerance criteria, "termtol":

```
maxiter = 100;
termtol = 0.005;
```

As the example problem will be simple a very small amount of memory will be required; hence, all results may be stored. Three arrays are therefore introduced: "x_arr" that will store the solutions; "fx_arr" that will store the function values; and "dfdx_arr" that will store the gradient values. There will only be one value to store in each of these arrays for each iteration. As the maximum number of iterations is determined by "maxiter," the size of the three arrays can be preallocated as:

```
x_arr = NaN(maxiter+1,1);
fx_arr = NaN(maxiter+1,1);
dfdx_arr = NaN(maxiter+1,1);
```

All the entries of the arrays are initially assigned as "not a number" (NaN), so that there can be no doubt as to whether or not the solver has entered values into the individual entries. The arrays will also store the initial guess and associated values; hence, the size of the arrays becomes maxiter + 1. Next the "startguess" will be stored in the x_arr array; the function and gradient values will be calculated and stored:

```
x_arr(1) = startguess;
fx_arr(1) = fx(x_arr(1));
dfdx_arr(1) = dfdx(x_arr(1));
```

Note that it is not required to change any of the above if *f*(*x*) changes as we have used a function to define *f*(*x*). At this point, we are ready to go to step 1 in the flowchart in Figure 3.12. We can see that there are iterations between steps 1 through 4; hence, a for loop, which will ensure that the program does not loop infinitely will be introduced:

```
for i =2:maxiter+1

end
```

Note that the iterations start at $i = 2$, simply because the first entry in the arrays is used to store iteration 0, that is, the starting position. Consequently, the iterations must end at maxiter + 1. The next steps are very straightforward; three individual lines inside the for loop covering steps 1, 2, and 3 (Figure 3.12):

```
for i =2:maxiter+1
    x_arr(i) = x_arr(i-1)-sf*dfdx(x_arr(i-1));
    fx_arr(i) = fx(x_arr(i));
    dfdx_arr(i) = dfdx(x_arr(i));
```

Finally, we will introduce an additional termination criterion, which simply will stop the iterations if the gradient becomes less than termtol:

```
if abs(dfdx_arr(i)) <= termtol
        break
end
```

Note that the absolute, that is, numeric value of the gradient is used, `abs()`, as this may be a negative number if, for example, dealing with a convex function. If the gradient, for example, was -1.0, the if sentence above would terminate the iterations if the absolute value was not used because $-1.0 \leq 0.005$ is true.

The program of the single variable GDM is now complete:

```
%Gradient Descent Method - Single Variable

%Clear memory
    clear
    clc
    tic
%Define problem
    sf = 0.1;
    fx = @(x) x^2;
    dfdx = @(x) 2*x;
    startguess = 1;
%Define termination parameters
    maxiter = 100;
    termtol = 0.005;
%Create arrays
    x_arr = NaN(maxiter+1,1);
    fx_arr = NaN(maxiter+1,1);
    dfdx_arr = NaN(maxiter+1,1);
    x_arr(1) = startguess;
    fx_arr(1) = fx(x_arr(1));
    dfdx_arr(1) = dfdx(x_arr(1));
%Iteration loop
    for i =2:maxiter+1
%Calculate xi solution
    x_arr(i) = x_arr(i-1)-sf*dfdx(x_arr(i-1));
%Calculate function value fxi
    fx_arr(i) = fx(x_arr(i));
%Calculate gradient value dfdxi
    dfdx_arr(i) = dfdx(x_arr(i));
%Check for convergence
    if abs(dfdx_arr(i)) <= termtol
        break
    end
  end
toc
```

The variables sf, fx, dfdx, startguess, maxiter, and termtol may all be changed to suit the problem being solved. Running the example problem provides the same results as determined in section 3.6.1, but just how quick is it? In real time solving the example problem takes less than 1 s. In fact, if `tic` is inserted immediately after the `clc` command and `toc` is inserted as the last line, MatLab will display the time elapsed. For this particular problem, the elapsed time was 0.004519 s, which is very low indeed; however, the time required to solve increases exponentially with problem complexity as will be subsequently discussed.

Before the focus of attention turns to the Newton–Raphson method, the changes required to the MatLab code to solve multivariable problems with GDM is worth mentioning. It is possible to utilize built-in functions in MatLab to refine the process by using cell arrays to store the function handles as well as symbolic expressions to complete the differentiation to reduce the amount of "manual labor" required to solve any given multivariable and multiexpression problem. Following this, it is a simple and straightforward process of "expanding" the iterative solver to reflect the relative differences between Equations (3.57) and (3.59).

3.6.3 Newton–Raphson method

The Newton–Raphson (NR) method, also known as Newton's method or Newton's iteration, is also a gradient-based root finding method that may be used to determine extreme points of a function, that is, optimization. The method is in many ways similar to the GDM method; there are, however, some subtle differences, as will be subsequently explained. Analog to GDM, NR can also be applied to optimization problems containing single, as well as multiple, variables. In connection to structural analysis and optimization, it is interesting to point out that NR is frequently employed in connection with explicit FEA as the numerical iteration tool to achieve equilibrium for the individual timestep, demonstrating the vast number of variables the method is capable of handling. For the purpose of initially presenting the method in this section, a single variable version will, however, be used. As was the case with GDM, NR requires a starting point to be defined; the method will thereafter iterate by updating the solution between iterations using Equation (3.60)

$$x_{i+1} = x_i - sf_i \frac{f(x_i)}{\nabla f(x_i)} \quad | \quad i \geq 0 \tag{3.60}$$

In (3.60) x_i is the current "solution," x_{i+1} is the "updated solution," sf_i is a scale factor that may be updated every iteration, $f(x_i)$ is the current functional value, and $\nabla f(x_i)$ is the functional value of the first derivative, i.e. the gradient. Comparing (3.60) to (3.57), it can be seen that the only differences between the two are that the current functional value $f(x_i)$ is part of the former, not the latter. Furthermore, $f(x_i)$ is divided by $\nabla f(x_i)$ meaning that NR uses a ratio of function value to gradient, whereas GDM only uses the actual gradient.

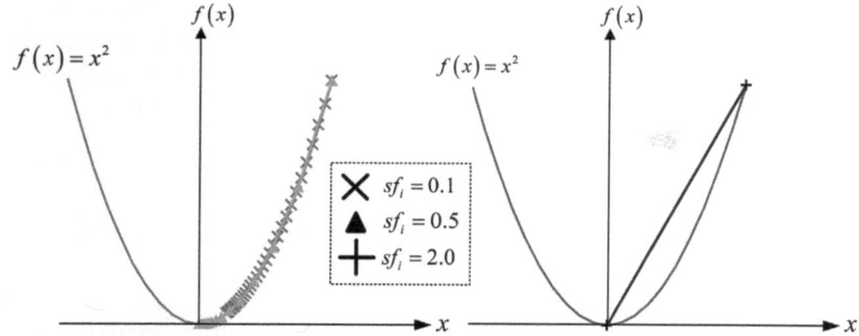

FIGURE 3.13 Newton–Raphson example problem.

In order to directly compare the GDM and NR methods, the simple example of minimizing $f(x)=x^2$ with a starting point $x_0=1$ will be repeated. As in the previous section, the problem will be solved using three (constant) sf_i values: 0.1, 0.5, and 2.0. The results obtained using NR are illustrated in Figure 3.13.

By comparing the results illustrated in Figure 3.11 to those of 3.13, the general difference between GDM and NR may be understood. Starting with the lowest scale factor, $sf_i=0.1$ it can be seen from Figure 3.11 that GDM takes larger steps at the start of the algorithm $(x_0=1)$ which decrease as the function values, that is, as $f(x)$ approaches the minimum point ($x = 0$). The NR results illustrated in Figure 3.13, on the other hand, show that the steps are more "evenly spaced" when compared with the GDM, for sf_i = 0.1; this is because the ratio of function value to gradient only varies marginally in the first three to four iterations. After 20 iterations the "solution" is that $x_{20}=0.35489$ which leads to $f(x_{20})=0.34056$ meaning that the proposed "optimum" solution found by NR is considerably inaccurate in this particular incidence after 20 iterations. In fact, the first time the solution becomes accurate to the second digit, that is, 0.00, is after 104 iterations. In comparison, the second graph illustrated in Figure 3.13 representing the sf_i = 0.5 solution initially displays a more rapid descent toward the true optimum solution, and reaches a solution that is accurate to the second digit after 19 iterations as $x_{19}=0.00423$. In this particular case, NR is inferior to GDM as the true optimum solution was achieved in a single step for sf_i = 0.5 using GDM, as illustrated in Figure 3.11. For sf_i = 2.0, NR is however superior to GDM as the true optimum solution is obtained in a single step. This particular example cannot be generalized into, for example, stating that GDM is superior to NR for sf_i = 0.5, as this will be entirely dependent upon the individual problem, and in particular the objective function. Although not clearly visible from the above example, the starting point, that is, initial solution, will also have a noticeable (if not significant) effect in many cases as will be subsequently demonstrated.

The above examples do, however, demonstrate the fundamental principles of GDM and NR, including the potential effects of the scale factor sf_i.

The above example only contained a single variable, namely x_i; it is, however, very straightforward to expand the NR algorithm to any multitude of variables simply by using a matrix and vector system. The iterative step for the single variable NR method in (3.60) can thereby be expanded to Expression (3.61)

$$\{x_{i+1}\} = \{x_i\} - 1 \cdot sf \cdot [J]^{-1} \cdot \{f(x_i)\} \quad | \quad i \geq 0 \tag{3.61}$$

In Expression (3.61), $[J]^{-1}$ is the inverted Jacobian matrix that contains the partial derivatives of the individual $f(x)$ functions.

3.6.4 MatLab example of Newton–Raphson method

To create a MatLab program for the single variable Newton Raphson method, the flowchart first needs to be defined; this is almost identical to that of the gradient descent method, seen in Figure 3.12. In fact, the only difference is the equation number used to calculate the updated solution, that is, point 1, needs to change from (3.57) to (3.60). This also means that only one specific aspect of the MatLab program created for the gradient descent method needs to be changed for the Newton–Raphson method, namely, the update of x_i inside the for loop:

```
x_arr(i) = x_arr(i-1)-sf*(fx(x_arr(i-1))/dfdx(x_arr(i-1)));
```

The single variable Newton–Raphson MatLab code therefore becomes:

```
%Newton Raphson Method - Single Variable

%Clear memory
    clear
    clc
    tic
%Define problem
    sf = 0.1;
    fx = @(x) x^2;
    dfdx = @(x) 2*x;
    startguess = 1;
%Define termination parameters
    maxiter = 100;
    termtol = 0.005;
%Create arrays
    x_arr = NaN(maxiter+1,1);
    fx_arr = NaN(maxiter+1,1);
    dfdx_arr = NaN(maxiter+1,1);
    x_arr(1) = startguess;
    fx_arr(1) = fx(x_arr(1));
    dfdx_arr(1) = dfdx(x_arr(1));
%Iteration loop
    for i =2:maxiter+1
        %Calculate xi solution
            x_arr(i) = x_arr(i-1)-sf*(fx(x_arr(i-1))/dfdx(x_arr(i-1)));
        %Calculate function value fxi
            fx_arr(i) = fx(x_arr(i));
        %Calculate gradient value dfdxi
            dfdx_arr(i) = dfdx(x_arr(i));
        %Check for convergence
            if abs(dfdx_arr(i)) <= termtol
                break
            end
    end
```

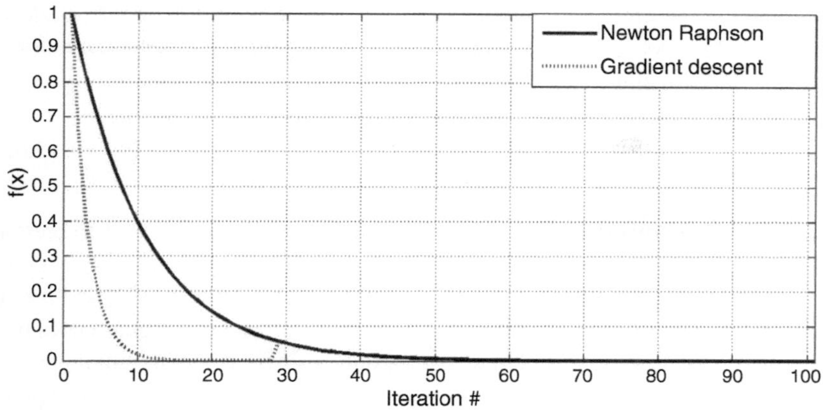

FIGURE 3.14 Function values vs. iteration #; for example, problem $f(x_i) = x^2$.

When solving the test problem using the Newton–Raphson approach, the time elapsed was 0.014032 s, percentage-wise considerably more than the gradient descent method of Sections 3.6.1 and 3.6.2. Part of the reason for this is the low *sf* value used, which dramatically reduces the steps of the NR solver. In order to aid the readers' understanding of the difference between the NR and GDM solvers, the function values $f(x_i) = x^2$ have been plotted direct from the MatLab results as illustrated in Figure 3.14.

Figure 3.14 reveals that the GDM initially converges significantly faster than the NR solver and seems to reach the optimum solution; $x = 0$ around the 20th iteration. At iteration 28 the GDM does, however, seem to diverge from the optimum solution and converge toward the NR solution and closely "follow" the NR solver results from this point onward. Reviewing the actual numerical values it is confirmed that the GDM and NR $f(x_i)$ values are identical from $i = 29$ up to the maximum number of iterations. Both solvers terminate at iteration 100, that is, immediately before the maximum number of iterations is reached because the `termtol` threshold has been violated.

As was the case with the GDM MatLab example of Section 3.6.2, the NR MatLab code developed in this section can relatively easily be extended to cater for multivariable problems by adapting the code to reflect the relative differences between Equations (3.60) and (3.61).

3.6.5 Steepest descent method

The steepest descent method (SDM) is also an LSM, just as the GDM and NR methods presented above, meaning that the method also utilizes the generalized expression of (3.55). There is, however, a more significant difference between the SDM and the preceding two methods, namely, the fact that the SDM contains an explicit definition of the scale factor sf_i which will be updated for each iteration. The general description and understanding of the SDM is not more cumbersome than for the GDM and NR methods. The presentation in this subsection will, however, be continued in the following subsections; an example function will therefore be introduced. A "classical" way of deriving the SDM is to initially observe the optimization problem of minimizing a general quadratic function as defined in (3.62)

$$f(x) = \frac{1}{2}x^T Ax - x^T b \tag{3.62}$$

The following derivation will not be comprehensive, as the main aim is to present the general steps of the method. For the subsequent derivation, x in (3.62) is a vector containing the variables, A is a symmetric positive definite matrix, and b also represents a vector. The fact that A is symmetric positive definite guarantees that the solution to the system is unique, that is, only a single global minimum exists; consequently, the function is convex. In general, a minimum (or maximum) point is global and unique if the second-order differential is greater than zero for all other points of the domain (less than zero for maximization problems), as indicated in Expression (3.10). For a multiple dimension function this is equivalent to ensuring that the Hessian matrix is consistently larger than zero as stipulated by (3.63)

$$H = f'' = \nabla^2 > 0 \tag{3.63}$$

It is clear that the solution x_{min} that minimizes the quadratic function of (3.62) is the one where the gradient is equal to zero due to the convexity of the function. Taking the gradient, that is, the first derivative of the quadratic function in (3.62) and setting equal to zero leads to Expression (3.64)

$$\nabla f(x) = Ax - b = 0 \Leftrightarrow Ax = b \tag{3.64}$$

The right-hand side of Expression (3.64) is a linear system of equations that, in principle, can be solved using any LSM. Note how the right-hand side of (3.64) is distinctly similar to the fundamental equation of FEA $\{F\} = [K] \cdot \{x\}$. The method of steepest descent follows the same ideology as the NR and GDM methods above, namely that the optimal search direction for any given minimization problem min($f(x)$) from any solution point x is the direction of the negative gradient $-\nabla f(x)$. As Expression (3.64) is the gradient of the generalized quadratic test problem in (3.62), the unique global minimum is obtained only when (3.64) equals zero. For all other "solutions" or iterations Equation (3.64) will not equal zero; in other words, $Ax - b \neq 0$. The "remainder," or the "residual" r, may be considered to serve as a measure for the relative error of the current solution and is defined as Equation (3.65)

$$r_i = Ax_i - b \tag{3.65}$$

The lower r becomes, the closer the current solution x_i is to the true optimum solution x_{opt} which is obtained when $r_i = 0$; consequently, $f(x)$ has been minimized. As the current solution x_i approaches x_{opt} it would be sensible to reduce the step size taken, that is, the relative change of x_i to avoid "skipping" or "overshooting" the optimum point as, for example, illustrated on the right-hand side of Figure 3.11.

In their basic form, neither the GDM nor the NR methods have any explicit "built-in automatic feature" to reduce the step size as the residual decreases; as previously discussed, sf_i may be manually changed once per iteration. SDM does, however, have a built-in function for decreasing the step size as a function of the residual as dictated by Equation (3.66)

$$sf_i = \frac{\nabla f(x_i)^T \nabla f(x_i)}{\nabla f(x_i)^T A \nabla f(x_i)} = \frac{r_i^T r_i}{r_i^T A r_i} \tag{3.66}$$

In Equation (3.66), the numerator is the dot product of the gradient and the transpose of the gradient; the denominator is in essence identical, apart from the inclusion of the matrix A in the dot product. The updated sf_i thereby becomes a function of the relative influence of A as it is multiplied onto the gradient, (3.64), and more importantly the size of the residual r. The iterative expression for the SDM can thereby be defined, shown in Equation (3.67)

$$x_{i+1} = x_i + sf_i \nabla f(x_i) \Rightarrow x_{i+1} = x_i - \frac{r_i^T r_i}{r_i^T A r_i} \nabla f(x_i) \Leftrightarrow x_i - \frac{r_i^T r_i}{r_i^T A r_i} r_i \tag{3.67}$$

In order to better understand the working principle of SDM, an example will now follow. At this point, it is important to remember that the system $Ax - b$ must be a set of linear equations as previously stipulated. As the method works on the gradient, that is, the first derivative, this means that a general quadratic function such as the one defined in (3.62) can be solved using SDM. For the purpose of this example, consider the problem of minimizing the system defined in (3.68)

$$Ax - b = 0: \begin{bmatrix} 3 & -1 & 1 \\ -1 & 3 & -1 \\ 1 & -1 & 3 \end{bmatrix} \begin{Bmatrix} x_{1i} \\ x_{2i} \\ x_{3i} \end{Bmatrix} - \begin{Bmatrix} -1 \\ 7 \\ -7 \end{Bmatrix} = \begin{Bmatrix} 0 \\ 0 \\ 0 \end{Bmatrix} \tag{3.68}$$

As with any LSM, a starting point must be defined; in this particular case, the starting point defined in (3.69) will be used.

$$x_0 = \begin{Bmatrix} x_{10} \\ x_{20} \\ x_{30} \end{Bmatrix} = \begin{Bmatrix} 0 \\ 0 \\ 0 \end{Bmatrix} \tag{3.69}$$

Strictly speaking, the first step should be to ensure that the matrix A is symmetric positive definite; in this particular case, it is straightforward to visually verify that the matrix A is symmetric by reviewing (3.68). Furthermore, as the determinant of A equals 20, the matrix is symmetric positive definite and SDM may be used. The first step is to calculate the residual r as a function of the initial solution x_0, that is, by inserting the information of (3.69) into (3.65) the current residual r_0 may be found, as seen in Expression (3.70)

$$r_0 = Ax_0 - b = \begin{bmatrix} 3 & -1 & 1 \\ -1 & 3 & -1 \\ 1 & -1 & 3 \end{bmatrix} \begin{Bmatrix} 0 \\ 0 \\ 0 \end{Bmatrix} - \begin{Bmatrix} -1 \\ 7 \\ -7 \end{Bmatrix} = \begin{Bmatrix} 1 \\ -7 \\ 7 \end{Bmatrix} \tag{3.70}$$

With the residual r_0 known, the scale factor sf_0 may be obtained using Equation (3.66), leading to Equation (3.71)

$$sf_0 = \frac{\overbrace{\begin{bmatrix} 1 & -7 & 7 \end{bmatrix}}^{r_0^T} \overbrace{\begin{Bmatrix} 1 \\ -7 \\ 7 \end{Bmatrix}}^{r_0}}{\underbrace{\begin{bmatrix} 1 & -7 & 7 \end{bmatrix}}_{r_0^T} \underbrace{\begin{bmatrix} 3 & -1 & 1 \\ -1 & 3 & -1 \\ 1 & -1 & 3 \end{bmatrix}}_{A} \underbrace{\begin{Bmatrix} 1 \\ -7 \\ 7 \end{Bmatrix}}_{r_0}} = \frac{99}{423} \approx 0.2340 \tag{3.71}$$

The updated solution x_1 may now be determined using the general expression of (3.67), as calculated in (3.72)

$$x_1 = \begin{Bmatrix} x_{11} \\ x_{21} \\ x_{31} \end{Bmatrix} = \begin{Bmatrix} 0 \\ 0 \\ 0 \end{Bmatrix} - 0.2340 \begin{Bmatrix} 1 \\ -7 \\ 7 \end{Bmatrix} = \begin{Bmatrix} -0.2340 \\ 1.6383 \\ -1.6383 \end{Bmatrix} \tag{3.72}$$

From (3.72) the next residual r_1 may be found, Expression seen in (3.73)

$$r_1 = Ax_1 - b = \begin{bmatrix} 3 & -1 & 1 \\ -1 & 3 & -1 \\ 1 & -1 & 3 \end{bmatrix} \begin{Bmatrix} -0.2340 \\ 1.6383 \\ -1.6383 \end{Bmatrix} - \begin{Bmatrix} -1 \\ 7 \\ -7 \end{Bmatrix} = \begin{Bmatrix} -2.9787 \\ -0.2128 \\ 0.2128 \end{Bmatrix} \tag{3.73}$$

sf_1 may now be found, as in Equation (3.74)

$$sf_1 = \frac{\overbrace{\begin{bmatrix} -2.9787 & -0.2128 & 0.2128 \end{bmatrix}}^{r_1^T} \overbrace{\begin{Bmatrix} -2.9787 \\ -0.2128 \\ 0.2128 \end{Bmatrix}}^{r_1}}{\underbrace{\begin{bmatrix} -2.9787 & -0.2128 & 0.2128 \end{bmatrix}}_{r_1^T} \underbrace{\begin{bmatrix} 3 & -1 & 1 \\ -1 & 3 & -1 \\ 1 & -1 & 3 \end{bmatrix}}_{A} \underbrace{\begin{Bmatrix} -2.9787 \\ -0.2128 \\ 0.2128 \end{Bmatrix}}_{r_1}}$$

$$= \frac{8.9633}{24.4455} \approx 0.3667 \tag{3.74}$$

Using the results of (3.73) and (3.74) the updated solution x_2 may now be obtained; seen in Equation (3.75)

$$x_2 = \begin{Bmatrix} x_{12} \\ x_{22} \\ x_{32} \end{Bmatrix} = \begin{Bmatrix} -0.2340 \\ 1.6383 \\ -1.6383 \end{Bmatrix} - 0.3367 \begin{Bmatrix} -2.9787 \\ -0.2128 \\ 0.2128 \end{Bmatrix} = \begin{Bmatrix} 0.8582 \\ 1.7163 \\ -1.7163 \end{Bmatrix} \tag{3.75}$$

The residual r_2 is calculated in (3.76)

$$r_2 = Ax_2 - b = \begin{bmatrix} 3 & -1 & 1 \\ -1 & 3 & -1 \\ 1 & -1 & 3 \end{bmatrix} \begin{Bmatrix} -0.8582 \\ -1.7163 \\ 1.7163 \end{Bmatrix} - \begin{Bmatrix} -1 \\ 7 \\ -7 \end{Bmatrix} = \begin{Bmatrix} 0.1418 \\ -0.9929 \\ 0.9929 \end{Bmatrix} \tag{3.76}$$

The updated scale factor is calculated in (3.77)

$$sf_2 = \frac{\overbrace{\lfloor 0.1418 \;\; -0.9929 \;\; 0.9929 \rfloor}^{r_2^T} \overbrace{\begin{Bmatrix} 0.1418 \\ -0.9929 \\ 0.9929 \end{Bmatrix}}^{r_2}}{\underbrace{\lfloor 0.1418 \;\; -0.9929 \;\; 0.9929 \rfloor}_{r_2^T} \underbrace{\begin{bmatrix} 3 & -1 & 1 \\ -1 & 3 & -1 \\ 1 & -1 & 3 \end{bmatrix}}_{A} \underbrace{\begin{Bmatrix} 0.1418 \\ -0.9929 \\ 0.9929 \end{Bmatrix}}_{r_2}}$$

$$= \frac{1.9919}{8.5106} \approx 0.2340 \tag{3.77}$$

The updated solution x_3 can now be determined, as shown in Equation (3.78)

$$x_3 = \begin{Bmatrix} x_{13} \\ x_{23} \\ x_{33} \end{Bmatrix} = \begin{Bmatrix} 0.8582 \\ 1.7163 \\ -1.7163 \end{Bmatrix} - 0.2340 \begin{Bmatrix} 0.1418 \\ -0.9929 \\ 0.9929 \end{Bmatrix} = \begin{Bmatrix} 0.8250 \\ 1.9487 \\ -1.9487 \end{Bmatrix} \tag{3.78}$$

The residual r_3 is calculated in (3.79)

$$r_3 = Ax_3 - b = \begin{bmatrix} 3 & -1 & 1 \\ -1 & 3 & -1 \\ 1 & -1 & 3 \end{bmatrix} \begin{Bmatrix} 0.8250 \\ 1.9487 \\ -1.9487 \end{Bmatrix} - \begin{Bmatrix} -1 \\ 7 \\ -7 \end{Bmatrix} = \begin{Bmatrix} -0.4225 \\ -0.0302 \\ 0.0302 \end{Bmatrix} \tag{3.79}$$

The avid reader may continue this example to obtain zero residual; however, it should be noted that the iterations may continue infinitely depending upon the number of significant digits under observation. The true solution to the example problem is listed in (3.80)

$$x_{opt} = \begin{Bmatrix} 1 \\ 2 \\ -2 \end{Bmatrix} \tag{3.80}$$

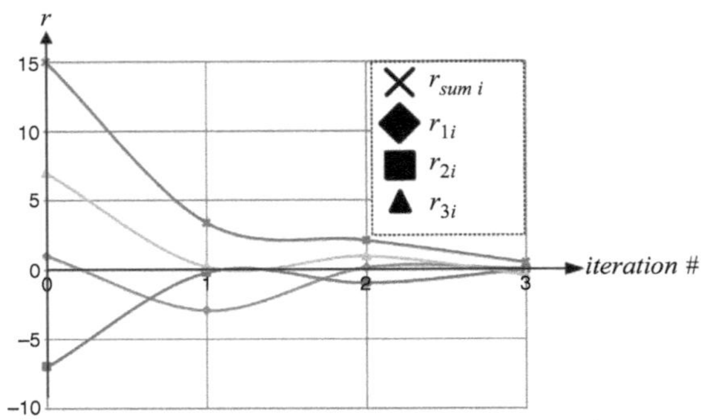

FIGURE 3.15 Residuals versus iteration #.

It can be argued that the two latter equations, that is, rows 2 and 3 of the $Ax - b$ system, are relatively close to the optimum solution as the residuals are 0 to the second significant digit. This may be substantiated by comparing the x_3 solution of (3.78) to the true optimum solution of (3.80). The first equation is, however, relatively far from its optimum with a numerical residual of 0.4225. The convergence of the solution may be tracked by reviewing the individual residuals as a function of the iteration number as plotted in Figure 3.15.

In Figure 3.15, r_{1i} refers to the first entry in the residual vector of Expression (3.68). The figure clearly displays that all three individual entries in the residual vector converge toward zero as expected. Consequently, the total residual, that is, the "length" of the residual vector r_i, also converges toward zero as expected. It is interesting to point out that both r_{2i} and r_{3i} are closer to zero for $i = 1$ than they are for $i = 2$, underlining the importance of solving the system of equations simultaneously. The graphs in Figure 3.16 plot the individual entries in the solution vector x_i as a function of the iteration number.

For the individual x_{2i} and x_{3i} variables, the graphs in Figure 3.16 demonstrate a continuous convergence toward the individual optimum values of 2 and −2, respectively. The x_{1i} variable does, however, diverge at $i = 1$ as it moves away from the true optimum value of 1; this is the main cause of the relatively large r_{1i} residual at $i = 1$ in Figure 3.15. Although the true optimum solutions x_i are most often unknown, it is pivotal to review/plot the convergence plots for the residual values/error estimations and variations of variables, as this gives clear indications about the progress of the optimization. Graphs such as the ones plotted in Figures 3.15 and 3.16 are often used to determine relative convergence of the specific iterative method/LSM and are very often used to control when the iterations terminate to avoid an infinite number of iterations, as the true optimum solution (down to an infinite number of significant digits) is most likely never obtained using LSM and numerical iterations. In certain instances, the final iteration may not be the "best" solution, as some problems may initially converge toward an optimum solution before diverging or at least fluctuating around it as will be subsequently discussed; this is, however, not the case when the matrix A is symmetric positive definite.

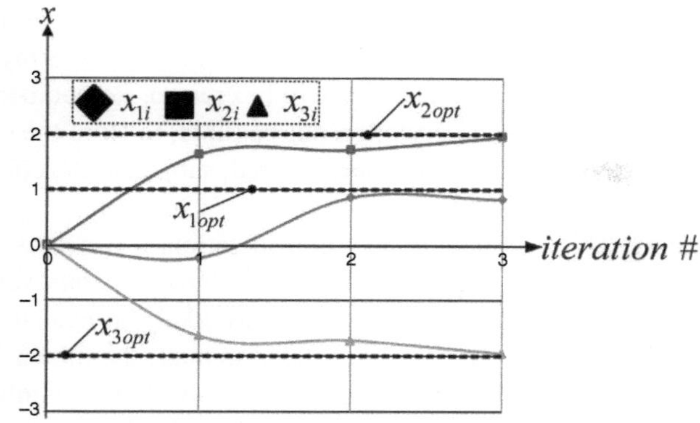

FIGURE 3.16 Solution vector entries versus iteration #.

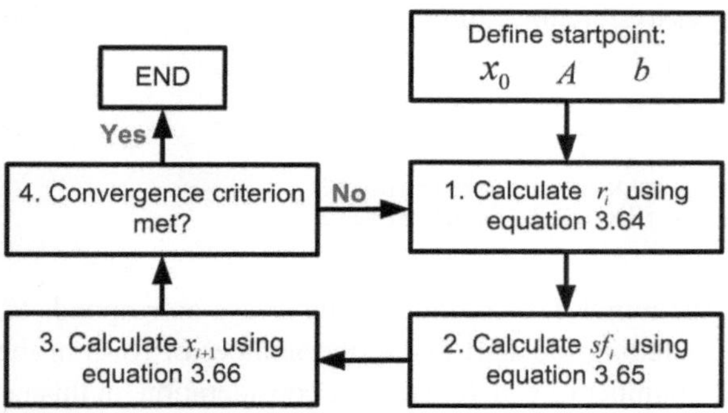

FIGURE 3.17 Flowchart of steepest descent method.

3.6.6 MatLab example of steepest descent method

Firstly, it is imperative to plot the flowchart for the SDM algorithm that contains one initial step and four iterative steps, as illustrated in Figure 3.17.

The MatLab program will start identical to the previous ones where the memory is cleared, the termination parameters are set, and the problem is formulated including the start guess. This leads to the following lines of MatLab code:

```
%Steepest Descent Method

%Reset memory
    clear
    clc
    tic
%Define problem
    A = [3,-1,1;-1,3,-1;1,-1,3];
    b=[-1;7;-7];
    startguess=[0;0;0];
%Define termination parameters
    maxiter=100;
    termtol = 0.005;
```

Remember the problem being solved in this example follows that of the generalized quadratic expression in Equation (3.62). Next, the arrays that will be used to store the entire solution history will be created. Remembering that in general it is not advisable, and very often not possible, to store the entire solution history due to the amount of memory required, subject to the complexity of the problem under consideration.

In the specific sample problem under consideration, it is necessary to store three separate values for x_i, Equation (3.68), and three separate values for r_i, Equation (3.70). This number is, however, highly likely to change if another problem is considered. It is, therefore, beneficial to make the dimensions of the x and r arrays "dynamic" for each problem to be solved. The number of rows or columns in A or the number of entries in the initial solution vector x_0 may, for example, be used to determine the number of entries required per iteration. Finally, the maximum number of iterations is defined using maxiter = 100 the size of the arrays to store the entire history of r_i and x_i can, therefore, be defined as follows:

```
%Create arrays
    dim1 = size(A,1);
    r_arr = NaN(dim1, maxiter+1);
    x_arr = NaN(dim1, maxiter+1);
```

Note that the number of columns in the arrays corresponds to the number of iterations, including the start guess; the number of rows in the arrays corresponds to the number of equations under consideration, in this particular case defined by the number of rows in A.

The only remaining variable to be stored is the scale factor sf_i, that is, a single value per iteration; hence, the dimensions of this array are only dependent upon the maximum number of iterations:

```
sf_arr = NaN(1,maxiter+1);
```

Before the iterative cycle commences, the start guess will be transferred into the `x_arr` array:

```
%Store initial solution
    for j = 1:dim1
        x_arr(j,1) = startguess(j,1);
    end
```

Next the iterative cycle needs to commence:

```
%Iteration loop
    for i = 1:maxiter+1
```

According to step 1 (Figure 3.17), r_i must now be calculated using Equation (3.65). Attention to detail is paramount at this stage, as the x_i values are stored in a much larger array than used in the previous example. Noting that the column entry of x_arr dictates the iteration number i, the x_i vector may be obtained using the iteration number and a ":" sign that in turn will collect the entire column from the MatLab array. Similarly, the calculated r_i values should be stored in a single column per iteration. The expression for calculating and storing the r_i vector in the MatLab program therefore becomes:

```
%Calculate Residual ri
          r_arr(:,i) = A*x_arr(:,i)-b;
```

According to step 2 (Figure 3.17), the scale factor sf_i is the next to be calculated and stored. Due to the way the r_arr has been defined there is no need to change any of the "transpose" setups of the original equation, (3.66), for the MatLab program:

```
%Calculate scale factor sfi
    sf_arr(1,i) = transpose(r_arr(:,i))*r_arr(:,i) /
                  (transpose(r_arr(:,i))*A*r_arr(:,i));
```

The next solution, x_{i+1}, should now be obtained, provided that the current iteration is not the last one:

```
%Calculate updated solution xi+1
      if i <= maxiter
          x_arr(:,i+1) = x_arr(:,i)-sf_arr(1,i)*r_arr(:,i);
      end
```

Finally, the last iteration step is to check for convergence. In this case, the absolute sum of the active column, that is, iteration of the r_arr vector will be used. It is imperative that the absolute sum is used as opposed to a simple summation to get an accurate overview of the current residuals. Take, for example, the r_0 vector of (3.70) which sums to 1, but the absolute sum is 15, indicating a very different level of residual. The convergence or termination criterion is therefore introduced as follows:

```
%Check for convergence
          if sumabs(r_arr(:,i)) <= termtol
              break
          end
```

The steepest descent MatLab program is thereby complete:

```
%Steepest Descent Method

%Reset memory
    clear
    clc
    tic
%Define problem
    A = [3,-1,1;-1,3,-1;1,-1,3];
    b=[-1;7;-7];
    startguess=[0;0;0];
%Define termination parameters
    maxiter=100;
    termtol = 0.005;
%Create arrays
    dim1 = size(A,1);
    r_arr = NaN(dim1, maxiter+1);
    x_arr = NaN(dim1, maxiter+1);
    sf_arr = NaN(1,maxiter+1);
%Store initial solution
    for j = 1:dim1
        x_arr(j,1) = startguess(j,1);
    end
%Iteration loop
    for i = 1:maxiter+1
        %Calculate Residual ri
            r_arr(:,i) = A*x_arr(:,i)-b;
        %Calculate scale factor sfi
            sf_arr(1,i) =
transpose(r_arr(:,i))*r_arr(:,i)/(transpose(r_arr(:,i))*A*r_arr(:,i));
        %Calculate updated solution xi+1
            if i <= maxiter
                x_arr(:,i+1) = x_arr(:,i)-sf_arr(1,i)*r_arr(:,i);
            end
        %Check for convergence
            if sumabs(r_arr(:,i)) <= termtol
                break
            end
    end
toc
```

Solving the example problem of Section 3.6.5 takes 0.02486 s and gives the solution listed in (3.81) after 11 iterations.

$$x_{\text{opt}} = \begin{Bmatrix} x_1 \\ x_2 \\ x_3 \end{Bmatrix} = \begin{Bmatrix} 0.9999 \\ 1.9999 \\ -1.9999 \end{Bmatrix} \tag{3.81}$$

When comparing Equation (3.81) to the optimal solution of (3.80), it can be seen that the results obtained are not entirely accurate; this is, however, dependent upon the termination tolerance set. If, for example, termtol is set to 0.0005, the solution found is accurate to the fifth significant digit after 13 iterations. The iteration history of the solution x_i is illustrated in Figure 3.18.

Figure 3.18 may be compared to Figure 3.15, as they represent the same problem; Figure 3.15 does, however, only display the first three iterations. Figure 3.18 reveals a smooth convergence for all three individual variables as expected.

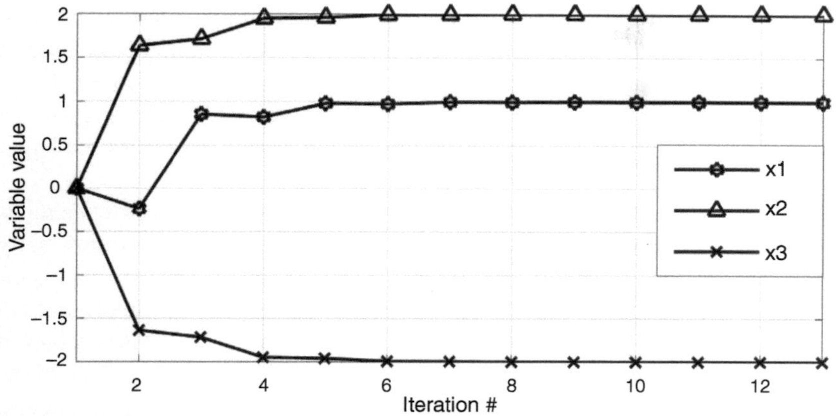

FIGURE 3.18 Solution vector entries vs. iteration #.

3.6.7 Conjugate gradient method

The conjugate gradient method (CGM) is perhaps the most cumbersome to explain relative to the ones presented in the preceding sections. CGM belongs to a number of methods known as *A-conjugate* methods. Remembering that conjugate in algebraic terms simply means to change the sign of a term, the conjugate of $3x + 1$ is simply $3x - 1$. In relation to the CGM, this means that a set of nonzero vectors p_i that are conjugate with respect to the A matrix of (3.62) are introduced, as dictated by (3.82)

$$p_i^T A p_j \quad \forall i \neq j \tag{3.82}$$

By introducing n linearly independent vectors $p_0, p_1, \ldots, p_{n-1}$ that span the entire $\mathbb{R}^n$ space, it can be proven that the generalized quadratic function of (3.62) can be minimized by successively minimizing the function along the direction of each p_i vector. The vectors p_i therefore span what in optimization is generally referred to as the solution space; the n-dimensional space that contains all possible solutions x to the minimization problem $f(x)$ in general, including the example defined in (3.62). An error estimation analog to the residual vector r_i, introduced in Section 3.6.5, may now be set up. This error estimation is based on the fact that the *A-conjugate* vectors p_i act as a basis for $\mathbb{R}^n$ and the conjugate property introduced in (3.82), which leads to the generalized solution expression of (3.83)

$$x_{\text{opt}} = x_0 + sf_0 p_0 + sf_1 p_1 + \ldots + sf_{n-1} p_{n-1} \tag{3.83}$$

The burning question at this stage is, of course, how are the *A-conjugate* vectors p_i actually obtained? These may be obtained in a variety of ways; the most "obvious" is to determine the eigenvectors of A, which may however be a lengthy and computationally heavy process. Another option is to use an orthonormalization process such as the Gram–Schmidt process that may be used to orthonormalize, that is, to create orthogonal unit length vectors, in the Euclidian $\mathbb{R}^n$ space. Taking this option will, however, require a relatively large usage of memory to store intermediate values for completing the process for larger systems.

There is, however another method in which each new conjugate vector p_i can be calculated using only the previous vector p_{i-1}, meaning that significantly less computational effort and memory is required compared with calculating eigenvectors or using the Gram–Schmidt process. The method repeatedly ensures that the new vector p_i is conjugate to the previous vector p_{i-1}. This is achieved by updating the search direction, that is, vector p_i, as a linear combination of the negative residual r_i and the previous search direction, that is, p_{i-1} vector; hence, the process was named the conjugate gradient method. Following the general expression for all LSM, Equation (3.55), Equation (3.84) dictates how the CGM determines an updated search direction, i.e. p_i vector.

$$p_{i+1} = -r_{i+1} + \beta_i p_i \tag{3.84}$$

β_i is a scale factor and is calculated in a similar fashion to the sf_i value of (3.66), as dictated by (3.85)

$$\beta_i = \frac{r_i^T A p_{i-1}}{p_{i-1}^T A p_{i-1}} = \frac{r_{i+1}^T r_{i+1}}{r_i^T r_i} \tag{3.85}$$

The residual r_{i+1} in Equation (3.84) is calculated using Equation (3.86)

$$r_{i+1} = r_i + sf_i A p_i \tag{3.86}$$

The scale factor (step size) sf_i in Equation (3.86) is calculated using Equation (3.87)

$$sf_i = \frac{r_i^T r_i}{p_i^T A p_i} \tag{3.87}$$

The updated solution may now be calculated using the general LSM method, seen in Equation (3.88)

$$x_{i+1} = x_i + sf_i p_i \tag{3.88}$$

The observant reader may have noticed that in order to start the iterative steps of the algorithm x_0, r_0 and p_0 are required. x_0 is of course user defined, r_0

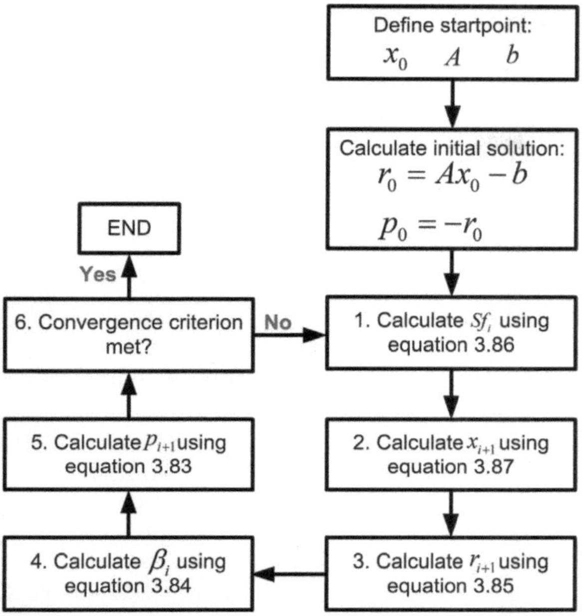

FIGURE 3.19 Flowchart of conjugate gradient method.

is computed using Equation (3.65), but p_0 cannot be directly calculated using Equation (3.86), as it requires information from the previous search direction, the very basis of conjugate. The initial search direction p_0 is, therefore, simply taken to be the negative residual r_0 as defined in Equation (3.89)

$$p_0 = -r_0 \tag{3.89}$$

The derivation of the steps above does not give a very clear or intuitive overview of the flowchart of the conjugate gradient algorithm, which is listed in Figure 3.19.

In order to obtain a detailed understanding of the CGM, the example from Section 3.6.5 will be repeated using the CGM flowchart of Figure 3.19, that is, the problem defined by Equations (3.68) and (3.69). As illustrated in Figure 3.19, the first step is identical to that of the SDM, namely to calculate r_0 using Equation (3.65), which in this particular example has been done in (3.70). For the CGM, the initial search direction p_0 is straightforwardly obtained using Equation (3.89). The starting point of the example problem may, therefore, be summarized, Expression (3.90)

$$A = \begin{bmatrix} 3 & -1 & 1 \\ -1 & 3 & -1 \\ 1 & -1 & 3 \end{bmatrix} \quad b = \begin{Bmatrix} -1 \\ 7 \\ -7 \end{Bmatrix} \quad x_0 = \begin{Bmatrix} 0 \\ 0 \\ 0 \end{Bmatrix} \quad r_0 : \begin{Bmatrix} 1 \\ -7 \\ 7 \end{Bmatrix} \quad p_0 : \begin{Bmatrix} -1 \\ 7 \\ -7 \end{Bmatrix} \tag{3.90}$$

The next step is to calculate the scale factor sf_0, as in Equation (3.91)

$$sf_0 = \frac{\overbrace{\begin{bmatrix} 1 & -7 & 7 \end{bmatrix}}^{r_0^T} \overbrace{\begin{Bmatrix} 1 \\ -7 \\ 7 \end{Bmatrix}}^{r_0}}{\underbrace{\begin{bmatrix} -1 & 7 & -7 \end{bmatrix}}_{p_0^T} \underbrace{\begin{bmatrix} 3 & -1 & 1 \\ -1 & 3 & -1 \\ 1 & -1 & 3 \end{bmatrix}}_{A} \underbrace{\begin{Bmatrix} -1 \\ 7 \\ -7 \end{Bmatrix}}_{p_0}} = \frac{99}{423} \approx 0.2340 \tag{3.91}$$

The updated solution x_1 may now be determined, Equation (3.92)

$$x_1 = \overbrace{\begin{Bmatrix} 0 \\ 0 \\ 0 \end{Bmatrix}}^{x_0} + \overbrace{0.2340}^{sf_0} \overbrace{\begin{Bmatrix} -1 \\ 7 \\ -7 \end{Bmatrix}}^{p_0} = \begin{Bmatrix} -0.2340 \\ 1.6383 \\ -1.6383 \end{Bmatrix} \tag{3.92}$$

At this stage there is no difference between the CGM and the SDM method, as the right-hand side of Equation (3.92) is identical to the right-hand side of Equation (3.72); the additional steps of the CGM have, however, not been completed yet. The next step is to calculate the residual r_1, Equation (3.93)

$$r_1 = \overbrace{\begin{Bmatrix} 1 \\ -7 \\ 7 \end{Bmatrix}}^{r_0} + \overbrace{0.2340}^{sf_0} \overbrace{\begin{bmatrix} 3 & -1 & 1 \\ -1 & 3 & -1 \\ 1 & -1 & 3 \end{bmatrix}}^{A} \overbrace{\begin{Bmatrix} -1 \\ 7 \\ -7 \end{Bmatrix}}^{p_0} = \begin{Bmatrix} -2.9787 \\ -0.2128 \\ 0.2128 \end{Bmatrix} \tag{3.93}$$

Next the scale factor β_0 is calculated, as in Equation (3.94)

$$\beta_0 = \frac{\overbrace{\begin{bmatrix} -2.9787 & -0.2128 & 0.2128 \end{bmatrix}}^{r_1^T} \overbrace{\begin{Bmatrix} -2.9787 \\ -0.2128 \\ 0.2128 \end{Bmatrix}}^{r_1}}{\underbrace{\begin{bmatrix} 1 & -7 & 7 \end{bmatrix}}_{r_0^T} \underbrace{\begin{Bmatrix} 1 \\ -7 \\ 7 \end{Bmatrix}}_{r_0}} = \frac{8.9633}{99} \approx 0.0905 \tag{3.94}$$

Finally, a new search direction can be calculated, Equation (3.95)

$$p_1 = -\overbrace{\begin{Bmatrix} -2.9787 \\ -0.2128 \\ 0.2128 \end{Bmatrix}}^{r_1} + \overbrace{0.0905}^{\beta_0} \overbrace{\begin{Bmatrix} -1 \\ 7 \\ -7 \end{Bmatrix}}^{p_0} = \begin{Bmatrix} 2.882 \\ 0.8465 \\ -0.8465 \end{Bmatrix} \tag{3.95}$$

The second iteration may now commence by calculating the scale factor sf_1, Equation (3.96)

$$sf_1 = \frac{\overbrace{\begin{bmatrix} -2.9787 & -0.2128 & 0.2128 \end{bmatrix}}^{r_1^T} \overbrace{\begin{Bmatrix} -2.9787 \\ -0.2128 \\ 0.2128 \end{Bmatrix}}^{r_1}}{\underbrace{\begin{bmatrix} 2.8882 & 0.8465 & -0.8465 \end{bmatrix}}_{p_1^T} \underbrace{\begin{bmatrix} 3 & -1 & 1 \\ -1 & 3 & -1 \\ 1 & -1 & 3 \end{bmatrix}}_{A} \underbrace{\begin{Bmatrix} 2.8882 \\ 0.8465 \\ -0.8465 \end{Bmatrix}}_{p_1}}$$
$$= \frac{8.9633}{20.9780} \approx 0.4273 \tag{3.96}$$

The updated solution x_2 may now be determined, as calculated in Equation (3.97)

$$x_2 = \overbrace{\begin{Bmatrix} -0.2340 \\ 1.6383 \\ -1.6383 \end{Bmatrix}}^{x_1} + \overbrace{0.4273}^{sf_1} \overbrace{\begin{Bmatrix} 2.8882 \\ 0.8465 \\ -0.8465 \end{Bmatrix}}^{p_1} = \begin{Bmatrix} 1.000 \\ 2.0000 \\ -2.0000 \end{Bmatrix} \tag{3.97}$$

The next step is to calculate the residual r_2, as in Equation (3.98)

$$r_2 = \overbrace{\begin{Bmatrix} -2.9787 \\ -0.2128 \\ 0.2128 \end{Bmatrix}}^{r_1} + \overbrace{0.4273}^{sf_1} \overbrace{\begin{bmatrix} 3 & -1 & 1 \\ -1 & 3 & -1 \\ 1 & -1 & 3 \end{bmatrix}}^{A} \overbrace{\begin{Bmatrix} 2.8882 \\ 0.8465 \\ -0.8465 \end{Bmatrix}}^{p_1} = 10^{-15} \cdot \begin{Bmatrix} 0.4441 \\ -0.8882 \\ 0.8882 \end{Bmatrix} \tag{3.98}$$

At this point, a new scale factor β_1 and search direction p_2 should be calculated; however, it is evident that the solver has already converged when reviewing the residuals in (3.98). In this particular case, the optimum solution is also listed in Equation (3.81), which is identical to the right-hand side of Equation (3.97) to the fourth significant digit as a minimum. In this particular case, the superiority of the CGM method compared with the SDM is clear due to the fast conversion of the former when compared with the latter.

3.6.8 MatLab example of conjugate gradient method

Based on the CGM example of the previous method, the focus will now be on developing a MatLab code of the method. As highlighted in the previous section, there are a number of similarities between the CGM and SDM algorithms;

the MatLab code for the CGM can therefore be obtained by adapting the SDM code developed in Section 3.6.6. Firstly, it is obvious that additional solver parameters are required for the GCM; as it is still desirable to store the complete solver history, an additional set of arrays will be defined for this purpose. In particular, two additional arrays to store the search direction p_i and the additional scale factor β_i needs to be created. By adding these to the SDM algorithm presented in Section 3.6.6, the following code is obtained:

```
%Conjugate Gradient Method

%Reset memory
    clear
    clc
    tic
%Define problem
    A = [3,-1,1;-1,3,-1;1,-1,3];
    b=[-1;7;-7];
    startguess=[0;0;0];
%Define termination parameters
    maxiter=100;
    termtol = 0.0005;
%Create arrays
    dim1 = size(A,1);
    r_arr = NaN(dim1, maxiter+1);
    x_arr = NaN(dim1, maxiter+1);
    p_arr = NaN(dim1, maxiter+1);
    sf_arr = NaN(1,maxiter+1);
    b_arr = NaN(1,maxiter+1);
%Store initial solution
    for j = 1:dim1
        x_arr(j,1) = startguess(j,1);
    end
```

The above code is all that is required prior to the iteration loop of the algorithm. At this stage, we will deviate slightly from the flowchart of Figure 3.19, as the first step will be to calculate the residual value r_i. The reason for this is that r_0 can be computed inside the iterative loop using the same lines of code as is required for step 3 (Figure 3.19), thereby avoiding "double programming". The first lines inside the iterative loop therefore become:

```
%Iteration loop
   for i = 1:maxiter+1
       %Calculate Residual ri
           r_arr(:,i) = A*x_arr(:,i)-b;
```

Next, the special case where $i = 1$, that is, iteration 0, must be considered, as there is no previous search direction the search direction must be set according to Equation (3.89), leading to the following lines of code:

```
%Calculate initial pi
            if i == 1
                p_arr(:,1) = -1*r_arr(:,1);
            end
```

The program now "returns" to step 1 (Figure 3.19), that is, the calculation of the scale factor sf_i, which is simple to adapt from the SDM code:

```
%Calculate scale factor sfi
            sf_arr(1,i) = transpose(r_arr(:,i))*r_arr(:,i) /
                          (transpose(p_arr(:,i))*A*p_arr(:,i));
```

The calculations of x_{i+1}, r_{i+1}, β_i, and p_{i+1} are all subject to the current iteration, not being the last one. As the variable arrays have previously defined according to the maximum number of iterations +1 to include the initial solution, it is necessary to include an if loop that only calculates the above variables if the current iteration number is not the last one:

```
%If not last iteration
            if i <= maxiter
```

The updated solution x_{i+1}, residual r_{i+1} scale factor β_i, and the updated search direction p_{i+1} can now all be computed as illustrated in the flowchart of Figure 3.19, which all reside inside the iteration if loop:

```
%If not last iteration
    if i <= maxiter
        %Calculate updated solution xi+1
            x_arr(:,i+1) = x_arr(:,i)+sf_arr(1,i)*p_arr(:,i);
        %Calculate updated residual ri+1
            r_arr(:,i+1) = r_arr(:,i)+sf_arr(1,i)*A*p_arr(:,i);
        %Calculate scale factor bi
            b_arr(1,i) = transpose(r_arr(:,i+1))*r_arr(:,i+1) /
             (transpose(r_arr(:,i))*r_arr(:,i));
        %Calculate new search direction
            p_arr(:,i+1) = -1*r_arr(:,i+1)+b_arr(1,i)*p_arr(:,i);
    end
```

Finally, the convergence check is completed:

```
%Check for convergence
            if sumabs(r_arr(:,i)) <= termtol
                break
            end
```

The MatLab program of the conjugate gradient method is thereby complete:

```
%Conjugate Gradient Method

%Reset memory
    clear
    clc
    tic
%Define problem
    A = [3,-1,1;-1,3,-1;1,-1,3];
    b=[-1;7;-7];
    startguess=[0;0;0];
%Define termination parameters
    maxiter=100;
    termtol = 0.0005;
%Create arrays
    dim1 = size(A,1);
    r_arr = NaN(dim1, maxiter+1);
    x_arr = NaN(dim1, maxiter+1);
    p_arr = NaN(dim1, maxiter+1);
    sf_arr = NaN(1,maxiter+1);
    b_arr = NaN(1,maxiter+1);
%Store initial solution
    for j = 1:dim1
        x_arr(j,1) = startguess(j,1);
    end
%Iteration loop
    for i = 1:maxiter+1
        %Calculate Residual ri
            r_arr(:,i) = A*x_arr(:,i)-b;
        %Calculate initial pi
            if i == 1
                p_arr(:,1) = -1*r_arr(:,1);
            end
        %Calculate scale factor sfi
            sf_arr(1,i) =
transpose(r_arr(:,i))*r_arr(:,i)/(transpose(p_arr(:,i))*A*p_arr(:,i));
        %If not last iteration
            if i <= maxiter
                %Calculate updated solution xi+1
                    x_arr(:,i+1) = x_arr(:,i)+sf_arr(1,i)*p_arr(:,i);
                %Calculate updated residual ri+1
                    r_arr(:,i+1) = r_arr(:,i)+sf_arr(1,i)*A*p_arr(:,i);
                %Calculate scale factor bi
                    b_arr(1,i) =
transpose(r_arr(:,i+1))*r_arr(:,i+1)/(transpose(r_arr(:,i))*r_arr(:,i));
                %Calculate new search direction
                    p_arr(:,i+1) = -1*r_arr(:,i+1)+b_arr(1,i)*p_arr(:,i);
            end

        %Check for convergence
            if sumabs(r_arr(:,i)) <= termtol
                break
            end
    end
```

The MatLab code produced took 0.000619 s to solve the example problem above, very fast compared with the steepest descent method as previously discussed.

There are many variations and adaptations of the conjugate gradient descent method presented in the previous two sections. Variations include the nonlinear conjugate gradient method (NLCGM), the biconjugate gradient method (BCGM), and the biconjugate gradient stabilized method (BCGSM). Although these additional algorithms may be very interesting for some readers, the above presentations cover the fundamental understanding of a conjugate gradient method sufficiently in the context of this book.

3.6.9 Comparison of gradient-based methods and line search methods

In the previous sections, the GDM, NR, SDM, and the CGM have all been presented including MatLab programs to demonstrate how they work through examples. But how do they actually compare from a more holistic viewpoint? As the reader may already be aware, it is not a straightforward task of identifying exactly where and when a specific algorithm should be applied. There is, however, a visual way of highlighting the very distinct difference between the CGM and, for example, the SDM methods as demonstrated by Figure 3.20.

Figure 3.20 represents a contour plot of a convex function, as well as the solution steps of the CGM and SDM algorithms. Note how the SDM method utilizes very different size steps and also how these clearly follow the steepest gradient. In the example of Figure 3.20, the SDM solver passes through all points between 1 and 5 before arriving at the optimum solution at 6. The CGM does, however, take significantly larger steps and only passes through points 1, 2, and 6. Note how the search direction vector p_2, that is, the line from 2 to 6, is orthogonal to the previous search direction p_1 represented by the lines 1 and 2. Finally, it can be seen that lines 1 and 2 are identical for both algorithms, as also explained in the previous examples in Section 3.6.7.

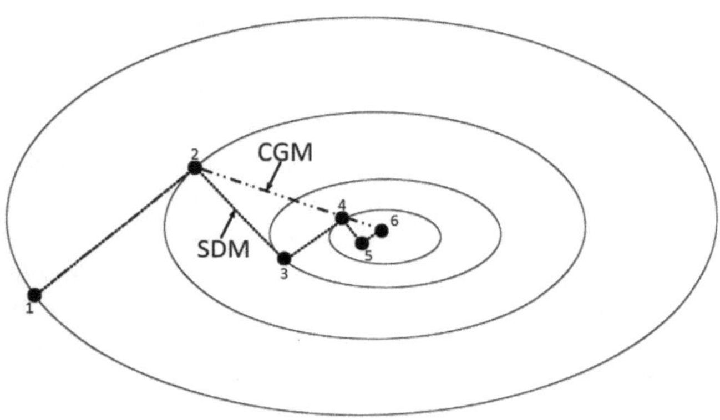

FIGURE 3.20 Visual interpretation of difference between CGM and SDM solvers.

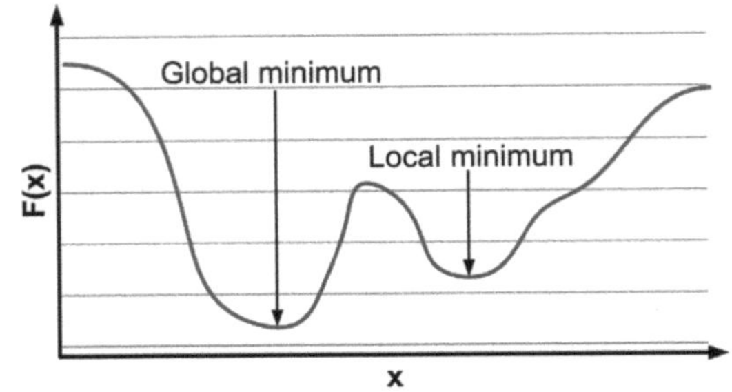

FIGURE 3.21 Example of continuous nonconvex function.

Gradient-based solvers are, in general, slow to converge for functions that have flat plateaus; one of the more general examples is the Rosenbrock function, Equation (3.99)

$$f(x_1,x_2)=(1-x_1)^2+c\left(x_2-x_1^2\right)^2 \tag{3.99}$$

In the Rosenbrock function, c is a constant often set to 100 and also sometimes set to 105. The function is typically used to test optimization algorithms because of its flat plateau toward the minimum point, which means that the gradients in the vicinity are relatively small which in turn leads to small step sizes, and ultimately longer computing time.

So far, the focus of attention has primarily been on convex or concave functions, as illustrated by Figure 3.4. But what happens when the functions are nonconvex (or nonconcave)? In short, the global extremum does become significantly more difficult to distinguish from the local ones. In order to appreciate this, observe Figure 3.21.

In Figure 3.21, the global minimum is once again distinguishable, but the graph also contains a local minimum and a local maximum; such points are often referred to as saddle points. What if the start guess is placed close to one of these saddle points, for example, the local minimum illustrated in Figure 3.21? In this case, the majority of gradient-based methods and line search methods will almost certainly converge toward the local minimum and ultimately terminate at that point. In the case of Figure 3.21, it is clear to see where the global minimum lies and so "steps" can be taken to ensure convergence toward the global minimum. But in most optimization scenarios, the function cannot be fully plotted; the global minimum is unknown – it is after all what is being searched for – and finally Figure 3.21 demonstrates a relatively simple case where only two additional saddle points or extrema are displayed. A nonconvex function does, in principle, add an infinite number of extrema, complicating the mathematical modeling as well as the optimization problem.

3.7 ADDITIONAL MATHEMATICAL OPTIMIZATION METHODS

This section could very easily dwarf the preceding ones, quite simply because of the vastness of algorithms and techniques that exist. The field is in essence infinite, simply because of the many variations and adaptations that exist of various "fundamental" techniques. The convex programming techniques, the gradient-based algorithms, and the line search methods presented in the previous sections were chosen because of their relative simplicity and their high relevance toward many engineering optimization problems. It should be highlighted that even though these methods have been presented the information supplied should only be considered an introduction. The in-depth mathematical derivations have been omitted quite simply because it is of limited interest to most practicing engineers, unless they are developing algorithms. Specific areas that have not been covered in detail in this chapter but are worth exploring further will be briefly covered in the following subsections.

3.7.1 Lagrangian multipliers

The method of Lagrangian multipliers (LM) is also a gradient-based approach, and was founded on geometric observations. The basic idea of the method, which applies to constrained optimization, is to "merge" the objective function $f(x)$ with the equality constraint(s) $c(x)$ as specified in Equation (3.100)

$$c(x) = cst \tag{3.100}$$

Note that at this point the constraint $c(x)$ must be an equality constraint, that is, equal to the constant *cst*. Initially, the two are of course separate, but in LM the two are combined as indicated in Equation (3.101)

$$f(x,\lambda) = f(x) - \lambda \cdot c(x) \tag{3.101}$$

The objective function thus becomes a function of both x and the Lagrangian multiplier λ. Analog to the methods presented in Section 3.6, the gradient is subsequently set to zero, shown in Equation (3.102)

$$\nabla f(x,\lambda) = 0 \tag{3.102}$$

The gradient in (3.102) is, however, more complex, as the second term in (3.101) has been introduced into the objective function; for example, the gradient now contains a partial derivative with respect to the Lagrangian multiplier λ. LM will be extensively used in Chapter 7, and will therefore be further introduced in that chapter.

3.7.2 Karush–Kuhn tucker optimality conditions

The Karush–Kuhn-Tucker (KKT) optimality conditions are a set of first-order conditions which partially ensures that a solution x_{opt} found for the optimization problem $f(x)$ is indeed an optimum solution. The remaining conditions that ensure a truly optimum solution are generally referred to as regularity conditions or constraint qualifications. These generally refer to aspects such as linearity, independence of variables, etc.

The KKT optimality conditions are defined as a set of equations; they are, however, most often employed simply as the KKT multipliers. Their specific application relates to the extension of LM, specifically when inequality constraint(s) of the form defined in Equation (3.103) are introduced.

$$d(x) \leq cst \wedge d(x) \geq cst \tag{3.103}$$

The principle is, in essence, a simple extension of that of the LM in (3.101); however, the inequality constraints are multiplied by the KKT multiplier μ. Using the LM principle of merging the objective function with the constraints the optimization problem of maximizing or minimizing $f(x)$ subject to the equality constraint $c(x) = cst$ and the inequality constraint $d(x) \leq cst$ becomes as listed in (3.104)

$$f(x,\lambda,\mu) = f(x) + \lambda \cdot c(x) + \mu \cdot d(x) \tag{3.104}$$

Equation (3.104) will be solved by setting the gradient $\nabla f(x,\lambda,\mu)$ equal to zero. The principle can straightforwardly be extended to include m equality constraints and n inequality constraints, as indicated in (3.105)

$$f(x,\lambda,\mu) = f(x) + \sum_{i=1}^{m} \lambda_i \cdot c_i(x) + \sum_{i=1}^{n} \mu_i \cdot d_i(x) \tag{3.105}$$

The use of KKT conditions will also be extensively explained in Chapter 7.

3.7.3 Trust regions

In Section 3.6, the use of line search methods and gradient-based methods was considered to determine sf_i and search directions. The general underlying idea of this was that the search direction and step size should be governed by the gradient. The basic principles of trust regions (TR) share a lot of similarities with line search methods; TR is a gradient-based method. The fundamental difference is that a quadratic model is trusted to be correct in the vicinity of the current solution x_i. The algorithm therefore ensures that the next step, that is, x_{i+1}, remains with the trusted region. The quadratic model $q_i(p)$ takes the form of Equation (3.106)

$$q_i(p) = f(x_i) + \nabla f(x_i)^T \cdot p + \frac{1}{2} p^T \cdot sf_i \cdot p \tag{3.106}$$

In (3.106), p is the search step that is updated for every iteration; x_i is the current solution and sf_i is a scale factor. An ellipse is most often used as the geometrical shape of the trust region; this may subsequently be used to increase or reduce the size of the trust region as a function of the relative convergence. The search step p_i is updated once per iteration; in essence, it represents the relative difference between the predicted function value obtained using $q_i(p)$ and the actual functional value $f(x_i)$. p_i is updated as defined in Equation (3.107).

$$p_i = \frac{f(x_i) - f(x_i + p_i)}{q_i(0) - q_i(p_i)} \tag{3.107}$$

p_i is thereby a measure for the relative precision or indirect convergence of the method. The closer p_i is to 1, the higher is the level of correlation between predicted $q_i(p)$ and actual $f(x_i)$ function values, indicating a high level of precision and accuracy of $q_i(p)$ meaning a high level of trust should be put into $q_i(p)$. This in turn means that the trust region can be expanded to increase the step size and thereby to reduce computational time. On the flip side, if p_i is far away from 1, the trust region should be reduced to increase the relative correlation.

TR methods can thus be used to solve the same type of problems as the previous examples explained; in fact, the method can be readily accessed in a multitude of commercially available programs such as MatLab.

3.8 ADDITIONAL ASPECTS OF STRUCTURAL OPTIMIZATION

The purpose of this section is to initially present some additional aspects, considerations, and phenomena that any engineer utilizing structural optimization should keep in mind. The aspects raised are, in general, applicable to any type of optimization; the specific argumentation will, however, be conducted in a mechanical engineering context. The discussions in this section are by no means exhaustive and should only serve as a starting point for considerations to be made in connection with optimization problems; the majority of the areas discussed in this section will be addressed throughout the remainder of this book.

3.8.1 Design space and design volume

One of the most important aspects of engineering and science in general is efficient and specific communication that leaves no room for interpretation. At this point, it is therefore important to present the difference between design space and design volume, two aspects that are very commonly mixed up. The design space is of dimension n, where n corresponds to the number of design variables under consideration in a specific optimization problem. In other words, if each design variable is represented by a vector $\{ds_n\}$, the span of all $\{ds_n\}$ vectors is the design space. If the boundaries of each design variable are now inserted, the remaining space is often referred to as the allowable,

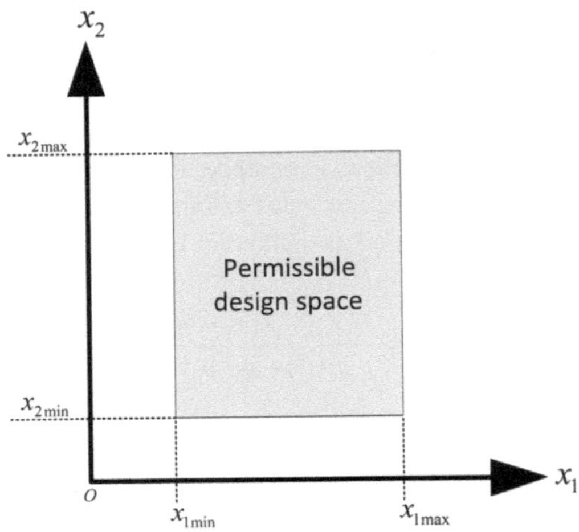

FIGURE 3.22 Design space example.

feasible, or permissible design space. Figure 3.22 illustrates a two-dimensional design space as spanned by the two vectors $\overrightarrow{Ox_1}$ and $\overrightarrow{Ox_2}$ where O is the origin of the coordinate system.

Figure 3.22 also illustrates the permissible design space as bounded by an interval constraint for each of the two variables. Figure 3.22 could, for example, represent an optimization problem relating to the cross-sectional geometry of a square beam where x_1 represents the width and x_2 represents the height of the beam. The space spanned by the design variable vector is also often referred to as the solution space as it also spans the solutions to the optimization problem. For the remainder of this book, the term design space will however generally be used. To complete the graphical interpretation of an optimization problem, it can be realized that the objective function that is a function of the design variables would be perpendicular to the plane illustrated in Figure 3.20 for a 2D problem, and thus a 3D illustration of the specific optimization problem could be created, as illustrated in Figure 3.23.

The term design volume is a space that is maximum three-dimensional, and it refers to a geometrical space most often representing a physical part. Consider, for example, the task of optimizing the conrod of an engine. To consider the full geometry of the conrod, one must utilize a 3D geometrical space. The volume that the boundary edges of the conrod geometry enclose is the design volume, assuming that all the dimensions of the conrod can be changed during the optimization. In this case, the design volume is 3D.

If a cross-sectional plane of the conrod was used for the optimization, the plane used would constitute a 2D geometrical space that would be the design volume, in this case 2D. Finally, if, for example, only the length of the conrod could be changed, the design volume would constitute a line and thus be 1D. The reader will become very familiar with design volumes as they will be utilized throughout the remainder of this book.

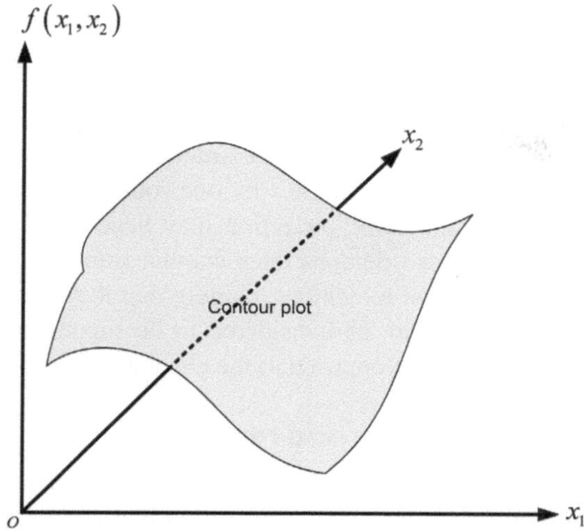

FIGURE 3.23 Design space and objective function contour example.

3.8.2 Uniqueness of solution

Throughout this chapter, it has been assumed that a single optimum point exists, that is, that a unique optimum solution exists. This is, of course, preferable as there can be no "doubt" that the optimum solution has been obtained; in other words, it is "certain" that the optimum solution is unique. However, most practical optimization problems do not possess this highly desirable feature as previously discussed. In fact, a single unique solution to an optimization problem is very seldom in the context of structural optimization. To demonstrate this point, consider the simple optimization task of minimizing the displacement of a single cantilever beam with a square cross section of height h and width w exposed to a single compressional force along its longitudinal direction. Assuming a uniform cross section, as well as linear elastic and isotropic material behavior, the problem may be solved using Equation (3.108)

$$\max: \frac{EA}{L} = \frac{E \cdot b \cdot h}{L} \tag{3.108}$$

Subsequently, two permissible solutions s_1 and s_2 are found as defined in Expression (3.109)

$$s_1 = \begin{Bmatrix} b \\ h \end{Bmatrix} = \begin{Bmatrix} 25 \\ 50 \end{Bmatrix} \quad S_2 = \begin{Bmatrix} b \\ h \end{Bmatrix} = \begin{Bmatrix} 50 \\ 25 \end{Bmatrix} \tag{3.109}$$

The question is which of the two solutions listed in (3.109) is the optimum solution to the optimization problem of (3.108)? The simple answer is both! In the context of this example problem, there is no reason why one solution would be preferred over another, as they will both provide the same functional value for the objective function. There may, of course, be "practical" preferences, for example, relating to manufacturing or even aesthetics; however, if these are not clearly defined and included in the optimization problem, definition no unique solution can be found. This is a key topic of structural optimization; very often it is not possible to "objectively" decide why one solution should be preferred over another. Hence, a "subjective" selection may have to be made. Furthermore, the above example also demonstrates another interesting dilemma that any engineer or designer must remember; namely that the results of structural optimization should, in general, be considered to be suggestions as to how a structure may be engineered, as opposed to the absolute answer.

3.8.3 Design sensitivity and robustness

The next general topic to consider is the sensitivity of a structural design. This is particularly important for vehicle optimization and even more so when it relates to safety critical structures. The fundamental question to be considered in this context is, can a structure be "over-optimized?" The short answer is yes. Consider the cantilever beam exposed to a single compressional load F, as illustrated in Figure 3.24.

Given the earlier scenario, the beam could be highly optimized, for example, with an aim of minimizing the mass and be able to fully withstand the load F exerted onto the beam. But how "certain" is it that the beam is exposed to a load that is exactly aligned with its longitudinal direction? What happens if the load is not actually applied in the direction of F, but more along the direction of F_1? If not carefully considered, the consequences could be very significant. The above issue does, of course, not only relate to structural optimization but also structural analysis, engineering, and even physics in general. There are countless parameters that may influence the structural performance of any given system; for example, impurities in the material leading to a reduced Young's modulus (or even an increase); manufacturing defects; installation errors; etc. From a practical engineering viewpoint, these types of errors or uncertainties are often catered for through the use of safety factors, for example, an increase in material usage, conservative estimations of mechanical material properties, exaggerated load magnitudes, etc. The use of safety standards, most often dictated through professional standards or legislation, is of course necessary, as there are inherent uncertainties about most of these parameters, which must be taken into account from the very start of the optimization process. If these aspects are overlooked

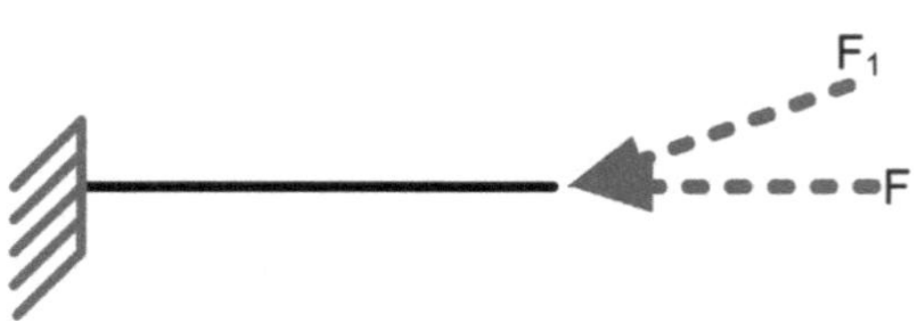

FIGURE 3.24 Cantilever beam example.

or ignored the benefits of conducting the optimization in the first place may be completely negated in the end result. Would you wash and wax your old car the day before you sell it as scrap metal?

There are specialized fields of optimization which attempt to deal with uncertainty or even randomness in the input variables; these are referred to as robust optimization, stochastic optimization, and design sensitivity analysis. In general, optimization that does not take the random variations of variables into consideration is referred to as deterministic optimization.

Even if the "practical" uncertainties discussed above were to disappear, there are still numeric uncertainties associated with the FE modeling discussed in Chapter 2 to consider. As most structural optimization is based on FEA, these uncertainties may be significant in terms of the optimization results, as the FEA provide the "information" to the structural optimization as will be subsequently discussed.

The remaining sections of this chapter will focus on general types of optimization.

3.8.4 Multiobjective optimization

The basic principles of multiobjective optimization are relatively easy to understand. As the name suggests this relates to scenarios where it is desirable to optimize more than one characteristic of a system; in other words, there will be more than one objective function, that is, multiple objectives. Consider, for example, the cantilever beam exposed to a compressional load in Section 3.8.2 where the objective was to minimize the displacement. At the same time it may be desirable to reduce the mass, in which case the expression for the displacement and the expression for the mass would not be objective functions. Note that both are functions of the design variables: the height h and the width w of the beam. Unfortunately, the two objectives are not ideal partners as a decrease in mass is likely to lead to an increase in compliance, that is, larger displacements. One of the preferred options for solving this type of problem is simply to "relegate" one of the objective functions to being a constraint. For example, if the objective remains to minimize the displacement, an inequality constraint specifying the mass to be less than a certain value could be introduced. Needless to say, multiobjective optimization problems are generally significantly more difficult to resolve when compared with their single-objective "counterparts." There are, however, a range of techniques, principles, and methods available for multiobjective optimization; the perhaps most simple being the weighted sum method where all objectives are assigned a weighting by the user. For example, if preference is given to minimizing displacement over minimizing mass the objective function for the displacement may be multiplied by 0.66, whereas the objective function for the mass is multiplied by 0.33.

Caution should be taken not to confuse multiobjective optimization with multicriteria optimization where the algorithm chooses from a series of alternatives, not numerical variables as in multiobjective optimization.

3.8.5 Multidisciplinary optimization and multiphysics optimization

Most engineering problems, particularly in the context of automotive engineering, involve more than one type of analysis. For example, to reduce the mass of an exhaust system, the structural performance of the system must be considered, but also the flow of gas through the system. As the exhaust gas passes through the system, the gas interacts with the boundaries, that is, it resonates of the surface forcing the walls of the exhaust system to vibrate which in turn changes the characteristics of the flow. How can a problem such as the above be solved? The most commonly used approach so far is to consider the problems separately and then "bring the information" together, resolving any conflicting issues at the later stages of the design. This is, however, not an efficient way of analyzing such problems, as there can be many unforeseen problems at the later stages of the design, which in many cases does not become obvious until a prototype is built. One of the potential answers to this is to use multidisciplinary of multiphysics simulation, including optimization. Multidisciplinary and multiphysics optimization are, however, often intertwined. Multiphysics analysis and optimization refers to using a single (numerical solver) to solve the problem; this will, however, use different physics to do so. For example, the fluid structure interaction (FSI) problem of the exhaust system may be solved using only FEA, as FEA is capable of modeling both the fluid flow and the structural response, hence the term multiphysics. On the other hand, FEA is not generally considered to be the best solver for fluid problems; computational fluid dynamics (CFD) solvers are most often based on the finite volume (FV) method. Hence, it may be desirable to use a certain solver for the CFD analysis, for example a commercial CFD package, and another solver, for example a commercial FE package, for the structural mechanics. Coupling the data from the two separate analysis packages and thereafter performing the optimization is referred to as multidisciplinary optimization. In general, the difference between multiphysics and multidisciplinary analysis and optimization, therefore, lies in the way the data for the optimization is obtained.

REFERENCES

Griffith, R. E., & Stewart, R. A. (1961). A nonlinear programming technique for the optimization of continuous processing systems. *Management Science*, *7*, 379.

Ravindran, A., Ragsdell, K. M., & Reklaitis, G. V. (2006). *Engineering optimization: methods and applications* (2nd ed.). New York: Wiley, ISBN: 978-0-471-55814-9.

Chapter | four

Introduction to Structural Optimization and Its Potential for Development of Vehicle Safety Structures

CHAPTER OUTLINE

Throughout the preceding chapters, aspects of vehicle design including architectures, structural and safety requirements, fundamental FE theory, and basic optimization principles have been discussed. In these chapters, a number of remarks have indicated the links between all of these aspects with limited detailed and specific information about how this works in the "real world." This chapter is the first step toward providing this information, as it primarily closes the "theoretical gap," whereas Chapter 8 (Applications of Linear Optimization to Concept Vehicle Safety Structures) will close the gap by means of structural optimization examples and case studies. The first step in this process is to understand the basic interaction between the FEA presented in Chapter 2 and the mathematical optimization techniques presented in Chapter 3.

In general, FE-based optimization is not a "standalone tool," as FEA only supplies the information required for the optimization, which in turn can be, but does not necessarily have to be, integrated within the FE solver. In order to better understand this, consider Figure 4.1.

Nonlinear Optimization of Vehicle Safety Structures. http://dx.doi.org/10.1016/B978-0-12-804424-7.00004-X

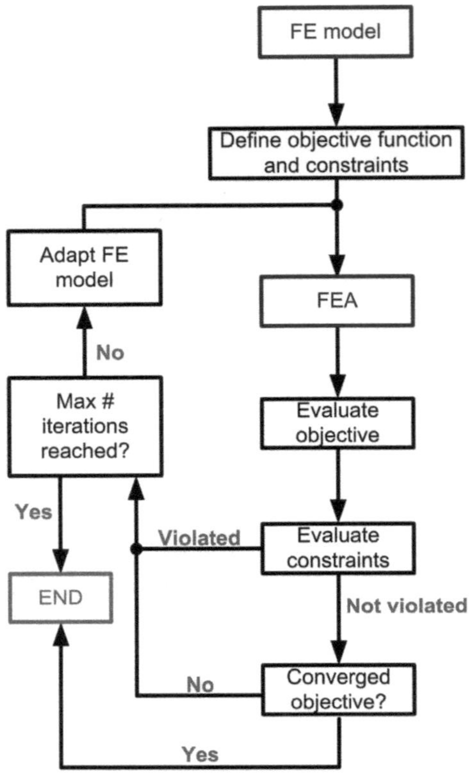

FIGURE 4.1 Basic FE-based optimization flowchart.

Figure 4.1 illustrates the overall steps of a general and fundamental FE-based optimization algorithm. As indicated in Figure 4.1, FEA merely provides the "information" required for the optimization. In Sections 3.1 and 3.2, the term structural optimization was defined and the fundamental principles of optimization were presented. The following sections will build on these definitions, starting with an overall presentation of the general types of structural optimization. The initial question is: what types of structural optimization exist?

The short answer is relatively simple as structural optimization may, in general, be divided into three major categories:

1. Topology Optimization
2. Shape Optimization
3. Size Optimization

Due to space limitations only the intricate details of topology and shape optimization will be presented in this chapter; size optimization will, however, be used in examples later on in this book. Before the different types of optimization are presented, it is beneficial for the reader to understand the general difference between these three types of structural optimization, which is perhaps best done by introducing an example. Consider the task of creating a 2D structure capable

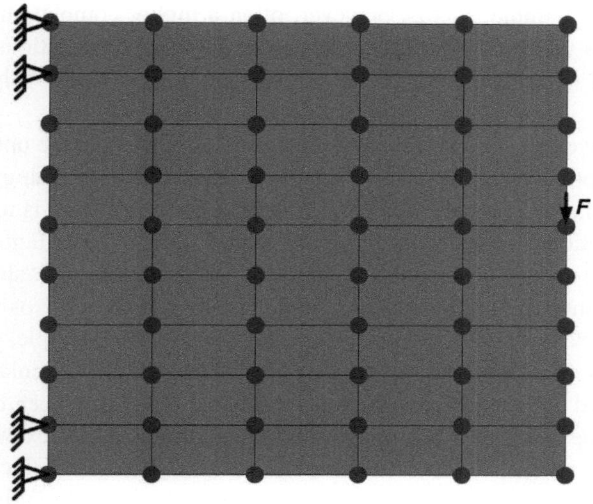

FIGURE 4.2 Example problem.

of withstanding a single-point load given certain constraints as illustrated in Figure 4.2.

For the purpose of solving the task, optimization will be employed to find the lightest structure, while meeting prespecified maximum displacement, as well as maximum stress constraints.

In order to solve the task, the optimum load paths are firstly to be defined. It is clear that the "box" structure illustrated in Figure 4.2 is not the ideal "shape" (or topology) for the prescribed boundary conditions. A topology optimization is conducted to determine the optimum load paths, that is, the topology of the structure, the results of which are illustrated in Figure 4.3.

The left-hand side of Figure 4.3 represents the actual outcome of the topology optimization; the right-hand side represents the "simplified" structure (using 1D elements). Note that the interpretation of topology optimization results and creating the associated optimum mechanical design may seem relatively trivial

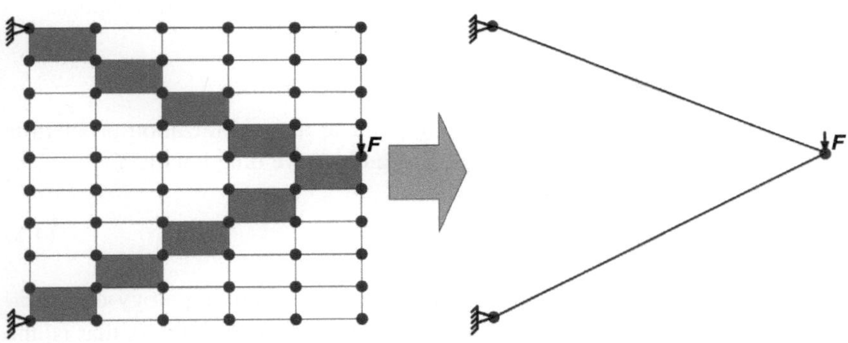

FIGURE 4.3 Topology optimization result.

in the above example; this is, however, often a highly complex and cumbersome task, as will be further demonstrated in the following sections as well as in Chapter 8 (Applications of Linear Optimization to Concept Vehicle Safety Structures).

Topology optimization can thereby be used to determine the optimum load paths, i.e. topology. On the basis of the interpretation of the topology optimization results, i.e. the right-hand side of Figure 4.3, the next task is to determine the cross sectional geometry of the two "beams" illustrated in Figure 4.3. The first step is to determine the cross sectional geometry in terms of "shape" as can be done by means of shape optimization, that is, should the cross-section be circular, a rectangular box section, an I-beam, etc. For this example, assume that the optimum cross-sectional geometry of both "beams" is a circular tube. The optimum load paths and cross-sectional geometry have now been determined. The remaining task is to determine the cross-sectional sizes; for example, the inner and outer radii of the circular tube, and indirectly the gauge thickness. This can be done by utilizing size optimization, thereby obtaining the fully optimized structure.

The previous example provides a fundamental understanding of the subtle differences between topology, shape, and size optimization, although it must be underlined that shape optimization is far wider than "merely" determining the cross-sectional geometry of a beam; indeed, the same could be said for shape optimization that also is much more applicable than simply determining radii of a tube. Nevertheless, these are classical applications encompassing a general definition of the three types of structural optimization. It should be noted that the sequence in which topology, shape, and size optimization were applied in the previous example is not set in stone; neither is it a general rule that all three must be applied to the same problem. It should also be noted that some of the methods may be applied to resolve the same task; topology optimization may, for example, also be used to determine the optimum cross-section of a beam.

The application of any of these structural optimization categories will be continuously discussed both in terms of theoretical applications but also using case studies and "real world" engineering examples throughout this book. The following sections will, however, primarily focus on the theoretical aspects.

4.1 TOPOLOGY OPTIMIZATION

As introduced in Chapter 3, the starting point of any optimization problem is most often to mathematically define a single-objective function (4.1)

$$f(x) \tag{4.1}$$

In (4.1), x can represent any quantity, for linear static topology optimization x often represents Young's modulus, E, or the mass density, ρ, that is, the objective function is often as defined in (4.2). Please note that the objective function $f(x)$ can, in general, contain more than one design variable, that is, x_1, x_2, etc. For the purpose of the current discussion, only one design variable will be included.

$$f(x) \Rightarrow f(\rho) \tag{4.2}$$

In Expression (4.2), the mass density is used as an example of a design variable; this is because many topology optimization algorithms utilize a material distribution concept (MDC), that is, the material distribution is the main parameter (variable) during the optimization. There are two main variations of MDC; these are:

- Variable density method (VDM)
- Homogenization-based optimization (HBO)

Broadly speaking, the difference between VDM and HBO is that the latter utilizes the material microstructure during the optimization, whereas the former does not. The VDM approach is generally more frequently used than HBO, and is available in many commercial FE optimization software packages, such as Altair HyperWorks.

The VDM method will be discussed in the following sections, whereas the HBO approach will be discussed from Section 4.1.5 onwards.

4.1.1 The variable density method

For topology optimization, the function f in Equation (4.2) often represents stiffness or compliance (strain energy). In these cases, it may seem odd to define, for example, the strain energy as a function of the mass density, as there is no direct link between these two with respect to mechanics of materials. Nevertheless, ρ is used as the design variable in many applications of topology optimization. The reason for this is essentially related to human intuition, as most engineers find it "easier" to interpret a plot of density, as opposed to a plot of stiffness. Initially, the relationship between density and (material) stiffness is defined through the VDM, as defined in Equation (4.3)

$$\begin{gathered} C_{ijkl}(y) = \rho(y) \cdot C^0_{ijkl} \\ \text{where} \\ \rho(y) \in \lfloor 0,1 \rfloor \end{gathered} \tag{4.3}$$

In Equation (4.3), $C_{ijkl}(y)$ represents the material properties of the individual element y. For a linear elastic isotropic material, C_{ijkl} only contains two independent variables: Young's modulus, E, and Poisson's ratio, ν. For the purpose of conducting linear static (isotropic) topology optimization, Poisson's ratio generally assumes a constant value, as it is sufficient to vary E. C^0_{ijkl} represents the "original" (and constant) material properties.

Finally, the relative element/material "density," $\rho(y)$, represents the design function, that is, the function that determines the scaling factor attributed to the individual element y. For the discrete formulation, $\rho(y)$ can therefore only assume the values 0 and 1.

Thus, Equation (4.3) states that a linear relationship exists between ρ and the constitutive tensor C_{ijkl}.

Due to the linear relationship between ρ and C_{ijkl} (E), the aforementioned respective plots of "stiffness" and "density" will be identical, provided that the 0–1 discretization is used.

Equation (4.3) is only valid in the linear elastic region of the Prandtl diagram (stress–strain curve), as Hooke's law is only valid in that region. This limitation is ignored for the time being, as the initial purpose is to discuss the governing optimization principles. It should, however, be noted that in general C_{ijkl} is applicable to linear isotropic and linear nonisotropic materials. In general, the constitutive tensor is not used to describe nonlinear material behavior; instead, this is often mathematically described by a strain energy function.

As discussed in Chapter 3, one or several constraint functions are often defined, Expression (4.4), in addition to the objective function of (4.2)

$$g_m(\rho) = \lim_m \wedge g_n(\rho) \le \lim_n \tag{4.4}$$

In Expression (4.4), g_m represents the equality constraint(s) with the limiting value(s) $\lim_m$, and g_n represents the inequality constraint(s) with the limiting value(s), $\lim_n$. As presented in Chapter 3, the total number n_c of constraints in (4.1) is in principle infinite. In most structural optimization applications, n_c is however most often a finite number, typically in the range of 1 to 20, and almost always relates to practical considerations such as displacement values, stress levels, or a maximum mass value of the component.

If n_c is set to zero, then no constraints will exist and the topology optimization will most likely remove all material from the model, leaving a "blank" design volume, which obviously makes no sense with respect to the "practical" overall aim of, for example, minimizing the mass of a given component subjected to external loading; hence, unconstrained optimization is most often of little or no interest with respect to topology optimization and structural optimization in general. On the other hand, if n_c is very large, the design space will so significantly reduce meaning that an "optimum" solution could be overlooked, or even no viable solution can be found due to the optimization problem being over-constrained.

With the objective function $f(\rho)$ and the constraints $g_m(\rho)$ and $g_n(\rho)$ defined, the optimization task can, for example, be defined as Expression (4.5)

$$\begin{gathered} \text{minimize } f(\rho) \\ \text{subject to} \\ g_1(\rho) \le \lim_1; \quad g_2(\rho) = \lim_2; \quad g_3(\rho) > \lim_3 \end{gathered} \tag{4.5}$$

If the optimization task is to maximize the stiffness of a structure, then the objective function in Expression (4.5) may, for example, represent the compliance (strain energy) that should then be minimized to maximize the stiffness.

For a linear elastic body not subjected to initial strains nor stresses, the linear strain energy, U, can be defined as Equation (4.6)

$$U = \frac{1}{2}\int \left(\{\sigma\}^T \cdot \{\varepsilon\}\right) dV = \frac{1}{2}\int \left(\{\varepsilon\}^T \cdot [E] \cdot \{\varepsilon\}\right) dV \tag{4.6}$$

In Equation (4.6), dV indicates that the function must be integrated over the volume of the body (component) to obtain the combined strain energy. Assuming linear and isotropic material behavior, the VDM principle of Equation (4.3) may be combined with Equation (4.6) to further define the objective function in Equation (4.7)

$$f(\rho) = U(\rho) = \frac{1}{2}\int \left(\{\varepsilon\}^T \cdot \left[E(\rho)\right] \cdot \{\varepsilon\}\right) dV \tag{4.7}$$

As previously explained under the assumption of linear and isotropic material behavior, C_{ijkl} may simply be represented by Young's modulus E.

It is important to note that Equation (4.7) is a linear function; if all of the constraint functions are also linear, the optimization problem can be described as linear. This is, in principle, independent of whether linear or nonlinear FE is utilized. Provided that all functions in (4.5) are linear, the optimization problem will per definition be linear. If any of the functions in (4.5) are nonlinear, then the optimization problem is inherently nonlinear, as previously presented in Chapter 3.

This means that there is a fundamental difference between nonlinear FE-based topology optimization and nonlinear optimization in general. The effects of this relate to how the specific optimization problem is formulated.

With Equations (4.5) and (4.7), the optimization problem becomes a "simple" mathematical problem of minimizing a function subject to given constraints. As discussed in Section 3.4, it is not feasible to calculate all possible permutations, simply because the number of these rapidly becomes infinite as the number of elements increase. By virtue of Equation (4.5), the problem can, however, seemingly be solved by means of differentiation, that is, using gradient and line search methods to solve Equation (4.8)

$$\nabla f(\rho) = 0 \tag{4.8}$$

Thereby, the optimization problem of (4.5) can theoretically be solved numerically by, for example, utilizing a simple Newton–Raphson algorithm. Unfortunately, Equation (4.8) does not immediately solve the overall problem at hand due to one particular assumption with respect to the gradient-based search methods presented in Chapter 3, namely that the function must be continuous.

It is immediately obvious that the objective function $f(\rho)$ is not a continuous function, as Equation (4.3) stipulates that the "scale factors," that is, ρ for the individual element, can only assume the value of 0 or 1. Consequently, $f(\rho)$ is merely defined as $f(0)$ and $f(1)$, making the objective function discrete, not continuous. Therefore, the gradient $\nabla f(\rho)$ only exist as defined in Expression (4.9)

$$\nabla f(\rho) \Rightarrow \nabla f(0) \wedge \nabla f(1) \tag{4.9}$$

As the objective function is not continuous, it is not possible to utilize the gradient-based approaches presented in Chapter 3 for obvious reasons; it is, however, highly desirable to be able to do so. This may be enabled by allowing $\rho(y)$ in Equation (4.3) to assume continuous values between 0 and 1; this is in mathematical terms known as "relaxation."

The introduction of relaxation may initially seem to be a disadvantage as the number of possible solutions thus becomes infinitely larger. However, as discussed in Section 3.4, the possible number of solutions to the discrete (initial) optimization problem rapidly becomes infinite anyway.

Thus, from a mathematical viewpoint, it can be argued that the increase in the number of possible solutions is of less importance when compared to the advantages of enabling the use of gradient-based solution methodologies.

It was previously mentioned that extra care must be taken with respect to "practical" aspects of the optimized design. An obvious example of this is the introduction of the continuous $\rho(y)$ function, which undoubtedly will lead to solutions suggesting element densities of, for example, 0.5. To better comprehend this issue, consider the FE model illustrated in Figure 4.4.

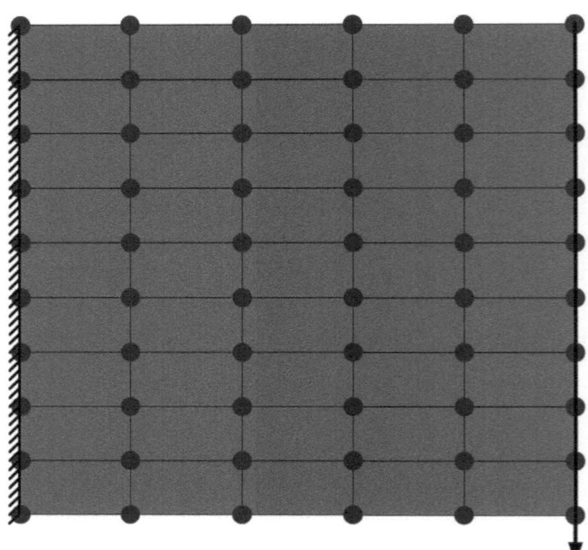

FIGURE 4.4 Example design volume for topology optimization.

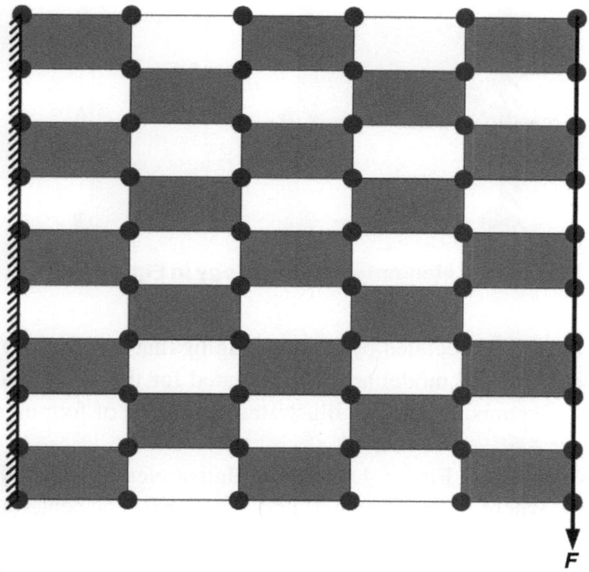

FIGURE 4.5 Potential solution to topology optimization problem.

The FE model illustrated in Figure 4.4 contains 45 elements; it is constrained at all nodes on the left-hand side and subjected to forces on all nodes on the right-hand side of Figure 4.4. All grey elements indicate where $\rho(y) = 1$, which initially is all of the elements.

Suppose the objective of the impending topology optimization is to minimize the mass of the model, whilst the constraints consist of a maximum displacement magnitude of the nodes where the load is applied, in addition to a maximum stress value applied to all elements in the model.

A possible solution to this topology optimization problem could be the one illustrated in Figure 4.5.

The white elements in Figure 4.5 denote elements where $\rho(y) = 0$.

The topology illustrated in Figure 4.5 is thus a mathematically viable solution; however, there are obviously some severe limitations from an engineering point of view.

Firstly, the structure in Figure 4.5 would be very difficult, if not impossible to manufacture. Furthermore, the "real world" version of the topology in Figure 4.5 would be inherently unstable. This is not necessarily the case of the FE model simply because of the FE discretization itself. In this context, all forces, displacements, etc., are transferred via the nodes. Hence, the compliance of the topology in Figure 4.5 is almost as low as that of Figure 4.4; however, the "mass" (i.e. number of gray elements) in Figure 4.4 is 45, whereas it is only 23 for Figure 4.5.

The problem illustrated by Figure 4.5 is a very common problem of topology optimization, and is often referred to as "checkerboard effect."

The checkerboard effect does, in principle, arise from the introduction of a continuous $\rho(y)$ function.

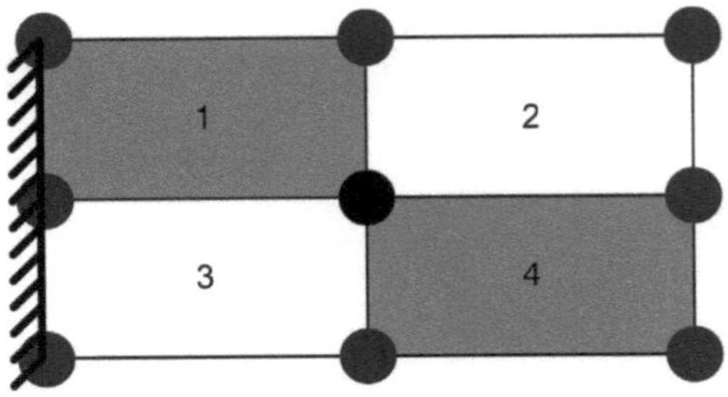

FIGURE 4.6 Close-up of elements from topology in Figure 4.4.

This fact can be appreciated by acknowledging that the displacements, stiffness values, etc. in a FE model are all calculated for the individual nodes, not the individual elements. Figure 4.6 illustrates a close-up of four elements of the model in Figure 4.5.

Elements 1 and 4 in Figure 4.6 have a relative element density value of 1, that is, $\rho(y) = 1$, whereas for elements 2 and 3, $\rho(y) = 0$. The center node belongs to all four elements; therefore, the $\rho(y)$ value in this node will be equal to the average from the four adjoining elements, that is, $\rho(y) = \frac{1+0+0+1}{4} = 0.5$.

Another example of the complications of introducing a continuous $\rho(y)$ function can be illustrated by using the original FE model illustration in Figure 4.4, but imagine that all elements have a slightly different shade of grey to indicate that for all elements in the model $\rho(y) = 0.5$. As this solution still contains a total of 45 elements, the "mass" of this solution is 22.5, that is, similar to that of Figure 4.5.

However, the feasibility of this alternative solution is equally "unfortunate" to that of Figure 4.5, as the suggestion in this solution is simply to replace the material with one that has a mass density, ρ, and a Young's' modulus, E, scaled by 0.5 to that of the "original" material in the model. This obviously defeats the purpose of topology optimization.

Thereby it can be concluded that using VDM for topology optimization with continuous scale factors is not always likely to return the "optimum" solution from an engineering viewpoint.

The introduction of a continuous $\rho(y)$ function may be "convenient" from a mathematical viewpoint; however, it conveys additional "practicality" problems such as the checkerboard effect.

The checkerboard effect can be relieved by using higher order elements, that is, elements with, for example, parabolic shape functions; this does, however, imply a substantial increase in CPU time, as discussed in Chapter 2. This renders this particular solution to the checkerboard effect infeasible, as the increase in CPU time is simply unacceptable for larger models. It therefore becomes necessary to introduce another method of "steering" the optimization towards a 0 to 1 solution whilst using the VDM approach, as shall be discussed in the next section.

It should be noted that other methods, such as the usage of continuous material distribution, have recently been proposed, suggesting that the mesh dependency and checkerboard effects can be completely avoided. This is in theoretically one of the main advantages of HBO optimization which will be discussed in Section 4.1.5.

4.1.2 The SIMP interpolation scheme

The gradient based algorithms of Section 3.6 can in principle be applied in principle to any optimization problem, as long as the objective function can be mathematically defined. The objective function for the topology optimization problem using the VDM method was defined in Equation (4.7) and relaxation was introduced in order to ensure continuity of the function, thereby enabling the use of the gradient-based methods from Section 3.6. In the case of FE based topology optimization, the approach is, however, not as straightforward as that. This is because Equation (4.7) contains the strain vector $\{\varepsilon\}$ which is found via FEA, that is, $\{\varepsilon(E)\}$. This means that an FE analysis must be conducted for each iteration of the optimization, somewhat "deviating" from the mathematical gradient-based approach to optimization, unless the optimization algorithm is directly embedded within the FE solver.

The focus will now return to the problems exposed in Section 4.1.1, such as the checkerboard effect that is a consequence of the relaxation, that is, the introduction of the continuous $\rho(y)$ function on the 0–1 interval. The solid isotropic material with penalization (SIMP) method is an interpolation scheme aimed at "guiding" the topology optimization toward the 0 to 1 solution by "penalizing" intermediate values of the design variable(s); ρ in Expression (4.5). Thereby, the number of unfeasible designs created by the checkerboard effect can be reduced. This is done by defining a mathematical "power law" linking ρ to E, an extension of the initial link defined in Expression (4.3). The SIMP interpolation scheme utilizes the "power law for representation of elasticity properties," as defined in Expression (4.10)

$$
\begin{gathered}
C_{ijkl}(y) = \rho(y)^p \cdot C^0_{ijkl} \\
\text{where} \\
p \geq 1 \wedge \rho(y) \in |0:1|
\end{gathered}
\tag{4.10}
$$

The optimization algorithms that utilize Expression (4.10) are generally referred to as VDM-SIMP; this is how they will be referred to throughout the remainder of this book. At first glance, Expression (4.10) is very similar to Expression (4.3). There are, however, subtle differences that hold significant influence.

Firstly, it should be noted that $\rho(y)$, that is, the design function, can now assume all values in the continuous interval between 0 and 1.

Secondly, Expression (4.10) contains the power factor p that theoretically can assume all values generally above or equal to 1.0. If p is indeed set to 1.0,

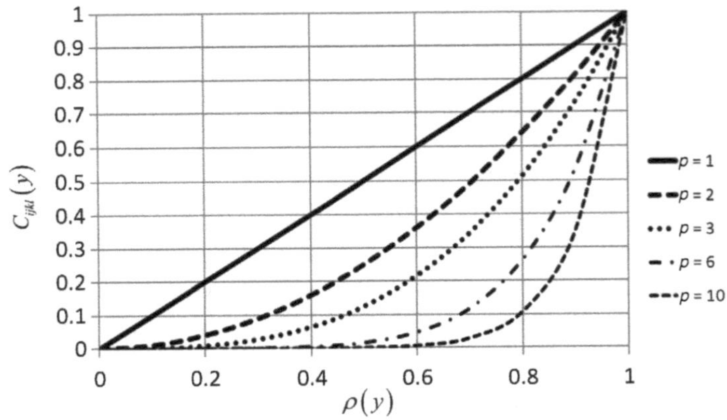

FIGURE 4.7 Graphical illustration of the influence of the penalization factor, *p*.

Expression (4.10) does essentially become equal to Expression (4.3), overlooking the fact that the former is defined for a continuous $\rho(y)$ interval, whereas the latter is not.

Figure 4.7 is a graphical illustration of the influence of the penalization factor, p, of the SIMP interpolation scheme, as defined in Expression (4.10).

From Figure 4.7, it can be seen that if p equals 1, there is no penalization for intermediate $\rho(y)$ values, as the relationship between ρ and C_{ijkl} is linear. As p increases, the penalization for intermediate densities becomes increasingly severe, thus "forcing" the optimization toward a 0 to 1 solution as it becomes "inefficient" to retain intermediate values.

For high p values, for example, 6 or 10, it is obvious that any element with a $\rho(y)$ value less than approximately 0.6 essentially has zero stiffness, which is in line with the overall purpose of the method. However, Figure 4.7 also displays a very steep slope between different $\rho(y)$ values in the interval 0.8 to 1 for the higher penalization factors. This is very likely to lead to severe instability problems, primarily from an optimization view point, but also potentially from an FEA view point. The underlying issue is that a relatively minor variation of $\rho(y)$ can lead to a significantly large change in stiffness, thus making the model very (if not extremely) sensitive to minor variations of $\rho(y)$.

A commonly utilized way of demonstrating the influence of the penalization factor, p, upon topology optimization results is by using a simple load setup such as the one illustrated in Figure 4.8.

For the following example, the model illustrated in Figure 4.8 is discretized using 5000 quad 2D shell elements with linear shape functions. The load, F, is applied at a single node and the model is rigidly clamped partially along the left-hand side, as illustrated in Figure 4.8. Utilizing the VDM-SIMP approach, a topology optimization problem is formulated with an objective of minimizing the mass subject to a single inequality constraint of a specified maximum displacement value at the point where the external load F is applied.

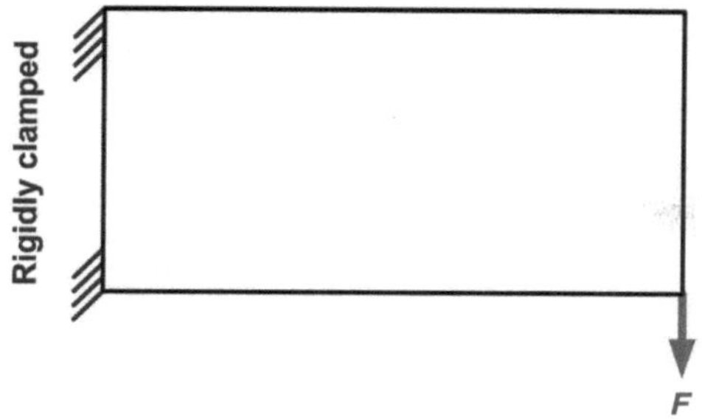

FIGURE 4.8 Example FE model.

Figure 4.9 shows the effects of varying the penalization factor, p, in Equation (4.10), using the integer values in the intervals 1 through 6.

The results illustrated in Figure 4.9 indicate that the value of the penalization factor may hold an influence upon the topology optimization results; it is evident that "higher" values of p may lead to a significant change of the resulting topology. Figure 4.9 may also be used to appreciate the "practical" effects of the penalization factor, as the topology optimization is "forced" toward a 0 to 1 solution. This can be substantiated by noting that for $p = 6$ there are many elements where $\rho = 1$, as indicated by the light grey elements, whereas $p = 1$, contains significantly more elements with darker shades of grey, indicating intermediate $\rho(y)$ values.

Figure 4.9 thereby demonstrates the influence of the penalization factor on quad elements, but what happens if tetra elements are used? Figure 4.10 may be used to answer that question, as it represents results from topology optimizations of the same basic problem as Figure 4.9 does. The only difference between

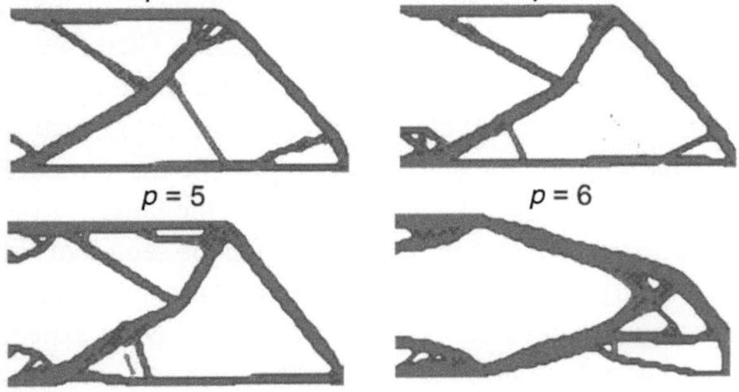

FIGURE 4.9 Topology optimization results, varying penalization factor, p - 5000 quad elements.

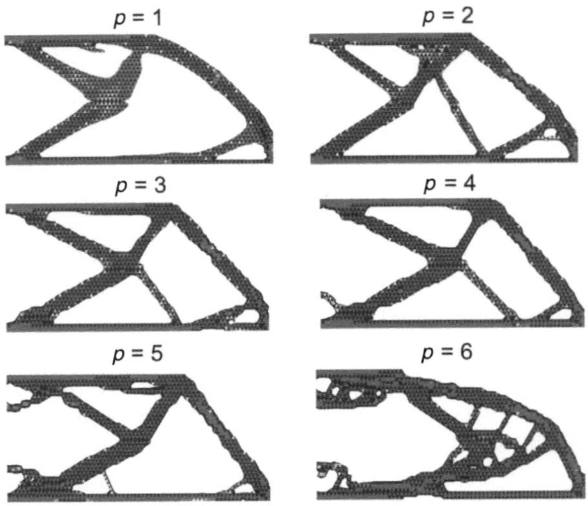

FIGURE 4.10 Topology optimization results, varying penalization factor, p - 4450 triangular elements.

the models represented in Figure 4.10 and the ones in Figure 4.9 is that the former represents models meshed with 4450 triangular (2D) elements with linear shape functions, that is, the results of Figures 4.10 and 4.9 represent models with approximately the same number of elements.

By comparing the results of Figure 4.10 to those of Figure 4.9, it can be seen that the choice of element type also has an effect upon the results; however, this is not likely to be directly linked to the SIMP interpolation scheme, but rather an effect of general FE discretization and modeling techniques.

Furthermore, it can be seen that the majority of load paths in Figure 4.9 are also present in Figure 4.10; the most significant differences occur in the $p = 6$ plots, which can be attributed to the previously discussed "instability" issues that can be appreciated by viewing the trends of the graphs in Figure 4.7.

It could be argued that the above comparison is biased, as the element size in Figure 4.10 is a factor of 1.5 times larger than that of Figure 4.9. In addition, the remeshing also implies that the node locations of the respective models have changed, possibly "skewing" the resulting topologies. Therefore, a third series of models is solved. These are based upon the same mesh used for the results in Figure 4.9; however, the quad elements have been split, resulting in 10,000 triangular elements with identical nodal locations for these models, that is, in Figures 4.8 and 4.10, excluding the added nodes in the latter.

By comparing Figure 4.11 to Figure 4.10, it can be seen that similarities between the respective topologies exist. However, significant variations also exist, thereby highlighting the mesh dependency issue, as may be expected since the input to the topology optimization depends upon FEA.

Furthermore, if Figure 4.9 is compared to Figure 4.11 it can also be seen that the respective topologies are similar in general. Further examination re-

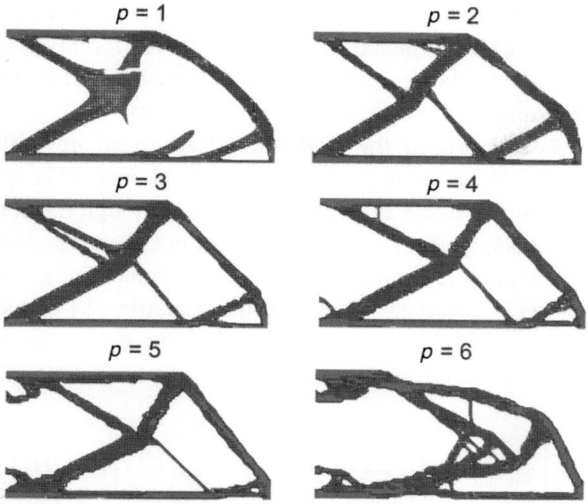

FIGURE 4.11 Topology optimization results, varying penalization factor, p - 10,000 triangular elements.

veals differences between the respective results, underlining the statement that the outcome of a topology optimization merely serves as a suggestion as to the location of the loadpaths, as it is highly unlikely that a single unique optimum exists, as discussed throughout Chapter 3.

The $p = 6$ results of Figures 4.9, 4.10, and 4.11 are all significantly different, thus substantiating the instability issue relating to "high magnitude" p values in connection with the SIMP interpolation scheme, as illustrated in Figure 4.7.

The above findings thus indicate that the selection of the p value is not always straightforward, as no single definite value can be determined in general. Guidelines such as the ones proposed by Bendsoe & Sigmund (2003) do however exist, as seen in Expression (4.11)

$$\begin{aligned} &\text{2D: } p \geq \max\left(\frac{2}{1-\nu^0}; \frac{4}{1+\nu^0}\right) \\ &\text{3D: } p \geq \max\left(15\left(\frac{1-\nu^0}{7-5\nu^0}\right); \frac{3}{2}\left(\frac{1-\nu^0}{1-2\nu^0}\right)\right) \end{aligned} \tag{4.11}$$

In Expression (4.11), ν^0 represents Poisson's ratio of the linear elastic isotropic material in question. According to (4.11), the minimum (integer) value of the penalization factor p should be equal to or larger than 2 for shell (2D) and 3 for solid (3D) elements, given ν^0 is equal to 0.3.

It is, therefore, seemingly beneficial to choose a relatively high value of p to further penalize the intermediate $\rho(y)$ values enforcing the 0 to 1 solution, thereby ensuring a more refined topology. As previously discussed, there are, however, also significant issues associated with choosing a large p value, as indicated in Figure 4.7.

With the general principle of the SIMP interpolation scheme presented, the focus can now turn to the implementation of VDM-SIMP into FE analysis.

The VDM-SIMP method is used in many state-of-the-art FE codes and software suites, an example of which is Altair HyperWorks.

The general strengths of the VDM-SIMP approach lie within the efficiency, versatility, and primarily stability of the scheme. This is particularly exploited in relation to linear FE-based analysis, although the scheme is also used for nonlinear analysis, as will be presented in Chapter 9. As presented in Chapter 2, the most efficient (and stable) way to conduct linear static FE analysis, in general, is to invert the global stiffness matrix $[K]$ and subsequently employ matrix and vector multiplication to solve the overall governing equation, as demonstrated in Expression (4.12)

$$\{F\}=[K]\cdot\{U\} \;\mapsto\; [K]^{-1} \;\mapsto\; [K]^{-1}\cdot\{F\}=\{U\} \tag{4.12}$$

The stiffness matrix $[K]$ contains data from the constitutive tensor C_{ijkl}, and is hence a function of material properties (as well as geometry).

The resulting global displacement vector $\{U\}$ in (4.12) is rapidly obtained once $[K]^{-1}$ has been calculated. The most CPU-intensive step of the process in (4.12) is to invert the stiffness matrix $[K]$ that theoretically needs to be repeated if the geometry or material of the model changes. Consequently, the stiffness matrix should, in principle, be reassembled and reinverted after each individual iteration of the topology optimization.

However, the VDM-SIMP approach is often implemented in a way that eliminates the need to repeat these steps, which may be done by exploiting the isotropic and, indeed, linear nature of the material model.

Due to this material behavior, combined with the previous statement that Poisson's ratio, ν, remains constant throughout the optimization, the inverted stiffness matrix $[\mathrm{K}]^{-1}$ can be defined as Equation (4.13)

$$[K]^{-1}=\left[E_y\right]\cdot\left[K\left(E_y\right)\right]^{-1} \tag{4.13}$$

In Equation (4.13), $[K]^{-1}$ is the "original" inverted stiffness matrix of (4.12); $[E_y]$ is an "elasticity matrix" containing Young's modulus for the individual (y), element and $[K(E_y)]^{-1}$ contains algebraic expressions including Young's modulus for the individual (y) element.

Thereby, an "updated" inverted stiffness matrix can swiftly be constructed for iteration $n + 1$ of the topology optimization by utilizing the $[E_y]$ matrix obtained from iteration n.

A similar approach to updating the (inverted) stiffness matrix could be taken for a nonisotropic material; however, it is not possible for nonlinear materials, as a simple scaling function such as (4.13) would not suffice. This is related to the fact that nonlinear material behavior is most often mathematically formulated by the usage of, for example, strain energy functions.

Thus, it can be avoided to reassemble and invert the stiffness matrix due to changes in material properties during the (linear FE-based) topology optimization. However, if elements are removed from the design volume, that is, if changes of geometry occur, the stiffness matrix will once again require reassembly and inverting.

This is nevertheless not the case for most VDM-SIMP-based optimization algorithms, as elements are in general not deleted; consequently, there is no need to reassemble the stiffness matrix, and the entire procedure of (4.12) only requires to be conducted once and it is therefore only required to redefine $[K]^{-1}$ in accordance with (4.13) between iterations of the topology optimization.

Thereby, the most CPU-intensive step of linear FEA is not repeated, and the CPU time required for the overall topology optimization process is significantly reduced.

There is, however, complications associated with this approach; namely that the resulting model will contain artificial stiffness, as the "geometry," that is, nodes, will remain within the model throughout the entire topology optimization process.

This is nonetheless only deemed to be a minor issue, as the vast majority of stiffness (in most cases) originates from the material properties, not the geometry itself.

Thus, the VDM-SIMP optimization method becomes very effective as the CPU-intensive step of inverting the stiffness matrix does not require repetition. In addition, the FE analysis is generally very stable and robust. Finally, the VDM-SIMP optimization method is, in general, also very stable, as no elements are deleted during the process.

Coincidentally, the algorithm is also very "flexible," if required, as the relative stiffness of any element can be extremely low in one iteration and very high in the next one, simply by varying $[E_y]$ in (4.13).

At this point, it should be noted that the stiffness value; the E_y value in (4.13), of any element is never set to zero for the purpose of conducting topology optimization in connection with the VDM-SIMP method. This is simply because if any E_y value is set equal to zero, $[K]$ effectively becomes singular, and consequently no inverted matrix exists and (4.12) cannot be used, requiring [K] to be reassembled.

Alternatively, E_y values may be set very low to represent a "0 element," meaning that a reassembly is not required, leading to the "practical" interval of $\rho(y)$ (in Equation (4.10)) to be as defined as Equation (4.14)

$$0 < \rho(y) \leq 1 \tag{4.14}$$

One of the associated issues with the 0 to 1 problem that is, of course, not solved by the introduction of SIMP is the lack of unique optimum solutions to the problem as discussed in Section 3.8.2. It is important to acknowledge that the 0 to 1 optimization problem of, for example, the BIW, is in principle a vast series of topology optimization problems, that is, an optimization problem to choose either 0 or 1 for each element, disregarding the relaxation introduced. In other words, if an FE model contains 100,000 elements, then the optimization may be perceived as 100,000 individual topology optimization problems that combined make up the design volume, for example, the BIW. However, these individual problems are necessarily related; therefore, the overall objective function that regards the problem as being global is introduced. This objective function is thus a generalized function, which is also often referred to as a distributed problem in mathematical terms. Solutions to this type of problem may, in many cases, not exist, as discussed in Section 3.8.2.

As a direct result of this, the topology optimization becomes mesh-dependent, as previously discussed. This fact can be further appreciated by observing Figure 4.12 that represents three "identical" topology optimization problems, where the difference between them is the actual mesh size. The models were solved using the VDM-SIMP method. The penalization factor of the models in Figure 4.12 was set to 2.0 in accordance with the rule of thumb in (4.11).

Figure 4.12 very clearly demonstrates the effects of mesh size upon the topology optimization results. It should be noted that the mesh dependency of topology optimization is a widespread problem, which theoretically is applicable to all FE-based optimization algorithms due to the discretization used for FEA.

Furthermore, Figure 4.12 also reveals that even though the SIMP interpolation scheme penalizes the intermediate $\rho(y)$ values, it does not eliminate the checkerboard effect. It can also be seen that the mesh size (i.e. average element size) has a noticeable influence upon the "severity" of the checkerboard effect. Therefore, extra care must always be taken when meshing the design volume prior to conducting topology optimization.

Thereby, the general presentation of the SIMP interpolation scheme is complete; the following section will present an alternative to SIMP, namely the summation of reciprocal values (SRV) constraint.

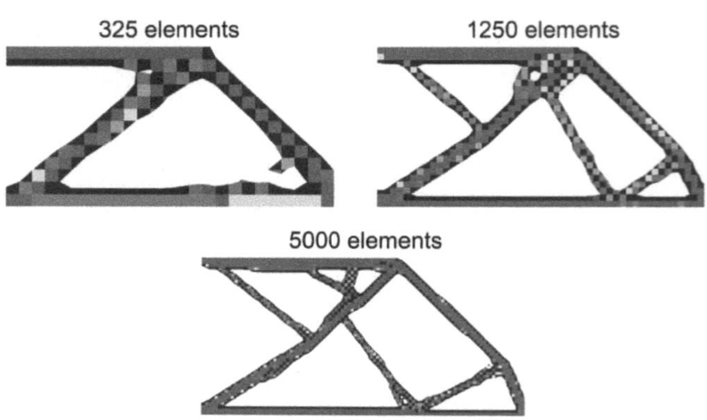

FIGURE 4.12 Example topology optimization mesh dependence.

4.1.3 SRV constraint

The summation of reciprocal values (SRV) constraint is another methodology that can be used to "steer" the continuous 0 to 1 optimization problem toward integer values. The SRV constraint can serve both as an alternative or an addition to the SIMP interpolation scheme, as presented in the previous section.

This is simply because the SRV is a constraint, which is imposed on the optimization problem, in the same manner as, for example, a maximum displacement constraint would be.

The initial governing theory relating to the definition of an objective function using continuous $\rho(y)$ values, etc., as explained in the preceding sections is thus also applicable to the SRV constraint.

The only potential difference is that the SIMP methodology is optional in this context. Alternatively, the penalization factor p in Equation (4.10) may be set to 1.0; either way, no penalization of intermediate $\rho(y)$ values is imposed.

For the purpose of clarity, identical notation to that used in the previous section will be utilized throughout this section. The starting point is Equation (4.10), more specifically $\rho(y)$, that represents the individual design variable attributed to the individual element. The basic form of the SRV constraint thus takes the form of Equation (4.15)

$$SRV = \sum_{y=1}^{m} \frac{1}{\rho_y}$$

where (4.15)

$$\rho_y \in \left| \rho(y)_{\min} : 1 \right|$$

In Equation (4.15), m represents the number of elements in the FE model, while $\rho(y)_{\min}$ defines the minimum value of $\rho(y)$. In order for Equation (4.15) to be valid for all elements, it follows that $\rho(y)_{\min}$ must be larger than zero, so to successfully apply the SRV constraint, Equation (4.10) needs to be redefined as Equation (4.16)

$$C_{ijkl}(y) = \rho(y)^p \cdot C_{ijkl}^0$$

where (4.16)

$$p \geq 1 \wedge 0 < \rho(y) \leq 1$$

By introducing Equation (4.16), it follows that the implementation of the SRV constraint means that the topology optimization problem is in essence no longer a 0 to 1 problem. However, this need not have a significant influence upon the overall solution. This is because even though there is a finite lower value for $\rho(y)$ in Equations (4.15) and (4.16), this can be set very low, as the only stipulation is that this value is nonzero. Furthermore, the same criterion was previously discussed in connection with the implementation of the SIMP scheme in the preceding section. By specifying that the values of $\rho(y)$ to either be equal to $\rho(y)_{min}$ or 1, all entries in the vector remain discrete. Additionally, if the number of unity $\rho(y)$ values in the model is defined as n, the $SRV_{discrete}$ constraint can be defined as Equation (4.17)

$$SRV_{discrete} = \frac{n}{1} + \frac{(m-n)}{\rho(y)_{min}} = n + \frac{(m-n)}{\rho(y)_{min}} \tag{4.17}$$

It can be mathematically proven that for any nondiscrete vector $\{\rho(y)\}$ Equation (4.18) is valid.

$$SRV\{\rho(y)\} < SRV_{discrete} \tag{4.18}$$

The optimum $\{\rho(y)\}$ vector is discrete (as the purpose is to solve a 0 to 1 problem); it therefore follows that SRV_{opt} is identical to $SRV_{discrete}$, Equation (4.19)

$$SRV_{opt} = SRV_{discrete} \tag{4.19}$$

Equation (4.18), in combination with Equation (4.19), thus states that SRV_{opt} must be a local (or global) maximum, as SRV_{opt} must be discrete. This can be defined as the topology optimization problem of Expression (4.20)

$$\begin{aligned}
&\min F(\rho(y)) \\
&\text{where} \\
&\sum_{y=1}^{m} \rho(y) = n + (m-n)\cdot \rho(y)_{min} \\
&\qquad \sum_{y=1}^{m} \frac{1}{\rho(y)} \geq n + \frac{m-n}{\rho(y)_{min}} \\
&\qquad 0 < \rho(y) \leq 1
\end{aligned} \tag{4.20}$$

In Expression (4.20), the equality constraint represents the total number of elements, $n + m$, and the inequality constraints represent the SRV constraint. Only one option exists to satisfy the latter and that is to satisfy the former, that is, the equality constraint. The only means of achieving that is to have $\rho(y)$ values of either $\rho(y)_{min}$ or 1. Thereby, the objective function may only contain discrete values.

The compatibility between the SIMP interpolation scheme and the SRV constraint is subtle. The primary point is the fact that both approaches require that $\rho(y)$ assumes nonzero values, as can be seen in (4.14) and (4.20) for SIMP and SRV, respectively.

Other alternatives to the SIMP approach also exist, such as the SINH method, which utilizes the hyperbolic sine function (SINH) to dictate the relationship between, for example, $\rho(y)$ and the compliance tensor C_{ijkl}. However, the SINH method essentially follows the same steps as the SIMP interpolation scheme, thus making many of the "issues" with SIMP and SINH identical with respect to application to nonlinear topology optimization.

The next section will briefly present the fundamental principles of design sensitivity analysis (DSA), which is not generally used as "stand alone" algorithm in connection with topology optimization. Instead, it is typically used as an "intermediate step" in the overall optimization algorithm, for instance, in connection with the SIMP interpolation scheme, as previously stated.

4.1.4 Design sensitivity analysis

The purpose of design sensitivity analysis (DSA), often also simply referred to as sensitivity analysis, is simply to obtain an overview of the sensitivities of the individual parameters of the objective function. This knowledge can then be used to "increase the efficiency" of the optimization by initially changing the magnitudes of the most sensitive parameters and subsequently "fine tune" the optimization by varying the "less sensitive" parameters. DSA can not only be used in connection with topology optimization, but is also applicable for shape, size, and any other form of optimization, including nonstructural optimization and does, of course, play a large role in terms of robust optimization as presented in Section 3.8.3.

The sensitivity of the individual parameters is found by means of differentiation, meaning that DSA is a gradient-based approach and therefore shares many commonalities with the approaches presented in Section 3.6. For example, if the objective function $f(\rho(y))$ in Equation (4.7) is used, the gradient vector $\nabla f(\rho(y))$ can be expressed as in (4.21)

$$\nabla f(\rho(y)) = \left\{ \frac{\partial f(\rho(y))}{\partial \rho(1)}, \frac{\partial f(\rho(y))}{\partial \rho(2)}, \ldots, \frac{\partial f(\rho(y))}{\partial \rho(n)} \right\} \tag{4.21}$$

In Expression (4.21), y represents the element number, and n represents the total number of elements in the model. The gradient vector of Expression (4.21) can be obtained by three different approaches:

1. *Analytical*, this requires that $f(\rho(y))$ is fully defined analytically. This is the most robust and efficient option;
2. Overall finite difference, if $f(\rho(y))$ is purely numerically defined;
3. *Semi-analytical*, if $f(\rho(y))$ is partially defined analytically.

In the particular case of linear FE-based topology optimization, $f(\rho(y))$ is generally fully analytically defined, such as Equation (4.7); thereby, the analytical option can be utilized. However, the strain vector $\{\varepsilon\}$ in Equation (4.7) will be a function of Young's modulus E, and thereby also ρ. Unsurprisingly, this means that the objective function $f(\rho(y))$ "becomes" dependent upon the FE analysis, seeking to obtain equilibrium for the fundamental Equation of FE:

$$\{F\} = [K] \cdot \{U\} \tag{4.22}$$

By combining Equation (4.22) with Expression (4.21) and using the chain rule of differentiation, an expression for the sensitivity with respect to design variable (i.e., the density of element y) can be found, as in Expression (4.23)

$$\begin{gathered} \frac{\partial\{F\}}{\partial\rho(y)} = \frac{\partial[K]}{\partial\rho(y)} \cdot \{U\} + [K] \cdot \frac{\partial\{U\}}{\partial\rho(y)} \\ \Updownarrow \\ \frac{\partial[K]}{\partial\rho(y)} \cdot \{U\} = \frac{\partial\{F\}}{\partial\rho(y)} - [K] \cdot \frac{\partial\{U\}}{\partial\rho(y)} \end{gathered} \tag{4.23}$$

The lower equation in Expression (4.23) thus represents the displacement sensitivities of the model, with respect to the relative mass density of element y. At this point, it should be noted that the CPU cost for solving Equation (4.23) is relatively low, and is directly proportional to the number of nodes/elements in the FE model.

In the case of linear topology optimization utilizing, for example, the SIMP interpolation scheme, Equation (4.23) will only have to be derived once, as the stiffness matrix [*K*] will remain "constant," that is, an algebraic function of ρ. This means that in a particular application and in line with the SIMP and SRV approaches, a large part of the efficiency of DSA can be attributed to the fact that the [*K*] matrix only requires assembly once.

The above sections have all assumed the usage of VDM, as presented at the start of Section 4.1; another general type of material distribution concept (MDC) also exists. HBO will be the focus of attention in the following section.

4.1.5 Homogenization-based optimization

In the preceding section, different approaches to topology optimization all using the variable density method (VDM) were presented and discussed. For linear elasticity the VDM approach assumes an explicit relationship between Young's modulus, E, and the relative mass density, ρ, of the material. The topology can subsequently be obtained by varying ρ and, thereby, E to obtain a topology and the associated changes of structural performance of the component in question. The VDM approach thus seeks to solve the topology optimization problem via the macroscopic material properties.

This approach works well for linear elastic problems; however, the FE-based VDM approach is also prone to the problem known as the checkerboard effect, leading to unfeasible designs, as illustrated by Figures 4.4, 4.5, and 4.6. Finally, the issue of mesh dependence was demonstrated in Figure 4.12.

As previously stated, an alternative to the VDM approach is denoted as homogenization-based optimization (HBO), which will be the focus of attention in this section.

The VDM approach is by far the most widely used in modern-day commercial FE software, despite the previously discussed issues. Possible reasons for this could be the implementation with FE combined with the proven robustness, efficiency, and the "relative simplicity" of the algorithms. HBO algorithms use a combination of macroscopic and microscopic considerations to perform topology optimization. The (theoretical) primary advantages of HBO as opposed to VDM are:

1. HBO can simultaneously conduct topology, shape, and size optimization;
2. HBO is mesh independent;
3. HBO completely avoids the checkerboard effect.

If the three advantages stated above can indeed be obtained, the potential for HBO-based algorithms will be significantly larger than VDM-based algorithms. However, given the overall aim, other aspects such as implementation with nonlinear material behavior, CPU cost, robustness, diversity, and accuracy must all be carefully considered before such a statement can be fully justified.

In mathematical terms, homogenization is the study of partial differential equations with rapidly oscillating coefficients. In order to relate this to continuum mechanics consider the following.

All materials are inhomogeneous (heterogeneous) at some scale; for example, the microstructure of steel varies significantly as a function of location. Nevertheless, for practical applications, it is most convenient to assume, for example, that structural steel has isotropic, that is, homogeneous, properties such as Young's modulus, Poisson's ratio, and volumetric mass density. For most (engineering) applications, this assumption is reasonable and provides a satisfactory level of accuracy, while significantly reducing the complexity of the engineering problem. For HBO-based optimization, these assumptions are removed. In order to demonstrate this in a mathematical context, suppose that Equation (4.24) describes Young's modulus, E, of a material as a function of location $u(x)$, where the (x) describes the location.

$$E = \frac{\partial}{\partial x}\left(A\left(\frac{x}{\delta} \right) \frac{\partial u}{\partial x} \right) \tag{4.24}$$

The value of E at a small distance δ from x will thus (rapidly) vary according to the magnitude of the $A\left(\frac{x}{\delta}\right)$ contribution, leading to a significant variation of E across the material/component in question. In order to "avoid" this variation, Equation (4.24) is redefined as Equation (4.25) according to the homogenization theory.

$$E = \frac{\partial}{\partial x}\left(A^{*} \frac{\partial u}{\partial x} \right) \tag{4.25}$$

Please note that E in Equations (4.24) and (4.25) is merely an example and could be substituted by any other material parameter.

In Equation (4.25), A^* is a constant representing the "effective property" of E, thus making Equation (4.25) "less fluctuating," which may, for example, be applied in a local region of the material/component in question.

The latter part of the last sentence is key to understanding the general principle of HBO.

The above explanation is merely a brief introduction into the general homogenization theory. In order to fully understand the principles and finer detail, it is imperative that the reader, as a minimum, possesses extensive knowledge about continuum mechanics and differential equations. Therefore, a more in-depth explanation of fundamental homogenization theory exceeds the scope of this chapter. Furthermore, while the in-depth understanding is desirable, it is not crucial to the continued discussion of the potential application to topology optimization.

The transition between Equations (4.24) and (4.25) can thus be used to represent the general homogenization theory. The "behavior" (response) of Equations (4.24) and (4.25) can thus be described as two "extremes," as the former dictates a rapidly oscillating function, while the latter dictates a significantly less rapidly oscillating function. In HBO-based algorithms, the material behavior is typically described as intermediate between these two extremes, and varies periodically. In other words, E will, for example, periodically obtain its constant value (A^*) at given points in the material/component and vary, Equation (4.25), in between these points. A graphical illustration of this variation can be seen in Figure 4.13 where "E" denote the "locations" of A^* (i.e. E), the remaining points are thus defined by Equation (4.25); note the periodicity illustrated in the figure.

As Figure 4.13 indicates, E has thereby become a "continuous variable" throughout the material (or component) in question. In principle, this means that the topology optimization problem can be described without the need for discretization; consequently, the checkerboard effect and mesh dependence will not occur.

At this point, it is, however, important to remember that in connection with topology optimization it is often highly desirable to apply constraints relating to,

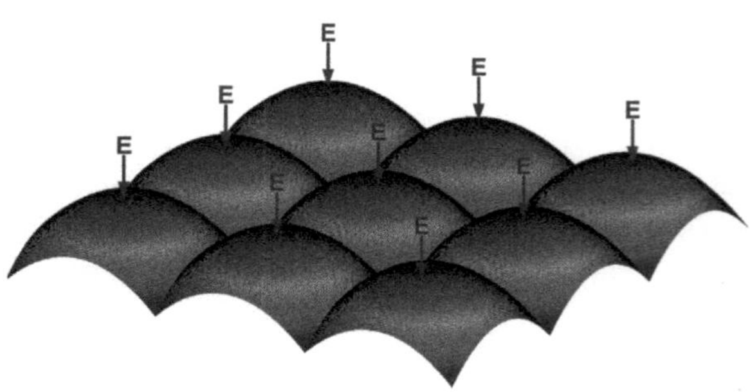

FIGURE 4.13 Illustration of HBO principle.

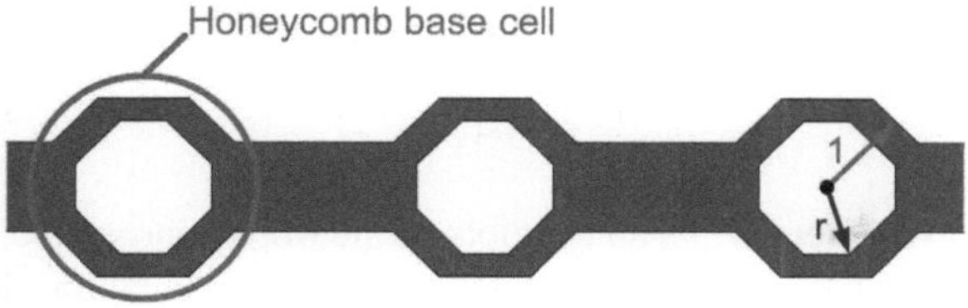

FIGURE 4.14 Honeycomb base cell for HBO.

for example, maximum displacement or stress values. These can, in principle, be obtained by the usage of mesh-free methods, that is, without the usage of FE. This approach is, however, often very cumbersome, if not impossible to define without the usage of discretization. Nevertheless, advances are made in this context, and the potential for future usage remains; if successfully solved, the use of FE for structural analysis will be obsolete. This will, however, not be further discussed in this book and the focus will return to FE-based optimization. Consequently, the structural analysis associated with the optimization will be mesh dependent.

The impending task is thus to define the periodic material properties in Figure 4.13 in relation to FE analysis. This could, for example, be done by introducing continuous FE approximation of design variables, as originally suggested by Matsui & Terada (2004).

The following methodology of implementing HBO is primarily based upon the approach of Zhao, Long, & Ma (2010).

In order to include the microstructure in the FE analysis, a unity honeycomb base cell could be employed as illustrated in Figure 4.14.

By utilizing the honeycomb base structure illustrated in Figure 4.14, the material microstructure is considered during FEA. Furthermore, the approach also allows for isotropic behavior, as the honeycomb base cell will mimic isotropic behavior on a macroscopic scale. In order to utilize the honeycomb base cell in connection with HBO and FEA, the "normal" interpolated shape functions used to construct the stiffness matrix $[K]$ are substituted with moving least squares approximation (MLS) shape functions. This method is, in essence, a curve-fitting technique and can briefly be defined as consisting of three primary components:

1. A weighting function
2. A basis function
3. A set of coefficients

Combined, the three components listed above works similar to, for example, a B-spline, that is, a weighted average which "decides" how the function acts in any given interval; this will be further explained in a subsequent example.

By introducing MLS, the design variable (e.g. the material density) becomes a continuous function over the design domain (in general), and the honeycomb base cell in Figure 4.14 may be used as a design variable, as in Equation (4.26)

$$r^e = \sum_{y=1}^{m} \Phi_y r_y \tag{4.26}$$

Equation (4.26) represents an example of the MLS approach to HBO.

The honeycomb base cell size r^e (Figure 4.14) at any location e in the design domain can be interpolated via Equation (4.26). m in Equation (4.26) represents the total number of design variables (e.g. the total number of elements in the FE model), while Φ_y represents the interpolation shape functions, that is, the ones that replaced the "original" FE shape functions.

From Equation (4.26), the density at location e can be derived as Equation (4.27). For details regarding this derivation, please see Zhao, Long, & Ma (2010).

$$\rho^e = \sum_{y=1}^{m} \Phi_y \rho_y \tag{4.27}$$

Equation (4.27) can thus be used in a similar manner to the general implementations discussed in Sections 4.1.1 and 4.1.2; however, the material microstructure has been taken into account in Equation (4.27).

The above presentations and discussions thus represent a very brief overview of the general principles and possibilities of HBO-based algorithms, but what is the feasibility of employing HBO algorithms for general topology optimization and, in particular, nonlinear topology optimization?

In order to complete a more in-depth discussion of HBO algorithms, it is, as previously mentioned, imperative that the reader possesses substantial knowledge of continuum mechanics and partial differential equations, thereby exceeding the scope of this particular chapter.

HBO-based algorithms hold very desirable potential with respect to nonlinear optimization and also for other areas under development such as the (efficient) inclusion of, for example, composites/nonisotropic materials into topology optimization. The overall potential advantages of HBO algorithms discussed in this section may be summarized as:

1. Mesh independency
2. Absence of checkerboard effect
3. Inclusion of material microstructure
4. Ability to cater for nonisotopic materials

At first glance, the HBO algorithms therefore hold many significant benefits when compared with the VDM-based algorithms discussed in Sections 4.1.1 through 4.1.4. However, many of these benefits become less valuable due to the inability of implementing these into actual code, especially in relation to structural analysis, where FEA remains the dominant method of analysis. The HBO algorithm thus holds significant potential for nonlinear topology optimization, but in order to fully enable this potential it seems that mesh-free analysis methods play a vital role, although partial implementation is possible as described above.

4.2 SHAPE OPTIMIZATION

As explained in Chapter 3, many of the optimization techniques available are highly adaptable, meaning that they can virtually be applied to any optimization problem, as an objective function can be used to represent essentially any system from economic forecast to structural analysis. This means that many of the techniques and methods described in Chapter 3 are not only applicable to topology optimization, but also shape and size optimization. This also extends to the fundamental principles described in Section 4.1, although the specific details will, of course, differ. With this in mind, the subsequent detailed mathematical descriptions of shape and size optimization will be minimal and will instead focus on general techniques.

As the name suggests, shape optimization is focused on the geometrical shape of the component to be optimized; the initial section of this chapter gave an example relating to a simple cantilever beam. As mentioned in that context, shape optimization does, however, stretch far beyond the "simple" task of determining an optimum cross-sectional geometry of a beam. A more "complex" example is presented in Figure 4.15.

The left-hand side of Figure 4.15 represents the initial model, whereas the right-hand side of Figure 4.15 represents the outcome of shape optimization. Initially, the change in shape illustrated in Figure 4.15 may seem similar to that of topology optimization and there is indeed a lot of similarity between the two; topology optimization is, however, able to completely change the topology, whereas shape optimization can merely change the shape, but not the overall topology. In the context of the example in Figure 4.15, this can be understood by the fact that during the shape optimization, only the shape of the holes has changed from a circular to a triangular shape; the number of holes, 3, and indeed the overall location of the holes, has however not changed as this is not within the remit of shape optimization; it is, however, within the remit of topology optimization. Shape optimization can be divided into two main approaches:

1. Parametric shape optimization
2. Nonparametric shape optimization

The parametric approach utilizes physical parameters, such as radii, height and length to optimize the shape of the structure in question. The design variables, that is, the geometric parameters chosen, are therefore the only optimization variables, meaning that the parametric optimization is simplified as it becomes

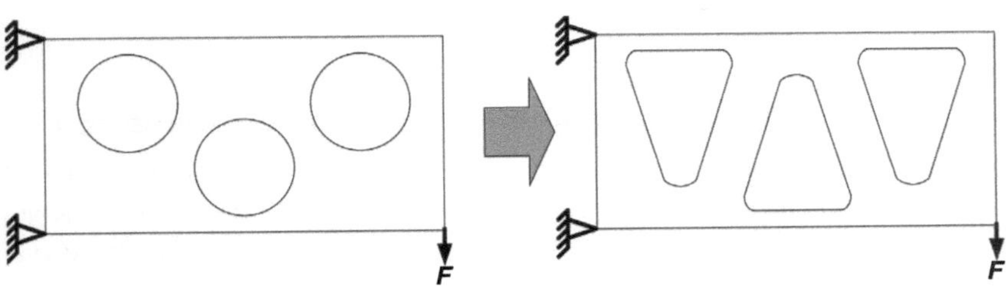

FIGURE 4.15 Shape optimization example.

more straightforward to predict/account for the variations of a shape due to the relatively simplicity of the parameters listed above. The parametric shape optimization is thereby performed simply by minimizing or maximizing the objective function within the solution space spanned by the design variables. This also means that in many cases FE is not required for parametric shape optimization, as the objective function and indeed constraint functions can be "easily" calculated using, for example, solid mechanics. If parametric shape optimization is to be performed, the model will need to be parameterized, which may be cumbersome if the previous step was, for example, derived through topology optimization. Consequently, the level of geometric complexity one can define with (a limited) number of design variables severely limits the detail of geometry which can be considered. This, in turn, indicates that in order to determine a truly optimum solution, *a priori* knowledge of the optimum solution is required. It should be noted that it is possible to use techniques such as nonuniform rational B-splines (NURBS) as used in computer-aided design (CAD) software to describe complex geometry in the context of parametric shape optimization. Nevertheless, the most often utilized method of performing shape optimization is through the use of nonparametric shape optimization. Nonparametric shape optimization operates at a more detailed FEA level, namely by observing and tracking change of nodes and elements, which also means that there is no need to (further) parameterize the model in question as an FE, or even a CAD model, can be directly used. In general, there are three main approaches to nonparametric shape optimization:

1. The Lagrangian approach
2. The Eulerian approach
3. Morphing

The three methods listed above may be familiar to the reader as two of them, the Eulerian and the Lagrangian, are also often used for fluid flow and fluid structure interaction problems in FEA. All of the approaches listed above utilize iterative methods in which the design is evolved from an initial start guess, as presented throughout Chapter 3; the overall flowchart does, therefore, not differ significantly from that illustrated in Figure 4.1, although the intricate details do. In fact, the difference between the three methods is not directly related to the overall optimization algorithm; it is, in actual fact, related to how the algorithm keeps track of the shape, i.e. geometry. In other words, the three approaches listed above can be considered to be the "memory" of the optimization and will briefly be described in the following.

4.2.1 The Lagrangian approach to shape optimization

The Lagrangian approach to shape optimization is essentially to keep track of the geometrical changes by monitoring nodes/points on the boundary of the geometry. As an example consider any of the circular holes on the left-hand side of Figure 4.15. This geometry could, for example, be defined and monitored by tracking the nodes illustrated in Figure 4.16.

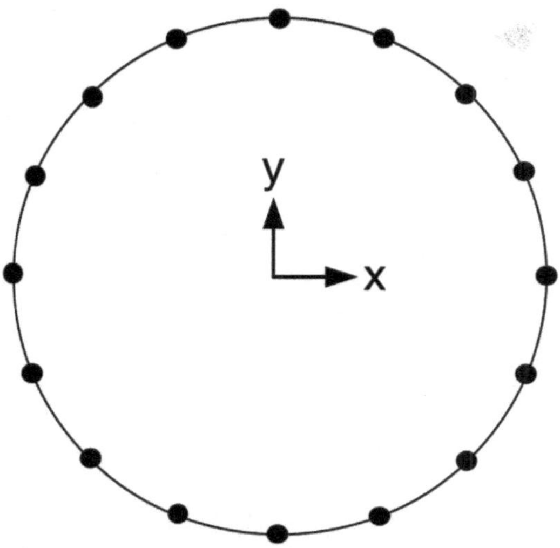

FIGURE 4.16 Lagrangian approach to shape optimization.

Based on the example in Figure 4.16, the subsequent shape optimization is simply performed by moving the coordinates of the nodes and tracking the response of the structure while storing the individual nodal coordinates. Any number of methods may be used to determine which nodes are to be moved, and in which direction(s). This could, for example, be based on strain, stress, or displacement information obtained from the FE analysis. The number of nodes will, of course, have an influence on the "complexity" of the impending shape optimization task as an increase in nodes will result in an increase of memory required to track the changes of the geometry. It may be difficult to see the difference between the Lagrangian approach illustrated in Figure 4.5 and the parametric shape optimization approach previously described. The main difference simply lies in the fact that the user or engineer is not required to specifically "lead" the parameterization process, as the nodes in Figure 4.16 are identical to the nodes of the FE model, meaning that the parameterization process is identical to the discretization process of FEA; meshing.

4.2.2 The Eulerian approach to shape optimization

The Eulerian approach to shape optimization differs from the Lagrangian approach by having nodes that do not move. Instead, the shape is contained within a rectangular "box" with an associated function describing and assigning values of "state" according to coordinates of the grid, which are fixed. Figure 4.17 illustrates the Eulerian approach to shape optimization.

On the inside of the shape a positive value is assigned; a value of 0 is assigned on the boundary of the shape, while a negative value is assigned on the outside of the shape. As the shape evolves, the function is updated to cater for the change of state at each node; however, the coordinates of the nodes do not

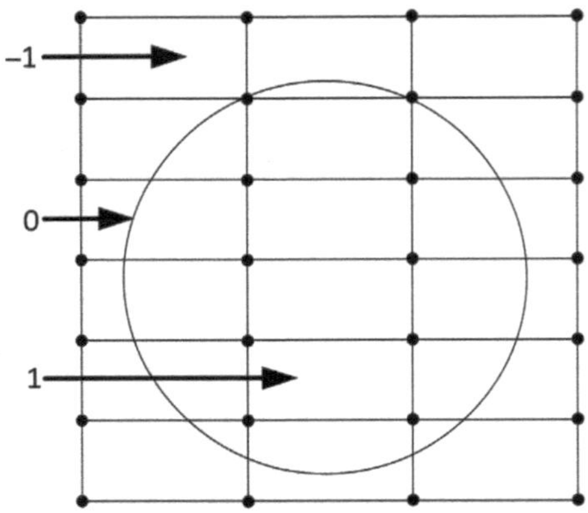

FIGURE 4.17 Eulerian approach to shape optimization.

change, as the nodes are fixed in space and "time" for all iterations. This approach is therefore more similar to the finite volume method (FVM) than FEM. The basic concepts of fixed nodes and functional values are the fundamental principles of the level set method (LSM), which is widely used for shape optimization, but will not be further discussed in this book. The choice of how the shape is moved may be determined in the exact same manner as that used for Lagrangian-based shape optimization (Figure 4.16).

4.2.3 Shape optimization using morphing

The use of morphing for shape optimization is to a very large extent based on the gradient-based methods presented in Chapter 3. The underlying principle of morphing is to consider the change of shape analogue to that of fluid flow. Imagine, for example, a casting process where the molten (and liquid) material is poured into a mold; the shape will gradually change and adapt according to the path of least resistance. By tracking the position of each particle inside the fluid flow a "time-history" map of the overall shape can be obtained, i.e. the shape optimization history. The term morphing stems from the mathematical background to the principle, that is, to consider the diffeomorphism which is an isomorphism of a smooth manifold, representing a topological space. The detailed explanation of this far exceeds the scope of this book, and is not essential in the context thereof. The heart of the principle is simply to create a function f_i of iteration i, which links the initial shape at iteration S_0 to the shape S_i at iteration i. Based on this function, a gradient method such as the ones presented in Section 3.6 may be utilized to determine the optimum solution in an iterative manner.

4.3 METAMODELING

Thus far in this book, it has been generally assumed that the optimization is conducted on the actual FE model; this generally comes at the expense of performing a number of FE analyses, which in certain instances may be acceptable and in certain instances may not, depending on the size of the FE model, the number of runs required, etc. This type of approach is termed direct optimization, as the optimization is performed directly on the FE model in this case. However, it is also possible to conduct indirect optimization by means of a surrogate model, often referred to as a metamodel. The main advantage of using a surrogate model for optimization is the reduction of CPU time, which can generally be significantly reduced. In contrast to direct optimization methods that determine the optimum solution from the "original" model, in this case an FE model, the optimum point in an indirect optimization is calculated from a model of the original model, which may seem counterintuitive with respect to accuracy and precision.

Several metamodel based optimization methodologies exist; the perhaps most commonly used approach is the response surface method (RSM), although other methods such as Kriging and neural networks (NN) may also be employed. In this book only RSM will be presented; instead, the emphasis will be on how an actual metamodel is constructed. It should also be noted that NN will be introduced in a wider context in Chapter 5. The ultimate aim of employing any of these approaches remains the same, namely to construct a surrogate model as accurately and adequately as possible based on a finite number of design points. Subsequently, the metamodel is used to determine the optimum solution.

The creation of a metamodel therefore requires a certain amount of information or number of design points. The exact number of design points is firstly determined by the algorithm of the point selection method, which will be discussed in Section 4.4 and secondly by the requirements of the specific metamodel technique.

4.3.1 Response surface methodology

Response surface methods (RSM), as introduced above, may be used as a statistical approach to optimization in the shape of a metamodel. This approach seeks to determine a relationship between the input variables and the response of, for example, the objective function. Using the knowledge obtained by conducting a series of numerical experiments, for example, FEA, a response surface is constructed and approximated, typically by a first-order polynomial or a second-order polynomial such as the one defined in Equation (4.28)

$$y = \beta_0 + \beta_1 x_1 + \beta_2 x_2 + \beta_{11}(x_1)^2 + \beta_{22}(x_2)^2 + \beta_{12} x_1 x_2 + \varepsilon \qquad (4.28)$$

In Equation (4.28), y represents the response, x_i represents the individual variable or regressors, k is the total number of variables, β_i represents the regression coefficients whilst, ε is the error. In this case, x_i spans a multiple dimensioned plane (hyper plane) in a k-dimensional space similar to the previously discussed solution space.

Generally, the objective of the impending optimization is to maximize or minimize one of the responses y, upon which the response surface is constructed. In order to create the response surface, it is of course necessary to have certain information available. This can be achieved in many ways, for example, via FEA. The number of FEA to be conducted will vary according to how many parameters, including the number of variables k and the point selection method used. Once the response surface has been created, the principle is to determine the maximum or minimum response of y. The overall steps of RSM can thereby be summarized as:

1. Determine the response(s) y to be considered and the number of variables k;
2. Collect sufficient data (see Section 4.4 for details);
3. Create the response surface, for example, using Equation (4.28);
4. Determine the maximum y value (using the steepest ascent method) or the minimum value of y (using the steepest descent method).

In order to better understand the underlying principle, a simple illustrative example with one variable, $k = 1$, is introduced. Figure 4.18 represents a function plot of the only response y under consideration in this example. Please note that this plot is, of course, not normally available; it is solely used for illustrative purposes in this example.

Imagine that the starting value of the only variable x_1 is A. Assuming that the task at hand is to minimize y it is necessary to determine which "direction" to take in order to minimize the response. As the function is "unknown," a "surrogate" must be created to decide whether to increase or decrease x_1; in other words, a response surface must be defined. In order to do so, the response is "sampled" at two points, P_1 and P_2, to get an idea of the "trends." With this information, a simple trend line is created as RS_1 in Figure 4.18;

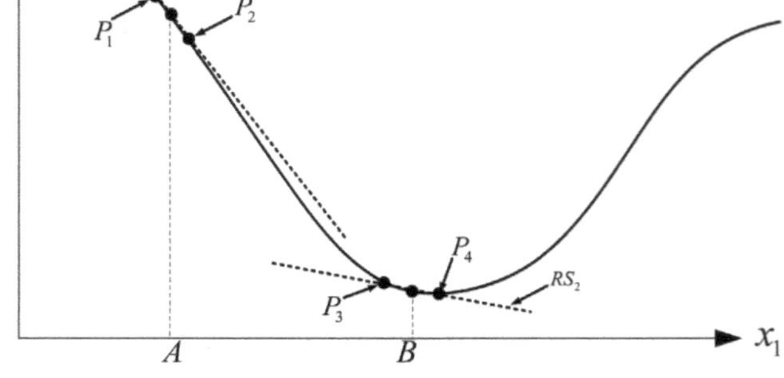

FIGURE 4.18 Single variable RSM example.

this line represents the response surface. In this simple case, the creation of the trend line is done via linear regression (based on points A, P_1, and P_2); a functional expression may thereby be obtained using the predicted intersection point with the y-axis, as well as the slope. In this case, the response surface is linear, meaning that a first-order polynomial is sufficient. In more complex cases, second-order polynomials may be employed as previously mentioned. With the response surface created, the optimization may proceed based on the trend line created; an optimum (minimum) may now be determined using SDM, as presented in Section 3.6.5. After a certain number of iterations, it may be required to "resample" the response y (using the original model, not the response surface) to check for errors. In which case, a new "starting point," B, and two new sampling points, P_3 and P_4, are used to inform and create the new response surface RS_2. If two or more response surfaces are created sequentially, as described above, the method is generally referred to as sequential response surface method (SRSM).

The above example is simplified, and very few real-world engineering problems are that simple. Firstly, it is, of course, highly likely that more than one variable exists; that is, $k \neq 1$, in which case more elaborate response surfaces will need to be created and the response "surface" becomes a hyper plane of dimension k. Another complication that is highly likely to occur is, of course, the fact that there is great variation or fluctuation in the responses from the individual sampling points, making it extremely difficult to create a "smooth" surface, in which case the linear regression step most often utilizes the least squares method (LSM) to efficiently and robustly complete the curve-fitting step of creating the response surface.

The major disadvantages of RSM and SRSM are the accuracy of the second-order polynomial with respect to a large number of design variables, for example, if the technique is to be used for topology optimization. In practice, RSM or SRSM methods are generally limited to a maximum of seven or eight variables due to the curse of dimensionality presented in Section 3.4. This makes RSM and SRSM well suited for, for example, shape and size optimization, but unsuited for topology optimization. In addition, the numerical experiments used to build the response surface must be carefully planned to obtain useful information, which will be the focus of attention in the next section.

It should be noted that only the RSM and SRSM methods have been briefly presented in this section; many other metamodeling techniques exist. Hence, the reader is encouraged to review other literature if extensive metamodeling is to be used. Many FE and mathematical software suites have built-in capabilities for creating metamodels using RSM and other methods; therefore, the overall purpose of this section was simply to provide the reader with an appreciation of how these methods generally work by means of curve fitting. The purpose of the following sections is analog to that of this section, namely, to provide the reader with a fundamental appreciation of the ideologies of each method as opposed to a thorough in-depth understanding.

4.4 POINT SELECTION METHODS FOR METAMODELING

As presented in the previous section, a suitable choice of design points is crucial for the accuracy of creating a metamodel, RSM or other metamodel. Just like an FE model, the metamodel is only as good as the input data upon which it is created. A number of methods to determine appropriate design points in a design space for the construction of a metamodel exist. Well-known and proven point selection algorithms such as factorial design, Koshal design, and composite design are in general meant for the deployment of polynomials such as RSM. The concept of the D-optimal design is well suited for linear regression and especially designed for this particular approach.

Space-filling designs determine the design points either randomly, according to a certain random method or using Latin hypercube sampling (LHS). Space filling designs are often chosen, assuming that a uniform distribution of the design points over the entire design space is desirable.

The following sections will briefly describe these various point sampling methods that, in general, may be referred to as design of experiments (DOE). For the following examples, it is beneficial to consider the design space, that is, the space spanned by x_i in Equation (4.28), to be represented by a square for $k = 2$ and a cube for $k = 3$.

4.4.1 Factorial design

The strategy of factorial design (FD) sampling is to determine the corners of the design space. The number of sampling points N_{SP} depends on the number of variables k and the degree G of the approximation method selected. The general formula for FD is as defined in Equation (4.29)

$$N_{SP} = G^k \tag{4.29}$$

For linear approximations, such as a first-order RSM G equals 2, and for a quadratic approximation such as the one defined in Equation (4.28), G equals 3. Figure 4.19 illustrates the "ideology" behind the FD sampling method for $k = 2$ and $k = 3$ using the square and cube geometries to illustrate the entire design space.

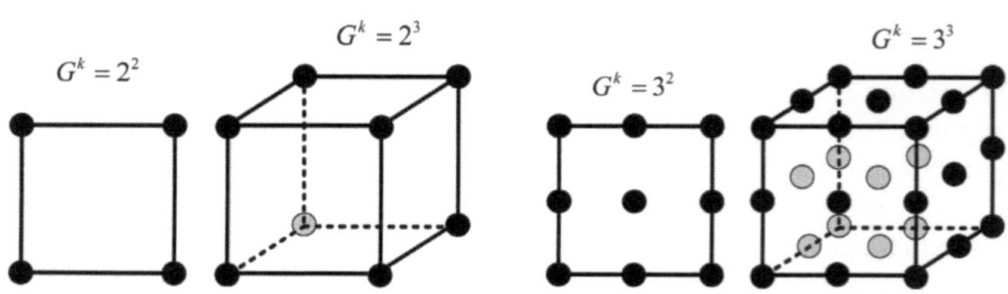

FIGURE 4.19 Factorial design example illustration for selected *G* and *k* values.

The two illustrations furthest left in Figure 4.19 illustrate the location of the design points for a first-order approximation, that is, $G = 2$, using two and three variables; $k = 2$ and $k = 3$. The two images to the far right of Figure 4.19 illustrate the experimental design for a second-order approximation, that is, $G = 3$, using two and three variables; $k = 2$ and $k = 3$. For $G = 3$, it can be seen that design points are added on the design space edges and the outer planes. The full factorial design (FFD), that is, the one utilizing Equation (4.29), has the characteristic that the distance between two adjacent design points is always identical for a constant approximation method, regardless of the number of variables. FFD does, however, "oversample" the design space, and often a fractional factorial design (FRFD) approach is utilized, which entails reducing the number of sampling points.

4.4.2 Koshal design

The ideology behind Koshal design is to provide the exact number of design points necessary to construct any given polynomial. Thus, for a linear approximation, Equation (4.30) is used.

$$N_{SP} = k + 1 \tag{4.30}$$

And for a quadratic polynomial, Equation (4.31) is used.

$$N_{SP} = k + 1 + \frac{k(k+1)}{2} \tag{4.31}$$

Figure 4.20 illustrates an example of the location of the design points for linear and quadratic approximations using $k = 2$ and $k = 3$.

The two illustrations furthest left in Figure 4.20 illustrate the location of the design points for $k = 2$ using first- and second-order approximations. The two images to the far right of Figure 4.20 illustrate the location of the design points for $k = 3$ using first- and second-order approximations.

4.4.3 D-optimal

The two sampling methods introduced in the previous sections were developed for regular design spaces, that is, design spaces that are only defined by the

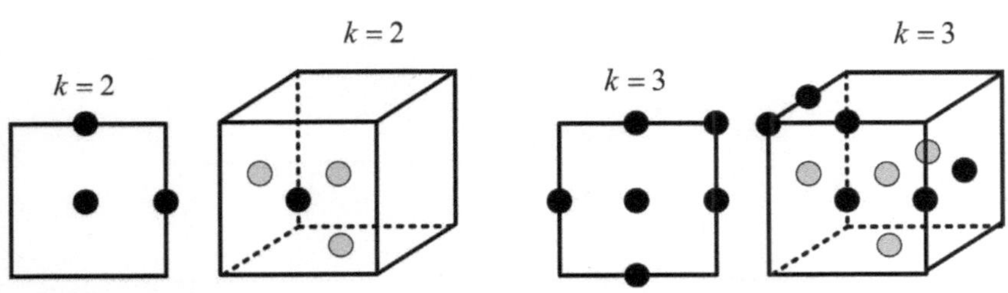

FIGURE 4.20 Koshal example illustration for selected *k* values using first- and second-order polynomials.

limitations of the design variables and thus form a hypercube. However, if the design space is irregular due to certain restrictions, these methods cannot be employed, for example, due to CPU limitations. The D-optimal design ideology was developed to cater for these instances. There are many variations of the method which is widely available in many FE (optimization) suites and mathematical programs such as MatLab; the specific definition of the model is dependent upon the chosen optimality criterion, as well as the metamodel chosen. The D-optimal method was compiled for linear regression models like the RSM and is, therefore, optimal for this specific metamodelling technique.

4.4.4 Space-filling designs

In Section 4.3, it was stated that for certain optimization problems or for probabilistic analysis it is desirable to achieve uniformly distributed design points over the entire design space. The point selection methods of the preceding sections are generally not suitable for this purpose for a variety of reasons. Space-filling designs are, however, well-suited for this specific purpose; the perhaps most widely used method is the Latin hypercube sampling (LHS) method.

In order to obtain a fundamental appreciation of the LHS method, Figure 4.21, which represents the design space of a two-variable problem, $k = 2$, is introduced.

The domain of each variable x_i is divided into m number of user-defined intervals or segments. In Figure 4.21, $k = 2$ and $m = 4$; hence, the two variables x_1 and x_2 are individually divided into four intervals leading to the overall design space being divided into 16 "segments." In order to create the meta model, LHS randomly chooses a value of x_1 and x_2 which, of course, uniquely determines one of the design space segments, that is, one of the 16 segments in Figure 4.21 in this particular example. Once the response has been sampled in a particular segment, it is "eliminated" and the remaining intervals are "randomly" examined until all segments have been sampled and the design space is filled. Figure 4.21 thus demonstrates that five segments have currently been sampled and

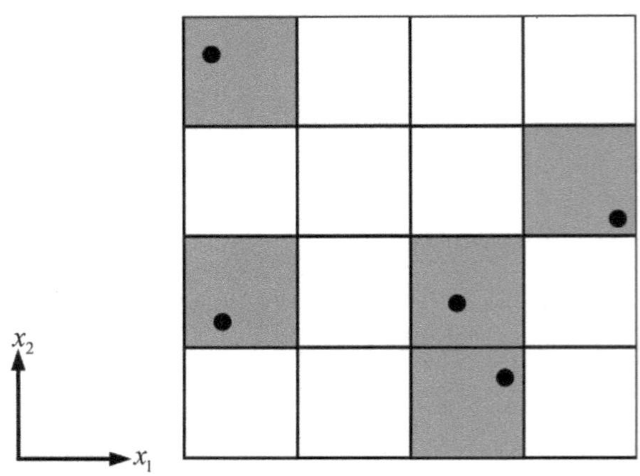

FIGURE 4.21 Latin hypercube sampling example illustration.

11 remain. LHS thereby ensures an "equal" treatment of each variable independent of its contribution to the response due to the "randomness" of selection variable values for the sampling.

4.5 OPTIMIZATION STRATEGIES FOR METAMODEL-BASED OPTIMIZATION

Section 4.3 introduced the overall concept of employing a metamodel with some of the associated advantages, and a brief overview of how a metamodel can generally be built using RSM. Section 4.3 introduced some basic examples of how information may be gathered through sampling and design points in order to create the metamodel, for example, a response surface. But what about the overall optimization and metamodel strategy? It is necessary to define an appropriate optimization strategy before commencing, as a particular DOE method may be preferential to use in a specific context. In this section, three general strategies of interest in connection with structural optimization will be introduced.

4.5.1 Single-stage optimization

In a single-stage strategy, all design points are sampled in only one stage, that is, design points are calculated once, the metamodel generated, and subsequently an optimization performed. This strategy generally works very well for radial-based functions which belongs to the neural network (NN) family and may be used for metamodeling. Although NN have not been covered in this chapter, they will be introduced in a wider context in Chapter 5. The radial basis function (RBF), which will be further introduced in Section 9.4.2 approach is generally used in connection with the space-filling design sampling method presented in Section 4.4.4.

4.5.2 Sequential stage optimization

In the RSM example of Section 4.3.1, a sequential optimization strategy was suggested meaning that a certain number of design points are calculated in each iteration. Compared with the single-stage strategy, the sequential method offers the advantage of terminating the optimization run once a satisfactorily accuracy of the metamodel is reached, although this does, of course, come at the disadvantage of increased CPU requirements.

4.5.3 Sequential stage optimization with domain reduction

The sequential strategy with domain reduction is an adaptation of the method introduced in Section 4.3.1. In fact, this is the method that is generally referred to as SRSM. The general principle does not deviate from the discussion in Section 4.3.1 or above. The only variation is that the design space is reduced from iteration to iteration, thereby making the current design space in any iteration a subspace of the entire or "original" design space. This is illustrated in Figure 4.22 where i represents the iteration number.

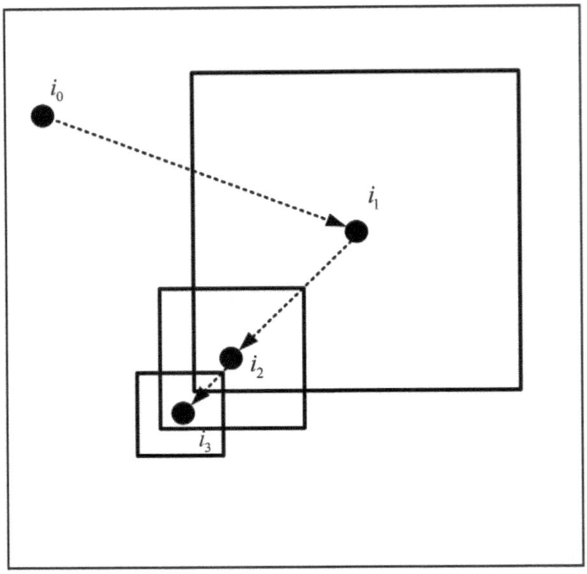

FIGURE 4.22 Domain reduction example illustration.

The benefit of SRSM is, therefore, that the design space is reduced in each iteration, making the optimization more efficient as the area is reduced. The potential disadvantage of this is, of course, that an optimum solution may be discarded as it lies outside of the current design space. The SRSM method, therefore, bears some resemblance to the trust region method presented in Section 3.7.3.

4.5.4 Model suitability and analysis of variance

Provided that a specific metamodel, sampling point method, and overall optimization strategy have been chosen, how is it known that the selection is actually appropriate? Given that metamodeling is in essence based on curve fitting, statistics, and probabilities, several well-known mathematical methods are often used to check the adequacy of the developed model. Generally, the model adequacy check is performed with methods known from stochastic/probabilistic mathematics such as error, root mean square (RMS) error, variance, standard deviation, or confidence interval calculations, although the specific implementations do vary slightly for the specific metamodel technique. Analysis of variance (ANOVA) is a further technique developed for probabilistic analysis. Prior to optimization, ANOVA can be used to rank the influence of different variables. Essentially, the process is carried out by analyzing the regression coefficients, for example, β_i in Equation (4.28). Variables with an insignificant influence upon the response y are thereby identified and can subsequently be eliminated from the optimization, significantly reducing the computational efforts of solving the optimization problem. ANOVA may initially seem similar to DSA introduced in Section 4.1.4; the profound difference is however that ANOVA is used in a statistical context, whereas DSA is used to determine sensitivity of outputs to input variables of a mathematical function.

REFERENCES

Bendsoe, M. P., & Sigmund, O. (2003). *Topology optimization – theory methods and applications*. Berlin, Germany: Springer Verlag.

Matsui, K. & Terada, K. (2004). Continuous approximation of material distribution for topology optimization. *International Journal for Numerical Methods in Engineering*, *59*(14), 1925–1944.

Zhao, H., Long, K., & Ma, Z. D. (2010). Homogenization topology optimization method based on continuous field. *Advances in mechanical engineering*, Volume 01/2010, Article ID: 528397, 7 pages.

Chapter | five

Applications of Linear Optimization to Concept Vehicle Safety Structures

CHAPTER OUTLINE

5.1 INTRODUCTION

Engineering a vehicle structure is a complex task to undertake. If vehicle architectures have the tendency to look similar nowadays new optimization programs have the capability to provide new opportunities to reconsider the principles of vehicle design. The opportunity of a holistic design approach could be investigated, using linear topology optimization, for the concept vehicle safety structures. The linear topology optimization method is robust for linear problems, even if some of the vehicle structural deformation is not occuring in the linear range. This chapter will investigate the opportunities and limitations of using such optimization techniques for BIW definition and the generation of optimized lightweight CAD geometry. One of the other challenges faced in the vehicle design phase is also to consider 'the vehicle external styling surface', as well as the formability of the panels, including reinforcements.

The chapter will base its argumentation using case studies relating to future hybrid electrical vehicle architectures.

Nonlinear Optimization of Vehicle Safety Structures. http://dx.doi.org/10.1016/B978-0-12-804424-7.00005-1

5.2 FULL VEHICLE STRUCTURE TOPOLOGY OPTIMIZATION

5.2.1 Introduction

In order to illustrate the capabilities of linear topology optimization, a HEV BIW with a 5 + 2 seating arrangement and a maximum length not exceeding 3900 mm is studied. The singular propulsion system of the vehicle will be electric. This requires that an electric motor will convert electric energy, originating from a battery pack, into mechanical motion to drive the wheels. The vehicle will be front wheel driven only. In addition to the battery pack, a range extender (RE), that is, a small displacement internal combustion engine (ICE) combined with a fuel tank (FT), was added in order to expand the operating range of the vehicle, as well as charging the battery pack.

The previous description clearly indicates that the setup of the vehicle in question differs significantly from the majority of "conventional" fossil-fueled vehicles currently manufactured. The intention was, therefore, to thoroughly investigate the possibilities of reducing the BIW mass, while accommodating the above design requirements. The starting point was essentially a blank canvas. The design volume for the topology optimization is defined, representing the cabin space and other packaging requirements. This design volume is illustrated in Figure 5.1, with approximate maximum exterior dimensions design volume of: (x, y, z) 3865 mm × 1850 mm × 1530 mm.

5.2.2 Crashworthiness criteria and design volume

For the purpose of the topology optimization, several loading scenarios need to be taken into account simultaneously. The reason for applying these various loading scenarios is to include a variety of possible impacts the vehicle might encounter during its lifespan into consideration.

Therefore, a total of six individual loading scenarios are defined:

1. Pole impact
2. Side barrier impact
3. Roof crush: "A" pillar

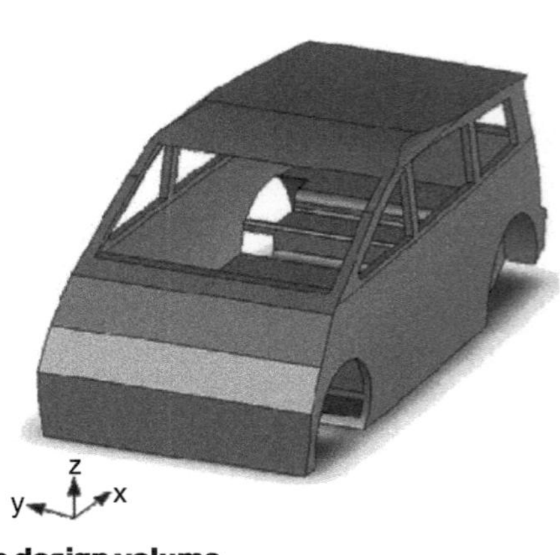

FIGURE 5.1 Vehicle design volume.

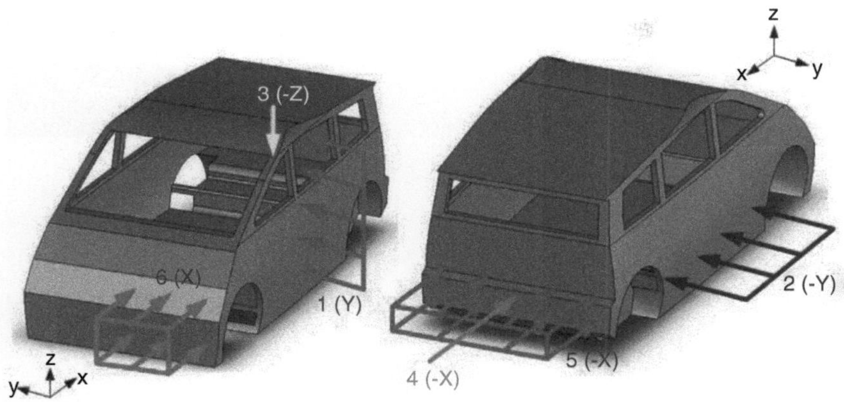

FIGURE 5.2 Location of the applied forces on a vehicle structure.

4. Low-speed centered rear impact
5. High-speed rear impact
6. Front impact: offset deformable barrier (ODB)

All loadcases are conducted as linear static. In addition to the above loadcases, a battery pack with a mass of 150 kg as well as a combined range extender and fuel tank with a mass of 110 kg are included. Illustrations of where and in which directions the various loads are applied can be seen in Figure 5.2. The bracket notations denote the direction of the applied loads relative to the coordinate system defined in Figure 5.2. For example, (*X*) denotes a load acting in the positive *x* direction.

The size and number of arrows for each loadcase does not depict the magnitude or the area over which a given loading is distributed. The magnitudes of the forces applied to each loadcase, as illustrated in Figure 5.2, are defined in Table 5.1.

TABLE 5.1 Applied Force Magnitudes

#	Loadcase	Applied Force	Applied Force Magnitude, *EVM* = 1200 kg
1	Pole impact	300 kN	300 kN
2	Side barrier impact	300 kN	300 kN
3	Roof crush ("A" pillar)	2.5·g·EVM	29.5 kN
4	Low-speed rear impact	150 kN	150 kN
5	High-speed rear impact	60·g·EVM	707 kN
6	Front impact (ODB)	60·g·EVM	707 kN

TABLE 5.2 Barrier Dimensions

#	Loadcase	Barrier Dimensions (mm) (Width · Height)
1	Pole impact	∅300
2	Side barrier impact	1560 · 475
3	Roof crush ("A" pillar)	250 · 250
4	Low-speed rear impact	250 · 250
5	High-speed rear impact	1375 · 250
6	Front impact (ODB)	1000 · 650

In Table 5.1, g is the gravitational acceleration (9.81 m/s^2) and EVM is the estimated vehicle mass that has been set to 1200 kg. The force magnitudes listed in Table 5.1 are distributed over areas with the dimensions specified in Table 5.2.

The loadcases listed in Table 5.3 are complemented by eight additional loadcases to include design robustness (Figure 5.3), as the crash tests may have some minor impact angle tolerances which have been estimated around ±5° for the offset deformable barrier (ODB) and ±10° for the rear impact barrier.

The impact values for robustness are listed in Table 5.3.

5.2.3 Discretization

The design volume was meshed using solid tetra elements with linear displacement functions and an average element size of 25.0 mm. This led to the generation of approximately 103,000 nodes and 527,000 elements.

TABLE 5.3 Loads for Robustness Design Analysis

#	Loadcase	Applied Force Magnitude
7	ODB +5°	707 kN
8	ODB −5°	707 kN
9	High-speed rear impact +5°	707 kN
10	High-speed rear impact −5°	707 kN
11	ODB +10°	707 kN
12	ODB −10°	707 kN
13	High-speed rear impact +10°	707 kN
14	High-speed rear impact −10°	707 kN

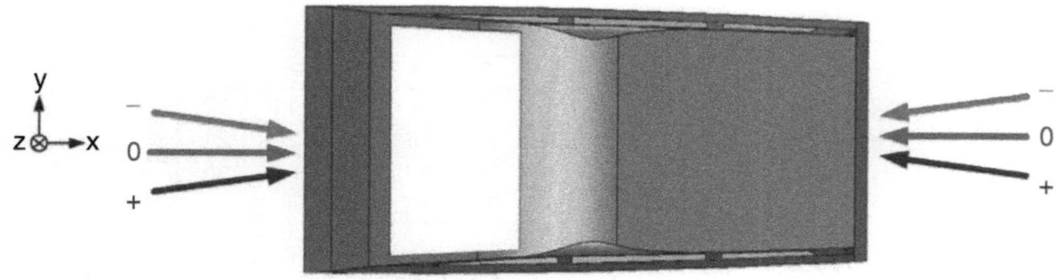

FIGURE 5.3 Impact sensitivity location.

5.2.3.1 PACKAGING OF ESSENTIAL COMPONENTS

In order to model the range extender as well as the battery box, it is decided to add two centers of mass (CM) to represent their inertial and structural influence.

The usage of two CM (i.e. battery pack and combined range extender/fuel tank) united with the discretization illustrated in Figure 5.4 led to a possible 1089 iterations (i.e. different models), which was not practical. Therefore, a series of considerations have been used to reduce the number of models. The first of such is the requirement for a symmetric BIW with respect to the *xz*-plane (Figure 5.1), that is, the *x*-axis in Figure 5.4. Due to the mass difference between the battery pack (150 kg) and the range extender/fuel tank (110 kg), a similar constraint cannot be applied parallel to the *yz* plane in Figure 5.1, that is, the *y*-axis in Figure 5.4. In addition, a series of considerations relating to, for example, general mass distribution have eliminated additional unfeasible configurations. A clear example of such a setup would be the CM of the battery pack in location 1 (Figure 5.4) and the CM of the range extender/fuel tank in location 2 (Figure 5.4). Thereby, the number of models has been reduced to 73, including a model containing only a single CM for all components located in position 14 (Figure 5.4) with a mass of 260 kg.

Two models (per BC type) intended to be used to assess the influence of the size of the battery have also been implemented. In Section 2.1, it was defined how the inertial effects originating from the mass of the components in question were implemented into the model. Equation 5.2 defines how these specific forces are calculated. The setup of model 69 is illustrated in Figure 5.5, where the respective locations of the battery pack as well as the range extender/fuel tank CM are indicated.

The roof was fitted with a thin sheet of steel, which will act as a 2D shear panel, with an element thickness = 0.7 mm.

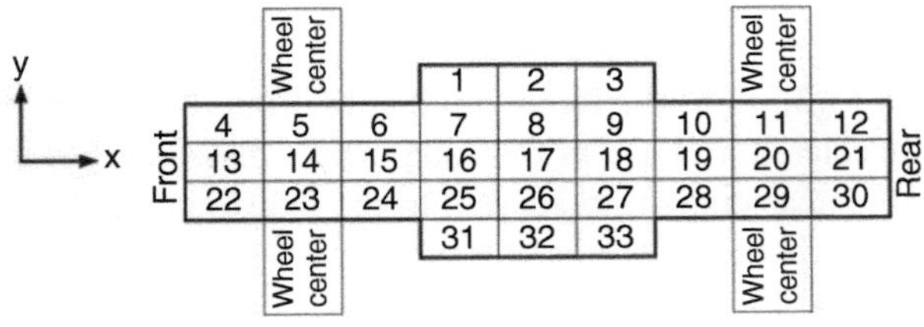

FIGURE 5.4 Possible locations to range extenders and battery box.

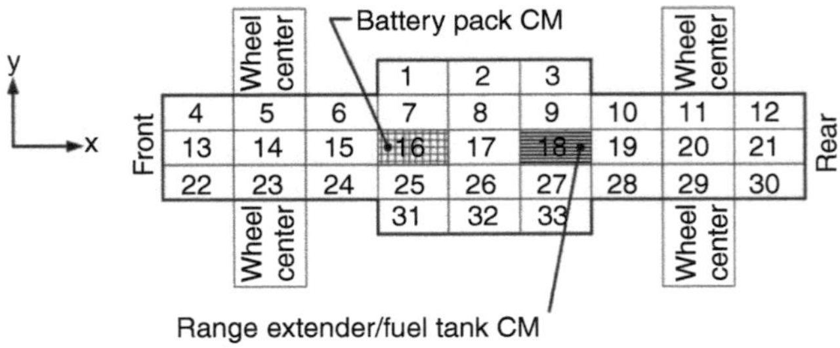

FIGURE 5.5 A configuration based on previous research (model 69).

5.2.4 Boundary conditions

In order to compute a linear topology optimization of this vehicle structure, two different types of boundary conditions can be considered, depending on the type of loading strategy adopted. The first method uses a standard method of restraining the model (single point constraints) and is based on restraining strategic nodes, while the second method does away with the restraining of nodes and uses inertia properties of the structure to compensate for its motion under loading (inertia relief).

5.2.4.1 SINGLE POINT CONSTRAINTS (SPC)

The usage of SPC requires definition of the allowable translational and rotational movements of specified nodes, that is, constraining certain degrees of freedom of these. In other words, it relates to solving Equation (5.1), which sets the relationship between stiffness matrix $[k]$, force $\{F\}$ and displacements $\{u\}$

$$\{F\} = [K] \cdot \{u\} \tag{5.1}$$

This time consuming reduction process needs to be repeated for every single variation of boundary condition using SPC. By utilizing SPC, the inertial effects originating from the mass of the battery pack and the combined mass of the range extender/fuel tank have been calculated as specified in Equation (5.2), that, is the relationship between applied forces, vehicle mass, and acceleration levels:

$$F_{\text{ie}} = cst \cdot g \cdot C_{\text{mass}} \tag{5.2}$$

In Equation (5.2), F_{ie} is the force (originating from inertial effects), g is the gravitational acceleration, C_{mass} is the components mass, and cst is a constant used to scale the acceleration, to adapt it to the individual loadcase. The individual values of the constant (cst) are listed as follows:

- 30 for loadcases 1 and 6 (high-speed impacts)
- 15 for loadcases 2 and 3 (side and barrier)

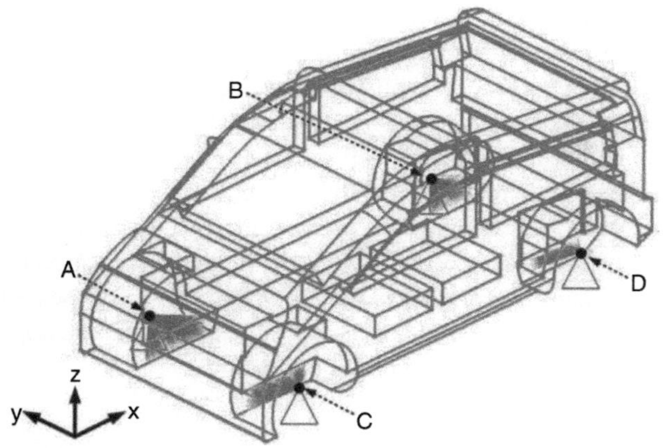

FIGURE 5.6 Location of the boundary conditions set for the SPC boundary scenario.

- 10 for loadcase 5 (low-speed rear impact)
- 0 for loadcases 4 and 7 (roof crush/torsion)

In order to compute using SPC, the model needs to be restrained in all the degrees of freedom to prevent any mechanism that would prevent the solver to compute a solution.

The four nodes (*A*, *B*, *C*, *D*) that unite the respective 1D rigid elements, as illustrated in Figure 5.6, are rigidly clamped, locked in all DOF (all translations and all rotations). These four nodes are located in the midpoint of the respective wheel wells in the *y*-coordinate and in the center point of the respective wheel wells in the *x*- and *z*-coordinates. The latter is illustrated in Figure 5.7, which is a close-up of the front left wheel well area.

5.2.4.2 INERTIA RELIEF (IR)

Inertia relief (Barnett & Widrick, 2005) utilizes a significantly different approach to obtaining load equilibrium of the models in question. In this approach,

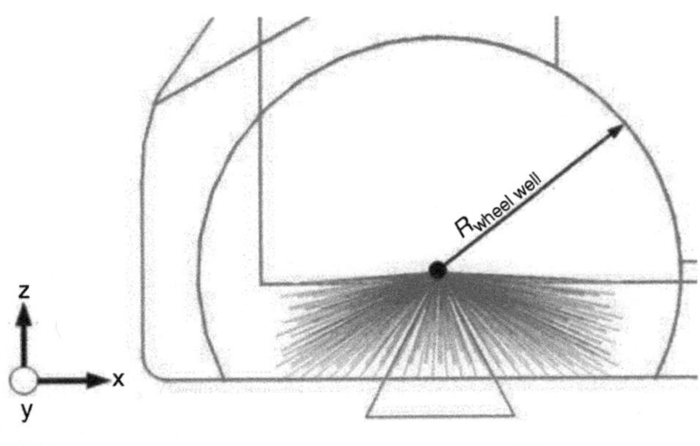

FIGURE 5.7 Wheel well area and SPC boundary conditions.

no DOF of any nodes are constrained (due to BCs). Instead, inertia relief works by balancing the external loading with inertial loads and accelerations within the structure itself. This is specifically done by "adding" an extra displacement-dependent load to the load vector $\{F\}$ in Equation (5.1).

This implies that if two "identical" models (where one utilizes SPCs and the other utilizes IR) are to be solved, then theoretically the inertia relief model should be slower to compute, simply because the stiffness matrix $[k]$ will contain additional terms. This is provided that the models in question only include a single loadcase and only one set of SPCs, or multiple loadcases utilizing one set of SPCs. The additional terms of the stiffness matrix can be appreciated by observing inertia relief formulation in Equation (5.3)

$$\{F\} = [K_{\mathrm{IR}}] \cdot \{u\} = \begin{bmatrix} [k] & 0 \\ 0 & [k_{\mathrm{add}}] \end{bmatrix} \cdot \{u\} \tag{5.3}$$

In Equation (5.3), $[k_{\mathrm{IR}}]$ is the stiffness matrix of the IR model, $[k]$ is the "original" stiffness matrix listed in Equation (5.1), and $[k_{\mathrm{add}}]$ represents the additional terms in the stiffness matrix, caused by the usage of IR. Thus, by comparing Equation (5.1) to Equation (5.3), the fundamental difference between the SPC models and the IR models can be appreciated.

The comparison indicates the existence of possibilities of (drastically) reducing the computation time of the topology optimization by utilizing IR compared with using SPCs, if models containing multiple loadcases (with different SPCs) are to be analyzed. This is because of the reduction process previously discussed. When computing the displacements due to the individual loadcases, the only parameter that will have to be adapted is the load vector $\{F\}$.

For the purpose of determining the stress distribution (or displacements) in connection with conducting a topology optimization, the stresses originating from the individual loadcases can simply be obtained by the method of superposition. The latter is, of course, only valid as long as the analysis is linear. Based upon the basic methodology of IR, it was not necessary to "transform" the inertial effects originating from the individual components, as is defined for the SPC models in Equation (5.2). Therefore, the representation of these masses was simply accomplished by means of nodal masses in the vicinity of the intended CM.

5.2.4.3 RUNTIME DIFFERENCES BETWEEN SINGLE POINT CONSTRAINTS AND INERTIA RELIEF

The two previous sections outline some of the possible differences with respect to CPU time when subsequently solving the FEM models (optimization). As the impending optimization was to be performed linear statically, the relationship between the stiffness matrix $[k]$ or $[k_{\mathrm{IR}}]$ and the volumetric mass density (ρ) was

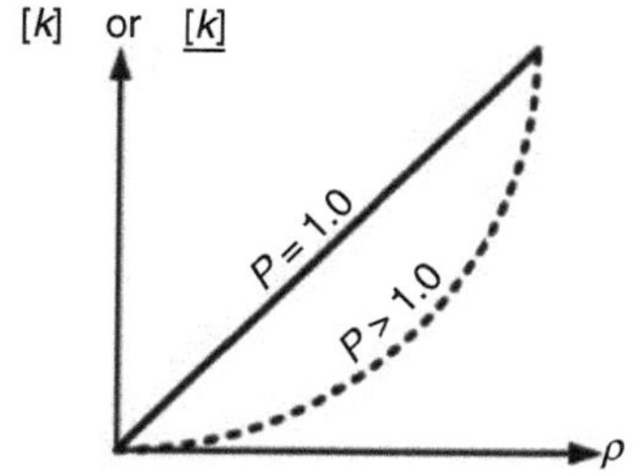

FIGURE 5.8 Relationship between [*k*] and *ρ*.

defined by the "power law for representation of elasticity properties," as formulated in Equation (5.4), that is, SIMP formulation

$$\underline{[k]}(\rho) = \rho^p [k] \tag{5.4}$$

In Equation (5.4), [*k*] is the penalized stiffness matrix and *p* is the penalization factor, which is used to determine the "type" of relationship between [*k*] and *ρ*. As long as *p* is equal to 1.0, the two are directly proportional, as illustrated in Figure 5.8.

This relationship can be adjusted, by varying *p* with the effects as indicated in Figure 5.8. The reason for adjusting this relationship is typically to penalize intermediate density values in order to avoid "vague" topology definitions, which is also sometimes referred to as "checkerboard effect."

However, subsequent analyses revealed that this was not a widespread problem for the models in question.

Therefore, in the remainder of this paper, the value of *p* will be set to 1.0, that is, a linear relationship between the stiffness matrix [*k*] and the mass density *ρ*. In addition to the above argument, the linear relationship also represented the shortest CPU time for a given model. With the above set, a series of trial runs were conducted on otherwise "identical" models, apart from the BCs, which are summarized in Table 5.4.

The above clearly indicated the significant differences in CPU time, as the usage of IR, as opposed to SPC, represented a reduction in CPU time of approximately 91%.

TABLE 5.4 Average Model Runtime Difference Between SPC and IR

	Runtime (s)	Runtime (h)
SPC	59249.3	16.5
IR	5163.5	1.4

The advantages of utilizing IR from a CPU point of view was obvious; however, it is very important to understand the different optimization outcomes of both methods, which is going to be discussed in the next section.

5.2.5 Post processing of optimization for SPC and IR

For the purpose of comparing the topology outcomes of SPC and IR, the design volume will be divided into "areas," as depicted in Figure 5.9. The following terms will henceforth be used when referring to the areas identified in Figure 5.9:

- Roof (1)
- Side (2)
- Front (3)
- Rear (4)
- Floor (5)

Starting with the roof: Throughout the variation of the individual models, where the main parameter is location of the CM, a general tendency of the roof topology has been found. This tendency is applicable to both the SPC and the IR models, thereby indicating that the location of the CM has only minor influence upon the topology of the roof area. This general tendency is illustrated in Figure 5.10.

An important factor to note when viewing the widespread triangulation of the roof structure is that it diverges significantly from conventional structures as illustrated in Figure 5.10, which represents the inclusion of a 2D shear panel, as defined previously.

The second area identified in Figure 5.9 was the side (2). Based on the study, it was found that a clearly defined topology for the side area proved difficult to obtain. However general tendencies are still found, even if some models do not display a clearly defined topology in this area.

A typical example of a vaguely defined topology of the side area is illustrated in Figure 5.11.

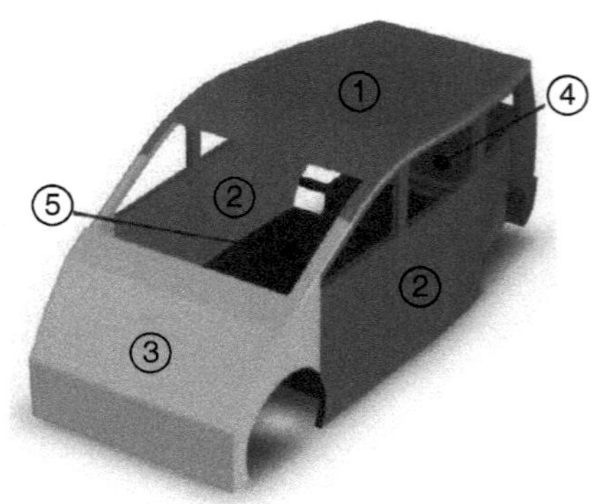

FIGURE 5.9 Design volume areas.

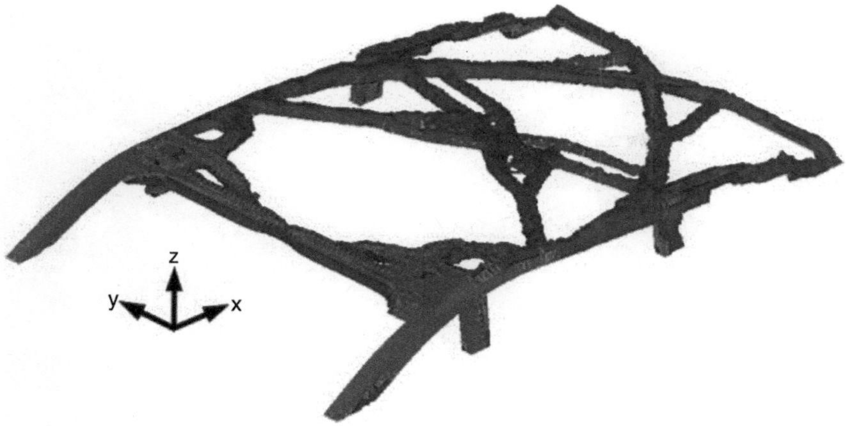

FIGURE 5.10 Roof topology.

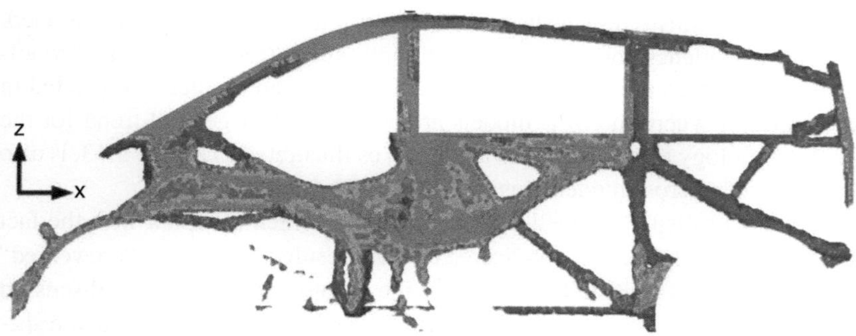

FIGURE 5.11 Vehicle side topology.

By observing Figure 5.11 and the associated stress plots/distributions, it is evident that additional topology in this area is desirable. A possible way of overcoming this problem is simply to adjust the penalization factor *p* and subsequently re-run the optimization. If the desired level of definition is still not achieved, then supplementary models are likely to be required, which might involve different approach altogether (Marur & Srinivas, 2008). However, a general tendency for the topology of the side area throughout this study exists, indicating that the topology of this area is influenced neither by the location of the components in question, nor then by usage of SPC versus IR.

In line with the previously discussed areas, the front did not show any significant tendencies between the SPC and IR models. In addition, the location of the CM also showed minor local changes in the topology of the individual models. Thus, a general tendency for the linear topology of the front area representing both the SPC and the IR models can be identified. This general tendency is illustrated in Figure 5.12.

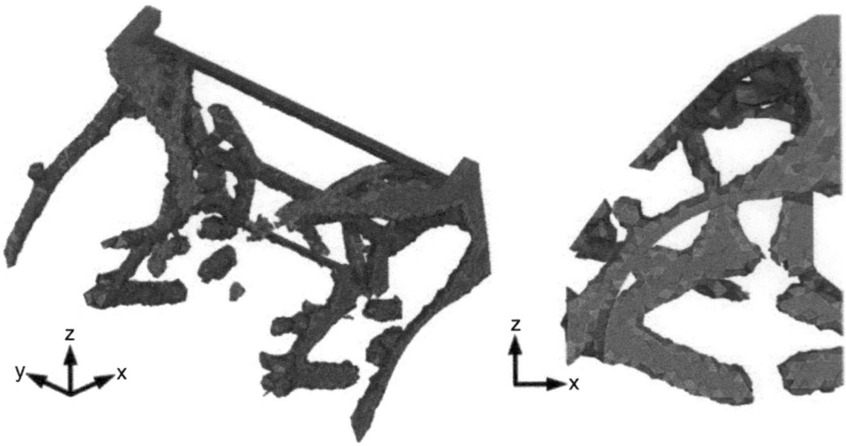

FIGURE 5.12 Vehicle front topology.

Figure 5.12 illustrates a well-defined topology that, as previously mentioned, is a general tendency for all models, with only minor (local) changes of topology between different models. These local variations are, in general, located in the vicinity of where the SPC models are constrained. A general trend for the rear area topology also exists. This tendency, as illustrated in Figure 5.13, is also distinguishable across models, regardless of BC type.

Figure 5.13 displays a well-defined topology, which (coupled with the fact that it is a general tendency) indicates that the results represent a "converged" topology for the rear area. The results of the topology optimizations discussed so far have only displayed minor (localized) changes of topology within the specific areas in question. These changes were caused by change of CM locations and/or a change in selected BC.

FIGURE 5.13 Vehicle rear topology.

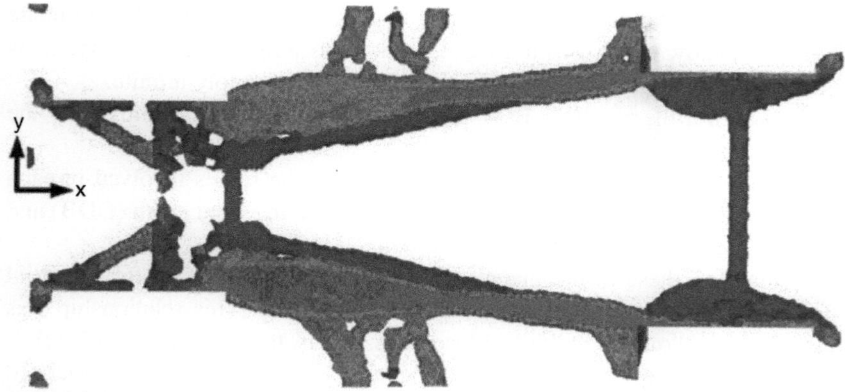

FIGURE 5.14 Floor topology, SPC.

The topology of the final area to be discussed is the floor (5), as indicated in Figure 5.9. The changes of topology of this area were significantly more substantial, with respect to a change of BC. Therefore, at this point, it was necessary to once again group the models by BC type, starting with the SPC models. In line with the previously discussed results, these models showed a general tendency that was applicable to all SPC models. Only very minor local changes of topology were found when the location of CM was altered. The generalized topology is illustrated in Figure 5.14.

An appreciation of the difference in results obtained by SPC and IR models can thus initially be understood by comparing Figure 5.14 to Figure 5.15.

Throughout this section, and culminating with Figure 5.15, it has been underlined that the effects of the location of CM upon the resulting topology were localized, with respect to the SPC models. This statement essentially summarizes the outcome of the topology optimization study, with respect to the SPC models; namely, that the location of the battery pack and range extender/fuel tank within the BIW had a trivial influence upon the topology optimization results.

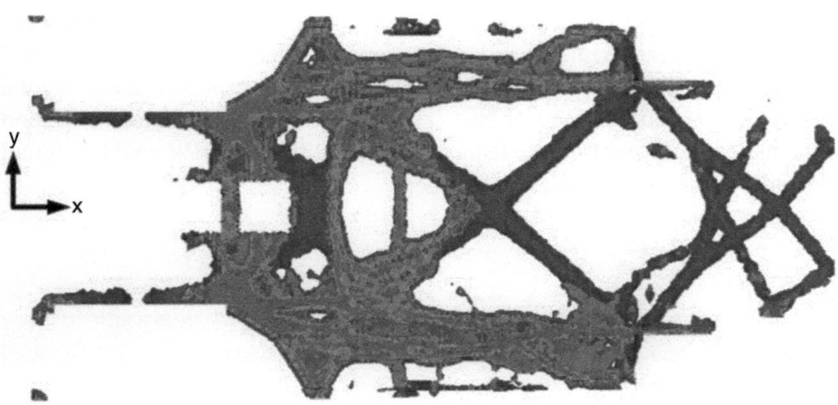

FIGURE 5.15 Floor topology, IR.

Based upon the above, the number of possible responses to the ultimate aim of this study, which was to find the optimal locations of these components within the BIW with respect to reducing BIW mass, is simply infinite.

So why did the resulting BIW topology not change significantly with the location of the CM for the SPC models?

A likely answer was simply the magnitude of the forces involved in each loadcase. The highest magnitude forces are found in the front crash (ODB) and the high-speed rear impact loadcases, as defined in Table 5.2.

Comparing these to the magnitude of the forces originating from inertial effects from the components, the difference was 90%. This relationship was repeated throughout the individual loadcases, that is, the force magnitude originating from component mass was significantly less than that of the associated "external" force. Therefore, it was thought that the optimization for the SPC models was primarily dictated by the external forces, drastically reducing the influence of the forces originating from component mass.

5.2.6 Realism of SPC and IR boundary conditions in linear topology optimization in vehicle architecture definition

Substantiation of the above statement was sought by studying the effect of doubling the mass of the battery box and the range extender (model 73), as specified in Table 5.5.

However, the results did not indicate any significant changes relative to the remaining SPC models.

In a final attempt to investigate the correctness of the above statements, the further model with the mass values scaled by a factor of 3 was computed and no significant mass increase was found. At this point, with this heavier model, the forces in question were still only approximately one-third of the magnitude of the "external" forces.

At this point, it could be argued to further increase the "mass scaling"; however, the relevance of such a model could be seriously questioned. Returning to Equation (5.2), model 73 represents a battery pack with a mass of 300 kg exposed to a deceleration of 30 g and a range extender/fuel tank of 220 kg exposed to 30 g. The relevance of increasing these values with respect to achieving the overall aim was questionable at best. Therefore, no specific (optimal) location of the battery pack nor range extender/fuel tank within the BIW could be found by means of the SPC models.

TABLE 5.5 Model Definition

Model #	Adjusted From Model 69 (Figure 5.5)
73	C_{mass} scaled by a factor of 2

With the above in mind, the focus turned to the floor topology of the IR models.

These results clearly showed significant differences between models when the locations of CM were altered. These differences were also reflected when the mass reduction (in %) of the individual models was calculated, using vehicle mass reduction evaluation, seen in Equation (5.5)

$$\text{Mass reduction} = 100 - \left(\frac{\text{FIMV}}{\text{ZIMV}}\right) \cdot 100 \tag{5.5}$$

In Equation (5.5), FIMV is the mass value of the final iteration and ZIMV is the mass value of the initial (0th) iteration. The above thus meant that the performance of the individual models, with respect to achieving the ultimate aim of the study, could now be specifically evaluated on a model-by-model basis. This was also the case with the SPC models, but the significant difference was that with the IR models the individual models stand out by displaying variety in floor topology and even more importantly in the value of mass reduction. The scope of mass reduction value for numerous IR models (73) ranged from approximately 81.2% to 91.2%. The corresponding range for the SPC models was 88.0% to 90.1%, thereby underlining the significant difference in results obtained.

Comparing the average mass reduction values of the SPC and IR models also revealed an interesting fact, namely that the average mass reduction values of the IR models were greater than those of the SPC models. The average mass reduction value of the IR models was 89.5% compared with 88.9% for the SPC models. Primarily based upon the above findings, seven models were selected for further investigation. The selection criteria for these seven models, which were all IR models, all showed a mass reduction in excess of 90.0%. These models are all listed in Table 5.6.

Model numbers 65, 68, and 72 are slightly different in setup than the other models, as previously explained. Model 65 had a larger "battery area" in the *xy*-plane (Figure 5.4) than the others; this covered the areas denoted in Table 5.2 and illustrated in Figure 5.6. Model 68 contained a coincident location for both components, that is, a mass value of 260 kg was located here. Finally, model 72 did not contain any mass for the components, and thus represented the BIW without the influence of the battery pack and range extender/fuel tank. Unsurprisingly, model 72 (which essentially is a replica of model 69) represented the largest mass reduction value, undoubtedly due to the absence of the component masses. Model 72 was, therefore, used as a benchmark to evaluate the remaining models, simply because it was unaffected by the components, that is, the influence upon topology of the additional mass from the components was eliminated. Comparing the 91.2% reduction of model 72 to the average mass reduction value of the remaining models listed in Table 5.6, it was found that the difference was approximately 1.1%, which was deemed insignificant compared with the additional 260 kg of mass included in these models.

TABLE 5.6 Models Selected

Model #	CM (Battery Pack)	CM (Range Extender/ Fuel Tank)	Mass Red.
19	16	15	90.0%
34	14	13	90.2%
35	15	13	90.3%
42	15	14	90.0%
43	16	14	90.1%
49	16	15	90.0%
65	1, 7, 16, 25, 31	20	90.0%
68	14	14	90.0%
72	N/A	N/A	91.2%

Another interesting fact, which can be found by studying Table 5.6, is that all models listed (with the exception of model 72) have located the CM of all components in the symmetry line (i.e. the x-axis) of Figure 5.9. At this point, the 73 initial models have been reduced to seven, including model 72, which was not a viable solution.

The next step was, therefore, to investigate the sensitivity of the remaining models.

5.2.7 IR models sensitivity study

The study was divided into three major parts:

- Angle sensitivity
- Stiffness sensitivity
- Combined angle and stiffness sensitivity

5.2.7.1 ANGLE SENSITIVITY STUDY

The angle sensitivity study was performed by including additional loadcases within the specific models. These loadcases were replicas of the front crash (ODB) and the high-speed rear impact loadcases, as defined in Section 5.2.2. Neither the application points nor the force magnitudes were altered during this study, only the angles, as depicted in Figure 5.3 (the "0" arrows representing the "original" setup).

The additional loadcases contained separate plus and minus loadcases. This was carried out for 5° and 10° angles, in addition to a combination of the two.

This led to a total of four different variations of each of the single models listed in Table 5.6, creating an additional (9 × 3) 27 models, specified as:

1. Original model (i.e. no additional loadcases)
2. 5° loadcase additions:
   a. 5° + front crash (ODB)
   b. 5° − front crash (ODB)
   c. 5° + high-speed rear impact
   d. 5° − high-speed rear impact
3. As specified in point 2 above, but with 10° angles
4. Both 5° (point 2) and 10° (point 3) loadcase additions

The results from these additional models showed a further change in (particularly) the floor area between the "original models" (point 1) and the models with additional loadcases (points 2–4). However, the individual differences between the latter models were not found to be significant. An example of this significant difference can be seen in Figure 5.16, which represents the results of the angle sensitivity study of model 19 (Table 5.6), also including parts of the front area and rear area topology.

The upper illustration in Figure 5.16 represents the original model, while the lower illustration represents the general topology of the remaining three models.

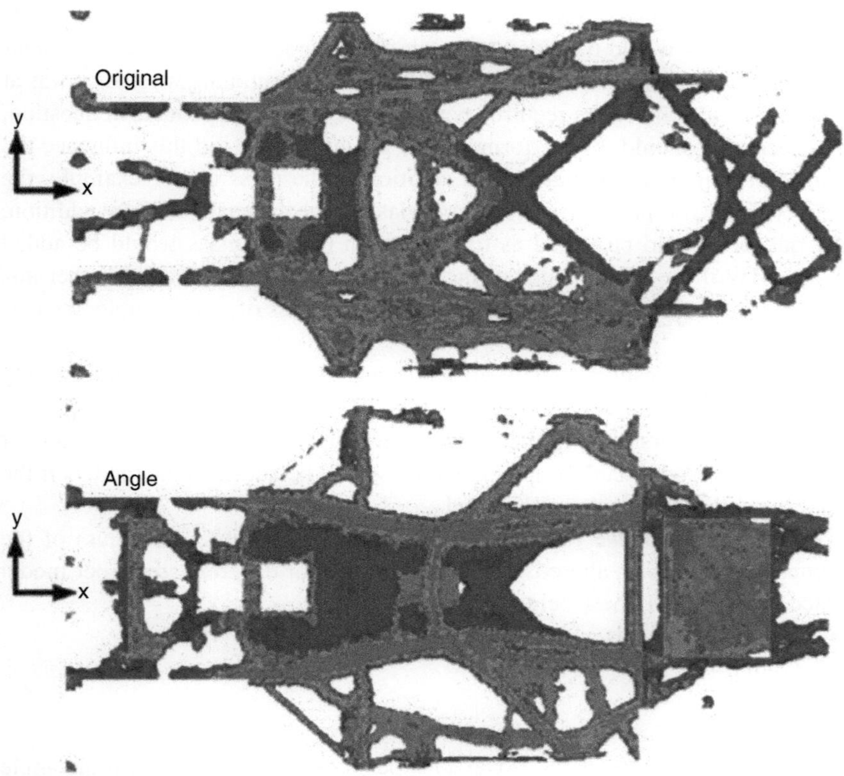

FIGURE 5.16 Floor topology, angle sensitivity study.

The differences in Figure 5.16 are clear, underlining the sensitivity of the models with respect to the angle of application for these particular loadcases. This result also substantiates the previous statement that the topology optimization was primarily dictated by the (largest) external forces. The above tendency was also found for the remaining models listed in Table 5.6. The next step in the IR models sensitivity study was the stiffness study.

5.2.7.2 STIFFNESS SENSITIVITY STUDY

The stiffness sensitivity study was based on the models listed in Table 5.6; omitting the angle study. The implementation of the battery pack and range extender/fuel tank within the actual models consisted of excluding specific elements from the design domain, while adding the required concentrated nodal masses.

These elements were solid elements with linear displacement functions, and the material characteristics of a mild grade steel. This was equivalent to having a "solid" steel plate in these locations, including the associated stiffness. Thereby, it was assumed that the performance of the components, for example, the battery pack, during loading was equivalent to the performance of the aforementioned steel plate, which was a significant performance requirement.

The reasoning behind this was that if these components were part of the main load-bearing structure (i.e. the crash structure), then no additional or unnecessary mass would be added to attach the components to the BIW. A counterargument to the above could be made by firstly questioning whether it was at all possible to achieve the required performance of the components in question, and secondly, should this performance be achieved, how did this influence the manufacturing and material cost in addition to the mass of, for example, the battery pack (or perhaps more appropriately, the battery box). In addition, the question could be raised as to how much (if any) mass would be added to the BIW if the performance of the components is lowered and whether this value would be in excess of the expected gain in mass of, for example, the battery pack.

The above questions cannot all be answered at this point in time, simply because of their complexity.

However, a first step in the effort to answer these was taken by means of this stiffness study; namely, how much mass will be added to the models if the battery pack compliance was lowered.

In order to achieve this, the stiffness value (i.e. Youngs modulus) of the component areas was altered. This was done in four different steps (per model listed in Table 5.6), these being:

- 210 GPa (original)
- 70 GPa
- 35 GPa
- 17.5 GPa

This also leads to an additional 27 models, as was the case with the angle study. The general outcome of this study was that the influence of the stiffness upon the topology optimization results was minimal. However, at this point, it must be underlined that the areas in question were (generally) 250 mm by 250 mm, as defined in Section 5.2.2, making them small compared with the overall dimensions of the design volume. This obviously had an effect on the results of the study. The exception of the above is model 65 (Table 5.5), which utilized a considerably larger area to represent the battery pack. The results of this model are illustrated in Figure 5.17, where the areas marked with an x represent the size of the battery pack. Please note that Figure 5.17 also includes parts of the front and rear area topology.

Figure 5.17 clearly shows a significant difference with respect to the floor topology, which was not found in the remaining models in the stiffness sensitivity study. This was, as previously mentioned, not unexpected as the area with the reduced stiffness was significantly larger in model 68 when compared with the remaining models listed in Table 5.6. The topology illustrated in the lower part of Figure 5.17 contains a less distinctive/defined topology than the upper illustration in Figure 5.17, which could possibly be resolved by adjusting the penalization factor p. However, to fully understand the implications of the reduced stiffness value, the global

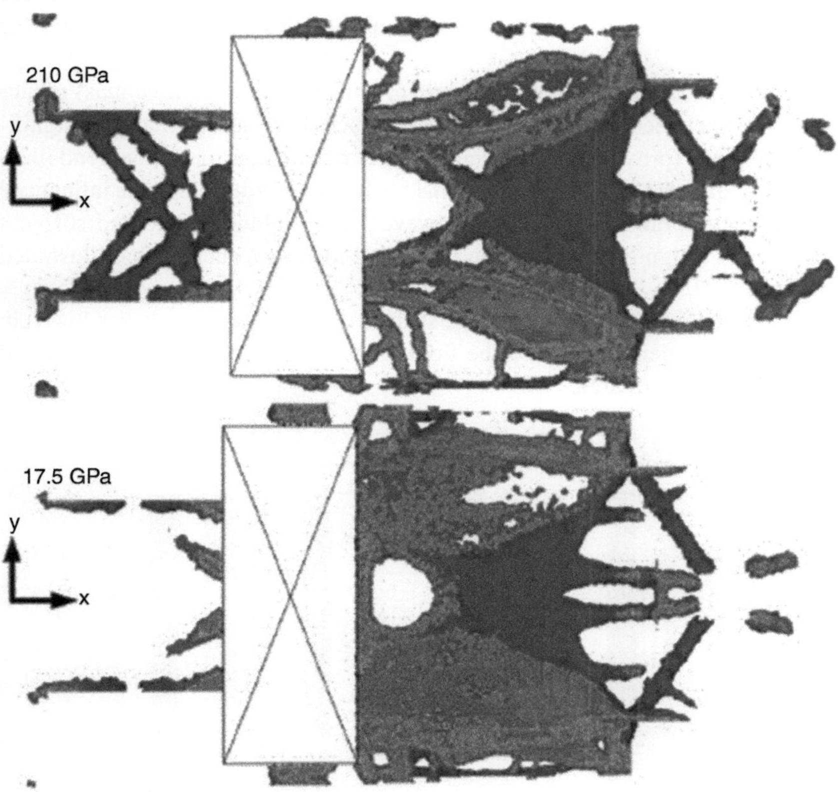

FIGURE 5.17 Floor topology, stiffness sensitivity study.

topology was reviewed because the changes were no longer localized to the floor area, although the most significant differences were still found there. An interesting fact was that the mass reduction of the 210 GPa model was exactly the same as the 17.5 GPa model, namely 90.0%. The above indicated that when the "external" forces, that is, the ones defined in Section 1.2, were applied, the distribution of these throughout the BIW had changed (between, e.g. the 210 GPa and the 17.5 GPa) model, which was also substantiated by the global change of topology.

It had been established that both the angle and the stiffness significantly influence the outcome of the topology optimization; therefore, these must be taken into account.

The next step was, therefore, to conduct a combined study incorporating both angles and stiffness in one study.

5.2.7.3 COMBINED ANGLE AND STIFFNESS SENSITIVITY STUDY

This combined study produced a total of 144 models (in addition to the original SPC/IR study). The results and implications of this study exceed the scope of this book; however, an important (and general trend) was identified. This can be understood by observing Figure 5.18, which represents the mass reduction in percentage as a function of angles and stiffness values for model 65.

An important trend to note in Figure 5.18 is that the individual mass reduction values between the individual grouped models do not display any significant difference (provided that the angle values remain constant). The trend illustrated in Figure 5.18 did not indicate that changing the stiffness value influenced the topology optimization results. However, the combined angle and stiffness study found a significant change in the global topology, such as that illustrated in Figure 5.17.

In general, it was found that if the stiffness of the battery pack was lowered, the relative element density of the floor area would decrease, while it would increase for the roof area.

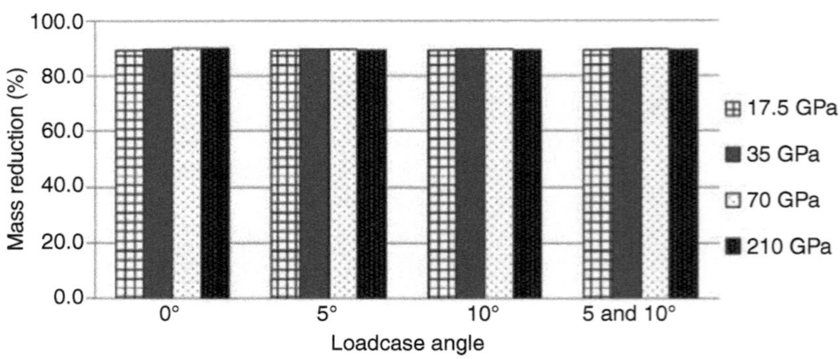

FIGURE 5.18 Combined angle and stiffness study results.

5.2.8 Discussion

The generalized topology of the roof area has remained very consistent throughout the entire study, while utilizing both SPC and IR, and also including the sensitivity studies, wherein only minor changes occurred. The simple conclusion is that the topology of this area has converged. This converged topology was, as previously mentioned, unconventional when compared with the roof bow structures of many modern-day passenger vehicles.

The side area topology has, in line with the roof topology, also remained consistent; however, a significant number of the models display a rather vague definition of the side area topology, as illustrated in Figure 5.11. Therefore, additional optimization studies will be required to achieve a more refined topology for this area.

The floor area topology was found to produce the most significant changes in topology during the studies. The implications of the battery pack and range extender/fuel tank CM were used to isolate nine models, from which eight are potential solutions to location of these CM. The effects of angles and stiffness upon the floor topology were investigated, leading to a suspected link between component (e.g. battery pack) compliance and BIW mass.

In general, the individual floor area topologies were found to be viable solutions that can be implemented to successfully withstand the dynamic crash-loading scenarios. However, this is solely based upon mechanical engineering judgements and is not at this point backed up by any calculations.

The results relating to the roof, floor, and partially the side area topologies, which in essence make up the safety cage of the vehicle, display a well-defined topology.

The front and the rear areas were also found to change significantly with angles and stiffness values, as indicated in Figures 5.17 and 5.18.

The response of the topology optimization with respect to the front and rear areas, primarily as a function of changing the angles, seems to be "triangulation," that is, the widespread use of triangles. This makes perfect sense from a linear static point of view, as the stiffest geometry in solid mechanics is indeed a triangle. However, this raises serious concerns when the subsequent step is taken into dynamic loading, primarily because of the triangles resistance to buckling, which undoubtedly will have a negative influence on the crushability, and therefore the dynamic crash performance of these very vital areas. This is evidently one of the major limitations of the linear solver and highlights the necessary further steps in the optimization procedure.

5.2.9 Conclusions and limitations

The discussions and the results presented in Section 5.2 have led to the definition of an alternative methodology/approach to BIW design. The results and methodology presented are only the initial steps.

However, it demonstrates the efficiency of this technique, while underlining the valuable outcomes that can be implemented in the continued BIW development process. These outcomes primarily determine the outline requirements for structural load path development, mass distribution, and location of crash structure.

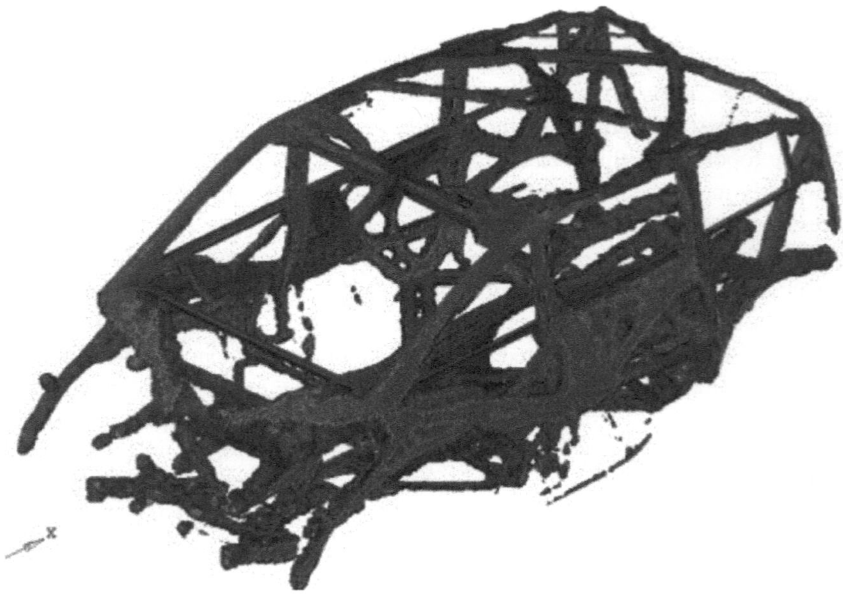

FIGURE 5.19 Isometric view, IR topology optimization example.

These were obtained by relatively simple modeling techniques, nevertheless implementing the required complexity in order to create a good starting point for the continued BIW design process. An example of this is illustrated in Figure 5.19.

Nonetheless, as discussed, there are some limitations of the linear (or implicit) topology optimization. This method cannot fully consider inertial effects, as well as other aspects such as detailed material property characteristics; strain rate effects, real-world effects, such as the interaction of the vehicle with its surroundings (ground and barrier), as well as the interaction of the assemblies within the vehicle itself. The particular topology optimization process presented in this chapter also utilizes simplified panel joining methodology, in addition to the joining of subassemblies.

To enhance the output of the linear topology optimization, an explicit optimization process could be developed (Soto, 2004). This process will likely use explicit crash analysis modeling techniques to generate a finite element representation of the linear structure and subject it to (dynamic) crash loading. However, this step is likely to be very extensive, and before commencing, the choice of methodology should be painstakingly considered (Duddeck, 2007; Zimmer et al., 2009; Acar & Solanki, 2009; Liao et al., 2008, Horstmeyer et al., 2009).

This process should include development of material properties where practicable, in addition to including relevant and available vehicle subassemblies (Richardson et al., 2003; Stephenson, 2010). The intention is that this process will optimize the shape and material properties of the crash structure to provide a suitable structure for high-speed dynamic crash impact.

5.3 FROM TOPOLOGY OPTIMIZATION TO COMPUTER-AIDED DESIGN (CAD) MODEL

5.3.1 Introduction

Following the previous chapter, it has been concluded that despite the linear topology techniques available, some limitations have been observed. Nevertheless, the linear topology computation could provide some interesting information that could be considered during the early stages of vehicle design. The topology results are, however, quite complicated to utilize, and provide a meaning to. Another typical topology optimization result is depicted in Figure 5.20, containing all the loadcases discussed in this chapter.

There are fundamentally two major questions:

1. How are the forces "flowing" inside the structure?
2. How can the result of the topology be converted into a structure that can be manufactured?

As an example, it can be observed that the front crash structure is quite complex and is vastly different from the "standard" front crash structures currently manufactured. It can be noted that the force path is triangulated to maximize the vehicle stiffness, which is incompatible with dynamic load dissipation during a crash event. During a crash event, the frontend of the vehicle is required to crush in a controlled manner, without catastrophic buckling. The structure circled in Figure 5.21 and Figure 5.22 is not compatible with this design requirement.

What can be done to overcome this topology optimization limitation?

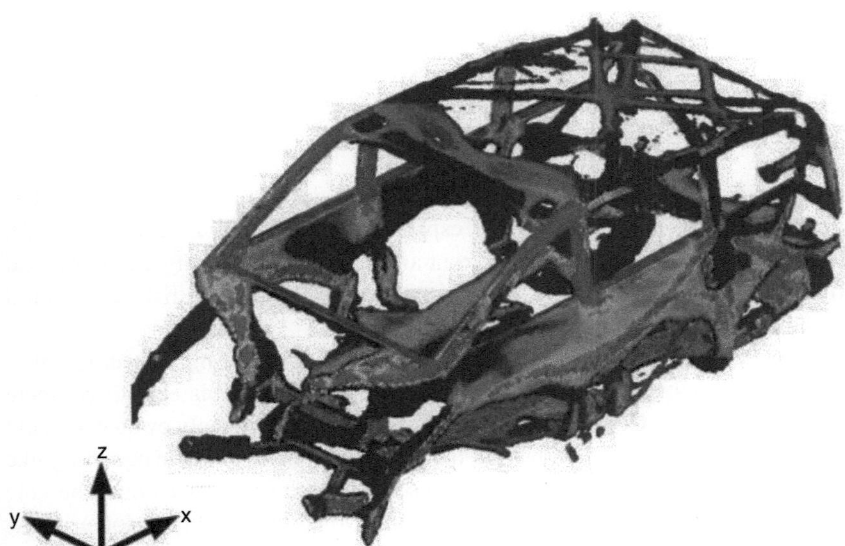

FIGURE 5.20 A typical topology optimization result.

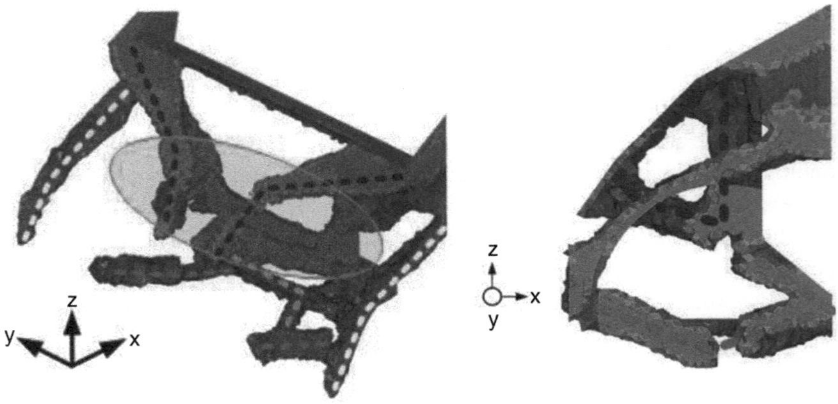

FIGURE 5.21 Vehicle frontend loadpath interpretation.

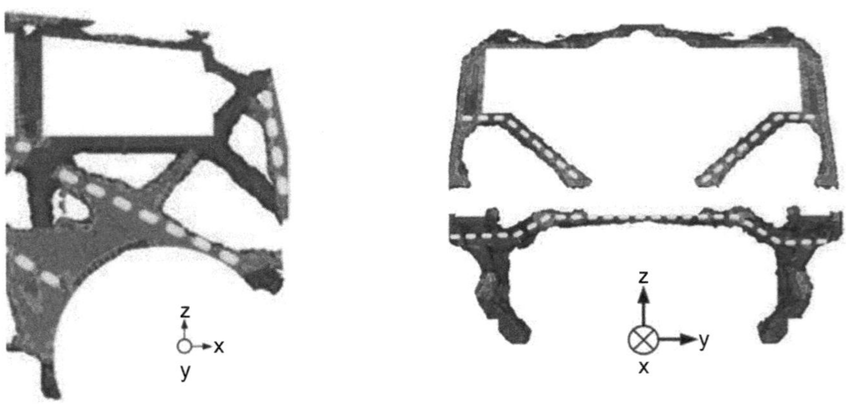

FIGURE 5.22 Vehicle rear end loadpath interpretation.

Considering other areas of the vehicle, such as the roof, for example, the interpretation can be quite cumbersome, especially in the header area (Figure 5.23). The roof is an intrinsic part of the front and side loadpath, as it carries the load to the other side of the vehicle in compression; hence, there is a risk of structural buckling which is undesirable.

The side structure of the vehicle works mostly in bending; hence, its interpretation is important, but less critical as the linear solver is able to compute bending scenarios. Consequently, the safety cage can be mostly evaluated using linear topology optimization but not the sections prone to extreme buckling, like the vehicle frontend and rearend. In these areas, the solver used must be able to compute bifurcation, which is not possible when using a linear solver (Bastien, 2010; Bastien & Christensen, 2011; Christensen et al., 2011; Christensen & Bastien, 2011a).

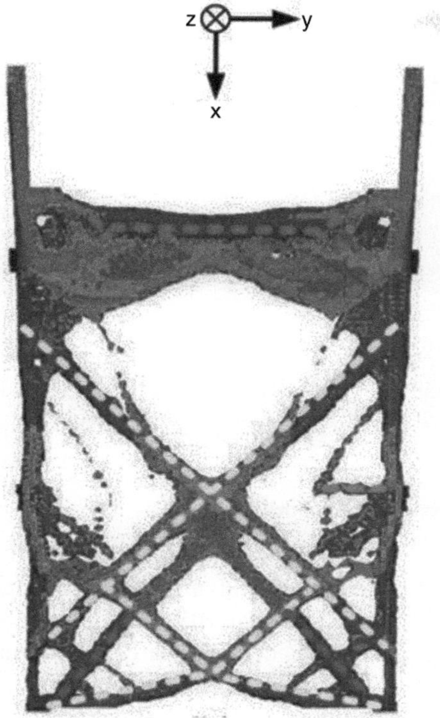

FIGURE 5.23 Vehicle roof loadpath interpretation (Figure 5.20).

Despite these limitations, the next part of this chapter will propose some methods to obtain a useful vehicle architecture based on linear topology optimization and provide useful guidance to the development of the vehicle body sections.

- The first method proposed in this chapter will be based on a holistic approach using beams and shear panels to size the full vehicle structure.
- The second method will use the fundamentals of the first method, and will also provide a useful design constraint that will aid in the sizing of the vehicle frontend structure.
- The third method will use the fundamental of the first method as an initial stage and will propose a useful method to develop manufacturable body panels.

5.3.2 Method 1: a holistic BIW sizing methodology

The Method 1 optimization technique in this section utilizes the design envelope of the Tata Beacon (LCVTP, 2011). The ultimate aim of this process is to complement the topology optimization phase by providing further detailed definitions of the local BIW structure (cross-sections), while keeping the vehicle architecture mass to a minimum. BIW development processes do, in general, not utilize both topology-, shape-, and size-optimization in succession to each other. Consequently, current BIW designs do not fully exploit the potential of structural optimization for minimizing their mass (Duddeck, 2007).

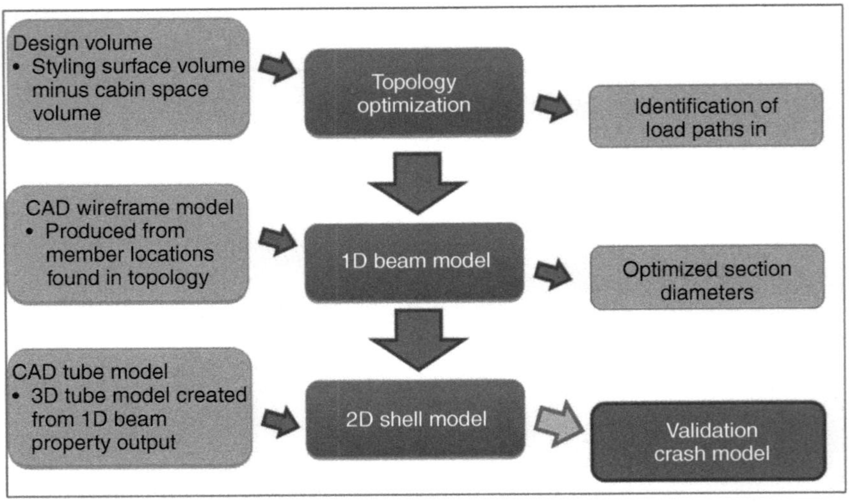

FIGURE 5.24 Holistic design process.

The flowchart of the proposed design process is illustrated in Figure 5.24, listing the necessary steps to minimize BIW mass. Future vehicle architectures begin with the styling envelope, defining the design volume for the topology optimization process. The design volume definition is necessary for the success of a truly lightweight architecture characterization (Christensen et al., 2011b).

The design volume is defined by removing the interior cabin volume from the volume created by the exterior styling surfaces. In locations where thin walls existed (i.e. roof and sides), a design volume was created (e.g. a 50 mm) to allow space for structural members to form during the topology optimization process.

All nonstructural components are excluded from the BIW design volume, such as the electric motor, the batteries, and the range extender, as these are not assumed to transmit any load originating from crash scenarios. Excluding these from the design volume ensures that no structural members are defined in these areas during the topology optimization, thus allowing sufficient space for these components and vehicle packaging in general.

Furthermore, vehicle apertures are only considered to be attached at hinges and locking points; hence, these are detached from the main body (design volume) and are fixed to the main body in such a way that they only provided longitudinal stiffness for the appropriate loadcases.

With the above considerations, the design volume for the BIW is defined and provides the starting point for the holistic optimization process (Christensen et al., 2011b).

The design volume is meshed with first-order tetrahedral elements and includes concentrated nodal masses to reproduce the inertial effects originating from, for example, battery mass. Vehicle apertures are constrained to the main vehicle body at the hinge and lock locations to represent longitudinal stiffness in connection with the relevant loadcases documented in Section 5.2.1.

A symmetry boundary condition is enforced in order for the topology to compute a vehicle symmetrical about its centerline.

The general load paths for the BIW architecture are then extracted by means of topology optimization, removing inefficient material with respect to the structural integrity exposed to the predefined loadcases (Figure 5.25).

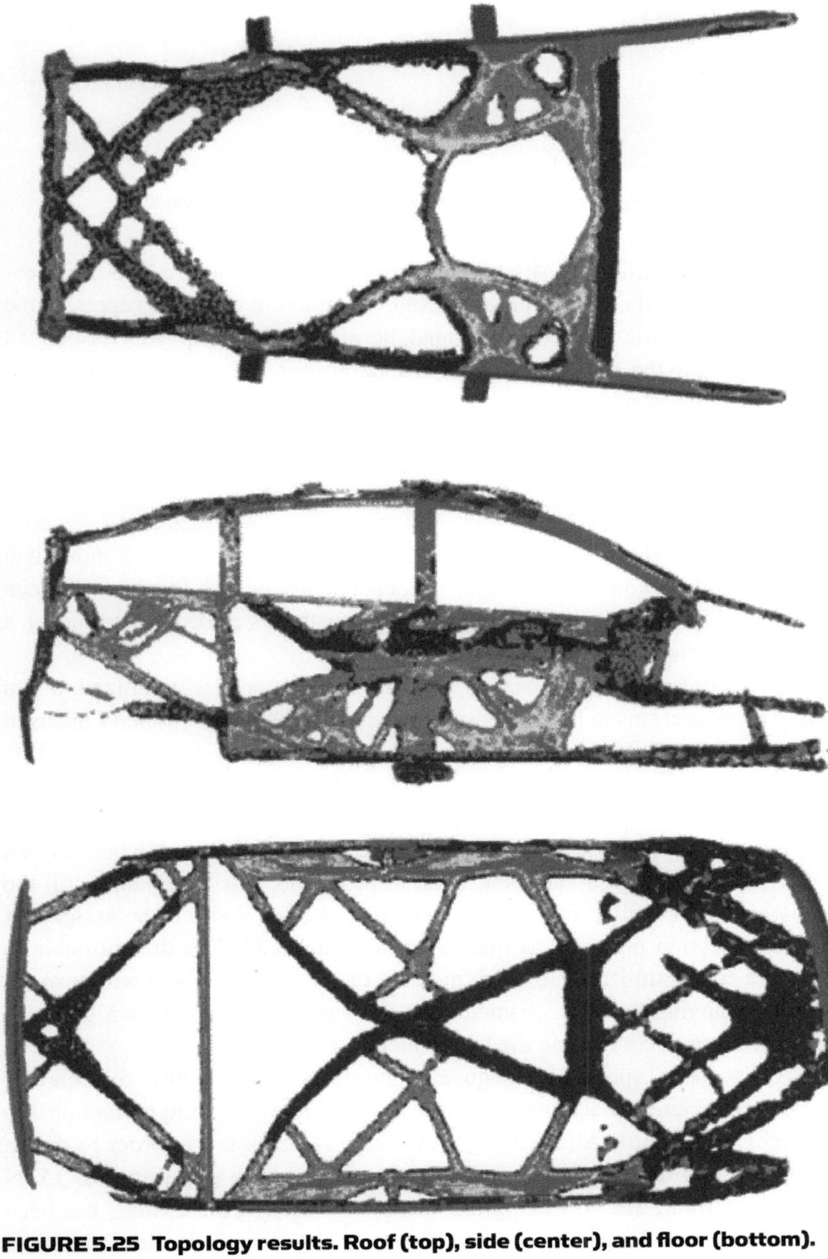

FIGURE 5.25 Topology results. Roof (top), side (center), and floor (bottom).

FIGURE 5.26 Topology optimization result with wireframe overlay.

In order to continue the BIW optimization process of Figure 5.25, it is necessary to simplify the BIW topology results by utilizing beams to represent the individual load paths. This step is manual, because of the complexity associated with interpreting the results, which requires a profound "engineering intuition."

In the future, it may be possible to automate this step using an engineering knowledge-based approach, which would output the location of these beams relative to the topology optimization results, as illustrated in Figure 5.26.

5.3.2.1 1D BEAM MODEL

The objective of the beam model is to create a model where each individual section (member) could be optimized independently of one another. This can be achieved within the preset optimization constraints aimed at producing a lightweight structure.

In order to transform the beams produced by the topology optimization into hollow beam sections to be further optimized, an initial beam property needs to be specified. A tubular section profile is chosen to find the ideal sectional stiffness and dimensions using a minimal number of design variables.

Since the topology optimization produced in excess of 100 individual members, the starting cross-sectional properties of each individual member, as well as the associated design variables, need to be defined. As an example, all the beams in the "A" pillar area will be gathered together and will be assigned a sole beam section property, as illustrated in Figure 5.27. It is then possible to perform a size optimization of each member of the vehicle, which will have the effect of changing the inner diameter and the outer diameter for each group of beams, leading to controlling the beam thickness.

Each property specified a sequential property ID for which the relevant beam is associated. Each property ID has the relevant design relationship (DVPREL input card (Altair, 2015)). This step is necessary in order to define the variables of the cross-sectional dimensions. A design equation (DEQATN input card (Altair, 2015)) linking the beam cross section dimensions has been implemented in order to prevent the crossing of the inner and outer radii and a

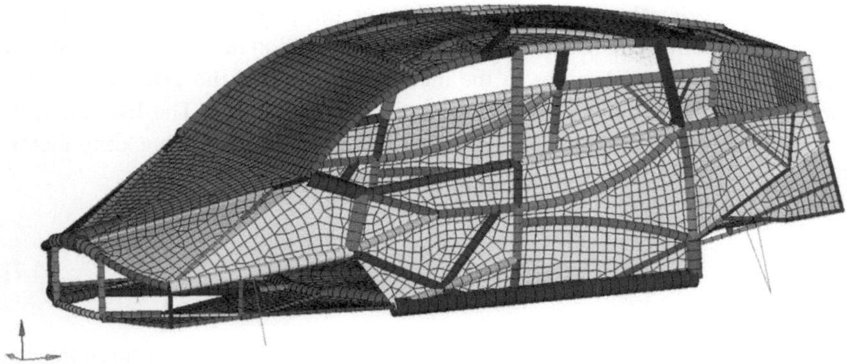

FIGURE 5.27 1D beam model with shear panels.

design variable (DESVAR input card (Altair, 2015)), after specifying the initial dimensions and the sizing limitations of the individual member.

The wireframe model is used to position beam elements at the locations produced from the topology optimization (Figure 5.26). Each line mesh is assigned relevant section property IDs that provided the corresponding dimensions and general optimization setup. Shell elements were used to represent shear panels between the beam elements, characterizing the outside body panels. The beam and shell elements in areas representing apertures are only attached at locations of the hinges and the lock points to represent the same fixing of that used for the topology optimization.

It is imperative to ensure that the same load steps of the 1D beam model are applied analog to the ones used for the topology optimization. In addition, the 1D model also utilizes inertia relief boundary conditions, as described in Section 5.2.

In addition to the optimization constraints outlined above, it is also crucial to include considerations with respect to buckling of individual members. The importance of this can be realized by considering the consequences of roof members buckling during roll over cases. Existing linear buckling analyses models using eigenvalue extraction are, however, unsuitable for the purpose of this model (Altair, 2015), as this is due to the incompatibility between inertia relief boundary conditions and eigenvalue models. Because severe crashes are a one-off event to the BIW, the deformation is only a concern when the bifurcation point limit load is reached.

Instead, the detection of buckling modes is based on Euler's buckling formula, Equation (5.6), which is calculating the critical buckling force F_{crit} for each member

$$F_{\text{crit}} = \frac{\pi^2 \cdot E \cdot I}{k \cdot L^2} \tag{5.6}$$

In Equation (5.6), E is Young's modulus, I is the second moment of area, k is the buckling mode factor, and L is the length of the beam member. The worst case buckling mode occurs when k in Equation 5.6 is equal to 1.

By monitoring the second moment of area, the critical buckling load of the individual member can be checked for each optimization iteration. Furthermore, by extracting the axial forces in the members caused by the external loading, the likelihood of buckling can be monitored and evaluated. The likelihood of buckling occurring can thus be controlled by ensuring that the buckling factor, Equation (5.7), remains true.

$$Buckling\ factor = \frac{Element\ axial\ force}{F_{\mathrm{crit}}} < 1 \tag{5.7}$$

A buckling factor under "1" leads to a reduced likelihood of bifurcation occurring. However, this may also be indicative of an over-dimensioned member, thus defeating the purpose of the optimization, and thus the maximum buckling factor had to be globally adjustable. Doing so also provided additional control for later stages of the optimization process, particularly if key members are found to buckle when utilizing dynamic crash modeling.

With the above considerations, the optimization of the beam (and shell model) is conducted to compute the structural integrity of the beam model including the buckling features.

5.3.2.2 2D SHELL MODEL

Following the beam optimization, the wireframe CAD model is reused to produce tubular surfaces over the wireframe. The radii of the tube sections are parameterized with the corresponding member beam ID. The CAD model is then linked to the property output file from the last iteration of the beam model optimization, thus automatically converting the 1D beam model into a 3D tubular CAD model.

The problem of generating such geometrical transfer is that adjacent beam sections will be different; hence, this will create a natural hinge. This is remedied by a loose shrink-wrap mesh that is produced based upon the 3D geometry, as illustrated in Figure 5.28.

The shrink mesh produced a single shell (thickness) structural mesh with hollow member sections, representing the optimized dimensions based on the beam optimization. The shrink wrap mesh produced blended connections between the individual members. The cross-sectional dimensions are created from the tube sizing and the sectional thicknesses from the beam model. Both are inserted into the shell element properties. This was automatically done by assigning an element located inside a "box" created by the coordinates of the member's line and radius.

It has to be noted that such organic vehicle structures are not seen to date; however, the aim of this exercise is to evaluate the optimization method.

5.3.2.3 VALIDITY OF METHOD 1 USING A CRASH MODEL

The 2D shrink map model is subsequently imported into commercial FE-software and a nonlinear dynamic crash analysis was completed. The purpose of this was to further validate the crashworthiness of the optimized design. Thus,

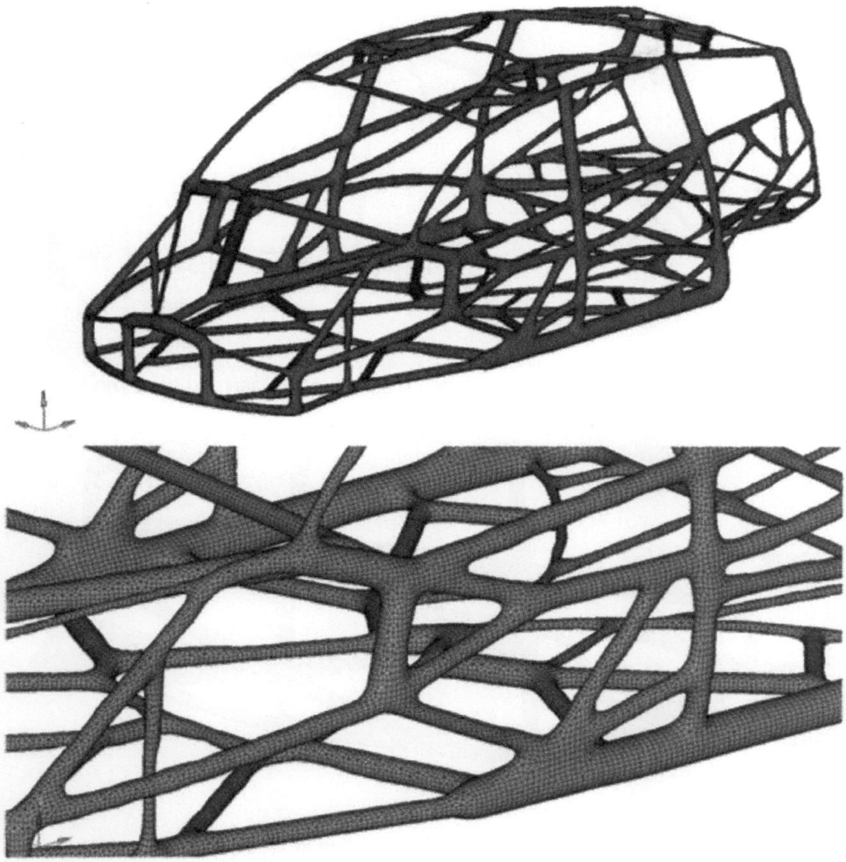

FIGURE 5.28 2D shell model using shrink-wrap meshing.

it was possible to ensure that buckling of key members did not occur during individual loadcases, hence confirming the validity of Method 1. Additional nodal masses across the entire structure were included in the model to reflect its total mass including drivetrain and seven occupants, thus realistically replicating the inertial effects during all loadcase scenarios.

For the front crash scenario, the vehicle collided with a rigid barrier following the FMVSS208 legislative requirements. As the BIW was developed based upon linear static topology optimization, it is anticipated that the front crash scenario would propose the largest challenges to the optimized BIW during dynamic crash modeling, due to the maximum buckling effect caused by the overhang of the front crash structure. The results from the front crash analysis are represented in Figure 5.29.

As indicated in Figure 5.29, the crash structure resisted the crash scenario, and the critical buckling of the safety cell was avoided. This indicates that the linear buckling sizing performed on the 1D beam elements was successful. The maximum deceleration level observed in Figure 5.29 is around 42 g, which is a reasonable value for general structures observed on the market (NHTSA, 2015).

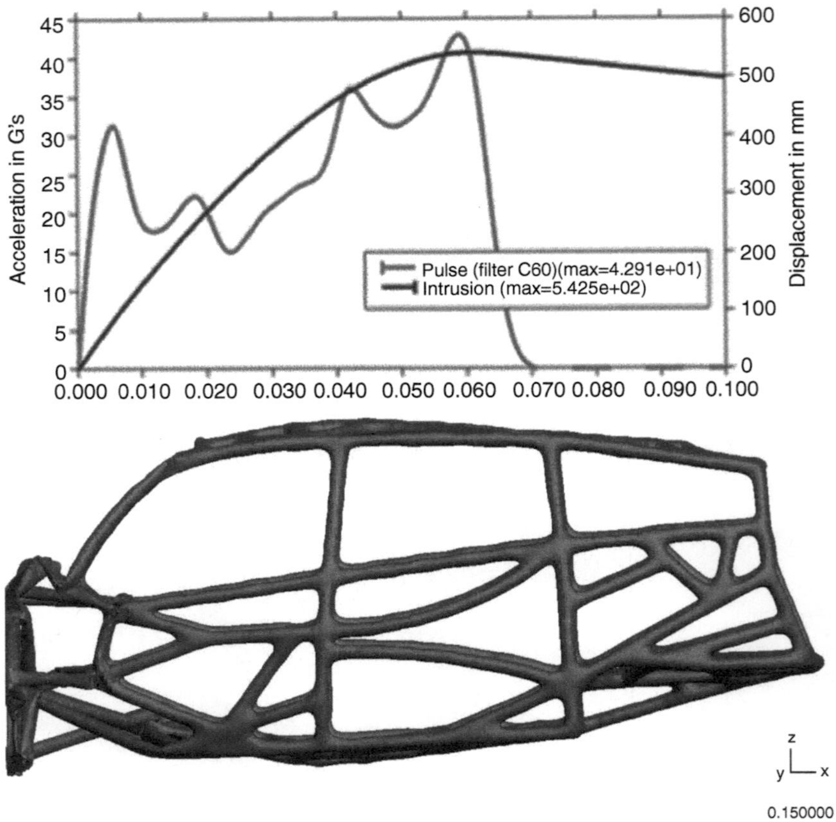

FIGURE 5.29 Crash model validation. Deformation at $t = 60$ ms; crash pulse characteristics.

The small (or inexistent) magnitude of the safety cell deformation substantiates the fact that no buckling is present.

It should, however, be noted that the acceleration levels and crush length results obtained from this analysis could not have been predicted based upon the linear topology or sizing optimization, as the solving was not transient dynamics. Therefore, it was necessary that the process includes a validation phase of the full structure, as illustrated in Figure 5.29, using a nonlinear explicit solving technique.

5.3.2.4 CONCLUSION OF METHOD 1: A HOLISTIC BIW SIZING METHODOLOGY

The skeleton of a semiautomatic process to generate an optimized BIW architecture has been demonstrated and represents all steps needed to develop a crash model based upon basic styling surfaces. This process can be completed within a very short time, realistically within one working day. The structure generated is suitable for evaluating a vehicle architecture and more particularly the safety cell. However, this process does not provide a manufactural solution.

Caution needs to be exerted, especially with respect to the front crash structure, where the linear topology optimization algorithm provides a load path that does not directly consider aspects such as buckling, bifurcation, material strain rates and damage. Therefore, it would be strongly advisable to review the front end design after the full process has been completed and converged. This is to extract the necessary loads in the "A" pillar and the seals to propose a perhaps more suitable (conventional) front end crash structure, which subsequently can be optimized for mass, utilizing the extracted loads as maximum loading objective functions. This leads to Method 2.

5.3.3 Method 2: detailed frontend design

Based on Section 5.2, it has been shown that linear topology optimization was adequate to define the safety cell loadpath, but not for areas where extreme buckling occurs, because of the limitations of the linear solver. Consequently, should only the safety cell definition be used after the topology, it is possible to "graft" a more typical vehicle front end (Figure 5.30), as long as the buckling loads limit do not violate in the "A" pillar area (Christensen et al., 2011b; Prem & Bastien, 2013).

A standard crash structure contains the longitudinal members and the shotguns which are the primary loadpath for the frontend. As discussed in Chapter 1, the longitudinal members provide most of the energy management of the impact and carry the load along the seals of the vehicle, while the shotguns tend to drive the impact load up to the upper part of the vehicle, through the "A" pillar. Figure 1.36 represents a typical frontend structure and is repeated as Figure 5.30 for convenience. The bumper beam and the crush can bracket are used for low-speed impact and are bolt-on sacrificial components (Figure 5.30). Usually, the turret, which is the home of the front suspension spring system, connects the longitudinal

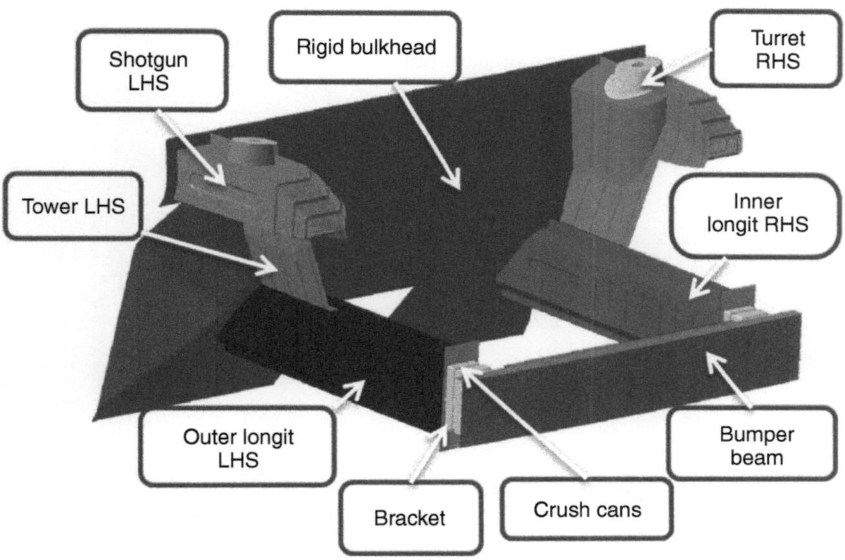

FIGURE 5.30 A standard vehicle frontend structure.

member and the shotgun (Figure 5.30). As explained in Chapter 1, the longitudinal members can be swaged to control the buckling mode and collapse force.

This vehicle buckling performance is met in Method 1, as it fulfills Equation (5.7). The steps are illustrated in Figure 5.31, where the compression load (negative normal to the structure) can be considered. The normal force values can be extracted from the postprocessor along various elements that are part of a structural member. In the case of Figure 5.32, the maximum load observed in the "A" pillar, after Method 1 optimization is 23.6 kN. This means that in order for the structure to be stable in buckling, no load should violate Equation (5.7) on which the topology optimization is based on (Method 1). However, when the vehicle styling is taken into account, the "A" pillar buckling value can be recomputed using Equation (5.7). In our case, the buckling load is raised from 23.6 kN to 300 kN, as the "A" pillar shape is given. Consequently, the load path of Method 1 is carried over, while the vehicle cross sections will be more subjected to standard CAD practice. This updated buckling value is higher than the tubular section computed in Method 1 because of the complex "A" pillar shape dictated

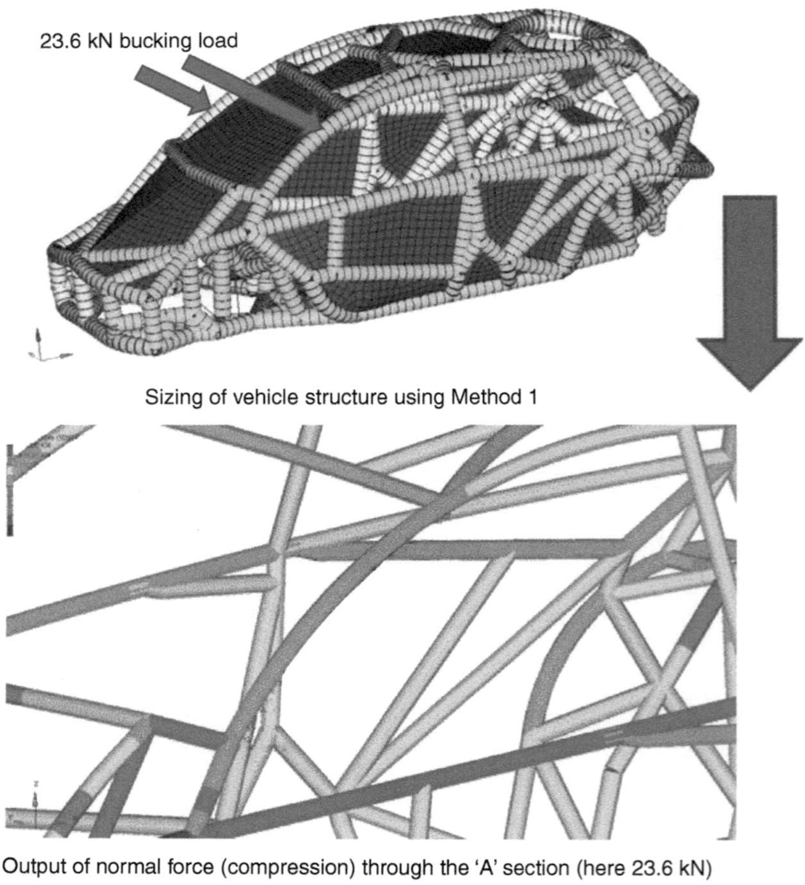

FIGURE 5.31 Reading of "A" pillar compression load requirement.

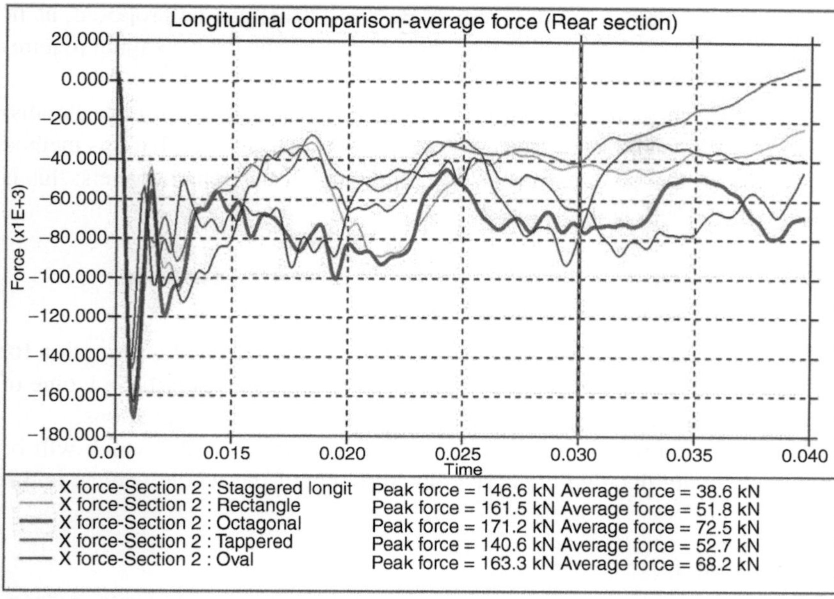

FIGURE 5.32 Collapse load of various crush can shapes aiming to meet the buckling requirement from the beam sizing.

by the styling surface and vehicle interior package. This updated buckling value is the maximum force allowed in the longitudinal member and hence is a design constraint that must not be exceeded.

As longitudinal member can be of different cross sections (rectangle, staggered rectangle, octagonal, oval, etc.), some nonlinear studies can be undertaken to find which section is able to withstand the maximum load 'permissible before' collapsing. Figure 5.33 suggests that most longitudinal members would meet the 300 kN normal force requirements, and would suggest that an octagonal-shaped longitudinal would transfer more load before buckling is initiated (Prem & Bastien, 2013).

Consequently, a standard CAD architecture could be "fitted" at the front of the vehicle structure based on topology results, whilst meeting vehicle package requirements.

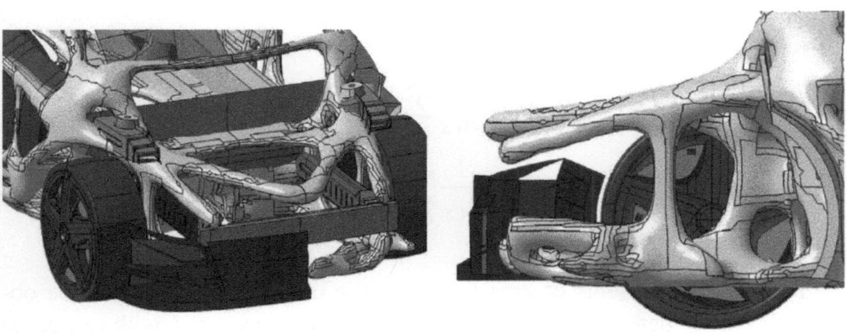

FIGURE 5.33 Overlay of typical structure over topology optimization results.

A more detailed structure (Prem & Bastien, 2013) can be proposed, as illustrated in Figure 5.30, meeting the topology results and the packaging requirement (Figure 5.33).

Method 2 is useful for the front end vehicle design and can provide a realistic structure. It can be optimized nonlinearly for shape and size, but this method does not provide a generic manner of optimizing all the vehicle panels: this is what Method 3 will aim to address.

5.3.4 Method 3: development of manufacturable body panel geometry

In order to generate the detailed local BIW cross sections in the preparation for manufacturing, an optimization method is created, building on the outcome of the topology optimization (Method 1) as well as the *I*-value estimations.

The first step in this process is to define a segment model which will be used to compute detailed cross sections throughout the whole vehicle architecture.

5.3.4.1 SEGMENT MODEL

As BIW members are typically manufactured from (metal) sheets, a "conventional" manufacturing process could be used in connection with the BIW design, using stamping processes. The cross section of each BIW member would consist of a series of segments. The cross section of a flat metal sheet would consist of one single segment, as illustrated in Figure 5.34. The segment example in Figure 5.34 is located in the *yz*-plane, and defined by the two endpoints A and B with the coordinates (Y_A, Z_A) and (Y_B, Z_B), respectively. The coordinates of the centroid C of the segment are (Y_C, Z_C). Finally, the thickness of the segment is denoted as t and the angle between the segment and the y-axis is denoted as α. The area of the segment, the angle α, and the coordinates for the centroid C can all be obtained, provided A, B, and t are known parameters.

Using the parallel axis theorem, the expressions for the second moment of area around the centroid C with respect to the global coordinate system axes can be defined in Equation (5.8), where l_{AB} is the length of the line segment between point A and point B

$$Y_C = \frac{1}{A_{CS}} \sum_{i=1}^{6} AS_i \cdot Y_{Ci}$$
$$Z_C = \frac{1}{A_{CS}} \sum_{i=1}^{6} AS_i \cdot Z_{Ci}, \tag{5.8}$$

The second moments of area for the entire cross section can thus be obtained by firstly determining these for the individual segment i, with respect to the global coordinate system using the parallel axes theorem, as expressed in

Equation (5.9), where d_Y and d_Z are the relative distances from the centroid of the individual segment to the centroid of the combined cross section

$$(I_{CY})_i = (I_{C_iY})_i + SA_i \cdot (d_Z)^2$$
$$(I_{CZ})_i = (I_{C_iZ})_i + SA_i \cdot (d_Y)^2 \qquad (5.9)$$

The second moments of area for the entire cross section can subsequently be obtained using Equation (5.10)

$$I_{CY} = \sum_{i=1}^{6} (I_{CY})_i$$
$$I_{CZ} = \sum_{i=1}^{6} (I_{CZ})_i \qquad (5.10)$$

Thereby, the second moment of area for any cross section can be calculated regardless of the number of segments within the cross section. The methodology presented above was subsequently programmed into a spreadsheet, which could be used to immediately calculate the second moments of area for any cross section (Figure 5.34). The calculations of the spreadsheet were subsequently validated using commercial CAD and FE software, as well as manual calculations. Following this, the spreadsheet was set up to be able to calculate and store the second moments of area for a series of cross sections in connection with each other, which could be used to calculate these for the entire BIW structure.

5.3.4.2 OPTIMIZATION OF VEHICLE ROOF STRUCTURE CROSS SECTIONS

In order to illustrate the application of crash section generation, the following optimization is based on roof assembly of a vehicle, as illustrated in Figure 5.35.

With the ability to calculate the second moments of area for all roof members, these could be equated to the outcome of the individual I-value estimations discussed in Section 5.2.2, as illustrated in Figure 5.31. It should also be noted that parts of the cross section will follow the outer profile, that is, the A-surface of the vehicle (vehicle skin). This meant that the I-value could be constrained in

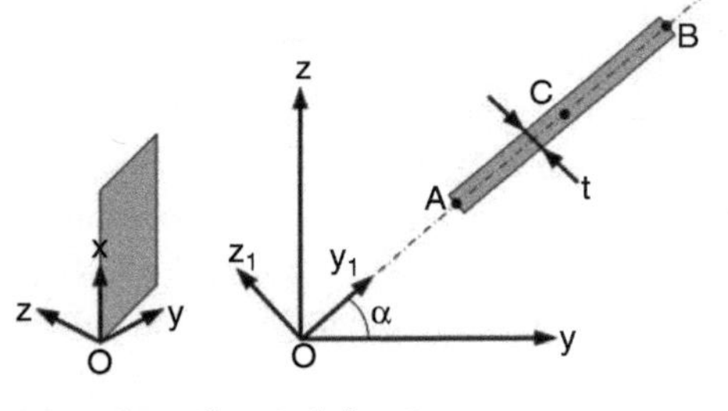

FIGURE 5.34 Cross section of metal sheet.

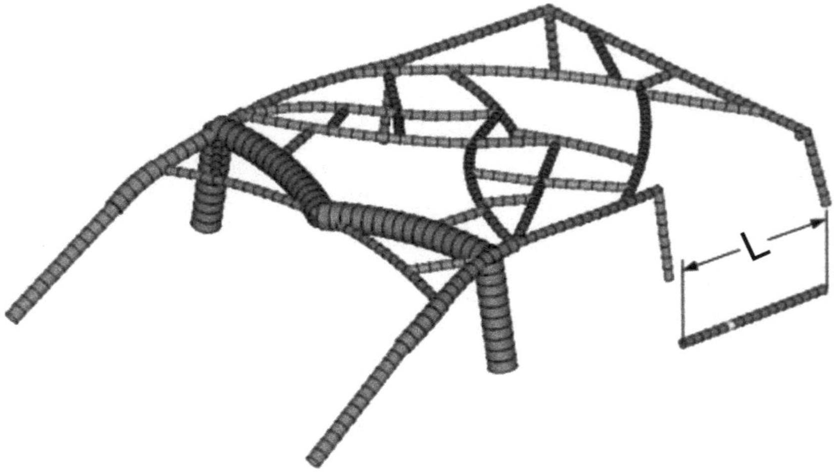

FIGURE 5.35 Roof subassembly with separate members of length "*L*."

connection with shape and size optimization with the objective of minimizing the BIW mass.

Alternatively, the objective could also be to maximize the *I*-values of the cross-sections, with the ultimate aim of defining manufacturable cross-sections.

The optimization was to be conducted using commercial software (Christensen et al., 2012), which aimed to use design of experiments (DOE) to build the response surface of the vehicle cross section. Unfortunately, due to the vast number of design variables and design variable levels, this method was judged unpractical. Indeed, computing a response surface from more than 1000 points was not an efficient method of optimization, as it used a significant amount of disk space and caused an excessive memory shortfall when DOE response points had to be reloaded for the computation of the response surface.

It was found advantageous to perform a direct optimization, eliminating the need to compute the response surface from experimental points to then extract the minimum section mass. The direct optimization method was mostly chosen because solving each optimization in an Excel spreadsheet took just a few seconds with acceptable results, as will be discussed later. To conduct the optimization, it is necessary to define the design variables, including initial values and constraint magnitudes. The majority of these would have to be set on a local level, that is, for each roof member (section); however, parameters such as the mass of the roof would be monitored on a global level.

Figure 5.36 represents a "typical" cross sectional roof profile and can be used as a starting point for an example to define key design variables. The coordinates of points *A* through *G* (excluding *C*) are potential design variables for the optimization process. In this particular case, *A*, *B*, and *D* are fixed, as these in essence define the external profile, that is, the A-surface of the vehicle. Thereby, three design variables remained, namely the coordinates of points *E*, *F*, and *G*. There are, however, some restrictions on how these would be able to change,

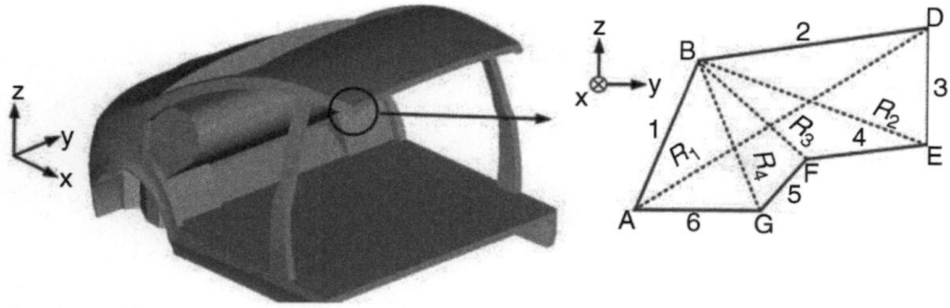

FIGURE 5.36 Initial design variables for shape optimization.

which was primarily related to the resulting interior cabin space (head room), as depicted in Figure 5.37.

In addition to the actual coordinates of E, F, and G, the gauge thickness of the individual segments (1–6) could also be design variables. However, as modern-day roof structures usually use a minimum steel gauge thickness of 0.7 mm for reasons other than crash performance, it was decided to set the gauge thickness of the six segments to 0.7 mm.

Figure 5.36 also displays four potential reinforcements (R_1, R_2, R_3, and R_4) that can be "inserted" or "removed" using simple Boolean operators. Furthermore, the gauge thickness of the four potential reinforcements, T_{R1}, T_{R2}, T_{R3}, and T_{R4}, are also included as design variables, which can either assume the discrete arbitrary values of 1.0 mm or 0.7 mm (standard BIW panel gauges). These discrete values may be adjusted to suit the metal coil thickness available to the vehicle manufacturer.

Finally, manufacturing constraints are taken into account during the optimization process; it is assumed that the roof structure would be tack welded and use a sheet metal forming process, meaning that blank holders and blend surfaces would have to be used. The optimization parameters are summarized in Table 5.7. Based on the information summarized by Table 5.7, the optimization problem was set up using Microsoft Excel coupled with Altair HyperStudy (Altair, 2015) to directly solve the optimization problem.

Using the roof rail as a case study, it is found that the most efficient solution, given the design variable in Table 5.7, use the reinforcement R_1 with a gauge thickness (T_{R1}) equal to 1.0 mm; this scenario meets all (mathematical) optimization constraints, including matching the I-value to that of the previous

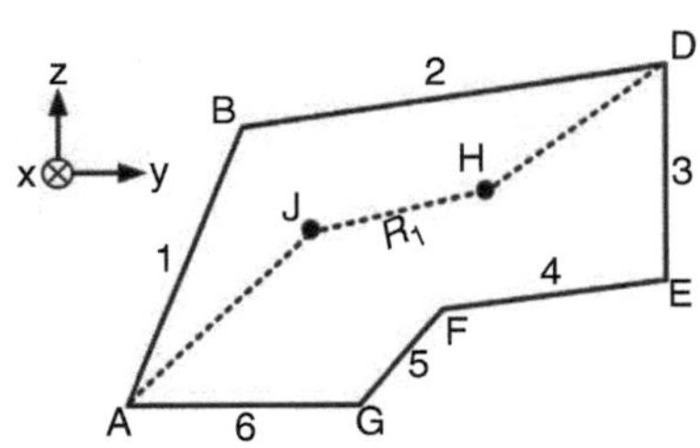

FIGURE 5.37 Updated design variables for shape optimization.

TABLE 5.7 Parameters for Initial Shape and Size Optimization

Design variables	Y_E, Z_E, Y_F, Z_F, Y_G, Z_G, R_1, R_2, R_3, TR_1, TR_2, TR_3, TR_4 R_3, TR_1, TR_2, TR_3, TR_4
Optimization responses	A_{CS}, I_{CY}, I_{CZ}
Optimization constraints	I_{CY} = fixed; I_{CZ} = fixed
Optimization objective	Minimize A_{CS}

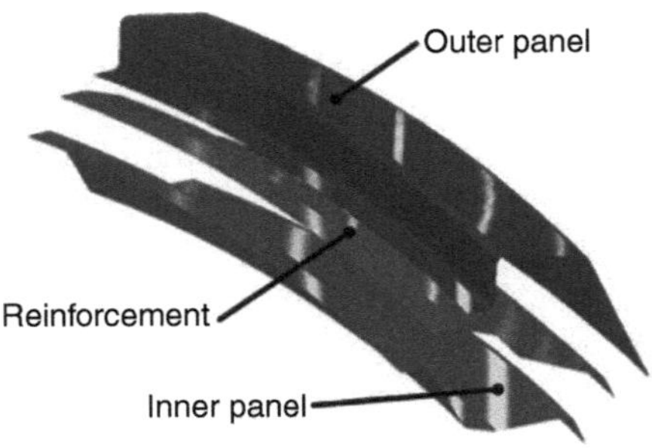

FIGURE 5.38 Optimized cross section.

estimation. Despite this, there are concerns about the manufacturability of the proposed design; hence, it is decided to insert two additional points (*H* and *J*) to guide the profile of the reinforcement R_1. The updated design variables are illustrated in Figure 5.37. With these changes, the optimization process was repeated and a manufacturable design is obtained. This consists of three parts: the inner panel, the outer panel, and the reinforcement, as illustrated in Figure 5.38.

To substantiate the feasibility of the proposed manufacturing methodology with respect to the design in Figure 5.38, the geometry is analyzed using FE-based sheet metal forming analysis (Altair HyperForm), using 2D (shell) elements with an average mesh size of 5 mm and the material parameters defined in Table 5.8.

TABLE 5.8 Material Parameters for Sheet-Forming Analysis

Parameter	Value	SI Unit
Youngs modulus (E)	210,000	MPa
Poissons ratio (ν)	0.3	
Yield stress value (σ_y)	315	MPa
Friction coefficient (FC)	0.125	

Figure 5.39 displays the relative thinning of the sheets after the forming process and relatively minor variation in thickness for all three parts analyzed. The part found to have the largest variation of gauge thickness, as a function of the forming process, was the outer panel. In this case, the maximum increase (of thickness) was 30% (maybe a risk of wrinkling) and the minimum decrease was −6%, meaning a maximum thickness of 0.91 mm and a minimum thickness of 0.66 mm, based on an initial gauge thickness of 0.7 mm. All elements of the analyses are below the failure level of the forming limit diagram. The results of the sheet forming analyses thus substantiate that the parts can be manufactured without any significant variations of thickness, indicating that the anticipated structural performance of the previously conducted FE analyses can indeed be obtained.

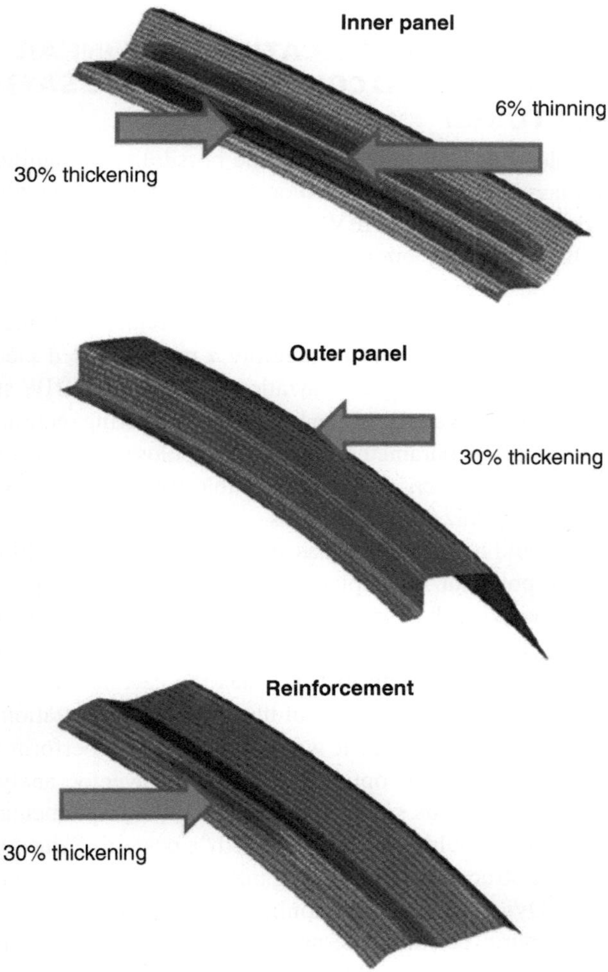

FIGURE 5.39 Results from sheet metal forming analysis of optimized cross section (relative thinning).

This method can be replicated for each panel member of the vehicle and as such will provide the main BIW section that will then need to be blended, which is still a challenging task. Maybe, in the adjoining sections, some extra geometrical constraints could be added to allow an easier transition. One of the obvious limitations is that most BIW panel shapes are rarely of constant shape; hence, some localized CAD adjustments may be needed. Nevertheless, coupling Method 1 followed by Method 3 will provide an excellent starting point for vehicle architecture, as it will provide very early in the vehicle program:

- Panel shapes that will meet external styling requirements
- Buckling performance to prevent vehicle intrusion inside the vehicle cabin
- Panel thicknesses
- Panel reinforcement shapes
- Panel stamp-ability

5.4 CONCLUSIONS: APPLICATIONS OF LINEAR OPTIMIZATION TO CONCEPT VEHICLE SAFETY STRUCTURES

The use of topology optimization enables the possibility of significant reductions in BIW mass, which has been demonstrated in connection with "alternatively" fuelled vehicles, such as HEV. The outcome of the study suggested that currently available topology optimization software can be used for BIW load path extraction, considering buckling as part of the optimization process. It was also observed that inertia relief (IR) provided more realistic results than using standard boundary conditions and is therefore a recommended method to use when performing linear topology optimization of full vehicle BIW structures.

Initially, it was investigated how the buckling engineering requirement could be included (as an active parameter) during the topology optimization process. This approach was, however, deemed unfeasible using current commercially available topology optimization algorithms.

As buckling could not be included as an "active parameter" within topology optimization, the potential for implementing it as a "passive" parameter was investigated. Recognizing that topology optimization uses linear static FEA, and a penalization factor of 1 (Christensen et al., 2012), the method of superposition was deemed applicable, meaning that the buckling calculations could be performed subsequently to the completion of the topology optimization.

As inertia relief was utilized, it was not possible to perform a buckling analysis during the topology optimization, as eigenvector analysis cannot be computed in this analysis setup. Consequently, the implementation of the buckling optimization could only be performed "a posterior" during the sizing phase, limiting the structural beam size against a linear Euler buckling critical load constraint. Dynamic (explicit) nonlinear FE analysis was used to substantiate the structural performance and the results of these analyses clearly indicated that the structures obtained were fully capable of withstanding the applied loading. It was observed that a vehicle model based on structural beams was useful; however, it did not contribute to addressing the manufacturing requirements. It was discussed that a standard crash structure could be implemented while meeting the maximum critical load to be transferred to the back end beam structure, which is an initial step to vehicle design from the topology results.

It was observed that there was no automatic method to extract the overall CAD geometry in a meaningful and manufacturable manner. To date (2015), the loadpath extraction requires a good engineering judgement and a trained eye and cannot be automated.

In order to achieve an optimized CAD model, an initial combined shape and size optimization study was set up in order to derive manufacturable cross sections based on the topology optimization and *I*-value estimations. Finally, the manufacturability of these cross sections was substantiated using FE-based sheet metal forming analysis to estimate the variations of gauge thickness, and indirectly the structural performance of the manufactured parts.

In conclusion, this chapter has demonstrated the potential for using current state-of-the-art commercially available optimization software to obtain a manufacturable BIW able to meet crash performance requirements, based on a styling (CAD) surface.

It also recognizes that revised (topology) optimization algorithms would be very beneficial in the effectiveness and validity of the overall methodology.

REFERENCES

Acar, E. K., & Solanki, K. (2009). Improving the accuracy of vehicle crashworthiness response predictions using an ensemble of metamodels. *International Journal of Vehicle Structures and Systems*, *14*(1), 49–61.

Altair, Altair Engineering, Inc. (2015). "*Altair HyperWorks manual*," Troy, MI.

Barnett, A.R, & Widrick, T.W. (2005). Closed-Form Static Analysis with Inertia Relief and Displacement- Dependent Loads Using a MSC/NASTRAN DMAP Alter. Available from: <ntrs.nasa.gov/archive/nasa/casi.ntrs.nasa.gov/19950013233_1995113233.pdf>. accessed on 01st March 2015.

Bastien, C. (2010). Topology optimization of a body in white for low carbon vehicle technology project. In: *Altair European conference,* Versaille, France., EHTC 2010, October 2010.

Bastien, C., & Christensen, J. (2011). 'Towards the light weighting of low carbon vehicle architectures using topology optimization'. In: *EHTC*, Bonn, Germany.

Christensen, J., Bastien, C., Blundell, M. V., Gittens, A., & Tomlin, O. (2011). Lightweight hybrid electrical vehicle structural topology optimization investigation focusing on crashworthiness. *International Journal of Vehicle Structures and Systems*, *3*(2), 113–122.

Christensen, J., Bastien, C., & Blundell, M. (2012). Effects of roof crush loading scenario upon body in white using topology optimisation. *International Journal of Crashworthiness 17*(1), 29–38. Available from http://dx.doi.org/10.1080/13588265.2011.625640.

Christensen, J., Bastien, C., Grimes, O., Appella, A., Bareham, G., O'Sullivan, K. (2011b). Generation of optimised hybrid electric vehicle body in white architecture from a styling envelope.

Christensen, J., Bastien, C., Blundell, M.V., & Kurakins, J. (2012). Lightweight body in white design using topology-, shape and size optimization. In: *Electric vehicle symposium (EVS26)*, Los Angeles, CA, 2012.

Duddeck, F. (2007). Multidisciplinary optimization of carbodies. Technical paper. Available online <www.springerlink.com/content/2741p870381l0l7>. London, UK. accessed on 01st March 2015.

Horstmeyer, M. F., Ren, X. C., Fang, H., Acar, E., & Wang, P. T. (2009). A comparative study of design optimization methodologies for side-impact crashworthiness, using injury-based versus energy-based criterion. *International Journal of Vehicle Structures and Systems*, *14*(2), 125–138.

LCVTP. (2011), Low Carbon Vehicle Technology Project. Re: BG/AM1011, partly funded by ERDF and AWM. Project based on Tata 'Beacon' vehicle layout. Participation of Coventry University for vehicle aerodynamics, HVAC and Vehicle Body Structure.

Liao, X., Li, Q., Yang, X., Li, W., & Zhang, W. (2008). A two-stage multi-objective optimization of vehicle crashworthiness under frontal impact. *International Journal of Vehicle Structures and Systems*, *13*(3), 279–288.

Marur, P. R., & Srinivas, S. (2008). A reduced-order finite element model for the simulation of automotive side structure crash response. *International Journal of Vehicle Structures and Systems*, *13*(2), 211–218.

NHTSA. (2015). Vehicle crash test database. Available from <www-nrd.nhtsa.dot.gov/database/VSR/veh/LatestTestInfo.aspx>. accessed on 01st March 2015.

Prem, A., & Bastien, C. (2013). Optimization of future lightweight multi-material vehicle crash structures. In: *Proceeding of the IMechE conference* held in Gaydon on 13 December 2013.

Richardson, S. A., Rechnitzer, G., Grzebieta, R. H., & Hoareau, E. (2003). An advanced methodology for estimating vehicle rollover propensity. *International Journal of Crashworthiness*, *8*(1), 63–72.

Soto, C. A. (2004). Structural topology optimization for crashworthiness. *International Journal of Vehicle Structures and Systems*, *9*(3), 277–283.

Stephenson, R. R. (2010). *The case for a dynamic rollover test*. Washington, DC: ICRASH, 2010.

Zimmer, H., Prabhuwaingankar, M., Duddeck, F. (2009). Topology and geometry based structure optimization using implicit parametric models and LS-OPT. Technical paper. Available online <www.dynalook.com/european-conf-2009/F-II-02.pdf>. accessed on 01st March 2015.

Chapter | six

Complications of Nonlinear Structural Optimization

CHAPTER OUTLINE

Throughout the previous chapters of this book, topics such as numerical simulation considerations, optimization principles, techniques, and examples have been presented. The presentation and discussions of the VDM-SIMP approach to topology optimization were explained from a theoretical viewpoint in Section 4.1; extensive examples of application were presented and discussed in Chapter 5. Although the approach was initially developed for optimization of structures utilizing isotropic material models exposed to linear static loading conditions, recent advances have attempted to adapt the method to include nonisotropic material models, for example, Sorensen & Lund (2012) and, in particular, nonlinear/dynamic loading scenarios through the introduction of the equivalent static load method (ESLM) as, for example, presented by Park (2010). Although there are certain limitations associated with the VDM-SIMP, it is seemingly one of the preferred approaches for nonlinear optimization; least of all with commercial FE software vendors such as Altair who have included ESLM into their most recent releases of HyperWorks. It is convenient to speculate as to why such a relatively large amount of focus and effort has been put into developing an algorithm that is deeply rooted in linear analysis, and in particular upon Young's modulus for nonlinear and dynamic optimization. The question is, how relevant and appropriate is this approach? As previously discussed, optimization algorithms and indeed FEA are only as good as the information that is provided as input; an obvious and critical question to be put toward the ESLM approach is therefore: "How representative is a linear

Nonlinear Optimization of Vehicle Safety Structures. http://dx.doi.org/10.1016/B978-0-12-804424-7.00006-3

static analysis based on Young's modulus of a nonlinear and dynamic loading scenario?" Although this is a crucial questions, there may be more than one answer; firstly, the context of the optimization problem at hand should be considered, and in particular the level of nonlinearity and levels of deformation. The BIW topology optimization examples of Chapter 5 were based on the VDM-SIMP approach, i.e. a linear static optimization algorithm, and did indeed demonstrate that it is possible to extract useful and beneficial information from such an algorithm in the context of large nonlinear deformations, which may be an indicator of the potential benefit of the ESLM algorithm. Indeed, many authors such as Witowski et al. (2014) have found ESLM to be useful; most publications do, however, include a "but" stating that more research is required to fully assess the feasibility of the method in particular where very large deformations such as full frontal vehicle impacts are concerned. The ESLM method is relatively new in mechanical engineering terms as it was firstly introduced by Shin, Park, & Park (2007), suggesting that the method is still in its infancy. Sceptics would, however, suggest that the method is fundamentally flawed due to its dependence on Hookes' law, that is, a linear relationship between loads and response, as presented in Section 4.1. In Chapter 5, the potential benefits of the VDM-SIMP approach were demonstrated, including a review of the limitations associated with this. It can be generally assumed that the majority of these pros and cons are "transferred" across to the ESLM method, as the basis of ESLM is VDM-SIMP based. The remaining sections of this chapter will investigate some of the "additional features" of ESLM with a view to highlighting some of the shortcomings of the method, thereby suggesting areas of further research and development to improve the credibility, reliability, and robustness of the approach for topology optimization of structures exposed to large (nonlinear) dynamic deformations. Furthermore, the sections will also highlight areas that would generally have to be addressed for any topology optimization algorithm. Although the following sections may seem very critical, it should be remembered that in this context the "benefits" of VDM-SIMP, for example, presented in Chapter 5 are also generally applicable to ESLM.

6.1 EQUIVALENT STATIC LOAD METHOD

As presented above, the ESLM approach does, in essence, use the "conventional" VDM-SIMP approach as successfully adapted for linear static (and isotropic) topology optimization. The primary difference between the linear static VDM-SIMP and ESLM lies in the generation of load steps, as well as an additional iteration loop; the overall flowchart of the ESLM optimization process is illustrated in Figure 6.1.

Figure 6.1 contains four primary stages, which will be subsequently explained, in addition to two iteration loops:

- Internal iterations (II): These are the optimization iterations contained within step 3 (Figure 6.1);
- External iterations (EI): These are the iterations relating step 1 to step 3 (Figure 6.1).

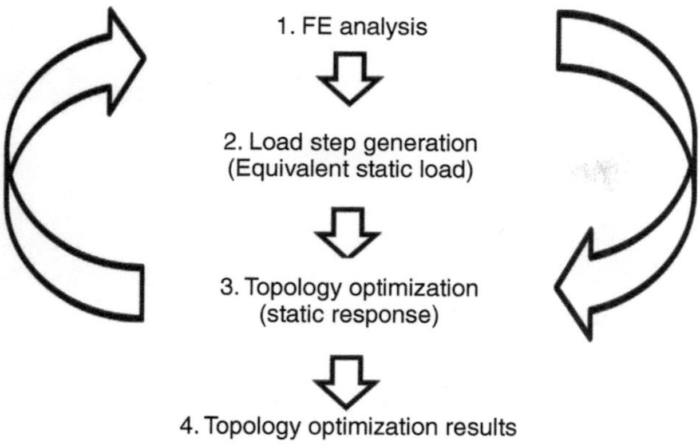

FIGURE 6.1 ESLM flowchart.

The two iteration loops above are important, as only the II are utilized in conventional linear static topology optimization; the II and EI acronyms will be utilized throughout the remainder of this chapter.

Step 1 in Figure 6.1 consists of an FE analysis, exactly the same as a "conventional," i.e. linear static, topology optimization algorithm. The difference is that the FE analysis in connection with ESLM is not confined to linear static modeling, but it may be dynamic and nonlinear. For the subsequent case studies of this chapter, three types of ESLM-based analysis are analyzed:

- Geometric nonlinear (implicit static)
- Geometric nonlinear (implicit dynamic)
- Geometric nonlinear (explicit dynamic)

Following the FE analyses, step 2 creates individual (linear static) load steps(s) to obtain a representative displacement profile to the analysis completed in step 1. The number of load steps created is automatically set equivalent to the number of timesteps. Once these equivalent static load steps (ESLS) have been created, step 4 conducts a "conventional" linear static topology optimization based on the outcomes of steps 1 and 2, while using the VDM-SIMP approach to complete the II. Pending convergence and/or termination criteria the algorithm will either return to step 1 and repeat the steps with the updated geometry via the EI loop, or terminate the analysis in step 4.

It has been documented by Altair (2013) that the ESLM method has been developed for "*Nonlinear responses from implicit static analysis, implicit dynamic analysis and explicit dynamic analysis.*" It is generally recognized that the ESLM approach is best suited for low levels of nonlinearity, for example small displacement nonlinear analysis.

The ESLM approach should, therefore, be well-suited for development of automotive roof structures, which generally experience a modest level of nonlinearity in connection with roll over events. Previously published research (Christensen, Bastien & Blundell, 2012) investigated the likely effects of

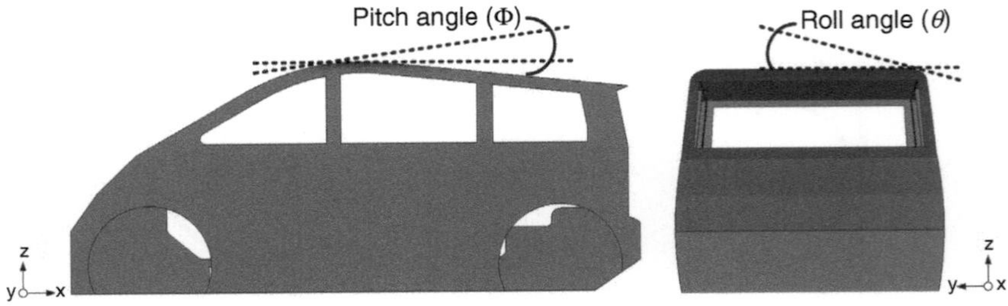

FIGURE 6.2 Pitch and roll angle definitions.

recently proposed changes to the Federal Motor Vehicle Safety Standards (FMVSS) 216 standard for roof crush upon roof structure geometry by means of linear static topology optimization. Where relevant, the outcomes of this previously published paper will be used in the current chapter, with an overall aim of drawing comparisons, highlighting significant differences, as well as evaluating the overall performance of ESLM versus linear static topology optimization based upon an automotive roof crush scenario. This will be completed to evaluate the feasibility, importance, and potential for using ESLM for nonlinear topology optimization in general. In this context, it is important to note the definition of the pitch angle Φ and the roll angle θ as defined in Figure 6.2, as they will be used throughout the remainder of this chapter.

The overall study will consist of three main phases documented in this chapter:

1. Initial study: This will contain the description of initial ESLM model setup and results;
2. Parameter sensitivity study: This will focus on the influence of optimization control parameters in connection with ESLM, as well as linear static topology optimization models;
3. Revised study: This will combine the lessons learnt from stages 1 and 2 above to refine the ESLM models before the final comparisons and associated conclusions draw the study to a close.

The individual parts described above will be the focus of attention in the following three sections of this chapter.

6.2 INITIAL OPTIMIZATION STUDY

The combined case studies to be presented in this chapter include a total of 131 topology optimization models. The general FE analysis setup is illustrated in Figure 6.3.

The general FE model illustrated in Figure 6.3 has the approximate external dimensions (x,y,z), 3000 mm, 1700 mm, 550 mm, consists of approximately 67,000 solid (tetra) elements with an average mesh size of 25.0 mm, and utilizes first-order displacement functions. These parameters are all specified based on the studies of Chapter 5. Using single point constraints (SPCs), all six degrees of freedom (DOF) are locked at the “base” of the roof pillars, as indicated in Figure 6.3. The external loading consists of a uniformly distributed load with a

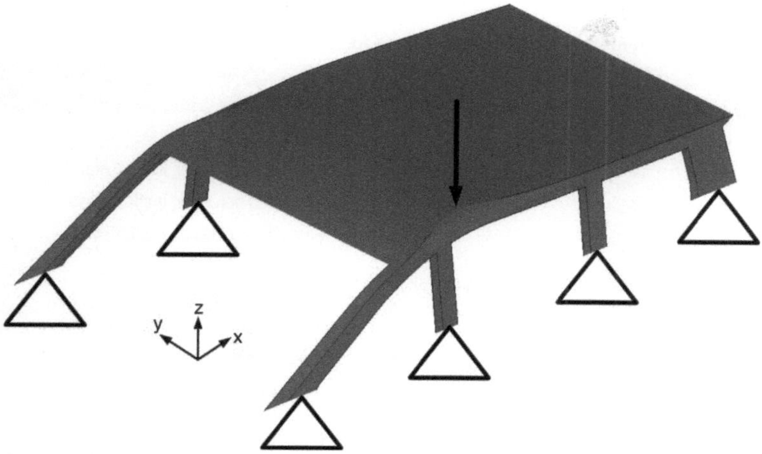

FIGURE 6.3 Illustration of general FE model.

combined magnitude of 45 kN spread across a total of 17 nodes; the application angle will be subsequently discussed. The topology optimization is set up with the main parameters listed in Table 6.1.

6.2.1 Initial model setup

With the above main parameters defined, the specific models to be included in the comparison case study need to be set up. As the purpose of the overall study is to draw direct comparisons between linear static topology optimization and the ESLM approach, an ideology of avoiding "over constraining" the optimization algorithms is adopted. This means that as few optimization control parameters as possible will be used in order to get a more "realistic" overview of the results from the individual optimization algorithms. Therefore, only the "global checkerboard control" and the minimum dimension of member size (MINDIM) parameters, which attempt to obstruct the creation of "small element clusters," are activated to reduce the likelihood of checkerboard effects occurring. Apart from these, the only other optimization control parameter initially activated is the penalization factor p for the SIMP interpolation scheme, as presented in Section 4.1. All parameters are applied to both the linear static and the ESLM models. In Chapter 5,

TABLE 6.1 Optimization Objective Parameters

Optimization Objective Parameter	Description	Value
Objective	Minimize mass	N/A
Constraint	Maximum compliance	1,000,000 mm/N
Constraint	Maximum nodal displacement	50 mm

TABLE 6.2 Parameters and Possible Values for Initial Model

Parameter #	Description	Possible Values
1	Load step type (LST)	Linear static Geometric nonlinear, implicit static Geometric nonlinear, implicit dynamic Geometric nonlinear, explicit dynamic
2	Maximum number of internal iterations per external iteration (MIE)	1 10
3	External load angle (ELA)	Pitch angle: 0°, roll angle: 0° Pitch angle: 10°, roll angle: 45°
4	Material model	Linear elastic Johnson–Cook elastic plastic
5	Nonlinear material parameters (NMP)	Mild grade steel High strength steel

it was found that a p value of 1.0 was appropriate for this specific application; despite the recommendations listed in Expression (4.11). With p determined, the remaining parameters to be varied between different models are chosen; these are listed and summarized in Table 6.2 and explained in the following.

1. *Load step type (LST)*
   The load step type (LST), listed as parameter 1 in Table 6.2, represents the possible options for linear and ESLM-based optimization in the software used and refers to step 1 of Figure 6.1. It is necessary to include all three options for nonlinear optimization to obtain a reasonable understanding of ESLM. The timestep for the models is set to 2.7E-9 s, which will remain constant throughout all the models detailed within this chapter.
2. *Maximum number of internal iterations per external iteration (MIE)*
   The maximum number of internal iterations per external iteration (MIE), defined as parameter 2 in Table 6.2, refers to the number of internal iterations (II) in step 3 (Figure 6.1) between the external iterations (EI), linking steps 1 and 3 in Figure 6.1. The ideology behind choosing 1 II per EI is that this should theoretically provide the highest possible "control" within the ESLM optimization process itself. This is because a (nonlinear) FE analysis would be conducted, step 1 (Figure 6.1), following each II optimization ensuring updated ESLS for every single EI, maximizing the accuracy of the ESLS. This should, however, also theoretically provide the highest central processing unit (CPU) requirements due to the potentially large number of (nonlinear) FE analysis required for each model. Nevertheless, the CPU time and cost is a secondary aspect in this context. The reason for choosing 10 II per

EI is to estimate the influence of this parameter, indirectly evaluating the potential tradeoff between accuracy and CPU requirements. Initially, it is not deemed suitable to further increase the number of II, as a very large value of this parameter will theoretically "revert" ESLM to linear static topology optimization, as the number of EIs completed will most likely be minimized.

3. *External load angle (ELA)*
   The external load angles (ELAs), listed as parameter 3 in Table 6.2, are defined as the angular rotation of the roof structure around the x-axis and the pitch angle is defined as the angular rotation around the y-axis in Figure 6.2. Roll and pitch angles of 0° represent the current 2015 European New Car Assessment Program (NCAP) roof crush test scenario, while the combination of a 10° pitch angle and a 45° roll angle represents the worst case scenario for roof crush as found by Chirwa & Peng (2010). The lessons learnt in the previous chapter demonstrate the necessity of incorporating sensitivity analysis for topology optimization. This is, however, of minor concern in this specific study, as the primary objective is the relative comparison of optimization algorithms. The loading for either of the two possible ELAs is assigned a time history curve, as illustrated by Figure 6.4.

   The loading is applied with a constant acceleration from 0 to 45 kN, as illustrated by Figure 6.4, representing an impact velocity of 2.22 m/s (8 km/h). It is chosen to extend the termination time to 0.085 s to ensure that a steady state is obtained within the models. It should be noted that the models do not contain any strain dependences with respect to the material parameters.

4. *Nonlinear material parameters (NMP)*
   The material models (parameter 4) and the NMP (parameter 5) used for the models of the initial studies, as listed in Table 6.2, are linear elastic and Johnson–Cook elastic–plastic. The latter represents an extension of the former, as the "only" difference occurs in the nonlinear region of the material curve. The nonlinear part of the Johnson–Cook elastic–plastic material model is controlled by the yield stress value, a plasticity hardening parameter as well as a plasticity hardening exponent. The material parameters used for the study are listed in Table 6.3.

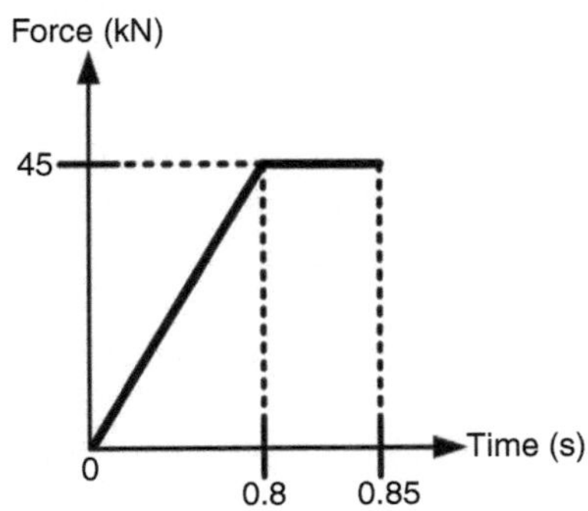

FIGURE 6.4 Force versus time history.

TABLE 6.3 Material Parameters

Material Model (Isotropic)	Material Description	Young's Modulus (MPa)	Yield Stress (MPa)	Plasticity Hardening Parameter	Plasticity Hardening Exponent
Linear elastic	Mild grade steel (linear)	210,000	N/A	N/A	N/A
Johnson–Cook	Mild grade steel (linear elastic–plastic)	210,000	276	0.076	0.643
Johnson–Cook	High strength steel (linear elastic–plastic)	210,000	430	0.017	0.51

On the basis of the above definitions and parameters, an initial study containing 26 individual topology optimization models is completed. This includes 2 linear static models and 24 ESLM models, consisting of 8 geometric nonlinear implicit static, 8 geometric nonlinear implicit dynamic, and 8 geometric nonlinear explicit dynamic models.

The models were all solved using symmetric multicore processing (SMP) on a high performance computer (HPC), utilizing 10 processors per node (PPN) on a single node of a Linux 64 bit based system.

6.2.2 Results of initial optimization study

Obtaining useful and relevant results from the 26 models of the initial study proved significantly difficult, apart from the two linear static optimization models that had already been refined, courtesy of Chapter 5. Both the implicit and explicit ESLM models seemed to have substantial stability problems. The root of the problems was initially thought to be the parameters controlling the implicit and explicit timesteps. These parameters were therefore reviewed, specifically the allowed maximum number of iterations to obtain equilibrium within each timestep increased in an attempt to stabilize the models. With this update, all 24 (ESLM) models were solved once again; the adjustment of implicit and explicit timestep parameters did, however, prove not to have a noteworthy effect on the stability of the models. Therefore, further investigation into the problem was undertaken. It was found that the initial FE; step 1 in Figure 6.1, were completed successfully. Additional investigation uncovered that the problem was rooted in the checkerboard effect, as can be perceived from Figure 6.5, representing a "typical" outcome of the 24 initial ESLM models.

The model illustrated by Figure 6.5 represents the outcome of the II returned to the EI in order to update the ESLS, in other words what is returned from step 3 to step 1 (Figure 6.1). Unsurprisingly, the subsequent FE analyses failed

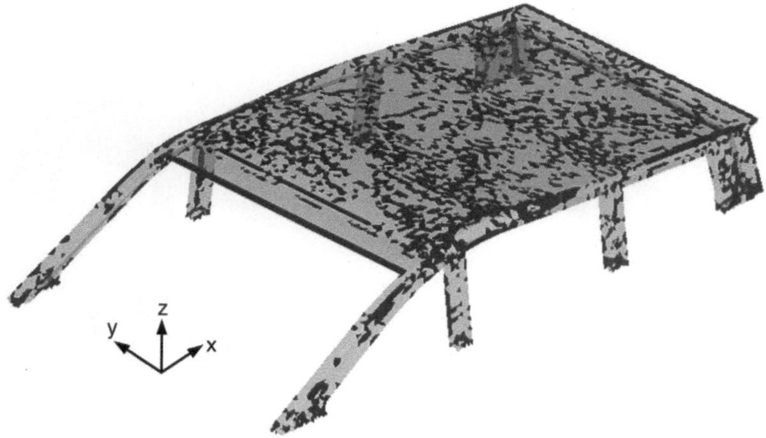

FIGURE 6.5 Checkerboard effect from initial models penalization factor = 1.0.

to achieve equilibrium during the implicit (or explicit) timesteps, which caused the models to error terminate. The maximum number of II (MIE), that is, 1 or 10 (Table 6.2), was not found to have a considerable effect on this "phenomenon," which most often occurred between the first and second EI loops.

The checkerboard effect, as illustrated in Figure 6.5, occurred despite the fact that the "built-in" global checkerboard control of the FE software was activated. In addition, the MINDIM optimization control parameter had also been previously activated. This aims to penalize the formation of small members, but does, however, not necessarily prevent the formation of checkerboard patterns during the optimization, as discussed in Chapter 4.

In an attempt to resolve the checkerboard issue, all 26 models from the initial study, including those utilizing linear static optimization, are updated with a penalization factor p of 3.0, which is the recommended maximum value for solid elements as defined in Section 4.1.

Initially, the results obtained from the 26 models with a p of 3.0 indicated that the changes had had the overall desired effect, as the average number of EI increased from 6 ($p = 1.0$) to 10 ($p = 3.0$). In addition, the average mass reduction, calculated as the ratio of the initial mass relative to the final mass, increased from 72% to 93%. Further postprocessing of the models does, however, reveal that the feasibility of the obtained structures has not drastically improved. Figure 6.6 illustrates an "identical" model to that displayed by Figure 6.5; the only difference between the two is that Figure 6.5 represents a p of 1.0, whereas Figure 6.6 represents a p of 3.0.

The model in Figure 6.6 successfully completed 1 EI and obtained a relative mass reduction of 30%; the corresponding values for the model in Figure 6.6 are 4 and 73%, respectively. Figure 6.6 is equivalent to the "type" of checkerboard effect illustrated in Figure 4.4.

The above example provides a very good indication of the overall effects of the increased penalization factor p found throughout the ESLM study. Although

FIGURE 6.6 Checkerboard effect, penalization factor = 3.0.

a significant number of models have indeed been stabilized by the increased p value, further inspection of the individual models reveal that the initial issue of a large number of models only completing 1 EI is still considerable. Approximately 40% of the 24 ESLM models with a p of 3.0 terminate during step 1 (Figure 6.1) when initiating the second EI. This is once again attributed to the checkerboard effect illustrated in Figure 6.5.

In an attempt to increase the stability of the models during these initial stages, it is decided to further "experiment" with p, by repeating all 26 initial models using a p of 0.5. This may initially seem controversial; however, it should theoretically increase the stability of the optimization models during the "early" EI, thus maximizing the number of EI, with the ultimate aim of obtaining an enhanced definition of topology. The influence of lowering p upon the results can be appreciated by observing Figure 6.7.

As illustrated by Figure 6.7, the effect of reducing p to 0.5 in essence corresponds to "reversing" the effects of the penalization in the SIMP interpolation scheme, as it "encourages" the use of elements with low relative element density

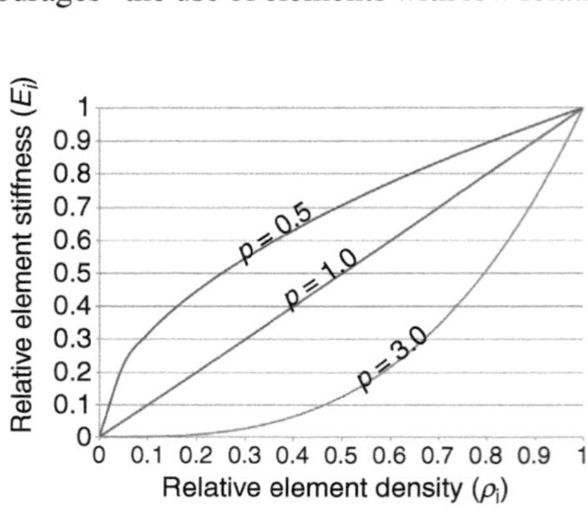

FIGURE 6.7 Effects of penalization factor.

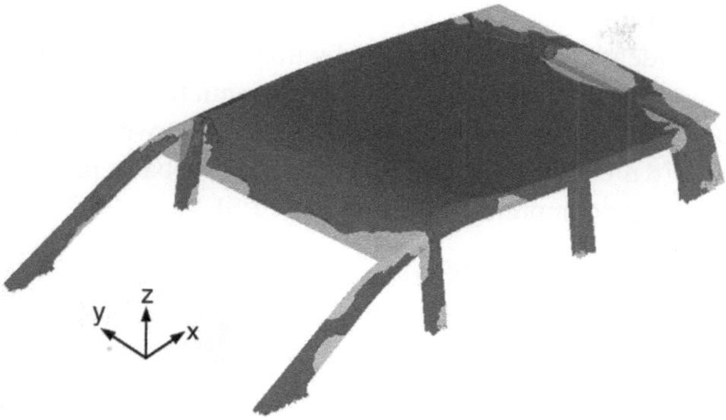

FIGURE 6.8 Checkerboard effect, penalization factor $p=0.5$.

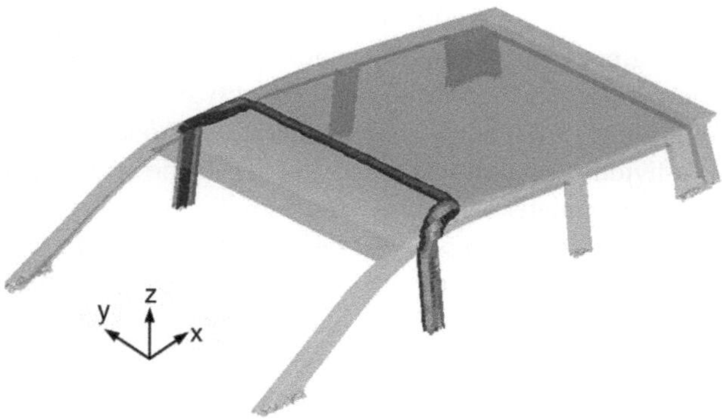

FIGURE 6.9 Corresponding linear static model.

and high relative element stiffness. Strictly speaking, this is also the "equivalent" of reversing the scale for the relative element density and will have an effect on the estimated mass of the resulting topologies. This means that the models utilizing a p of 0.5 cannot be directly compared to the relative mass reduction values of the ones utilizing p of 1.0 or 3.0. On the other hand, the models utilizing a p value of 0.5 should enhance the stability of the ESLM optimization models. With the above in mind, all 26 models from the initial study are recomputed using a p value of 0.5.

Figure 6.8 displays the results of the model equivalent to the ones displayed in Figures 6.5 and 6.6, apart from the fact that Figure 6.8 utilizes a p of 0.5.

As indicated by Figure 6.8, the results obtained by lowering the p to 0.5 does not significantly change the resulting topology, as can be seen by comparing Figures 6.8 and 6.6. The change of p did, however, have the anticipated effect on the number of EI, which increased to an average of 19 per model. Nevertheless, the feasibility of the topologies obtained is, as previously discussed, questionable. The results of the "corresponding" linear static models to the ESLM models of Figures 6.5, 6.6, and 6.8 are represented by Figure 6.9.

The results of all three linear static models are identical to the naked eye and obtained an identical relative mass reduction of 90.6 % (within 0.1%). This includes the model with p equal to 0.5, thus substantiating the initial hypothesis that a p value of 0.5 can indeed be used. However, the overall results obtained from the initial 78 models (26 times 3) indicate that the basic model setup has to be reviewed to better comprehend the effects of ESLM.

6.2.3 Parameter sensitivity study

When the initial study, as documented above, was set up, the underlying ideology was to use as few optimization control parameters as possible to better comprehend the "true" effects of ESLM. Nevertheless, the results from the 78 models of the initial study demonstrated that it is necessary to review this approach.

Therefore, a parameter sensitivity study involving a total of 10 ESLM models, using the LST geometric nonlinear implicit static (Table 6.2) only, is conducted. The main aim of the parameter sensitivity study is to enhance the stability of the models as well as to increase the feasibility of the topology results.

The starting point of the study is the general FE model illustrated in Figure 6.3. The individual subsections below highlight the main parameters addressed during the parameter sensitivity study.

1. *Aggressiveness of topology optimization algorithm (ATOA)*
   Analyzing the result files that contain the detailed specification of the optimization process, it soon becomes clear that the "aggressiveness" of the topology optimization algorithm will have to be reduced. This is the equivalent of the scale factors introduced and discussed in connection with the gradient-based methods and the line search methods of Chapter 3. In this particular case, it simply means reducing the maximum allowable percentage of material (mass) to be removed in any single optimization iteration from 50% to 10%. This was found to have a significant (positive) effect on the stability of the ESLM models.
2. *Minimum dimension of member size (MINDIM)*
   The MINDIM parameter, as previously explained, attempts to prevent the formation of small "clusters" of material, thus reducing the likelihood of widespread checkerboard effect, and is increased from three times the average mesh size to four times the average mesh size. This was found to have a positive effect on the feasibility of the results obtained by ESLM.
3. *Maximum number of internal iterations per external iteration (MIE)*
   During the setup of the initial study, it was postulated that having only one single internal iteration (II), step 3 (Figure 6.1), per external iteration (EI) would provide the best possible control of the equivalent static load (ESL) models, as the loadcases will be updated for each II of the topology optimization. However, as illustrated in Figure 6.5, the checkerboard effect is often widespread after a single II. By reducing the ATOA, as listed above, this tendency is likely to decrease. If the MIE is increased simultaneously, the likelihood of widespread checkerboard effect will be further reduced, as

the material will be allowed to be increasingly stressed in specific locations prior to recalculating the ESL.

Excessively increasing the MIE would mean that the ESLM models will "converge" toward a "standard" linear static topology optimization, as the influence of the nonlinear (and potential dynamic) effects will be minimized due to the reduced ratio of EI to II.

It could also be argued that the main underlying reason for the widespread checkerboard effect in Figure 6.5 is the fact that the displacement and the stress levels are considerably low in the initial analysis step 1 of EI 1 (Figure 6.1). This means that the initial analyses are, in essence, linear, as the magnitudes of the applied loading are not sufficiently large to induce nonlinear responses. This was known prior to the initial study, as the thickness of the design volume was approximately 70 mm where the external loading was applied, i.e. in the *z*-direction in Figure 6.3. However, as both the design volume and the applied loading were carefully selected, it is not deemed feasible to adapt these to evoke a nonlinear response from the model during the initial iteration. This finding seems to suggest that the ESL method has stability problems if no nonlinear responses are obtained during step 1 (Figure 6.1). A possible method of overcoming this problem would be to "refine" the design volume prior to the optimization; this will, however, (partially) counteract the overall purpose of conducting topology optimization.

In order to obtain a more in-depth understanding of the stability issue, the model represented in Figure 6.6 is reanalyzed. This reveals that in step 1, EI 1 (Figure 6.1), the maximum plastic strain value obtained is approximately 0.01. During the analysis in EI 7, the maximum plastic strain value obtained is 0.27, indicating that the structure has entered the nonlinear region. All the results obtained at this point clearly suggest that an MIE value of 1 or even 10 is not sufficient; the MIE value is therefore set to 100.

4. *Relative convergence tolerance (RCT)*
The effects of this parameter are the last to be investigated during the parameter sensitivity study. The reasons behind investigating the RCT parameter are linked to the ATOA and MIE parameters discussed above. The RCT parameter controls when the II terminates by comparing the relative convergence of the objective function between successive iterations, as previously presented. In other words, it terminates the II when the optimization becomes "ineffective"; the default RCT value is 0.5%. Increasing this value to 5% means that the II will most likely terminate more rapidly, reducing the overall CPU time. This was found to have a positive effect on the overall stability of the ESLM models, in combination with the ATOA and MIE parameters discussed above.

6.3 REVISED OPTIMIZATION STUDY

On the basis of the results obtained by the initial study and the information gathered during the parameter sensitivity study, a revised study is defined. The general FE model illustrated in Figure 6.3 is "reused," including the optimization

TABLE 6.4 Optimization Control Parameters for Revised Study

Description	Value
Aggressiveness of topology optimization algorithm (ATOA)	10%
Minimum dimension of member size (MINDIM)	100
Relative convergence tolerance (RCT)	5%

constraints listed in Table 6.1. The variable optimization parameters used for the study remain identical to the ones listed in Table 6.2, apart from parameter 2, MIE, which was set to 100. In addition, a sixth parameter p that can assume the values of 0.5, 1.0, and 3.0, as presented in Section 6.2.3, is set. Finally, the additional optimization control parameters discussed in Section 6.2.2, and listed in Table 6.4, are implemented. With the revised values, a total of 43 additional models are created.

The models from the revised study proved to be significantly more stable than the ones from the initial study, with approximately 70% of the models successfully completing in excess of 1 EI, clearly demonstrating that the application of the additional control parameters listed in Table 6.4 had the desired effects.

In the following subsections, the effects of individual parameters upon the topology optimization results will be compared and discussed. The main topics covered in these sections will be the influences of the penalization factor p, load step type (LST), nonlinear material parameters (NMP), external load angles (ELA), and scalability of ESLM.

6.3.1 Penalization factor *p*

As discussed throughout Section 4.1, the penalization factor p is always a parameter of interest when dealing with optimization algorithms utilizing the SIMP interpolation scheme. Although the main purpose of the parameter is to reduce the checkerboard effect, it may also have a significant effect upon the overall resulting topology, as presented in Chapters 4 and 5. The effects of this parameter were studied on a smaller scale in connection with the parameter sensitivity study presented in Section 6.2.2. An expansion of this was included within the revised study, which included a total of 13 models with a p value of 0.5, 16 models with a p value of 1.0, and 13 models with a p value of 3.0. However, only 13 p 1.0 models were included in this comparison; the reason for the additional 3 p 1.0 models will be presented and discussed in another subsection. Furthermore, the models being discussed in this section will be ESLM models only. Table 6.5 provides an overview of the results obtained using the three different p values.

TABLE 6.5 Overview of Results Based on *p*

	Average Value		
Penalization Factor *p*	Mass Reduction (%)	# of External Iterations	CPU Time (s)
0.5	95.6	9	19,150
1.0	88.6	7	1,280,141
3.0	82.3	8	784,630

At this point, it should be reiterated that the mass reduction listed in Table 6.5 for p 0.5 is not straightforwardly comparable to the ones for the 1.0 and 3.0 models, due to previously discussed reasons relating to the ratio of relative element stiffness to relative element density, as illustrated in Figure 6.7. Initially disregarding p 0.5, Table 6.5 reveals a minor difference in the average mass reduction between the p 1.0 and p 3.0 models, where the former displays an approximate increase of 7% relative to the p 3.0 models. This is obtained via an average of 7 EI for p 1.0, as opposed to a comparable 8 EI for p 3.0, indicating that a p value of 1.0 is more effective. The average CPU time between the two p groupings is also different, with the p 1.0 models using an average CPU time 1.6 times that of the p 3.0 models. The p 0.5 grouping clearly represents the lowest CPU time, averaging less than 1% and 2.5% of the CPU time relative to the p 1.0 and 3.0 groupings, respectively. One of the primary reasons behind this significant difference originates from the explicit models, as will be discussed later. In addition, the influence of the checkerboard effect is once again significant. Figure 6.10 represents the results from the revised study of the model "identical" to the one illustrated in Figure 6.5 with a p value of 1.0.

FIGURE 6.10 Model with updated optimization control parameters, penalization factor = 1.0.

Coincidentally, the results to the "corresponding" p 0.5 and 3.0 models, relative to Figure 6.10, are now well represented by the topology illustrated in Figure 6.5. This tendency is occasionally observed; however, no general tendencies can be directly deducted from it. The overall conclusion of the study is, therefore, that the influence of p could be very significant; however, it is not a "stand-alone" parameter, as found in the "initial study." It is, in fact, very dependent upon other optimization control parameters. With the current setup, the trend is, however, that the models with a p of 1.0 are the most stable and efficient (in terms of CPU time), in addition to providing the most feasible results, as illustrated in Figure 6.10.

6.3.2 Load step type (LST)

The load step type (LST) is, as previously presented, divided into three possible categories: nonlinear geometry (implicit static), nonlinear geometry (implicit dynamic), and nonlinear geometry (explicit static). The comparison to the linear static models will be subsequently completed.

Figure 6.11 represents the results of the nonlinear geometry (explicit dynamic) LST with a p of 1.0. The model illustrated by Figure 6.11 is thus "identical" to the model represented in Figure 6.10, which represents the nonlinear geometry (implicit static) model. The results of the "remaining" model with the nonlinear geometry (implicit dynamic) LST are well-represented in Figure 6.6. Therefore, by comparing Figures 6.6, 6.10, and 6.11, an initial understanding of the influence of LST can be obtained.

By comparing Figure 6.10 to Figure 6.11, it is obvious that the topology of the latter is better defined than the former, with a reduced checkerboard effect. This indicates that the nonlinear geometry (explicit dynamic) LST models provide clearer definitions than the nonlinear geometry (implicit static) models do. This trend is found to be largely applicable throughout all models included in this study.

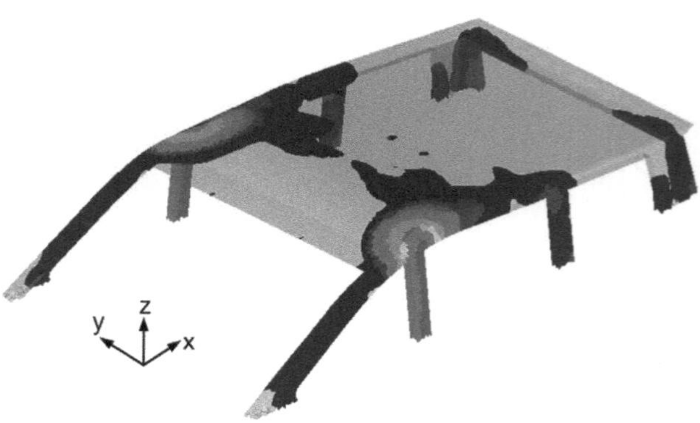

FIGURE 6.11 Nonlinear geometry (explicit dynamic) LST.

TABLE 6.6 Overview of Results Based on LST

Load Step Type (LST) (Nonlinear Geometry)	Average Value Mass Reduction (%)	# of External Iterations	CPU Time (s)
Implicit static	75	8	4,061
Implicit dynamic	79	8	16,607
Explicit dynamic	91	6	2,936,834

Table 6.6 contains the average values obtained across all ESLM models when divided into groups according to LST applied.

Table 6.6 supports the trend identified from Figures 6.10 and 6.11, as explicit dynamic models obtain an average mass reduction value considerably larger than those obtained via the implicit static and implicit dynamic models. Furthermore, the explicit dynamic models obtain this considerably larger mass reduction value using 6 EI on average, as opposed to 8 EI for either of the implicit LST. In other words, the explicit dynamic models seem to be the most "efficient" per EI.

At this point, it should however be noted that, as listed in Table 6.6, the implicit static LST models were completed using an average CPU time of 4061 s, the implicit dynamic models were completed using 16,607 s, and the explicit dynamic models were obtained using 2,936,834 s. In other words, the implicit static and implicit dynamic models used 0.1% and 0.5% of the CPU time used by the explicit dynamic models, respectively.

Throughout the study not a single "set" of models, that is, "identical" models apart from the LST were found, where all three models provided feasible/comparable results to an extent beyond that demonstrated by Figures 6.6, 6.10, and 6.11, as discussed above. Generally, the explicit dynamic models are found to be the most stable, but at the same time also the most CPU intensive. In general, it is also found that the topologies obtained from comparable models using different LSTs do not display significant differences beyond those illustrated by Figures 6.10 and 6.11.

The overall conclusion upon the effects of the LST is, therefore, that the choice of LST should predominately be made as a function of the type of load scenario desired, as would be customary for a "conventional" standalone FE analysis, not as a function of the optimization itself.

6.3.3 Nonlinear material parameters (NMP)

The nonlinear material parameters (NMP) discussed in this section purely relate to those listed as parameter 5 in Table 6.2; a mild grade steel (linear elastic–plastic) or high strength steel (linear elastic–plastic), both using the Johnson–Cook material model; the specific NMPs are listed in Table 6.3.

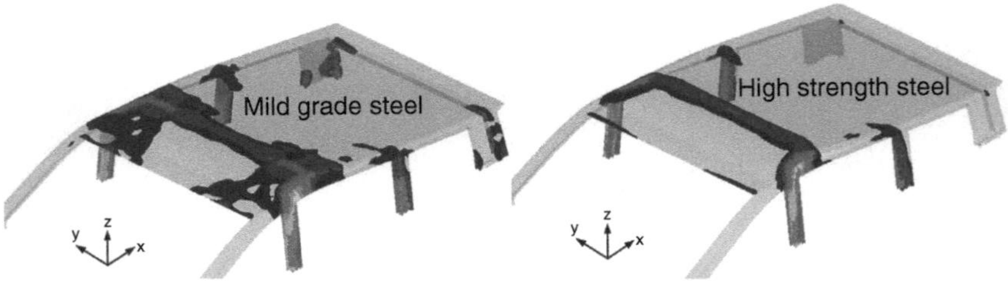

FIGURE 6.12 Effects of nonlinear material parameters (NMP).

Figure 6.12 represents an example of the "typical" difference found between two "identical" models utilizing different NMP. As illustrated in Figure 6.12, the models utilizing the high strength steel values generally display more "distinct" topology results. The mild grade steel model in Figure 6.12 is obtained after 7 EI, achieving a mass reduction of 93.3%, and a combined CPU time of 13,825 s. The corresponding values for the high strength steel model are 6 EI, a mass reduction of 92.8%, and a CPU time of 9,252 s.

The above results are obtained using implicit dynamic LST and indicate that using a high strength steel model increases the "efficiency" of the optimization by obtaining a similar mass reduction using approximately 30% less CPU time, which is generally found to be the case using implicit dynamic LST.

Dividing the models into NMP groupings and including all LST does, however, not support this trend, as documented by Table 6.7.

Table 6.7 clearly shows that the average CPU time consumed in connection with mild grade steel is only approximately 52% of that used for the high strength steel models. In addition, the mild grade steel models obtained an average mass reduction value of 80% using 8 EI, as opposed to 71% using 5 EI for the high strength steel models.

Based on the above, no clear trend or tendency could be found with respect to the influence of the NMP. Trends in subsets of models, such as the one for the implicit dynamic models discussed above, were, however, identified, indicating that the influence of the NMP is not a "stand-alone" parameter, but is heavily dependent upon other factors such as the LST used.

At this point, it should be noted that there are many factors that could potentially influence the above, one of which would be the degree of nonlinearity of

TABLE 6.7 Overview of Results Based on NMP

	Average Value		
Nonlinear Material Parameter (NMP)	Mass Reduction (%)	# of External Iterations	CPU Time (s)
Mild grade steel	80	8	244,756
High strength steel	71	5	468,508

the individual models, for example maximum displacement values. These are likely to have an effect on the above results, but are not further investigated during the study documented in this chapter.

6.3.4 External load angles (ELA)

The reason for including the external load angles (ELAs) as a parameter within the study is in essence to obtain an indication of the sensitivity of the models, as this is of primary concern with respect to crash structures and, to a lesser extent, topology optimization in general. The underlying intention in this particular case is to establish whether any significant variations of the "level" of sensitivity exist between different LST sets. In other words, the intention is, for example, to determine whether the "nonlinear geometry (explicit dynamic)" models are more sensitive to the ELA than the "nonlinear geometry (implicit static)" models.

Throughout the study, a very clear tendency emerged, strongly suggesting that there are no distinct differences in the level of sensitivity between different groupings of LST. This is not surprising given the fact that all models utilize identical optimization parameters; the minor differences found between sets of LST are most likely caused by the difference in "starting point," that is, the difference in ESLS obtained during step 1 (Figure 6.1).

Before summarizing the findings of the ESLM study, a study devoted to exploring the scalability of the ESL method will be presented in the following subsection.

6.3.5 Scalability of ESLM

The purpose of this section is to document a subset of the case study devoted to the scalability of ESLM. The underlying reason for conducting this study is based on suspicions arising in connection with the investigation into the stability problems previously discussed. This study contains a total of three nonlinear geometry (implicit static) models, accounting for the additional three models using this LST type, as previously mentioned. It should be noted that this particular study may, of course, be software biased; nevertheless, the issue of optimization scalability is general and must be carefully considered in the context of high performance computing (HPC).

All optimization results presented so far in this chapter were obtained using HPC computing utilizing a single node and 10 processors per node (PPN). Initially, it was found that a significant number of models error terminated after a very short duration in the region of 10 s of effective CPU time. The only error returned from the FE solver indicated that the specific solver was unable to achieve equilibrium. It was suspected that the issue was related to the amount of memory available; the PPN was, therefore, reduced from 12 to 10 prior to when the studies included in this chapter were completed. This did relive the problem to a certain extent. Nevertheless, it was decided to compare the results of two completely identical models, one solved using 1 PPN and another one solved using 10 PPN. This was found to have a very dramatic effect upon the results, as the model that utilized 1 PPN completed a total of 30 EI (which was the maximum allowed value), whereas the model with 10 PPN completed a total of 8 EI prior to terminating due to excessive checkerboard effect, as illustrated in Figure 6.5. The mass reduction between the two models is, of course, significant due to the significant difference in EI. Nevertheless, the 30 EIs were completed using a total of 3,132 s, as opposed to 1,562 s to complete the 8 EIs on 10 PPN. The FE software utilized sequential multicore processing (SMP); all models were solved on the same Linux 64 bit-based HPC. The results obtained in this limited study suggest that the models will be more stable if run on 1 PPN. This suspicion has, however, not been further investigated.

6.3.6 Overall conclusion of revised optimization study

The above subsections have highlighted the majority of key findings in the revised optimization study. It is clear that not all results provide conclusive tendencies, as there are many factors to be considered, many of which proved not to have significant effects individually but proved to be considerably influential in combination with other parameters. The most obvious example of this is the influence of the nonlinear material parameters (NMP), as previously discussed.

The penalization factor p, the load step type (LST), and the NMP were predominantly found to be influential upon the overall "definition" of the resulting topology. In other words, these were not found to significantly change the topology in general, but were, in essence, found to reduce the checkerboard effect, hardly surprising in terms of p. This also had a direct effect upon the stability of the models, and consequently upon the number of EIs completed, CPU time used, and mass reduction obtained.

The theoretical ideology of a maximum internal iterations per external iteration (MIE) value of 1 initially adopted proved not to be practically feasible due to significant checkerboard and associated instability problems experienced. It could be argued whether this specific problem was related to the ESLM algorithm or indeed to the specific software used. Regardless of the root of the problem, the fact remains that the stability of the majority of models was significantly increased by the introduction of additional optimization control parameters as listed in Table 6.4.

The overall study did only include one design volume and two different sets of external load application angles (ELAs) that could potentially also have a significant effect upon the stability of the models. The main reason behind this argument is that the initial analyses, that is, step 1 (Figure 6.1), did neither display high stress nor strain levels, indicating that the structure is initially "underloaded," thus experiencing linear behavior only. This may lead to significant checkerboard effects in subsequent iterations, as the relative material density and therefore element stiffness is significantly reduced, inadvertently destabilizing the models. Although this problem was overcome by increasing MIE and decreasing the aggressiveness of the topology optimization algorithm (ATOA), it could still be argued that this "un-favored" the ESL approach. In the same context, the question of whether or not the ESL models should be exposed to a certain degree of nonlinearity, for example, significant displacement values at step 1, EI 1 (Figure 6.1), to be effective could also be asked. If this is the case the structure, i.e. design volume, would need to be relatively "well-defined" prior to the optimization commencing, meaning that the fundamental advantages of conducting a topology optimization study would be diminished. In order to fully answer this question, additional studies using multiple software suites would have to be completed.

6.4 ESLM VERSUS LINEAR STATIC TOPOLOGY OPTIMIZATION

With the overall conclusion of the ESLM-based optimization study completed, it is now possible to compare these to "conventional" linear static VDM-SIMP-based optimization. A total of 12 linear static optimization models were included in the overall study; however, valuable knowledge was also drawn from Chapter 5 and previous research that contained in excess of 300 linear static topology optimization models. The results obtained from all 12 models contained in this chapter did not vary significantly, although minor differences did appear. The results obtained from linear static topology optimization are thus well represented by Figure 6.9, and so a direct comparison of the ESL models in Figures 6.10, 6.11, and 6.12 will provide an indication of the difference in results between ESL and linear static optimization. By doing so, it is immediately clear that the singular load path extracted in Figure 6.9 is also noticeably distinct in all ESL-based results. In fact, it could be argued that Figures 6.9 and 6.12 (high strength steel) display the most refined topologies of the four results depicted. These are all very good indications that the primary (and optimal) load path should be located as indicated in Figure 6.9, regardless of whether the dynamic effects are taken into account or not.

On the basis of the model results and statistic information included within the study presented in this chapter, the use of ESL-based optimization for development of structures exposed to large nonlinear deformations is limited. This is primarily related to the excessive increase in CPU time and the lack of increased information (detailed load path generation) obtained therefrom.

In addition, the ESL models experienced significant stability issues that could be attributed to a number of factors, the majority of which were investigated and documented in this chapter. It could also be argued that the stability issues may have been caused, or increased, as a function of applied loading and/or the design volume, as the preliminary analysis indicated that the initial loading coupled with the specific design volume did not invoke a nonlinear response from the models in question. According to Altair (2013), the ESL method "*can be applied at the concept design phase as well as design fine tuning phase*." This means that the use of ESL throughout the studies of this chapter is well within the realm of the ESL method. Nevertheless, the "optimization engine" of the two methods remains identical: VDM-SIMP.

The examples presented in Chapter 5 did, in essence, utilize a "manual" version of ESL, as the dynamic crash loads were represented by "equivalent" static loadcases defined by using "engineering intuition and experience." This did, however, allow for the use of linear static modeling, with the additional benefit of significantly lower CPU time relative to the ESL models of this chapter. The fact does, however, remain that linear static modeling does not consistently provide feasible results for nonlinear (and highly dynamic) crash scenarios, such as a head-on vehicle collision. The results from this chapter indicate that the use of ESL-based models did not significantly enhance the feasibility of the topology results. The primary reason behind this is most likely that the two approaches to optimization are in essence identical, as discussed by Park (2011), which states that *"we cannot use design variables or functions that cannot be defined in linear static response optimization."* It could, nevertheless, be explored whether ESL might be advantageously utilized in a later stage of the design process once the design has initially "matured," that is, initial load paths have been extracted. Regardless of whether or not this approach was to be successful, the fact still remains that truly nonlinear topology optimization of crash structures requires substantial improvements to provide highly efficient and feasible results.

Furthermore, the previously highlighted possible "limitations" of the design volume and applied loading adopted throughout this chapter could be further investigated. This could, for example, be completed via a case study focusing on the development of an automotive front crash structure that would be exposed to very large nonlinear deformation, as well as dynamic loading. Such a study could follow the lines previously set out in Chapter 5, which utilized linear static topology optimization.

In a wider context, many of the issues highlighted through the use of ESLM in this chapter any FE-based topology optimization algorithm for structures exposed to large (nonlinear) deformations must address. Regardless of whether or not the individual engineer or scientist "believes" in the potential of ESLM for nonlinear topology optimization, the status quo remains that a significant amount of development and improvement in terms of obtaining feasible topology optimization results for vehicle crash structures in a CPU "efficient" manner is needed. The following chapters will focus on optimization algorithms that do not use the VDM-SIMP or ESLM methods in the context of nonlinear optimization.

REFERENCES

Altair Engineering, Inc. (2013). *Altair HyperWorks manual*. Troy, MI.

Chirwa, E. C., & Peng, Q. (2010). Modelling of roof crush using the newly updated FMVSS 216. In: ICRASH 2010. Washington DC, September 2010.

Christensen, J., Bastien, C., & Blundell, M. V. (2012). Effects of roof crush loading scenario upon body in white using topology optimization. *International Journal of Crashworthiness, 12*(1), 29–38.

Park, G. J. (2010). Equivalent static loads method for non linear static response structural optimization. LS-Dyna Forum, Bamberg, 2010, Available from: <http://www.dynamore.de/de/download/chapters/forum10/chapters/G-I-02.pdf>

Park, G.J. (2011) "Technical overview of the equivalent static loads method for non-linear static response structural optimization" (online) available from: <http://link.springer.com/article/10.1007%2Fs00158-010-0530-x> Accessed 03.09.14.

Shin, M. K., Park, K. J., & Park, G. J. (2007). Optimization of structures with nonlinear behavior using equivalent loads. *Computer Methods in Applied Mechanics and Engineering, 196*(4–6), 1154–1167.

Sorensen, R., & Lund, E. (2012). Ply-based optimization of laminated composite shell structures under manufacturing constraints. Available from <http://www.escm.eu.org/eccm15/data/assets/1870.pdf> Accessed 23.08.14.

Witowski, K., Mullerschon, H., Erhart, A., Schumacher, P., & Anakiev, K. (2014). Topology and topometry optimization of crash applications with the equivalent static load method. In: *Proceedings from the 13th international LS-Dyna users conference*.

Chapter | seven

Heuristic and Meta-Heuristic Optimization Algorithms

CHAPTER OUTLINE

The purpose of this chapter is to review some of the more "unconventional" optimization algorithms with respect to optimization of mechanical structures exposed to large (nonlinear) deformations. The idea of this chapter is to further enlighten the reader of optimization principles which are not currently considered to be "main-stream" for structural optimization but may be further developed to resolve some of the inherent problems relating to nonlinear optimization as discussed throughout previous chapters. This chapter not only presents and evaluates some of these "alternative" algorithms; it also serves as a foundation for Chapter 8, in which the idea of developing a new optimization algorithm for topology optimization is further presented.

In the previous chapters of this book, various optimization techniques have been presented. These include the general techniques and principles presented in

Nonlinear Optimization of Vehicle Safety Structures. http://dx.doi.org/10.1016/B978-0-12-804424-7.00007-5

Chapter 3, as well as the more specific ones relating to structural optimization in terms of shape, size, and topology optimization algorithms presented in Chapter 4. In this chapter, additional optimization algorithms will be presented. Although the principles are applicable to almost all optimization applications, the presentations and discussions will be conducted in the context of topology optimization.

Optimization algorithms may be categorized in a number of different ways. In this chapter, the focus will be aimed at two main categories: mathematical and heuristic/meta-heuristic algorithms, as will be subsequently presented. Moreover, optimization algorithms may also be divided into approximation and exact algorithms (Vazirani, 2001). Exact algorithms work on the principle that all permutations of a given optimization problem be computed; the optimization problem thus reduces to a "simple" selection problem. However, as discussed in Chapter 3, the curse of dimensionality rapidly eradicates the feasibility of employing exact algorithms for most topology optimization and indeed structural optimization applications. Hosny (2010) states:

> *The term approximation algorithm is often used to refer to an optimization algorithm which provides a solution that is guaranteed to be within a certain distance from the optimum solution every time it is run, with provable runtime bounds.*

All algorithms discussed in this book may, therefore, generally be considered to belong to the approximate algorithm family, although certain limitations will apply, as will be discussed throughout the remainder of this book.

7.1 MATHEMATICAL ALGORITHMS

The algorithms presented in Chapter 3 may generally be considered to represent a small selection of the overall mathematical algorithms available. From a "practical" engineering viewpoint, mathematical-based algorithms are "ideal" for structural optimization, as these generally have a clear and systematic structure, enabling a relatively straightforward implementation into numerical analysis software, such as commercial FE programs. Gradient-based methods such as the optimality criteria method (OCM), moving asymptotes (MA), and sequential linear programming (SLP) are most often employed, although only the latter was presented in Chapter 3. From a scientific viewpoint, the perhaps most pivotal feature of mathematical algorithms is that they (generally) allow it to be proven (within certain limitations) that the optimization is converging toward a (global) optimum solution, as opposed to a (local) "random" solution, provided a unique global optimum solution indeed exists, as discussed in Chapter 3. This "proof" is, of course, dependent upon a number of parameters governing the specific optimization problem. As mentioned above, Chapters 3 and 4 were focused on mathematical algorithms including homogenization-based optimization (HBO). As concluded in Chapter 4, the use of approaches such as HBO has extremely limited applications within engineering in general, as it is not possible to define a continuous function to describe, for example, the stress state throughout a component. The opportunities of utilizing such an approach thus address a much wider aspect of analytical and numerical modeling, and would (if solved) make the use of finite element obsolete for stress analysis; this approach will therefore not be further discussed.

Chapter 3 also contains detailed discussions relating to the VDM method and the SIMP interpolation scheme, which are perhaps the most widely used and accepted approaches to topology optimization of linear elastic and isotropic problems. Indeed, many adaptations and variations of the SIMP interpolation scheme have been used for a wide variety of applications outside of the linear elastic and/or isotropic remits of the original formulation. Sorensen and Lund (2012) did, for example, utilize the scheme for optimization of composite ply lay-up. Chapter 4 also contains suggestions as to how the SIMP scheme could be "modified" to cater for (minor) levels of material nonlinearity; however, the general issues related to SIMP as highlighted throughout Chapter 3, and in particular the ESLM complications discussed in Chapter 6, are still applicable. The overall conclusion was that the SIMP interpolation scheme is not suited for large deformation nonlinear topology optimization. In light of this information and the infeasibility of formulating continuous functions (for example, stress functions) for most mechanical engineering structural analysis/optimization problems, the potential scope for "pure" mathematical algorithms to be successfully and efficiently applied to nonlinear optimization of vehicle safety structures is limited. The focus in this chapter is therefore turned to heuristic algorithms.

7.2 HEURISTIC AND META-HEURISTIC ALGORITHMS

If the mathematical optimization algorithms presented in Chapter 3 and Section 7.1 are considered capable of providing the "exact" (optimum) solution to a given optimization problem, heuristic algorithms may (in the extreme) be considered to take the most "practical" approach to optimization. Pure heuristic algorithms are most often experience-based and utilize a rational and, above all, practical approach to optimization and problem solving in general.

The scenario of a structural engineer reviewing and modifying an initial design based upon personal experience and numerical analysis results may thus be defined as a heuristic "algorithm." Hence, pure heuristic algorithms have inherent uncertainties with, for example, repeatability (imagine that the structural engineer is replaced) and relevance of solution (how can it be justified that the solution is indeed an optimum?). Heuristic algorithms thus generally have many undesirable factors of uncertainty associated with them.

Meta-heuristic algorithms are an adaption of heuristic algorithms and overall address many of the common issues associated with heuristic algorithms. Meta-heuristic algorithms most often adapt ideologies from other sciences, such as biology and evolution, and combine these with physical experimentation results, as well as experience. The following subsections will introduce some of the competing meta-heuristic algorithms.

7.3 EVOLUTIONARY ALGORITHMS

The first types of meta-heuristic algorithms to be introduced are generally referred to as evolutionary algorithms or more specifically evolutionary structural optimization (ESO) algorithms.

The general methodology of the ESO algorithms is very simple and is loosely based upon the theory of evolution, that is, the general principle of "survival of the fittest." In other words, this type of algorithm ensures that the "fittest elements" survive by gradually removing "inefficient" material from the structure in question. The first book on ESO algorithms was published by Xie and Steven (1997).

7.3.1 Evolutionary structural optimization (ESO)

The ESO algorithms solely rely on FE analysis to obtain the data necessary for conducting the topology optimization. Thereby, the general flowchart of an evolutionary-based algorithm can be defined as illustrated in Figure 7.1.

As illustrated in Figure 7.1, the flowchart of the evolutionary-based algorithms is slightly different than that of the mathematical-based algorithms presented in Chapter 4. The governing theory behind the evolutionary-based algorithms is, in general, less complicated than that of the mathematical algorithms.

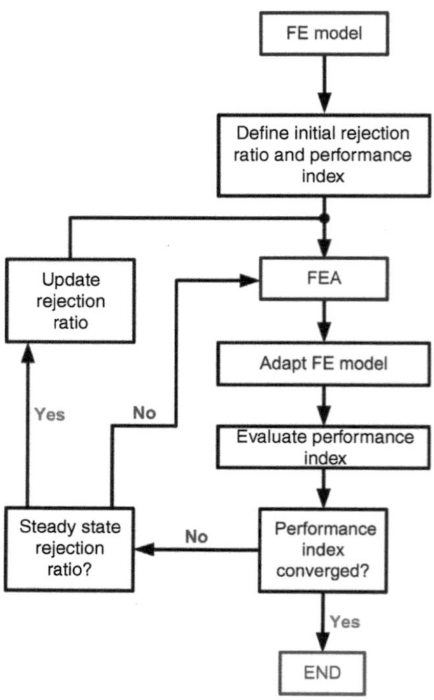

FIGURE 7.1 Flowchart of evolutionary algorithm.

An evolutionary structural optimization (ESO) algorithm may work as described by the following seven steps.

1. Define an FE model
2. Define initial rejection ratio (RR) and performance index (PI).
   The RR is the methodology used in ESO to determine which elements are "inefficient." This could, for example, be based on strain or stress values; for instance, the RR could be defined as Equation (7.1)

$$RR = \frac{\sigma_y}{\sigma_{\max}} \tag{7.1}$$

   In Equation (7.1), σ_i is the stress of the individual element, y, and $\sigma_{\max}$ is the maximum allowed stress; for linear applications this could, for example, be the yield stress value. Typical values of RR range between 1% and 25%.
   The performance index (PI) is a recent addition to ESO, and was initially suggested by Querin (n.d.). The ideology behind the PI is, in essence, the same as the objective function of the mathematically-based algorithms, namely to provide a single number "ranking" the performance of the topology in that specific iteration, enabling the "optimum" structure/topology to be identified via this single number.
3. Conduct an FE analysis (linear or nonlinear);
4. Adapt FE model. This step simply consists of deleting those elements which have RRs below that specified in Equation (7.1);
5. Evaluate the performance index and check if a significant change has occurred since the last iteration. This step is very similar to that used in the mathematical-based algorithms as previously presented;
6. If the PI has not converged, check if the RR has reached a steady state, that is, if no elements (or very few) have been deleted in the current iteration, compared with the preceding one(s);
7. If the RR has reached a steady state, increase it according to Equation (7.2)

$$RR_{\text{updated}} = RR + ER \tag{7.2}$$

In Equation (7.2), RR_{updated} will replace the current RR value by adding the evolutionary ratio (ER). The ER is also a predefined value to be selected from model to model. A typical value of ER is 5%.

The above steps are repeated until the performance index has converged. The overall steps of the ESO and the mathematically-based algorithms are thus very similar. However, the way in which these steps are conducted is quite different.

The most important thing to note is that ESO does not reduce the stiffness (or density) of any element; it straightforwardly deletes the "inefficient" ones.

This strategy has its advantages as well as its disadvantages. The main advantage lies in the fact that the geometry is updated according to the optimization, that is, no "artificial" stiffness remains in the model as it does in, for example, the SIMP scheme (where E of an element is lowered while the "geometrical stiffness" remains). This also means that ESO can utilize linear as well as nonlinear FE with only very minor changes to the algorithm definition. The disadvantages of deleting elements between iterations primarily relate to instability issues and the "irreversibility" of the process. If an element is deleted, it cannot be "recreated" at a later stage in the process if required. This is possible using SIMP, as the "scaling" factors of $[K]$ can simply be adjusted.

Furthermore, ESO algorithms need to reassemble $[K]$ for every iteration, as the topology (geometry) changes. As previously mentioned, this is not required for the VDM-SIMP approach. Variations of the ESO algorithm have been proposed; these have primarily been developed to overcome some of the above issues relating to the algorithm.

The two main variations of ESO are denoted:

- Additive evolutionary structural optimization (AESO)
- Bidirectional evolutionary structural optimization (BESO)

7.3.2 Additive evolutionary structural optimization (AESO)

The AESO algorithm works very similar to the ESO algorithm, except instead of deleting elements, it adds them. This means that the optimal structure is not found from "a full design volume," but rather an empty one. This can be better understood by observing Figure 7.2, which may be used to illustrate the starting points for an ESO and an AESO algorithm, respectively.

Please note that Figure 7.2 is merely an illustration; a "minimum" number of elements is required in order to carry out the initial FE analysis. Due to the overall similarities and the "identical" approach of ESO and AESO, the two types of algorithm experience the same difficulties, namely that of the inability to "reverse" the steps, that is, recreating (ESO) or deleting (AESO) elements. On this basis, the bidirectional evolutionary structural optimization (BESO) algorithm was developed. BSO essentially "combines" ESO and AESO so that the algorithm is capable of creating, as well as deleting, elements to a certain extent.

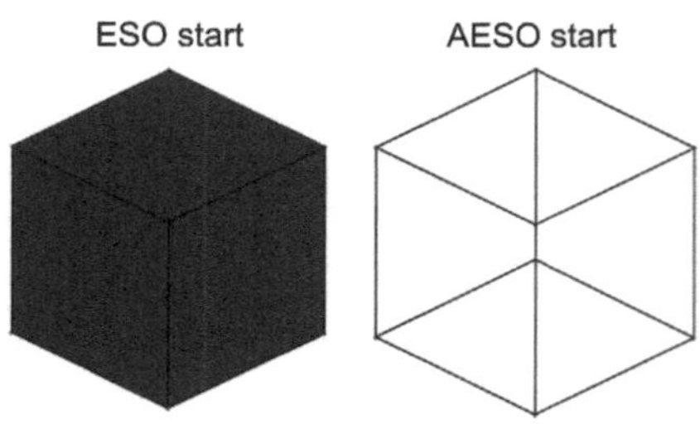

FIGURE 7.2 Illustration of ESO and AESO starting points.

7.3.3 Bidirectional evolutionary structural optimization (BESO)

In theory, the BESO algorithm works well and does resolve some of the obvious disadvantages of the ESO and AESO algorithms. However, some significant practical issues severely restrain the BESO algorithm. These primarily relate to the CPU cost of the algorithm which, in general, is high, as a full FE analysis is required for each iteration (see Figure 7.1). Furthermore, a number of research papers have rigorously questioned the validity of ESO topology optimization results; "discussions" through journal and conference paper publications have been ongoing. At this stage, it is important to remember the stumbling blocks for all topology optimization algorithms, such as the curse of dimensionality and the "accepted trade-off" between speed and accuracy of the optimization algorithm, as indicated in Section 7.2. The main point to raise here is that the BESO algorithms are "not perfect," but neither are the remaining available algorithms; nonlinear optimization is an area of immense research activity which is very much still in its infancy.

ESO-based optimization can, in general, be classified as belonging to the "greater family" of optimization algorithms known as genetic algorithms (GA). Completing a 2015 literature review of GA and ESO optimization did provide a number of publications, but none directly related to nonlinear topology optimization. The most recent paper from the "fathers" of BESO, Y.M. Xie and X. Huang, directly focusing on nonlinear topology optimization is from 2008 (Researchgate, 2014), indicating that the development of BESO for nonlinear topology optimization has not been widely pursued recently.

Given the general similarities between GA and BESO and that the literature review did not disclose any publications from other authors specifically relating to the development of GA or BESO for nonlinear topology optimization post-2008, it is presumed that Huang and Xie (2008) contains the most recent version of BESO for nonlinear topology optimization; the following will, therefore, utilize the information of this publication, as this is representative of GA algorithms in general.

The paper focuses on both geometrical and material nonlinearity under displacement loading, and focuses on the robustness of the algorithm, as defined in Chapter 3. Based on information in the paper, an overall flowchart for the revised BESO algorithm has been created (Figure 7.3).

The paper highlights three main aspects that have been improved in the "updated" BESO algorithm as:

1. Filtering scheme (2)
2. Improved sensitivity analysis (1, 3)
3. A new procedure for adding and removing elements (5)

The brackets above relate to the algorithm steps illustrated in Figure 7.3. It can be noted that the "filtering scheme step" falls in between the two parts of the "sensitivity analysis," and indeed the two are intertwined. The algorithm in Huang and Xie (2008) aims to maximize the structural stiffness of the structure

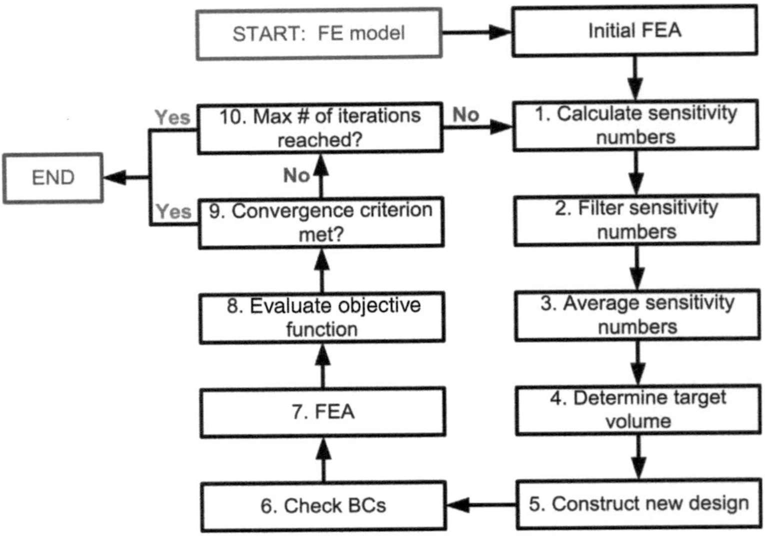

FIGURE 7.3 BESO flow diagram.

with a constraint based on maximum allowed displacements. The objective function thus seeks to maximize the external work that is set equal to the total strain energy under the caveat of a quasistatic load condition. While this condition may not be of major concern to certain topology optimization problems, it is certainly an aspect that must be thoroughly considered if dealing with, for example, high-velocity crashworthiness scenarios. This potential concern could possibly be circumnavigated by introducing a very low timestep in the FE analysis; this would, however, have a significant effect upon the overall CPU run time.

The revised sensitivity analysis and filtering scheme aims to determine the influence upon the individual (FE) element upon the response of the objective function, that is, the sensitivity. Overall, this is achieved by observing the (strain) energy levels within the individual element and the method used to summarize the total strain energy. Equation 15 in Huang and Xie (2008) denotes the variation of the objective function as a function of the individual element x

$$\Delta f(x) = \lim_{n\to\infty} \sum_{i=1}^{n} \left(E_i^e - E_{i-1}^e\right) = -E_n^e \tag{7.3}$$

In (7.3), E denotes the strain energy, e denotes that the definition is carried out on a single element, i denotes the specific interval on the design displacement, and n denotes the total number of intervals on the design displacement. E_n^e is thus the total (numeric) strain energy of the removed element. There are two important aspects to denote in relationship to the above: Firstly, the i index can be considered to be the equivalent of the timestep t in an FE analysis, as the two are coincident. With this in mind, the second and more general significance of (7.3) becomes clearer; the algorithm now utilizes time-history information, that is, it utilizes the value of the previous time step to assess the impact upon the objective function. Although this information is perhaps of limited use at present, it may be worth analyzing the potential of expanding this "time-history" further to enhance the credibility of the algorithm. Huang and Xie (2008a) state:

> *The Eq. (15) means that the decrease of the total external work due to removing one element is equal to the total strain energy of the element in its final deformed state and irrelevant to the size of displacement intervals.*

There are two main implications of the above statement. Firstly, it is the basis for the claim in Huang and Xie (2008) that the algorithm is mesh independent. On the face of it, this statement is quite bold, but in terms of technical significance it is of minor importance, as all FEA per definition is mesh dependent, as discussed in Chapters 1, 2, and 4. This means that despite the above statement, the "input information" is arguably mesh-dependent, questioning the overall importance of the "mesh independence" statement.

The second main implication of the above citation lies within the intricate details in the wording, more specifically the term "final deformed state." This implies that ultimately only the final deformed state is of relevance and the intermediate steps are not. The validity of this statement can be justified in combination with the previous caveat of quasistatic loading, but arguably warrants further investigation in connection with optimization of, for example, crash structures. A more holistic approach to this topic is, of course, related to the overall aim of any specific topology optimization study. For example, the design of a front end crash structure (where acceleration levels during the impact must be taken into consideration) is significantly different from a support bracket where only the final deformation is of relevance. In other words, the significance of the dynamic nature of any given problem must be carefully considered.

Returning to Equation (7.3) Huang and Xie (2008a) defined the sensitivity of the (existing) element i as

$$\alpha_i = E_n^i \tag{7.4}$$

Equation (7.4) thus confirms the above statement that the sensitivity α_i of element i is only dependent upon the last state n, that is, the final deformed state. Huang and Xie (2008a) also state:

> *Note that the sensitivity number is a direct measure of the variation of the objective function due to element removal or addition. To maximize the total external work, the evolutionary process will be conducted by removing elements with the smallest sensitivity numbers and adding the elements with the highest ones. Mathematically such a procedure is known as the "hill-climb" method or the "steepest descent" algorithm.*

The above citation thus indicates that the BESO algorithm fundamentally uses the hill climbing (HC) optimization technique that will be subsequently presented in Section 7.8. This also reinforces the statement relating to the lack of "diversification" of the BESO optimization algorithm significantly increasing the likelihood of converging toward local optimum solutions, as previously discussed in Chapter 3.

The focus of attention is now turned to the "filtering scheme" improvement in Huang and Xie (2008); step 2 (Figure 7.3). As per Equation (7.4), the sensitivity number for existing elements is extracted directly from FEA. This is, of course, not possible for void elements, but it may be necessary to add elements to the model, for example, in highly stressed areas. At this point, it is convenient to expand the explanation of what constitutes a void element. It is important to remember that BESO optimization works from a full design volume (DV) and subsequently voids (deletes) elements in each iteration, as illustrated between Figures 7.4a and 7.4b. Finally, void elements may be added as illustrated in Figure 7.4c.

Two things should be noted when observing Figure 7.4: firstly, that the elements added in (c) were originally part of the full DV in (a); hence, the nodal coordinates are predefined. Secondly, it should be noted that (b) and (c) in Figure 7.4 may occur within the same iteration step, that is, elements may be both voided and added. The above may seem identical to the VDM-SIMP approach; it is, however, important to remember that the VDM approach does not delete elements from the FE model between iterations, as it merely reduces Young's modulus. In BESO, the elements are completely removed from the stiffness matrix of the FE model; however, the nodal coordinates and element definitions remain in the memory of the optimization software enabling the transition between Figure 7.4b and c; in other words, the element is marked as void. It should also be noted that it is possible to initially define void elements. As an example, the structure included in the FE analysis at iteration 0 is Figure 7.4c, but the full DV of the optimization is as illustrated in Figure 7.4a.

The mathematical description for adding elements may be obtained by redefining the sensitivity numbers and simultaneously introducing the filtering

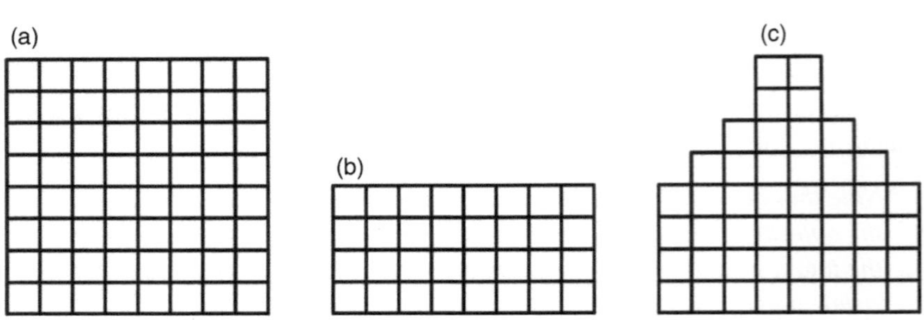

FIGURE 7.4 (a) Full design volume, (b) elements void (deleted), and (c) elements added.

scheme. Huang and Xie (2008b), therefore, define the sensitivity number for adding elements as

$$\alpha_i = \frac{\sum_{j=1}^{N} \omega(r_{ij})\alpha_j}{\sum_{j=1}^{N} \omega(r_{ij})} \tag{7.5}$$

In Equation (7.5), N is the total number of elements in the FE mesh; $\omega(r_{ij})$ is a weighting function defined as Equation (7.6) (Huang & Xie, 2008b)

$$\omega(r_{ij}) = \frac{\exp\left(-\left(\frac{(r_{ij})^2}{2\left(\frac{r}{3}\right)^3}\right)\right)}{2\pi\left(\frac{r}{3}\right)} \quad \Bigg| \quad \begin{matrix} \{i \in N \mid r_{ij} \le r\} \\ j = 1,2,\ldots,N \end{matrix} \tag{7.6}$$

In Equation (7.6), r_{ij} is the direct distance between the centers of elements i and j; r is a user selected filtering radius. Equations (7.5) and (7.6) thus underpin the subsequent discussions of Section 7.5, which highlights some of the similarities between Cellular Automata (CAA)/Hybrid cellular automata (HCA) and BESO; in this case, the influences of the "neighborhood." In relation to Equations (7.5) and (7.6) Huang and Xie (2008b) state:

> *The above sensitivity filter serves two purposes: (1) to extrapolate sensitivity number within the full design domain and (2) to smooth the sensitivity number in the neighborhood of each element and prevent the problem of checkerboard patterns and mesh-dependencies in topology optimization.*

With the above in mind, the focus of attention can now turn to the third improvement of the BESO algorithm proposed in Huang and Xie (2008): the new procedure for adding or removing element, step 5, Figure 7.3. The intricate details of the algorithm are of less relevance in the current context; what is important is the overall use of 0 (void) and 1 (full) element values. The revised procedure ranks and treats all elements in the model, that is, both 0 and 1 elements together. The main reason why this is significant is that boundary conditions (BCs) are not initially considered at this step (5, Figure 7.3), when determining the elements to void or add. This could, for example, lead to computational instability or even a significant change in the overall physics of the optimization problem if, for example, an external load or constraint is removed. In an attempt to avoid this scenario, step 6, Figure 7.3 is introduced. In this context, Huang and Xie (2008c) state:

> *Once such a problem occurs, using a finer mesh or replacing the removed elements with "soft" elements to preserve the boundary or load can effectively circumnavigate problem.*

Huang and Xie (2008) do not further specify the definition of a "soft" element, but it can be deduced that this is an element with lower stiffness or indeed lower volumetric mass density. Parallels can thus be drawn between the BESO algorithm and the VDM-SIMP algorithm discussed in Chapter 4.

The final sections of Huang and Xie (2008) contain topology optimization examples from both linear and nonlinear models using BESO. The conclusions contain two separate statements of particular interest (Huang & Xie, 2008d):

> *Due to partial unloading of material, the defined optimization problem for nonlinear material is ill-posed and the solution may not always convergent because oscillation between designs of two different deformation modes.*

Convergence is always a key aspect of any topology optimization algorithm; the above statement reveals a very undesirable potential pitfall of utilizing BESO in connection with nonlinear material behavior. Questions can be raised as to why this problem seemingly has not been further addressed since 2008; regardless of what the answers to these are, it is certain that if BESO is to be further refined for topology optimization of nonlinear structures, the above must be a key area of investigation.

The second main finding according to Huang and Xie (2008d) is:

> *The examples illustrate substantial differences in the topologies of the linear and nonlinear optimal designs. Unlike the linear design which is independent of the design displacement, the nonlinear optimal design is highly dependent on the specified design displacement. The results show that the nonlinear design at its design displacement is always better than the linear design in terms of the total energy, which is the objective function.*

This is a very interesting statement that underpins the previous statement relating to the significance of the "final deformed shape," but the above also clearly indicates that the final deformed shape is heavily dependent upon the previous stages of the deformation, which is hardly surprising from a mechanical engineering viewpoint. It does, however, also emphasize that the need for analysis into the direct importance and influence of the "intermediate deformation steps" upon the results of the topology optimization must be explored further. In other words, should the intermediate energy levels, E_i, in Equation (7.3) be further implemented into the optimization algorithm? This will be further addressed in Chapter 8.

TABLE 7.1 Advantages and Disadvantages of BESO Algorithm

Advantages	Disadvantages
Strong, direct, and relatively simple implementation into FEA.	Quasistatic loading assumption.
Local and global constraint control.	Nonconvergent and potential oscillation between "equal" design solutions.
Direct link between objective function and FEA.	No direct consideration of "intermediate" deformation modes/levels.
Neighborhood influence.	Relatively high computational cost.
Inclusion of gradient information (time-history).	Significant issues with high levels of nonlinear material behavior.
	Low dispersion of search within potential solution space, high probability of "converging" toward local optimum points.

The final remark to be noticed in the above citation is that using the BESO algorithm and test cases in Huang and Xie (2008), the nonlinear topology optimization results always outperform the linear counterparts. If this is indeed the case, it is suggested that applying nonlinear topology optimization algorithms to linear topology optimization problems will lead to more optimal solutions. It is understood that this potential advantage comes at the cost of increased CPU time.

The presentation of the BESO algorithms is thereby complete. Despite a number of significant issues relating to the application and results from the BESO algorithm, it also raises a number of potential benefits and general areas of interest which should be considered, whichever nonlinear topology optimization algorithm is chosen going forward. The main advantages and disadvantages of the BESO algorithm are summarized in Table 7.1.

7.4 REQUIREMENTS FOR OPTIMIZATION OF STRUCTURES EXPOSED TO LARGE (NONLINEAR) DEFORMATIONS

All discussions throughout this book have predominately been focused on algorithms that are currently being used for structural optimization on a large scale, for example, in commercial FE software. All the previous discussions, models, examples, and results have clearly indicated that none of these approaches fully meets all the requirements of a topology optimization algorithm for structures exposed to large (nonlinear) deformations. The overarching requirements for an algorithm/methodology catering for structures exposed to large (nonlinear) deformations can, in general, be listed as:

1. Proof of convergence toward an optimum solution
2. Reasonable overall computing requirements (CPU time, memory, etc.)
3. Appropriate level of ability to cater for the natural complexity of the load scenario such as:
   a. Time-dependent loading
   b. Nonlinear system response
   c. Variations in material characteristics as a function of deformation
   d. Overall system response versus localized effects
   e. Relevance and feasibility of results to "real world" engineering

The above list of "requirements" is by no means exhaustive and is, in general, open to interpretation and discussion. It is, however, clear that all individual algorithms discussed so far have severe limitations with one or multiple aspects of the above list. Furthermore, it is unlikely that any one algorithm will be able to fully meet all requirements; ironically, this problem can in itself be considered to be a multiobjective optimization problem. With this in mind, the algorithms presented in the subsequent sections are not generally applied to topology optimization problems in connection with structural analysis, crash scenarios, and vehicle safety structures. This means that the majority of algorithms presented will not have been widespread applied to topology optimization of structures exposed to large deformations using a rigorous and well-documented basis. The potential application of the algorithms and the feasibility to meet the above list of requirements will be discussed throughout each of the following sections.

7.5 HYBRID CELLULAR AUTOMATA

Hybrid cellular automata (HCA) is the optimization algorithm incorporated in Livermore Software Technology Corporations' (LSTC) commercially available LS-TaSC optimization software capable of handling large deformation nonlinear problems, such as vehicle crashworthiness (LSTC, 2011). According to LSTC (2011a), the algorithm is a heuristic one. However, according to LSTC (2011) the algorithm was first derived by Tovar (2004) which clearly utilizes bone biology to define the HCA algorithm, arguably making it a meta-heuristic algorithm. The HCA algorithm is an adaptation of the cellular automata (CAA) model originally developed by Stanislaw Ulman and John Von Neuman in the 1940s, originally known as Cellular AutmatoN (CAN). The general principle behind CAA is as follows. Imagine a series of "individual cells" that may have finite states, for example, "1" or "0," as is the basic principle of FE-based topology optimization. Each individual cell is "related" to a set of cells denoted as its "neighborhood." An initial state of each cell is set (at $t = 0$), a new overall state (at $t = 1$) is then achieved using a fixed rule, for example, the outcome of an FE analysis; the new state of each cell (at $t = 1$) is then obtained based upon its current state (at $t = 0$) and "the responses" from its neighborhood. The neighborhood may be defined in slightly different ways; one of the most often used neighborhoods is the Von Neumann neighborhood (Barati, 2014).

Comparing the above description to that of the evolutionary structural optimization (ESO) algorithms presented in Section 7.3, the parallels are very distinct, as the ESO algorithms use a "snapshot" of the current state, that is, FE analysis results to choose "initial states" and thereafter use "neighborhoods" to determine the state of each element (cell).

The basic topology optimization problem in LS-TaSC is solved using a uniform internal energy density as the objective function (Roux, 2011). The topology optimization problem in LS-TaSC is formulated as Equation (7.7) (Roux, 2011a)

$$\left.\begin{array}{l} \min\limits_{x} \sum\limits_{i=}^{N} \sum\limits_{j=1}^{L} \left(w_j U_j(x_i) - U_j^* \right) \\ \text{subject to: } \sum\limits_{i=1}^{N} \rho(x_i) V_i \le M^* \\ \qquad C_j^l \le C_j \le C_j^u, j = 1, 2, \ldots, J \\ \qquad x_{\min} \le x_i \le 1.0 \end{array}\right\} \tag{7.7}$$

In (7.7), w is a weighting factor, U is the internal energy density of element I, V_i is the element volume, U^* is the internal energy density set point, and C_j is constraint j with lover (l) and upper (u) subscript boundaries. The change in the individual design variable i at time t and the updated design variable at $t + 1$ are defined in Expression (7.8) (Roux, 2011a)

$$\begin{array}{l} \Delta x_i^t = \dfrac{K\left(U_i^t - U^*\right)}{U^*} \\ x_i^{t+1} = x_i^t + \Delta x_i^t \end{array} \tag{7.8}$$

In (7.8), K is a scale factor. According to Roux (2011), LS-TaSC is controlled by three user-defined parameters: mass fraction (MF), convergence tolerance parameter (CTP), and proximity tolerance parameter (PTL). LS-TaSC utilizes a "full design volume" for a starting point, as does ESO (see Figure 7.2). In Chapter 4, the objective function and constraints for the variable density method were formulated, and are repeated here for convenience:

$$\begin{array}{c} \text{minimize } F(\rho) \\ \text{subject to:} \\ g_1(\rho) \le Val_1, \quad g_2(\rho) = Val_2, \quad g_3(\rho) > Val_3 \end{array} \tag{7.9}$$

$$F(\rho) = U(\rho) = \frac{1}{2} \int \left(\{\varepsilon\}^T \cdot [E(\rho)] \cdot \{\varepsilon\} \right) dV \tag{7.10}$$

Comparing (7.9) and (7.10) to (7.7), it is clear that there are distinct similarities between the two "algorithms." The main difference lies in the fact that (7.7) utilizes the strain energy density, while (7.10) uses the (total) linear elastic strain energy. It should be noted that (7.10) is an example of how VDM works in general; the actual implementation will vary from software to software. Altair (2009) does, however, support that the above explanation is indeed the way that topology optimization is implemented into Altair HyperWorks' optimization module Optistruct. Altair (2009) also confirms that Optistruct utilizes a gradient-based approach to perform the actual optimization, as discussed in Chapter 3. Despite the similarities discussed above, HCA and LS-TaSC do not utilize a gradient-based approach (LSTC, 2011a):

> *This method updates the density of elements based on information from its neighbors. No gradient information was required.*

This means that LS-TaSC/HCA performs the optimization based on "current" information and is not influenced by "time-history," that is, previous iterations are not considered as part of the optimization process. In other words, the optimization process has no "memory" as, for example, Optistruct does (Altair, 2009). The above statement thus reveals a distinct difference between CAA and HCA, as the latter does not use "time-history," i.e. gradient information, whereas the former (CAA) does.

From an overall perspective, this means that HCA/LS-TaSC is, in essence, very similar to evolutionary algorithms that were discussed in Section 7.3. This also suggests that LS-TaSC "shares" some of the inherent problems of evolutionary algorithms with respect to proven convergence toward a truly optimum solution as also discussed in Section 7.3. As previously discussed, this is one of the major advantages of mathematical-based algorithms; however, when large levels of nonlinearity are introduced, the "predictive" (linear) nature of the systems and the associated benefits diminish. The lack of "time-history" influence, for example, gradient-based information, is a potential area of concern, as it is very likely that both the HCA and ESO algorithms will rapidly move toward a specific solution (topology). For relatively "simple" topology optimization problems using relatively, small design volumes, this may in fact be advantageous, but for larger more complex problems, it indicates that neither algorithm is able to "diversify" its search pattern and avoid local extremum/optimum points (see Figure 3.21).

The general advantages and disadvantages of the HCA/CA-based algorithms are listed in Table 7.2.

7.6 COMBINATORY OPTIMIZATION PROBLEMS

A combinatorial optimization problem (COP) can, in general, be perceived as an optimization problem containing only discrete variables. FE based topology optimization falls into this category, as each individual element should either

TABLE 7.2 Advantages and Disadvantages of CAA/HCA Algorithms

Advantages	Disadvantages
Strong, direct, and relatively simple implementation into FEA.	No time-history data considered.
Local and global constraint control.	No convergence justification.
Direct link between objective function and FEA.	Relatively high computational cost.
Neighborhood influence.	Low dispersion of search within potential solution space, high probability of "converging" toward local optimum points.

be included or excluded from the structure; this could, for example, be done using the Boolean operator 1 (included) or 0 (excluded). As presented in Chapter 4, the introduction of VDM turns the problem into a continuous one. A further general requirement of COPs is that the solution space should contain a finite number of entries, that is, discrete or alternatively a countable infinite set (Schrijver, 2003).

Due to the curse of dimensionality (Chapter 3), it is clear that the reference to a "finite solution space" above rapidly becomes violated with respect to topology optimization, but this may not necessarily be the case for the "countable infinite set." It is clear that most "real world" topology optimization problems will have an infinite number of solutions, and so the "reduction" to a "countable number of infinite solutions" will most likely influence the results of such an algorithm. With this in mind, the study into COPs and the applicability to nonlinear topology optimization may continue.

COPs may further be divided into two main categories: deterministic combinatorial optimization problems (DCOP) or stochastic combinatorial optimization problems (SCOP). The difference between the two types of problems is easily understood through the following example. Consider the optimization problem of creating packed lunches for a set upper sale price with maximum nutrition. The packed lunch may consist of eggs, meat, and bread. Knowing the cost, as well as the nutritional value, of each item, the DCOP can be solved relatively easily by creating an objective function of nutrition and a single constraint relating to the cost. The optimization problem may subsequently be solved by, for example, using one of the gradient-based methods presented in Chapter 3. Now imagine that the "packed lunch business" will be selling lunches over the coming school year. No single supplier is, however, able to commit to a set price or even a consistent product. This means that both the price and the nutritional value of each item will vary. The optimization problem is now more complex, as a certain level of "randomness" or "uncertainty" has been introduced; the nature of the problem has become stochastic, and the overall optimization problem is now a stochastic combinatory optimization problem (SCOP).

Leaving the world of "packed lunches" and returning to the world of structural optimization, it is obvious that there are direct parallels between the above example and that of linear versus nonlinear topology optimization. Although there are several practical differences, the essence of the DCOP problem is (in its basic form) similar to that of linear optimization; it is possible with great accuracy to "predict" the outcome of the optimization and it can be justified that an optimum solution has been obtained, for example, via the use of gradient-based optimization. The essence of the SCOP problem also holds parallels to the nonlinear topology optimization problem; it is very difficult if not impossible to predict the variations in the response of the system. Consider, for example, the task of consistently and repeatedly predicting the changes of structural response of an FE crash model when adding or deleting elements; this is the equivalent of adding uncertainty into the system. The stochastic nature of this system is most likely not completely random, which indicates that it may be possible to introduce probabilities into the optimization algorithm. With this in mind, it is now useful to review algorithms that in general focus on SCOP; for simplicity, it is, however, more convenient to introduce these in terms of DCOP and subsequently address the additional complications with respect to SCOP. Bianchi et al. (2008) completed a significant survey of meta-heuristic algorithms. This survey forms the basis of the explanations presented as follows.

7.7 ANT COLONY OPTIMIZATION

Ant colony optimization (ACO) is, as the name suggests, based on the natural behavior of ant colonies and their individual worker ants. When ants forage, they naturally seem to find a "logical" and "effective" route between their nest and the food source; in other words, they seem to determine an optimum route. This observed behavior is the basis of ACO (Bianchi et al., 2008). Imagine two ants walking from the nest to a food source using two different routes. As the ants walk they release pheromones which naturally decay over time. The ant which (randomly) selected the short route will commence the return leg of the journey quicker than the other ant, thus reinforcing the pheromone trace on the shorter route. Other ants will instinctively follow the stronger pheromone path, reinforcing and even adding to this. According to Bianchi et al. (2008), the ACO algorithm contains three major steps that constitute the central optimization loop of the algorithm:

1. Construct Ants Solutions. This is the procedure for which "ants" incrementally and stochastically construct paths, that is, solutions in a wider optimization context;
2. Evaporate Pheromone. This is the process in which pheromone for certain solutions are decreased using "local" information; therefore, this step is also often referred to as local update. This step is pivotal to ensure that the ACO algorithm does not prematurely converge to a single solution;

3. Deamon Actions. This step refers to decisions made based on global information relating to the optimization problem. Note the difference between local in step 2 and global in step 3. Analog to step 2, step 3 is also often referred to as global update.

The three steps highlighted above are repeated until the optimization problem has converged or is otherwise terminated via a prespecified termination condition.

Even though the above description is general and perhaps somewhat abstract in relation to nonlinear topology optimization and the much higher level of details in the previous sections, the basic ideology of the ACO algorithm can be understood. Based on the above, it is clear that significant development and analysis work is required before ACO could be applied to the nonlinear topology optimization problems of concern. It is also clear that it is unlikely for the ACO algorithm to be a "stand-alone solution" to the nonlinear topology optimization problem. It is perhaps more appropriate to see the ACO algorithm as something which could be merged into other structural optimization algorithms, for example as an "evolution" of BESO, that is, an algorithm that could be combined with that presented in Section 7.3. Based on this relatively short review of ACO, it is not feasible to detail a specific list of advantages and disadvantages of ACO for nonlinear topology optimization, quite simply because additional analysis and development work is required. It should, however, be noted that ACO has the potential to be combined with, for example, BESO to overcome the previously stated concerns with respect to the diversity of the search area for BESO algorithms.

7.8 STOCHASTIC HILL CLIMBING

Stochastic hill climbing (SHC) is a variation of the more general optimization search technique denoted as hill climbing (HC). As presented in Section 7.3, HC is the fundamental search method used in the BESO optimization algorithm. HC is generally considered to belong to the "local search" techniques, analogue to the trust-regions presented in Chapter 3. The basic principles of HC are as follows. A starting solution is required, that is, an initial guess for a viable solution is needed; in the case of BESO, this constitutes an initial FE analysis. Based on this initial guess, variations to the current solution are introduced based on some predefined rules and the newly "updated" solution is then analyzed. The performance, i.e. objective function, of the "updated" solution is then compared to the "original" solution; if the former outperforms the latter, it will be accepted as the "current" solution and the iterative process will repeat. If the "original" solution outperforms the "updated" solution, the former will remain the "current" solution and further variations of the "original" solution will be made. It is, of course, necessary to introduce a termination criteria, for example, maximum number of iterations. As discussed in Section 7.3, the specific implementation of the "predefined rules" in terms of BESO constitutes the filtering and averaging of the element sensitivity numbers; steps 2 and 3 in Figure 3.3. As highlighted in Table 7.1, questions about the "low dispersion of search within the solution space" for this type of algorithm could be raised. This statement is supported by Bianchi et. al. (2008a):

> *Such a simple and very local general search behaves quite poorly in practice, particularly because when a local optimum is found, the algorithm stops improving, and combinatorial problems often have a local optima whose objective values are much worse than that of the global optimum.*

SCH attempts to reduce this problem by not only "obeying" the predefined rules for variation, but also introduce a certain level of "randomness," hence stochastic HC, by including larger (global) variations not governed by the local variation laws. The ideology behind this is an attempt to avoid converging toward local optimum points. It should, however, be noted that this "randomness" will still be governed by some rules to enable the implementation into numerical software. A very crude example of SHC adaptation of the BESO algorithm would be to "at random" select a predefined number of elements or perhaps percentage of the design volume to remove per iteration.

Although the SHC adaptation introduces a potential method for avoiding local optimum solutions, it also has the unfortunate likelihood of introducing numerical instability into the FE model, let alone the optimization problem in general. A perhaps more refined and systematic variation of HC is denoted as Tabu search (TS).

7.9 TABU SEARCH

As mentioned in the previous section, Tabu search (TS) is in essence a variation of HC. The ideology of TS is very much the same as SHC, to avoid premature convergence toward a local optimum solution. There are three general TS concepts (Bianchi et al., 2008):

1. Best Improvement. Consider a cluster of solutions, which are variations of each other, as was the case with HC in Section 7.8. Imagine a simple minimization problem, where the current solution A provides the objective value 5. Variations of the first solution A, that is, the "neighbors" B and C, provide the values 6 and 7, respectively. B is then chosen as the solution to replace A because it is the "best neighbor," despite the fact that A is overall a better solution. The reason behind this is simply to avoid local optimums by expanding the search area inside the solution space.
2. Tabu Lists. Following on from point 1, B is now the current solution and the search is repeated. The neighbors of B are A and C. Under the rules of "best improvement," A will once again be chosen as the new solution quite simply because it is the best neighbor to B. This is, of course, detrimental to the efficiency of the algorithm, as the solution will now simply enter into a cycle of swapping between A and B. This general phenomenon was also observed in Huang and Xie (2008) where the convergence of the BESO algorithm was hindered by the oscillation between two separate solutions,

as discussed in Section 7.3. TS attempts to avoid this by storing information about each solution, in this case A, B, and C, in a so-called Tabu list. Due to the scale of information involved and limitations of available memory, it is unfeasible to store each solution in its entirety, and therefore the "changes" from one solution to another are typically stored. By utilizing a Tabu list, revisiting previous solutions can be prohibited.

3. Aspiration Criteria. The introduction of Tabu lists may, therefore, reduce or even remove the likelihood of oscillation between two solutions; it may, however, also prevent solutions that have not yet been explored. In extreme events this could lead to a better local, or even the global optimum solution, being discarded from the search space. This may occur as it might not be possible to distinguish a previous solution from an "unexplored" one due to the fact that not all details of the individual solutions are stored in the Tabu lists. Therefore, an "aspiration criteria" is introduced, a typical example of which is to stipulate that any new solution found must be better than the best solution found from the start of the algorithm (Bianchi et al., 2008).

Based on the above, the introduction of TS into the BESO algorithm holds the potential to reduce some of the disadvantages of the algorithm, namely the dispersion of the search as well as the oscillation between competing solutions.

7.10 SIMULATED ANNEALING

Simulated annealing (SA) is an optimization method that has received a lot of recent attention and general development within a wide range of different optimization applications, for example, Bureerat and Limtragool (2008), Garcia-Lopez et al. (2011) and Barati (2014). The principle of SA-based optimization was first presented by Kirkpatrick, Gelatt, and Vecchi (1983) in order to physically model the temperature-controlled annealing process. In other words, the SA optimization principle has its roots in statistical mechanics, a commonality it shares with entropy, which will be presented in Section 7.14. The SA algorithm may be classed as a "random search method" (Garcia-Lopez et al., 2011), indicating that the fundamental algorithm has similarities to, for example, ACO, SHC and TS as previously presented. Analog to, for example, SHC, SA commences with a "random solution,"; an initial "guess" of a solution. SA then attempts to gradually obtain a more optimum solution by searching for alternatives within the neighborhood of the current solution, similar to the trust region method presented in Chapter 3. This approach is equivalent to the controlled cooling required for the physical annealing process. The first step is also analog to many other optimization algorithms including SHC and TS. The method in which alternative solutions may be accepted as the "current" solution is, however, what sets SA apart from, for example, SHC and TS. There are two paths in which this can happen:

1. Improved fitness value. In this rule a new solution B is accepted to replace the current solution A if the objective function value of B is less than A (for a minimization problem);

2. Reduced fitness value. In this rule a new solution B is accepted to replace the current solution A if the objective function value of B is larger than A (for a minimization problem), given a certain probability.

The rule of point 1 above is clear and intuitive, but if used in isolation may lead to premature convergence toward a local optimum, as extensively discussed throughout the previous chapters.

The rule in point 2 above is perhaps counterintuitive. Firstly it is important to recognize that this rule is introduced to avoid premature convergence, as per point 1 above. The probability mentioned in point 2 is where the direct link to the annealing process lies, as the probability $P(t)$ is calculated using Equation (7.11)

$$P(t) = \exp\left(-T^{-1}\Delta F_t\right) \tag{7.11}$$

In Equation (7.11), ΔF_t is the difference in objective function value between the two steps (A and B, using the above notation). T is the current "temperature" or scaling parameter calculated as Equation (7.12)

$$T(t) = \alpha^j T_0 \big| 0 < \alpha < 1 \tag{7.12}$$

In Equation (7.12), α is referred to as the cooling parameter, j is the number of steps completed since the start of the algorithm, and T_0 is the initial temperature value.

As presented above, the fundamental principles of SA are relatively simple and easy to understand. The inspiration from the physical annealing process is clear, as are the rules introduced in an attempt to circumnavigate the likelihood of premature convergence toward an inferior local optimum point.

As stated at the start of this section, several recent applications of SA in combination with other optimization principles have recently been published.

Bureerat and Limtragool (2008) proposed to use SA in combination with multi resolution design variables (MRDV) in an attempt to reduce overall CPU time and checkerboard effects. The fundamental principle was to introduce a "translation function," altering the number of design variables between the actual FE analysis and the optimization problem. This enables the number of elements in the FE analysis to be larger than the number of design variables in the optimization problem reducing the computational effort of this step. Bureerat and Limtragool (2008) conducted this for linear topology optimization problems and concluded:

> *From the numerical test, it can be concluded that the use of MRDV results in the enhanced performance of the SA optimization method. The optimum results obtained using SA are said to be comparable to those obtained from using the classic gradient-based approach.*

The general principle of this paper is interesting; however, based on the above, the results are questionable both in terms of engineering feasibility and in terms of optimization algorithm efficiency. Furthermore, the task of defining a robust translation function for nonlinear analysis must be labeled as extremely difficult, at best. Despite the above, the general principle remains intriguing and the potential benefits most relevant to nonlinear optimization, in general. The ideology is, therefore, worth further exploration once a suitable nonlinear topology optimization algorithm has been identified and developed; this does, however, fall outside the initial scope of this book.

Barati (2014) investigated the opportunities for utilizing a combination of CA and quasi-simulated annealing (QSA) to solve a multiobjective optimization problem relating to a nuclear reactor. Overall, there is no difference between SA and QSA in terms of the definition of the two rules previously presented. The difference lies in where the rules are applied. SA applies the rules on a global level, where QSA applied the rules on a local level. Using the above example, SA would apply the rules using the objective function, whereas QSA would apply the rules on the element level. This is where the CA method comes into play, as it is used to define the neighborhood influence of the individual element, as described in Section 7.4. Barati (2014) did not use this approach for a "conventional" structural optimization problem; the emphasis was on a fuel element for a nuclear reactor. The overarching task was to estimate the performance of the proposed algorithm in direct comparison with previously published work on genetic algorithms (GA) (see Section 7.3) and neural networks (NN) (see Section 7.12). The results obtained by Barati (2014) were not decisive, and merely refer to the potential and future work as opposed to drawing any detailed conclusions. The work is, therefore, interesting and may certainly serve as inspiration for future research directions, but does not provide significant detailed information for future work.

Garcia-Lopez et al. (2011) investigated the effects of combining SA with SIMP for linear topology optimization using "classic" optimization examples such as cantilever beams, with an overall aim of reducing checkerboard effects. The methodology developed is referred to as SA-SIMP. Although the SIMP methodology is not of primary interest in terms of nonlinear optimization as discussed in Chapters 4 and 5, the overall strategy of the SA-SIMP approach is intriguing. The process of SA-SIMP defined in Garcia-Lopez et al. (2011) commences with an initial solution obtained using "conventional" SIMP in connection with FEA. Following this, the algorithm contains three major aspects:

1. Selection Method. This is the method in which "intermediate element densities" of the SIMP scheme are selected to be "upgraded" to 1 or "downgraded" to (nearly) 0 in the fixing method step. Garcia-Lopez et al. (2011) suggest four different ways of doing this: random selection, probability-based selection, random cluster selection, or probability-based cluster selection. The relative difference between the first two and the latter two is, in essence, whether or not the selection is carried out on an individual element basis or on a cluster of elements (using a neighborhood approach).
2. Fixing Method. As mentioned, this is the step in which elements are "upgraded" or downgraded." Garcia-Lopez et al. (2011) suggest two different ways of doing this: random assignment or rounding assignment.

3. Look-Ahead Strategy. Following steps 1 and 2 described above, a single "conventional" SIMP optimization iteration is completed. The results of the SIMP run are then included within the SA algorithm selection basis as described at the start of this subsection.

The results of Garcia-Lopez et al. (2011) found that the checkerboard effect was indeed reduced when using SA-SIMP when compared with gradient-based SIMP. Furthermore, the results indicated that using minimum compliance as the objective the results of SA-SIMP were consistently less compliant than the SIMP-based counterparts. Even though this algorithm and associated results are not directly applicable to nonlinear topology optimization, the overall methodology is certainly of interest for future work.

Based on the information within this subsection, the use of SA or a variation thereof could be combined with other optimization techniques such as BESO, and holds great potential to reduce some of the disadvantages of the BESO algorithm, namely the dispersion of the search, as well as the oscillation between competing solutions.

7.11 PARTICLE SWARM OPTIMIZATION

Particle swarm optimization (PSO) is another "nature inspired" meta-heuristic optimization algorithm. It was first proposed by Kennedy and Eberhart (1995) as a solution methodology to continuous nonlinear problems. The fundamentals of PSO are based upon the social behavior of animals hunting for food, such as a school of fish or a flock of birds. According to Kennedy and Eberhart (1995a), the original intent was:

> *To graphically simulate the graceful but unpredictable choreography of a bird flock.*

PSO is a population-based algorithm; in the initial formulation, the population (or swarm) consisted of a number of "collision proof-birds," denoted as agents or particles (Kennedy & Eberhart, 1995). As a starting point, consider a simple 2D coordinate system wherein each particle (bird) is given a random initial (x, y) location. Imagine now that each bird represents a potential solution to an optimization problem. For every iteration, each individual particle is given the x and y velocity of its nearest neighbor and thus the system of solutions is updated. This simple rule did, however, quickly result in the following observation (Kennedy & Eberhart, 1995a):

> *Unfortunately, the flock quickly settled on a unanimous, unchanging direction.*

Kennedy and Eberhart (1995) therefore introduced a stochastic variable that was initially labeled "*craziness*," which randomly changed x and y velocities of random individual particles in each iteration. Since the original introduction in 1995, there have been many minor adaptations and refinements of PSO, but the fundamental principles remain as described above. One of the more recent adaptations and implementations of PSO in connection with structural optimization is published in Luh et al. (2011), who introduced an adaptation denoted binary PSO (BPSO). This adaptation predominately consisted of introducing an alternative type of particles which changed the rules for how particles "interact." Luh, Lin, & Lin (2011a) focused upon linear multimodal topology optimization problems and concluded:

> *The results show that the modified BPSO algorithm and part of the structures show very well defined truss-like members. Moreover, parts of the derived structures have the same weight but quite different topologies.*

Based on the information above, PSO also seemingly offer an extended search area when compared with the standalone BESO algorithm. When compared with, for example, SA, CA, and TS, it is worth noting that PSO computes multiple solutions (particles) per iteration, whereas the former three only compute one solution per iteration of the optimization loop.

7.12 NEURAL NETWORKS

Neural networks (NN), also sometimes referred to as artificial neural networks (ANN), is yet another nature inspired methodology, which is based upon the study of the central nervous system and, in particular, the brain. According to Marsalli (2006), the concept of NN was first devised in the early 1940s by W.S. McCulloch and Walter Pitts who:

> *...tried to understand how the brain could produce highly complex patterns by using many basic cells that are connected together. These basic brain cells are called neurons...*

The original principle was not denoted as NN, but threshold logic. As indicated above, NN was not initially meant as an optimization tool, rather a modeling tool for complex nonlinear problems and has recently been adapted for structural optimization on a small scale, for example, Lin and Lin (2005). The central theme of NN is a pattern recognition methodology, in keeping with the original research question stated above. Just like the brain NN are capable of learning systems (patterns) which enables them to make predictions about the "future state" of the system under observation (Alavala, 2008), in mathematical terms, an NN may be perceived as a single function $f(x_i)$ which can be used to evaluate the system based on input parameters (design variables) x_i. In order to "define" $f(x_i)$, the NN needs to be trained or learn. The learning mode of an NN is generally referred to as "machine learning" (or pattern recognition) and is commonly divided into three main categories:

1. Supervised learning. In this mode, a set of example pairs y, z and the results $f(y)$ and $f(z)$ are known;
2. Unsupervised learning. In this mode, input data are given along with an objective function; the results are, however, not available;
3. Reinforcement learning. In this mode, input data are normally not given but established via, for example, another function.

In the ideal situation, supervised learning would always be applied. However, as these data are most often not available, so one of the other learning modes must be utilized. In relation to nonlinear (structural) topology optimization, it is perhaps most convenient to use reinforced learning mode. This is simply because the NN can be trained using other optimization techniques, such as ESO, PSO, or SA (Alavala, 2008). NN should, therefore, not be considered a stand-alone optimization solution, but rather as a potential "additional refinement." The principle of NN is certainly captivating from an overall algorithm efficiency viewpoint; it is, however, obvious that there are significant obstacles to overcome if an NN is to be "trained" for nonlinear topology optimization. It is also worth noting that NN algorithms are widely available in numerical software packages such as MatLab.

With this section, the overall review of meta-heuristic optimization algorithms is complete. This is by no means an exhaustive review; however, the general field of optimization algorithms is so extensive that it would be impossible to cover every single algorithm in detail. The above review can, therefore, be considered to be a "selective" review of relevant and potential algorithms for nonlinear topology optimization. The next section of this chapter will focus upon other general aspects that should be considered in connection with nonlinear topology optimization algorithms.

7.13 GENERAL PRINCIPLES

As previously presented, Chapter 8 will discuss the opportunities for developing a topology optimization algorithm for structures exposed to large (nonlinear) deformations. The purpose of this section is to present and discuss general principles and parameters that should be taken into account during the subsequent definition and development of the nonlinear topology optimization algorithm in Chapter 8.

7.13.1 Robust optimization

In Chapter 5, the term "robustness" was used to describe the "stability" or "sturdiness" of an algorithm; in other words, its ability to avoid "error termination," for example, due to numerical instability during FE analysis between iterations. The term "robust optimization" does, however, have a very specific definition, originally presented in Chapter 3, which is beneficial to reiterate at this point (Garcia-Lopez et al., 2013):

> *Robust design optimization consists of finding structural solutions such that the variability of the response – i.e. performance measure – is relatively insensitive to the uncertainty of the material properties and the demands. Most approaches to structural robustness evaluate the structural response in terms of displacement; however, others use the first and second moments of the response function.*

In other words, the principle of robust optimization relates to an optimization problem that considers the "noise factors" leading to uncertainty of the results. These are most often not parameters that can be controlled, and may include aspects such as the numerical inaccuracy of FE modeling. Although robust optimization is a field that has seen a lot of recent development, as seen in Garcia-Lopez et al. (2013a) and Chen (2012), no specific developments for nonlinear optimization have been found. Furthermore, the implementation of this principle will not be a top priority in the initial stages of the algorithm development and implementation. It may, however, be beneficial to consider potential avenues for implementation in the later stages of algorithm refinement in an attempt to enhance the overall algorithm viability as well as feasibility of results.

7.13.2 Fuzzy logic

In relation to the above introduction of robust optimization and the repeated discussions relating to numerical instability in connection with topology optimization and FE analysis in general, it may be beneficial to consider the principles of Fuzzy Logic (FL). FL may be considered a form of "many-valued logic." Consider the simple question: "is it Thursday today?" The obvious two answers to that question are "yes" or "no." In FL, other (fuzzy) answers such as "maybe" are also allowed. In the context of the above example, the advantages of FL are perhaps not clear; nevertheless, the fundamental principle of FL is. Now consider a "conventional" topology optimization problem, such as a cantilever beam. To solve such a problem could be defined as "asking" a simple question on a single element basis: "should this element be a part of the structure?" The obvious two answers are again "yes" or "no," or "1" or "0" using Boolean operators. This strict regime may lead to numerical instability (of FE models in particular), and is one of the reasons why the variable density method (VDM) is used for topology optimization, where relative element densities of, for example, 0.2 are allowed. A further example is the introduction of "soft" elements discussed in Section 7.3. In Section 7.6, where COPs were introduced, one of the pretences was that the answers should be discrete, which initially rules out the concepts of FL. In addition to this, one of the main drawbacks of VDM-SIMP is the checkerboard effect, an inherent side-effect of introducing FL (VDM). Despite these shortcomings, it may be necessary to introduce a certain level of FL into an optimization algorithm to ensure numeric stability.

7.13.3 Monte Carlo simulation

In Section 7.13.1, the definitions of robustness were discussed. This included the difficulty in predicting the change of results as a function of varying the design variables, in other words foreseeing the change in the system response (objective function) due to adding or removing elements from the FE model. A commonly used method of solving systems containing a large number of variables numerically is denoted as the Monte Carlo method (MCM). MCM is widely applied from engineering to financial forecasts, stock rates, and manufacturing strategies. Many different versions of MCM or Monte Carlo Simulation (MCS) exist and are readily available within commercial software packages such as Microsoft Excel and MatLab. According to MatLab (2014), MCS can be defined as:

> *Monte Carlo simulation is a method for exploring the sensitivity of a complex system by varying parameters within statistical constraints.*

In its basic form MCS is, therefore, not an explicit optimization tool, but rather a tool for prediction of sensitivity (robustness) of a particular system. MCS could, therefore, be an interesting "additional tool" for exploration of the robustness of results obtained from the nonlinear topology optimization algorithm, or alternatively could be incorporated into the optimization algorithm; this will, however, not be extensively investigated in this book.

The above citation also contains the phrase "statistical constraints" in connection with MCS. It may be possible to include these within the actual optimization algorithm. This could, for example, be implemented in connection with the "intermediate element energy levels" discussed in Section 7.3, as will be further elaborated in Chapter 8. Before this is done, it is, however, convenient to introduce some statistical principles. The reason for introducing these is the desire to reduce the level of "randomness" in the optimization by applying a more robust selection method for parameter variation, for example, element deletion or addition in connection with BESO. In Section 7.3, it was stated that only the final deformed shape (element energy level) is used to inform the optimization process. Expanding this to include intermediate levels may result in a more robust and accurate optimization process, as will be further investigated in Chapter 8. The methodology used to implement these is, however, pivotal to the overall success of the algorithm, as a poor choice of implementation may add additional uncertainty, numerical instability, and perhaps even cause divergence of the optimization algorithm in general, an example of which could be an element subjected to high levels of hourglass energy in isolated timesteps. It may, therefore, be beneficial to introduce the principle of entropy from statistical mechanics into an optimization algorithm.

7.14 ENTROPY

Different types of entropy exist; throughout this book the term will refer to Shannon entropy unless otherwise specified. The phrase Shannon entropy relates to the fact that Claude Shannon developed and expanded E.T. Jaynes' theory on entropy throughout several years. The term entropy is perhaps somewhat misleading as in this case it does not directly relate to the thermodynamical entropy property. In this context, entropy should be interpreted as an expression for the relative level of uncertainty of a system under consideration; this will subsequently be further expanded.

The principle of entropy (in the statistical sense) was first introduced by Jaynes (1957), where the aim was to use probability to determine the state (thermodynamical entropy) of a physical system. Jaynes' ultimate goal was to predict the final state of a system based upon sample data (from a much larger population) while avoiding making any "assumptions" about the system state not supported by the sample data; this was done using the entropy as an expression for the level of uncertainty. The idea was that maximizing the uncertainty would minimize the "unfounded assumptions"; hence, Jaynes' method was entitled the maximum entropy method (MEM), also known as the maximum entropy principle (MEP). A number of maximum entropy (ME) models have been derived, such as Gibbs and Boltzmann distributions, Markov random fields, logistic regression, exponential families, and Boltzmann machines, all serving different types of problems. A Layman's description of MEP can be found in Penfield (2003a):

> *The Principle of Maximum Entropy is a technique that can be used to estimate input probabilities more generally. The result is consistent with known constraints expressed in terms of averages, or expected values, of one or more quantities, but is otherwise unbiased.*

Furthermore Penfield (2003a) also states:

> *Information theory, particularly the definition of information in terms of probability distributions, provides a quantitative measure of ignorance (or uncertainty, or entropy) that can be maximized mathematically to find the probability distribution that is maximally unbiased.*

The above two citations capture the essence of MEP in the context of which it will subsequently be used in connection with nonlinear topology optimization, as will be explained throughout Chapter 8. The derivations and explanations throughout this section generally follow those of Jaynes (1957) and Penfield (2003), although the notation has been adapted to avoid confusion. According to Penfield (2003), a simple form of entropy can be expressed as Equation (7.13)

$$S = \sum_i \left[p(D_i) \log_2 \left(\frac{1}{p(D_i)} \right) \right] \tag{7.13}$$

In Equation (7.13), S is the entropy, that is, uncertainty, and $p(D_i)$ is the probability distribution for the individual states representing the design variables D_i of the system under consideration. As the natural logarithm with base 2 is used, the entropy is measured in bits. Also, please note that as a consequence of utilizing the natural logarithm in (7.13), the resulting probabilities assume a normal distribution. It is possible to use other distributions, but for the remainder of this book a normal distribution will be used. As per the rules of probability theory, the sum of probabilities must equal 1, Equation (7.14)

$$\sum_i p(D_i) = 1 \tag{7.14}$$

In order to aid the understanding of the principle of entropy S as well as detailing how S can be found in practice, it is convenient to introduce a relevant example. This will serve two main purposes: firstly, to explain the fundamental mathematics behind calculating the entropy, but will also indirectly highlight the strategy for the subsequent implementation into a nonlinear topology optimization algorithm.

Consider a simple FE model consisting of four elements exposed to predefined boundary conditions; the task is to obtain the lightest possible structure. Firstly, four design variables, D_i, are defined; in this example, these will be interpreted as Boolean operators defining whether an element should be included (1) or excluded (0) from the structure, analog to the VDM methodology. Initially, there is no reason for penalizing one element as opposed to another; hence, the initial design variable values are all set to 1, as in Expression (7.15)

$$D_1 = D_2 = D_3 = D_4 = 1 \tag{7.15}$$

At this stage it makes no sense to "select" the elements, as any such selection will be completely objective and heavily biased, but what about from an MEP viewpoint? As it is impossible to distinguish between the four elements, they should be assigned an equal probability, that is, $p(D_i) = 0.25$, which leads to a nonbiased solution with the highest possible entropy, which can be calculated using Equation (7.13). This scenario/type of solution is generally referred to as Bernoulli's principle of insufficient reason, and is of course not a relevant engineering solution. Additional information is, therefore, required, enabling the determination of a better probability distribution in the sense of a lower entropy, i.e. uncertainty, and a feasible structural solution. The MEP is a direct solution to this scenario and may, in general, be considered to be an extension of Bernoulli's principle of insufficient reason.

The additional information should enhance the ability to distinguish the individual design variables from each other. In other words, introducing an overall (global) quantity Q for which there are local values associated (with each design variable). Consider, for example, the total strain energy Q of the model with the individual element strain energy values $q(D_i)$. A global constraint can thus be set up as Equation (7.16)

$$Q = \sum_i p(D_i)q(D_i) \tag{7.16}$$

Continuing the above example an FE analysis is now conducted; as a result, the strain energy levels $q(D_i)$ are now known, as shown in Expression (7.17)

$$\begin{aligned} q(D_1) &= 0.1 \\ q(D_2) &= 0.2 \\ q(D_3) &= 0.3 \\ q(D_4) &= 0.6 \end{aligned} \tag{7.17}$$

An average strain energy level Q of the entire model is now chosen, as defined by Equation (7.18)

$$Q = 0.3 \tag{7.18}$$

Based on the information in (7.17) and (7.18), which of the four elements should now be "selected" to remain in the model and indeed how many should be selected for the structure with the lowest mass? It is obvious that based on the information in (7.17) there is still no unique solution to the constraint in (7.18). The constraint could, for example, be met by only selecting element 3, or alternatively by selecting elements 1 and 2. With the given information there is no objective justification for choosing one solution over the other, so which should be chosen? According to the MEP, the solution with the highest entropy should be chosen as to minimize any bias.

The above scenario is arguably a somewhat "idealized" scenario; firstly, the constraint of 0.3 can "conveniently" be obtained by the element combinations as listed above, but what if the constraint was 0.34 or the $q(D_i)$ were "less neat" than listed in (7.17)? Although the constraint of (7.18) can be manually set, it is clear that more "versatility" needs to be implemented into the MEP. By introducing the Boolean operators above, it was (indirectly) assumed that the elements could only be fully included or alternatively fully excluded from the structure. For calculation and FE stability purposes, this may be undesirable. It may, therefore, be convenient to introduce continuous variables to define the inclusion or exclusion of the individual design variables; in other words, allowing fuzzy logic into the solution of the system, analog to the VDM method. The primary potential negative effect of introducing fuzzy logic is, of course, the checkerboard effect. The discussion of whether or not this should be included will be continued throughout Chapter 8. Setting this discussion aside for the time being, maintaining the constraint in (7.18), and the element strain energies of (7.17), the problem can now be formulated using (7.14) and (7.16) through (7.18); two expressions for the system can now be defined in (7.19)

$$\begin{aligned} &p(D_1)+p(D_2)+p(D_3)+p(D_4)=1 \\ &0.1p(D_1)+0.2p(D_2)+0.3p(D_3)+0.6p(D_4)=0.3 \end{aligned} \tag{7.19}$$

The system in (7.19) is an indeterminate system, as there are two equations and four unknowns. The amount of entropy S associated with (7.19) can be obtained using Equation (7.13) leading to Equation (7.20)

$$S = \left\{ \begin{aligned} &p(D_1)\log_2\left(\frac{1}{p(D_1)}\right)+p(D_2)\log_2\left(\frac{1}{p(D_2)}\right) \\ &+p(D_3)\log_2\left(\frac{1}{p(D_3)}\right)+p(D_4)\log_2\left(\frac{1}{p(D_4)}\right) \end{aligned} \right. \tag{7.20}$$

For simpler problems (with, for example, three elements) it is possible to solve the above analytically by expressing one of the variables in terms of the other two. In general, it is, however, more convenient to solve the system numerically by the utilization of Lagrangian multipliers, as originally presented in Chapter 3; this can be done using Equation (7.21)

$$L = S - (\alpha - \log_2 e)\left(\sum_i p(D_i) - 1\right) - \beta\left(\sum_i q(D_i)p(D_i) - Q\right) \tag{7.21}$$

In (7.21), e is the base of the natural logarithm, α is a Lagrangian multiplier measured in bits, and β is a Lagrangian multiplier measured in bits per unit of G. In the above example, β is therefore measured in bits per Joule. The task at hand is to maximize L, thereby maximizing the entropy S. This can be done by differentiating (7.21) with respect to $p(D_i)$ and assuming S, α and β to be constants the result is Equation (7.22)

$$p(D_i) = 2^{-\alpha}2^{-\beta q(D_i)} \tag{7.22}$$

According to Penfield (2003), the entropy can be calculated without determining the probability distributions $p(D_i)$, as in Equation (7.23)

$$S = \alpha + \beta Q \tag{7.23}$$

Equation (7.23) can, therefore, be used to determine the level of uncertainty of a given problem, a number that could potentially be used as a parameter in the overall topology optimization algorithm. In order to get to this stage, it is necessary to determine the values of α and β; the former, that is, α, can be calculated as defined by Equation (7.24)

$$\alpha = \log_2\left(\sum_i 2^{-\beta q(D_i)}\right) \tag{7.24}$$

Penfield (2003) states that in general the highest computational efforts associated with MEP relate to solving Equation (7.24). There are, however, some additional caveats that should be noted:

1. If the system at hand contains a moderate number of states (potential solutions) and only one constraint, solving Equation (7.24) is not difficult. At this stage, it should be noted that the constraint referred to does not necessarily equal the "structural" constraints, for example, a maximum displacement constraint applied at the topology optimization level, as will be further discussed in Chapter 8.
2. The level of difficulty rises considerably as the number of constraints increases. If there are a very large number of solutions, the specific solution

cannot be obtained using Equation (7.24), although the overall solution may still be obtained using Lagrangian multipliers.

The two points above are critical to ensure that the MEP is applied in an appropriate context and will be further addressed throughout Chapter 8. Assuming that the conditions of point 1 above are met, Equation (7.24) is valid to determine α. In order to determine α, it is necessary to know β that can be found using Equation (7.25)

$$f(\beta)=\sum_i\left[\left(q(D_i)-Q\right)2^{-\beta(q(D_i)-Q)}\right] \tag{7.25}$$

Equation (7.25) is the fundamental equation to be solved; with one unknown, β. L (7.21), and thereby S (7.20), are maximized when Equation (7.26) is true

$$f(\beta)=0 \tag{7.26}$$

The overall MEP thereby "reduces" to solving Equation (7.25) subject to Equation (7.26). Please note that (7.25) does not contain any probability distributions; it is merely a function of the "sample values" $q(D_i)$ and the single constraint G. Solving Equation (7.25) numerically under these conditions is relatively straightforward and may be conducted using a number of techniques, including the gradient-based methods and line-search methods presented in Chapter 3.

Returning to the four-element FE model example introduced at the start of this section, it is now possible to calculate the entropy and individual probabilities associated with each element. This can be achieved in a number of ways, for example, via MatLab programming; this has been done for the following example, where the code is based upon the MEP principles and equations presented in this section in combination with a Newton–Raphson iterative solver, as presented in Chapter 3. The results can be found in Table 7.3.

TABLE 7.3 Results From First Numerical FE Example

Parameter	Value	Unit
α	2	Bits
β	5.0376e-8	Bits per Joule
Entropy (S)	2	Bits
$p(D_1)$	0.25	
$p(D_2)$	0.25	
$p(D_3)$	0.25	
$p(D_4)$	0.25	

When reviewing the results in Table 7.3, the numerical values of α and β are of little interest; the primary interest lies in the actual entropy S and the probability values $p(D_i)$. The numerical value of the entropy on its own can perhaps be somewhat cumbersome to interpret. Remembering that the entropy is a measure of uncertainty, the value becomes easier to comprehend; the higher the value, the higher the level of uncertainty. The probability values $p(D_i)$ denote the likelihood of the individual design variable D_i being included in the system to meet the single constraint Q; Equation (7.18), in this example. The example thus suggests that based on the input information there is no reason (probability) to use one element over another; this is indeed the least biased solution. The solution is undeniably identical to the one initially obtained by calculating average values. The MEP does, however, provide a repetitive and systematic approach to obtaining such as a solution. Furthermore, the solution is provided per design variable (element) easing the implementation into topology optimization. As previously discussed, the above example is rather "idealistic" and could certainly be solved manually. Complications of this were previously discussed, such as changing the average strain energy value Q. In order to investigate the effects of this Q is now updated as defined in Equation (7.27)

$$Q = 0.34 \tag{7.27}$$

With the change in Equation (7.27), it is now much more difficult to manually assign probabilities to the individual design parameters, avoiding any particular bias. Using the MEP approach, the updated system has been solved, the results of which are listed in Table 7.4.

Comparing the results in Table 7.4 to those of Table 7.3, it can be seen that the entropy has decreased in Table 7.4, indicating that there is less uncertainty in the result. This claim is supported by the probability values that now distinguish between all four design variables. The probabilities $p(D_1)$, $p(D_2)$, and $p(D_3)$ have

TABLE 7.4 Results From Second Numerical FE Example

Parameter	Value	Unit
α	2.2762	Bits
β	−0.88794	Bits per Joule
Entropy (S)	1.9743	Bits
$p(D_1)$	0.2195	
$p(D_2)$	0.2335	
$p(D_3)$	0.2483	
$p(D_4)$	0.2987	

TABLE 7.5 Results From Third Numerical FE Example

Parameter	Value	Unit
α	−0.12196	Bits
β	9.0767	Bits per Joule
Entropy (S)	0.89547	Bits
$p(D_1)$	0.5376	
$p(D_2)$	0.2866	
$p(D_3)$	0.1527	
$p(D_4)$	0.0231	

decreased, while $p(D_4)$ has increased. The results thus suggest that the "first" element to be removed should be D_1 and the "last" element to be removed should be D_4. This is in accordance with "engineering intuition," which would remove the element containing the least deformation energy to maximize the efficiency of the structure (in terms of energy absorption per unit mass). To draw a general conclusion based on this observation is, however, not as straightforward as it may seem. To understand why the average strain energy value is now changed according to Equation (7.28)

$$Q = 0.1 \tag{7.28}$$

The results from the change in (7.28) are listed in Table 7.5.

Firstly observing the entropy in Table 7.5 reveals that the level of uncertainty has considerably reduced when compared with the first two examples. According to Table 7.5, the "element ranking order" in which the elements should be chosen for inclusion in the FE model to meet the single constraint in (7.28) should now be: D_1, D_2, D_3, and finally D_4. From a purely probabilistic viewpoint, this makes perfect sense; from a purely structural engineering viewpoint, this is counterintuitive, as removing the element with the largest strain energy will inadvertently have an effect on the remaining elements. Hence, the implementation of MEP in topology optimization is not necessarily a case of choosing the elements with the highest probability values, as will be further discussed in Chapter 8. For the next example, the average strain energy value is now defined as Equation (7.29)

$$Q = 0.6 \tag{7.29}$$

The results from the updated constraint value are listed in Table 7.6.

The entropy in Table 7.6 is in effect 0 bits, indicating that there is no uncertainty in this particular example; only one element should be used: D_4. Even

TABLE 7.6 Results From the Fourth Numerical FE Example

Parameter	Value	Unit
α	46.1277	Bits
β	−76.8795	Bits per Joule
Entropy (S)	1.6534e-7	Bits
$p(D_1)$	0.0000	
$p(D_2)$	0.0000	
$p(D_3)$	0.000	
$p(D_4)$	1.0000	

though it must be concluded that this is the "correct" solution according to the MEP, it may not necessarily be a useful solution in a structural optimization context, as will be further discussed in Chapter 8.

With this, the initial presentation of MEP completes the starting point of Chapter 8 to initially define a nonlinear topology optimization algorithm.

REFERENCES

Alavala, C. R. (2008). Fuzzy logic and neural networks Available from: <https://www.academia.edu/1435724/Fuzzy_Logic_and_Neural_Networks_by_Chennakesava_R._Alavala> Accessed 01.09.14.

Altair. (2009). "blog" Available from: <http://blog.altair.co.kr/wp-content/uploads/2011/03/optistruct_optimization_10-0.pdf> Accessed 15.10.14.

Barati, R. (2014). A novel approach in optimization problem for research reactors fuel plate using a synergy between cellular automata and quasi-simulated annealing methods Available from: <http://www.sciencedirect.com/science/article/pii/S0306454914001091> Accessed 14.09.14.

Bianchi et al. (2008). A survey on metaheuristics for stochastic combinatorial optimization. Available from: <http://link.springer.com/article/10.1007%2Fs11047-008-9098-4> Accessed 27.08.14.

Bianchi et al. (2008a). A survey on metaheuristics for stochastic combinatorial optimization Available from: <http://link.springer.com/article/10.1007%2Fs11047-008-9098-4> page 246. Accessed 27.08.14.

Bureerat, S., & Limtragool, J. (2008). Structural topology optimisation using simulated annealing with multiresolution design variables. Available from: <http://www.sciencedirect.com/science/article/pii/S0168874X08000681> Accessed 22.09.14.

Chen, W. (2012). Robust shape and topology optimization under uncertainty. Available from: <http://www.google.co.uk/url?sa=t&rct=j&q=&esrc=s&source=web&cd=3&cad=rja&uact=8&sqi=2&ved=0CDsQFjAC&url=http%3A%2F%2Fwww.altairhyperworks.com%2F(S(abosnynyy2lsvkdbzowbzg05))%2FResLibDownload.aspx%3Ffile_id%3D1434%26from%3DResourceLibrary.aspx%253Fkeywords%253DRobust%252BShape%252Band%252BTopology%252BOptimization&ei=yP_ZU4akPMf17AbOwIHQBA&usg=AFQjCNE440sQsyqlLmJH77yf3--XRjz5-w&sig2=jsVrHwXIRksfY1YIDVlQsQ&bvm=bv.72185853,d.ZGU> Accessed 03.11.14.

Garcia-Lopez, M., et al. (2011). A hybrid topology optimization methodology combining simulated annealing and SIMP. Available from: <http://www.sciencedirect.com/science/article/pii/S0045794911001210> Accessed 10.11.14.

Hosny, M.I. (2010). page 16. Investigating heuristic and meta-heuristic algorithms for solving pickup and delivery problems. Available from: <http://users.cs.cf.ac.uk/C.L.Mumford/Research%20Topics/PDPTW/papers/Manar.pdf> Accessed 03.09.14.

Huang, X., & Xie, Y. M. (2008). Topology optimization of nonlinear structures under displacement loading. Available from: <http://www.sciencedirect.com/science/article/pii/S0141029608000205> Accessed 02.05.14.

Huang, X., & Xie, Y. M. (2008a). Topology optimization of nonlinear structures under displacement loading. Available from: <http://www.sciencedirect.com/science/article/pii/S0141029608000205> page 2059 Accessed 02.05.14.

Huang, X., & Xie, Y. M. (2008b). Topology optimization of nonlinear structures under displacement loading. Available from: <http://www.sciencedirect.com/science/article/pii/S0141029608000205> page 2060 Accessed 02.05.14.

Huang, X., & Xie, Y. M. (2008c). Topology optimization of nonlinear structures under displacement loading. Available from: <http://www.sciencedirect.com/science/article/pii/S0141029608000205> page 2061, Accessed 02.05.14.

Huang, X., & Xie, Y. M. (2008d). Topology optimization of nonlinear structures under displacement loading. Available from: <http://www.sciencedirect.com/science/article/pii/S0141029608000205> page 2065, Accessed 02.05.14.

Jaynes, E. T. (1957). Information theory and statistical mechanics. Available from: <http://bayes.wustl.edu/etj/articles/theory.1.pdf> Accessed 18.07.14.

Kennedy, J., & Eberhart, K. (1995). Particle swarm optimization. Available from: <http://www.cs.tufts.edu/comp/150GA/homeworks/hw3/_reading6%201995%20particle%20swarming.pdf> Accessed 30.07.14.

Kennedy, J., & Eberhart, K. (1995a). Particle swarm optimization. Available from: <http://www.cs.tufts.edu/comp/150GA/homeworks/hw3/_reading6%201995%20particle%20swarming.pdf> page 1943. Accessed 30.07.14.

Kirkpatrick, S., Gelatt Jr., C. D., & Vecchi, M. P. (1983). Optimization by simulated annealing. Available from: <http://minds.jacobs-university.de/sites/default/files/uploads/teaching/share/KirkpatrickSimulatedAnnealing.pdf> Accessed 30.07.14.

LSTC. (2011). Topology and shape computations for LS-Dyna. Available from: <http://www.lsoptsupport.com/documents/manuals/ls-tasc/ls-tasc-2.0-manual.pdf/view> Accessed 06.14.

LSTC. (2011a). Topology and shape computations for LS-Dyna. Available from: <http://www.lsoptsupport.com/documents/manuals/ls-tasc/ls-tasc-2.0-manual.pdf/view> page 11, Accessed 06.14.

Luh, G. C., Lin, C. I., & Lin, Y. S. (2011). A binary particle swarm optimization for continuum structural topology optimization. Available from: <http://www.sciencedirect.com/science/article/pii/S1568494610002905 Accessed 08.09.14.

Luh, G. C., Lin, C. I., Lin, Y. S. (2011a). A binary particle swarm optimization for continuum structural topology optimization. Available from: <http://www.sciencedirect.com/science/article/pii/S1568494610002905 page 2844, Accessed 08.09.14.

Marsalli, M. (2006). McCulloch Pitts neurons. Available from: <http://www.mind.ilstu.edu/curriculum/mcp_neurons/mcp_neuron_1.php> Accessed 6.09.14.

MatLab (2014). Monte Carlo Simulation. Available from: <http://www.mathworks.co.uk/discovery/monte-carlo-simulation.html> Accessed 08.08.14.

Penfield, P. Jr. (2003). Principle of maximum entropy. Available from: <http://www-mtl.mit.edu/Courses/6.050/notes/chapter9.pdf> Accessed 08.07.14.

Penfield, P. Jr. (2003a). Principle of maximum entropy. Available from: <http://www-mtl.mit.edu/Courses/6.050/notes/chapter9.pdf> page 80 Accessed 08.07.14.

Querin, O. M. *Evolutionary Structural Optimisation: Stress Based Formulation and Implementation,* Ph.D. Thesis, Available from: <http://www.mech-eng.leeds.ac.uk/menomq/papers/omquerin_phd_thesis.pdf>, site visited 25.05.11.

Researchgate. (2014). Yi Min Xie – publications. Available from: <http://www.researchgate.net/profile/Yi_Xie4/publications> Accessed 28.07.14.

Roux, W. (2011). Topology design using LS-TaSC Version 2 and LS-Dyna. Available from: <http://www.dynamore.de/de/download/papers/konferenz11/papers/session4-paper4.pdf> Accessed 12.06.14.

Roux, W. (2011a). Topology design using LS-TaSC Version 2 and LS-Dyna. Available from: <http://www.dynamore.de/de/download/papers/konferenz11/papers/session4-paper4.pdf> page 2, Accessed 12.06.14.

Schrijver, A. (2003). *Combinatorial optimization*. London: Springer.

Sorensen, R., & Lund, E. (2012). Ply-based optimization of laminated composite shell structures under manufacturing constraints. Available from: <http://www.escm.eu.org/eccm15/data/assets/1870.pdf> Accessed 23.08.14.

Vazirani, V. (2001). Approximation algorithms. Available from: <http://www.cc.gatech.edu/fac/Vijay.Vazirani/book.pdf> Accessed 23.08.14.

Xie, Y. M., & Steven, G. P. (1997). *Evolutionary Structural Optimization*. London, UK: Springer Verlag.

Chapter | eight

Definition, Implementation, and Partial Validation of a Nonlinear Topology Optimization Algorithm

CHAPTER OUTLINE

This chapter thereby serves multiple purposes:

1. To explicitly demonstrate some of the challenges facing a nonlinear topology optimization algorithm; these are applicable to nonlinear optimization in general;
2. To demonstrate that optimization does not have to be completed using expensive commercial software packages;
3. To demonstrate that optimization tools can be customized to specific purposes with the proper background knowledge;
4. To inspire all members of the academic and engineering communities to explore avenues of researching and ultimately solving a highly complex engineering problem; namely, that of nonlinear topology optimization.

This chapter will contain three major parts. Firstly, the proposed topology optimization algorithm will be outlined, discussed, and justified. Secondly, the programming of the optimization software will be presented. Finally, a series

Nonlinear Optimization of Vehicle Safety Structures. http://dx.doi.org/10.1016/B978-0-12-804424-7.00008-7

of small-scale fundamental case studies benchmarking the performance and potential for further development of the algorithm will be presented. The first step is, therefore, to overall compare and rank the optimization techniques and principles presented in the previous chapters.

8.1 ALGORITHM DEFINITION

The task of defining a topology optimization algorithm can, in effect, be divided into two main sections: firstly, to choose/define the overarching principle of the algorithm, and secondly, to add/change "minor" details relating to specific attributes of the algorithm, for example, to determine how neighboring elements affect each other. The first part of this section will, therefore, be dedicated to defining the overarching algorithm. As per previous discussions, these may initially be grouped into two categories, mathematical and meta-heuristic. The benefits of mathematical algorithms, in particular, the VDM-SIMP approach, have been discussed in previous chapters. This also includes the severe limitations of the pure mathematical algorithms in connection with optimization of structures exposed to large (nonlinear) deformations. On this basis this type of algorithm is, therefore, dismissed as a potential backbone for the optimization algorithm of this chapter. The second major group of algorithms discussed in Chapter 7 were the meta-heuristic algorithms, which have already been used in commercial software such as LSTCs LS-TaSC. The fact that it has been used in commercial software does not necessarily substantiate the suitability of the algorithm for structures exposed to large (nonlinear) deformations, as has been shown for the ESLM method. Many specific meta-heuristic algorithms were presented and discussed in Chapter 7, remembering that the overall "trade-mark" of this type of algorithm is the combination of physical theories and practicality in terms of computational efficiency. At this point, it is convenient to remember the list of requirements set forth in Section 7.4, which included the need for "reasonable overall computing requirements." The potential tradeoff between precision and algorithm efficiency (speed) makes the meta-heuristic algorithms attractive and is the main reason why the overall methodology of the optimization algorithm in this chapter will be based on meta-heuristics. As presented in Chapter 7, there are already (at least) two reasonably well-established meta-heuristics-based nonlinear topology optimization algorithms, HCA and ESO including variations thereof, such as CAA and BESO. As these algorithms have been analyzed and developed over a number of years, they could provide a good starting point for further algorithm development; this is, of course, subject to scrutiny and careful assessment of the advantages as well as the disadvantages of the individual algorithm in addition to feasibility of improving the algorithm. Starting with HCA as presented in Section 7.5, one of the main drawbacks of this algorithm is its lack of time-history data; the optimization is, in essence, "isolated" iterations with no data directly shared between optimization iterations. In addition to this, the optimization is based upon the final deformed shape of the component or system being optimized. These are significant shortcomings of the algorithm, as the time-data history between iterations is important for algorithm stability and control, as will be further discussed in Section 8.4.6. In addition to this, it is suggested that considering the intermediate stages of deformation is pivotal to obtaining an appropriate level of accuracy for optimization scenarios of structures exposed to large levels of nonlinear deformation. Based on these statements, the principle of HCA is therefore deemed to be fundamentally flawed in terms of suitability for nonlinear topology optimization.

This claim is substantiated by Singh (2014) and Chuang and Yang (2012) who stated:

> *Several issues and limitations were discovered during the release of LS-TaSC. One main concern is the HCA algorithm accuracy and robustness during optimization process. It has been seen through published examples that numerous iterations are required in order to achieve better visible topology optimization results. The non-periodic oscillatory behavior also found on the objective function (total internal energy density) history is an issue. These raise the concern of the robustness of HCA algorithm and eventually lead to computation inefficiency when applying to the large scale industry application problems.*

Chuang and Yang (2012) also includes case studies using other approaches including ESLM and a method denoted as inertia relief method (IRM) which is very similar to that presented in Chapter 5. Based on the above statements, it was chosen to abandon the HCA algorithm as a starting point for the nonlinear topology optimization algorithm in this chapter. This leaves only one type of "developed" meta-heuristic algorithm, the evolutionary-based algorithms. As previously presented, this includes three main derivations; the bidirectional ESO (BESO) was determined to be the most suitable one.

Throughout Sections 7.2 and 7.3, some of the similarities between HCA and BESO were highlighted in addition to the differences. To summarize, the BESO algorithms did resolve one of the fundamental issues of HCA, the gradient information (time-history) link between iterations. There were, however, some disadvantages repeated between both algorithms, such as the low dispersion of search within the potential solution space, the relatively high computational cost, and the lack of consideration for intermediate deformation modes, in additional to the over arching issue of the inability to prove that the solutions obtained are indeed optimum. Despite these BESO is considered a good starting point for a nonlinear topology optimization algorithm, as it is deemed feasible to resolve or reduce many of the disadvantages by implementing various principles and techniques.

Acknowledging that it might not be possible to fully resolve all the disadvantages of BESO, the importance of each disadvantage to the overall success of the revised algorithm was prioritized; the initial thoughts on how to improve the specific points are listed in Table 8.1. Note that Table 8.1 does not contain any reference to the proof of optimum solution; this is because addressing this problem is much more general and unfeasible to solve in the current context. Although very interesting and important, it is unlikely that this task will ever be fully resolved.

TABLE 8.1 Importance of Resolving BESO Disadvantages

"Original" BESO Advantage from Section 5.4	Importance of Resolving Disadvantage	Initial Thoughts for Improvement
Quasi-static loading assumption.	Medium	The quasistatic load assumption is overall consistent to that of explicit FEA; an adjustment in timestep size may be sufficient.
Nonconvergent and potential oscillation between "equal" design solutions.	High	Merging BESO with other optimization techniques, for example, SA or TS.
No direct consideration of "intermediate" deformation modes/levels.	High	Introduction of statistical measures such as entropy to enable the inclusion of intermediate values in a systematic and repeatable manner.
Relatively high computational cost.	Low	The accuracy of the algorithm is the main priority; it may be possible to reduce the optimization iterations to FEA ratio by the use of, for example, meta-modeling or an outer entropy-based loop.
Significant issues with high levels of nonlinear material behavior.	Medium	The sensitivity to high levels of nonlinear material behavior must be reanalyzed following the algorithm moderations; it is believed that the measure introduced to reduce the oscillation between solutions will also reduce this issue.
Low dispersion of search within potential solution space, high probability of "converging" toward local optimum points.	High	Merging BESO with other optimization techniques, for example, SA or TS.

Figure 8.1 illustrates the updated BESO flowchart of Figure 5.3, demonstrating in which steps the proposed changes will be incorporated.

As indicated in Table 8.1 and Figure 8.1, the only two optimization techniques chosen are SA and TS. The primary reason for this is that these two approaches are believed to be the only two simultaneously "offering" a solution to the "low dispersion" and the "oscillation" disadvantages listed in Table 8.1. Many of the algorithms presented in Chapter 7 offer potential solutions to the low dispersion disadvantage, but not to the oscillation disadvantage; hence, all other algorithms presented have been discarded. Please note that if the ranking presented in Table 8.1 was revised, other algorithms may once again be taken into consideration. Another aspect that may not be immediately clear from the above is that at this stage FL has not been introduced into the optimization algorithm; hence, the optimization

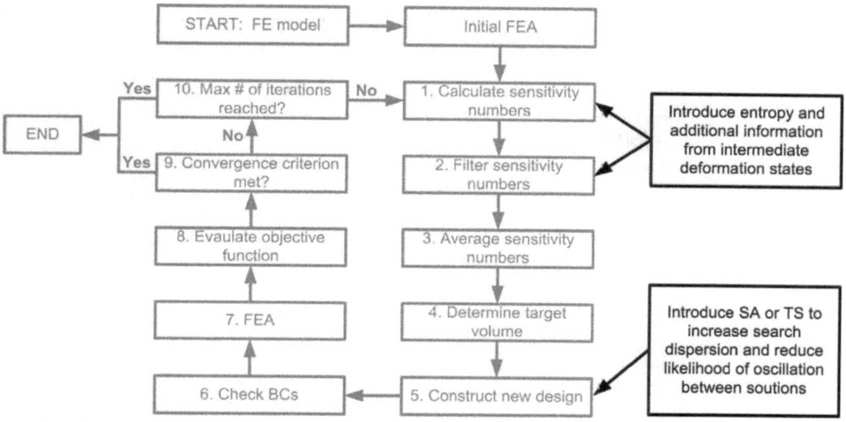

FIGURE 8.1 Proposed algorithm flowchart.

problem remains discrete, as the element inclusion or exclusion may be defined using the Boolean 0 1 operator. The following sections will present and discuss the changes to the individual algorithm steps outlined in Figure 8.1. It is important to note that the following presentations and discussions will only consider full design volumes, that is, no initial void elements are defined.

8.1.1 Calculate sensitivity numbers

As presented in Section 5.3.3, there are two different ways in which the element sensitivity numbers are calculated. For "existing" elements, the sensitivity number is directly obtained from the FEA data, as defined in Equation (5.4). For void elements, that is, elements that could potentially be added, the sensitivity numbers are calculated as defined in Equation (7.5); in the current context there is no need to adapt the sensitivity calculation for void elements. As discussed throughout Chapter 7, it is, however, suggested that the consideration of intermediate deformation states (energy) is key to enhancing the overall algorithm robustness and accuracy, meaning that Equation (7.3) should be revised. There are, of course, many different ways to do this; the simplest would be to utilize a mean or median value for each element, although this might lead to numerical instability if the strain energy as a function of time is somewhat exponential for example.

Other statistical principles such as the root mean square (RMS) value could also be an option. In this context, there are two main factors that should be taken into account when defining how the element sensitivity numbers are calculated:

1. The calculation should be as objective as at all possible, i.e. unbiased;
2. The discretization of the overall physical system should be considered, as FEA is merely an approximation of the "real world structure"; strain energy levels may be high in individual elements but considerably lower in the overall "local area" of the structure. The real-world structure does not experience the "peak" element levels, but rather the overall local area level.

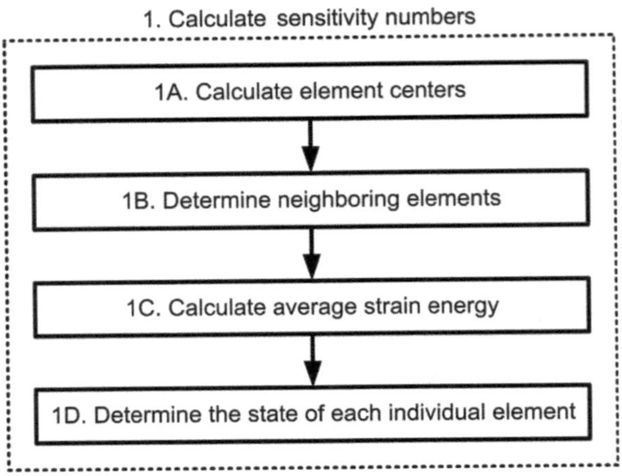

FIGURE 8.2 Detailed flowchart of step 1.

Point 1 above thus suggests using an unbiased selection procedure to statistically represent the state of the structure; an "ideal" choice would be the MEP as presented in Section 7.14.

Point 2 suggests that the element sensitivity should not solely be determined by the individual element, but also be influenced by its immediate neighbors. As presented throughout Chapter 7, there are many different ways of establishing neighborhood influence such as the Von Neuman neighborhood. The BESO algorithm does, however, already contain a neighborhood weighting function, Equation (7.6), which is deemed more suitable as it takes the distance between element centers into account, thereby considering the mesh size.

The intended changes to this step therefore consist of two main aspects: firstly, to introduce entropy as presented in Section 7.14 and secondly, to introduce the weighting function of Equation (7.6). Figure 8.2 illustrates the intricate steps of the step 1 of the overall algorithm in Figure 8.1.

Steps 1A and 1B of Figure 8.2 are relatively straightforward; they simply consist of determining the element centers as a function of nodal coordinates and element type, in addition to determining neighboring centers using the calculated element center coordinates and the user-defined radius r_{ij}, previously introduced in connection with Equation (7.6). Figure 8.3 illustrates how varying

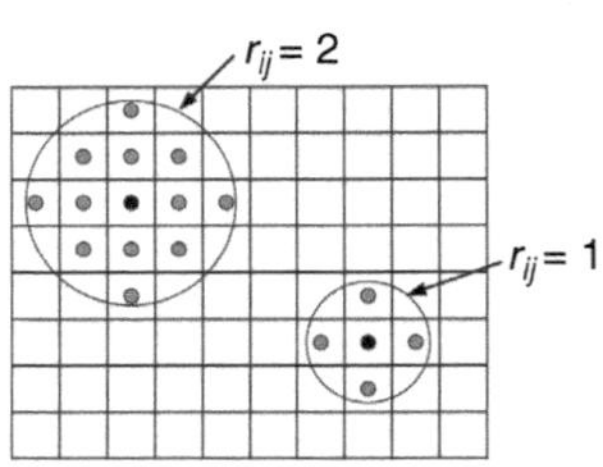

FIGURE 8.3 Illustration of neighborhood influence and strain energy calculation.

r_{ij} values will determine the size of the "neighborhood" and indirectly influence the subsequent average strain value calculations.

The first change of the BESO algorithm is incorporated into step 1C, which involves calculating the average strain energy of each "neighborhood." Please note that it is possible to use other quantities than strain energy; however, it is deemed that the strain energy is a good representation for the "condition" of the element, and will therefore be used throughout the remaining derivations and discussions of this chapter. If it was subsequently desired to use a different quantity, this would only constitute in minor changes to the software/algorithm being developed.

The average strain energy is calculated using the data from 1A and 1B in addition to the strain data obtained via FEA. The average strain data calculations are therefore indirectly a function of the user-defined r_{ij}value, as indicated in Figure 8.3. As previously discussed, the influence of each neighboring element will be weighted as a function of distance $\omega\left(r_{ij}\right)$, as defined in Equation (7.6). The calculation of average neighborhood strain energy ANS_i to be used in connection with the determination of the sensitivity number of the existing element i can, therefore, be defined as Equation (8.1)

$$ANS_i = \frac{\sum_{t=\Delta t_{\text{start}}}^{\Delta t_{\text{end}}} \left(E_i(t)\right)}{N\Delta t} + \sum_{j=1}^{NN_{\text{els}}} \left[\frac{\sum_{t=\Delta t_{\text{start}}}^{\Delta t_{\text{end}}} \left(E_j(t)\right)}{N\Delta t} \cdot \omega_{ij} \right] \tag{8.1}$$

In (8.1), NN_{els} is the number of neighborhood elements, Δt is the timestep from FEA, $E_j(t)$ is the strain energy of element j at time t, and $N\,\Delta t$ is the total number of timesteps included in the calculation. Please note that Equation (8.1) allows for the number of timesteps to be included to be user defined; per default this value will be set to include all timesteps from the FEA. The effects of this will be subsequently investigated in Section 8.4. With the average neighborhood strain energy ANS_i determined, it is now possible to determine the "probable" state of element i using MEP. This will be conducted analog to the Lagrangian multiplier single constraint example of Section 7.14. Using Equation (8.1), Equation (7.25) can now be updated to Equation (8.2), where $ESE_i(t)$ is the strain energy of element i at time t.

$$f(\beta) = \sum_{t=\Delta t_{start}}^{\Delta t_{end}} \left[\left(ESE_i(t) - ANS_i\right) \cdot 2^{-\beta\left(ESE(t) - ANS_i\right)} \right] \tag{8.2}$$

Using Equation (8.2), probabilities for a representative value of the element strain energy can now be calculated, which incorporates neighborhood as well as time history information, that is, intermediate deformation stages. The value with the highest probability will be chosen as the representative strain energy for the specific element.

This means that steps 1 and 2 of the original BESO algorithm flowchart of Figure 7.3 are incorporated into a single step using the flowchart in Figure 8.2 for "existing" elements and the "original" BESO sensitivity calculation of Equations (7.5) and (7.6) for void elements.

8.1.2 Average sensitivity numbers and determine target volume

Steps 3 and 4 of the original BESO algorithm in Figure 7.3, "average sensitivity numbers" and "determine target volume," remain unchanged at this stage. It is, however, necessary to describe how these are implemented, as this was not fully presented in Section 7.3.3.

Step 3 in Figure 7.3, "average sensitivity numbers," may initially seem obsolete following the significant changes introduced in how the sensitivity numbers are calculated as defined in Section 8.1.1. However, the averaging of step 3 serves a different purpose than that introduced in step 1. In step 1, the "averaging" was based on values obtained from timesteps from a single FE analysis of a defined structure within one optimization iteration. The "averaging" of step 3 refers to sensitivity values between separate optimization iterations, that is, between "different structures." The average element sensitivity number (ESN) is therefore a function of element number i as well as optimization iteration number j, as seen in Equation (8.3)

$$ESN_i = \frac{ESN_i(j-1) + ESN_i(j)}{2} \tag{8.3}$$

Equation (8.3) thus introduces a time-history link between optimization iteration loops.

Step 4 in Figure 7.3 "determine target volume" inadvertently decides the number of elements to void and/or add to the structure for the next optimization iteration. The first step is simply to calculate the individual volume of each solid element or alternatively the area of each shell element; this can be straightfowardly completed using nodal coordinates. The volume or surface area V_j of the structure for iteration j can thus be obtained by summation of the individual elemental volumes/areas. The volume is controlled by two user-defined parameters: target volume fraction (TVF) and evolutionary ratio (ER). TVF simply dictates the target fraction of the DV remaining after topology optimization. For example, a TVF value of 0.1 means that 10% of the origininal DV or area at iteration 0, $V(0)$, should remain at the end of the optimization process. Thus, the rule in Expression (8.4) must be followed.

$$V_{\text{end}} \simeq TVF \cdot V_0 \tag{8.4}$$

In most cases, it will of course not be possible to strictly adhere to Expression (8.4); hence, an acceptable residual value should be defined. Expression (8.4) and the residual thus also represent the "convergence criterion" of step 9,

Figure 7.3. Provided that the convergence criterion has not been met, the target volume TV_{j+1} for the next iteration is calculated using V_j and ER as defined in Equation (8.5)

$$TV_{j+1} = V_j \cdot (1 \pm ER) \tag{8.5}$$

The ± in (8.5) depends on whether V_j is larger or smaller than V_{end}. ER does, in essence, determine the "aggressiveness" of the algorithm and can be considered analog to the scale factor *sf* of line search methods introduced in Chapter 3. This means that a large number will allow a relatively large fraction of the DV to be removed or added in a single iteration, while a relatively small number will have the opposite effect. The ER value is, therefore, a key factor in determining the effectiveness of the algorithm, as well as ensuring the robustness of it. As presented above, the ER parameter is a constant once initially defined by the user. It is, however, possible to investigate the effects of making this parameter a variable, for example by using the probability function $p(B)$ principle from SA as defined in Equation (5.11). This will, however, not be further explored for the time being. The new design, that is, the structure at $j+1$, can therefore now be determined using the ESN and TV parameters.

8.1.3 Construct new design

Following the revised calculation of the ESN and TV parameters presented in the previous sections, the new design for the next iteration can now be created. This is a multistep process, step 5 in Figure 8.1; the overall flowchart of this specific step is illustrated in Figure 8.4.

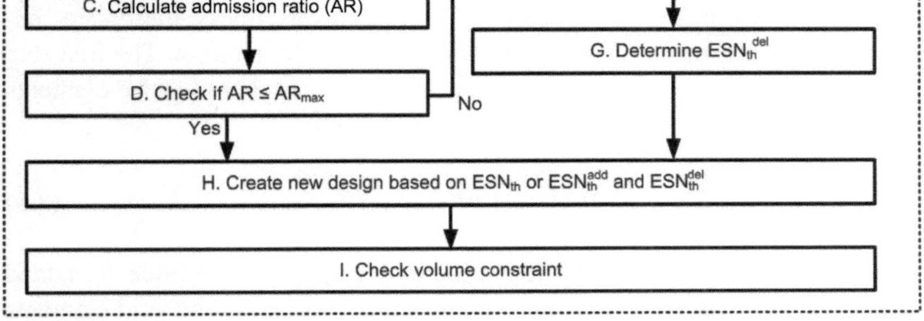

FIGURE 8.4 Detailed flowchart of step 5.

At this point, it is important to remember that in the revised BESO algorithm presented in Section 7.3.3, existing and void elements are evaluated simultaneously. Step 5A may, therefore, be considered to be a matrix with one row per element in the DV and the following header: element number i, $\mathrm{ESN_i}$ element volume and current element status (existing or void). In step 5A, this matrix is sorted according to descending $\mathrm{ESN_i}$ number, regardless of element status. In Step 5B, $\mathrm{ESN_{th}}$ is determined by summarizing the individual element volume EV_i according to the descending $\mathrm{ESN_i}$ values until Expression (8.6) becomes true

$$\sum EV_i \geq TV_{j+1} \tag{8.6}$$

$\mathrm{ESN_{th}}$ is then obtained by setting it equal to the $\mathrm{ESN_i}$ value of the element that makes Expression (8.6) true. In step 5C an admissions ratio (AR_j) is calculated, as defined in Equation (8.7)

$$AR_j = \frac{\mathrm{NEA_j}}{\mathrm{CNE_j}} \tag{8.7}$$

In (8.7), NEA_j is the number of elements added and CNE_j is the current number of non-void elements in the model. NEA_j can be determined by counting the number of elements with a current void status for which Expression (8.8) is true

$$ESN_i > ESN_{\mathrm{th}} \tag{8.8}$$

Next a maximum admission ratio (AR_{max}) is introduced; this is a user-defined value that indirectly dictates the maximum number of elements allowed to be added in any single iteration j. The primary reasons for introducing this factor are numerical stability and overall algorithm robustness. Step 5D is a simple comparison of AR to AR_{max}, thereby checking whether or not too many elements have been added in the current optimization iteration. If this is not the case, the process will proceed to step 5E; if this is the case, the process will proceed to step 5H.

The loops consisting of steps 5E, 5F, and 5G in Figure 8.4 are, as previously mentioned, introduced to enhance algorithm robustness; these steps become active if too many elements are added in a single iteration. The first step in this loop, 5E, is simply to determine the maximum number of added elements $MNEA_j$, and is calculated as defined in Equation (8.9)

$$MNEA_j = AR_{\mathrm{max}} \cdot CNE_j \tag{8.9}$$

$MNEA_j$ can now be used to determine a revised threshold value for added elements $ESN_{\mathrm{th}}^{\mathrm{add}}$ in step 5F. This value is found by ranking the void elements only according to descending ESN_i value and by choosing the entry equal to

$MNEA_j + 1$. The total volume of the added elements VAE_j can then be calculated by summation of the void elements for which Expression (8.10) is true

$$ESN_i > ESN_{\text{th}}^{\text{add}} \quad (8.10)$$

With the $ESN_{\text{th}}^{\text{add}}$ value determined, a corresponding value must be found for the deletion (void) of elements, step 5G; this is carried out according to the previously defined TV_{j+1} value, ensuring that the volume of removed elements VRE_j is as defined in Equation (8.11)

$$VRE_j = V_j + VAE_j - TV_{j+1} \quad (8.11)$$

Once VRE_j has been found, a revised threshold value for deleted elements $ESN_{\text{th}}^{\text{del}}$ can be found by sorting all nonvoid elements according to descending ESN_i values and subsequently summation of the EV_i values until Expression (8.12) is satisfied

$$\sum EV_i \geq VRE_j \quad (8.12)$$

$ESN_{\text{th}}^{\text{del}}$ is thus equal to the ESN_i value of the element that fulfils (8.12).

The next step of the flowchart in Figure 8.4 is 5H, "create new design." The specifics of this step will either use ESN_{th} or $ESN_{\text{th}}^{\text{add}}$ and $ESN_{\text{th}}^{\text{del}}$, depending on the outcome of step 5D. Regardless of the outcome of step 5D, the "input" to step 5H is, in essence, the same, a list of elements that will compose the structure at iteration $j+1$; hence, the following explanation will only utilize the ESN_{th} notation as the differences between the two possibilities are minute.

In the original BESO algorithm presented in Section 7.3.3, this step is relatively trivial as it merely consists of "reassigning" 0 or 1 values to the individual elements, thus creating an updated FE model for the following iteration. However, as identified in Section 7.3.3 and summarized in Table 7.1, one of the weaknesses of the original algorithm is its lack of ability to disperse the search area within the solution space, as well as its oscillation between two potential solutions. Step 5H presents an obvious opportunity for implementing SA or TS to alleviate these tendencies. The introduction of either of these will most likely have various effects upon the effectiveness and indeed robustness of the algorithm. SA does perhaps represent the most "diverse" of the two proposed changes, as it is classed as a "random search method," as presented in Section 7.10. The potential positive effects of this are that it ensures a significantly improved dispersion of search within the potential solution space; this may also be one of the negative aspects of the methodology, as "inferior" solutions may be accepted as a function of the "temperature," potentially causing divergence of the optimization as opposed to convergence. In addition, SA offers no immediate solution to the oscillation phenomenon previously discussed. TS, on the other hand, does offer a relatively straightforward way of preventing oscillation by using the Tabu lists and aspiration criteria discussed in Section 7.9. On the other hand, TS does perhaps not offer as wide a search pattern as can be expected from an SA algorithm. This may be a disadvantage but the effects of this are presently unknown. One of the general assumptions throughout this book has been the notion that the level of accurate structural performance can only be obtained using FEA. Hence, it follows that FEA provides good estimations of the current performance of the structure and indirectly provides good indications of load paths, regardless of the unpredictive nature of the overall problem. Based on this, as well as weighting the criteria listed in Table 8.1, it has been chosen to implement TS as an addition at stage 5. As stated in Section 5.9, TS is, in essence, a variation of HC, the "original" solution methodology of BESO, which may indicate a smoother transition and implementation of TS.

The revised algorithm will therefore utilize aspects of bidirectional evolutionary optimization, entropy, Tabu search, and simulated annealing methodologies and will therefore be denoted as BEETS.

8.1.4 Tabu search implementation

The focus is now returned to the specific task of implementing TS in step 5H, Figure 8.4. This can be done in many different ways; however, emphasis has been added to ensure that the number of iterations and thereby FE analyses does not become excessive, significantly increasing computing requirements. It is therefore decided to implement TS with parameters that can subsequently be adjusted by the user. These are number of iterations per Tabu search (NITS), variations per Tabu search (VTS), as well as variation ratio (VR). Figure 8.5 illustrates a flowchart providing an overview of the Tabu search iteration implementation into the BEETS algorithm.

In Figure 8.5, a "standard iteration" is analog to a normal optimization iteration, that is, all steps included in Figure 8.1 with the modifications discussed above. The NITS parameter decides how many "standard iterations" are to occur per Tabu search iteration (TSI). For example, if NITS equals 2, every other

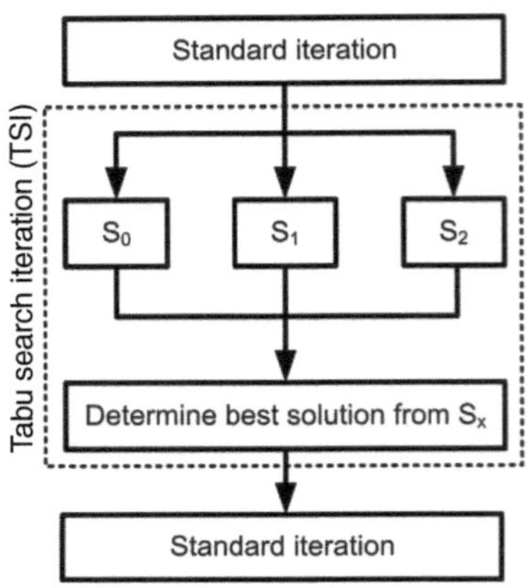

FIGURE 8.5 Tabu Search iteration implementation.

optimization iteration will be a TSI; if NITS equals 1, all iterations will be TSI. The TSI loop is in essence identical to the standard iteration up until step 5H. In step 5H, a TSI a solution S_0 is created based on ESN_i as explained above. However, a fixed number of solutions equal to VTS with "random" variations are also created: $S_1, \ldots, S_{VTS}$. These random variations will simply be obtained using a random number generator to identify nonvoid element numbers whose element status will subsequently be set to 1 subject to their current element; S_0 status.

The number of elements to be changed (NEC) will be calculated as a fraction of the current number of nonvoid elements (CNE) using the VR parameter as defined in Equation (8.13)

$$\text{NEC}_\text{j} = \text{VR} \cdot \text{CNE}_\text{j} \tag{8.13}$$

The Tabu lists and aspiration criteria will "ensure" that no previously analyzed solution will be reanalyzed. All variation models, $S_1, \ldots, S_{VTS}$, will thereafter be solved and steps 1 through 5 repeated. Following this, the "best" solution S_best of the $S_0, \ldots, S_{VTS}$ solutions will be identified and used to create the model for the following iteration $j + 1$. The determination of the "best" solution will follow a relatively simplistic principle; namely the ratio of volume to total strain energy (VTSE). The VTSE value for any VTS iteration will, therefore, be calculated as defined by Equation (8.14)

$$VTSE_k = \frac{\sum_{i=1}^{CNE} VE_i}{\sum_{i=1}^{CNE} ESE_i} \tag{8.14}$$

In (8.14), VE_i is the individual volume of the (nonvoid) elements and ESE_i is the total strain-energy of nonvoid elements (summed over Δt) and k represents the random variation run iteration number.

The above implementation method should help to maintain the robustness of the algorithm, because if any (or perhaps all) of the random variation solutions $S_1, \ldots, S_{VTS}$ are numerically unstable and error terminates, these can simply be disregarded and the algorithm will continue using S_0 if no other or better solution is available.

The Tabu list implementation into the BEETS software will use the following principles. Firstly, the use of a Tabu list will be optional, and can be switched on or off using a simple Boolean operator, as will be further explained in Section 8.2.2. As discussed in Section 7.9, it is not generally possible to store entire solution sets in Tabu lists, mainly due to memory limitations. The Tabu lists of the BEETS software will store the element status values; as it is not feasible to store all element status values, a fraction of the elements will be selected for the Tabu lists. This will utilize a Tabu list fraction (TLF) value specified by the user.

The number of elements stored in the Tabu list (NESTL) can thereby be calculated as defined in Equation (8.15);

$$NESTL = NE_0 \cdot TLF \tag{8.15}$$

NE_0 in (8.15) is the number of elements at iteration 0 (where all elements in the model are nonvoid). With the definition of *NESTL*, a list of element numbers whose statuses will be stored in the Tabu list can now be created. This will be done by randomly selecting *NESTL* number of elements.

In order to reduce the computational efforts associated with the Tabu list the first parameter used to detect potential "identical" iterations will be the volume of the FE model. If the volume of the current FE model (at iteration i) is identical to any of the volumes of preceding models (from iteration 0 to iteration i −1), the Tabu list will be activated to check if the element statuses of the elements are also all identical. If this is the case the software will recognize that two potential iterations are identical and prompt the user to decide whether or not the optimization should continue or terminate.

8.1.5 Simulated annealing implementation

The focus is now turned to the implementation of Simulated Annealing optimization principles into the BEETS algorithm and software. With the implementation of Tabu Search presented in the previous section, it is decided not to include all aspects of the SA optimization principles; the only aspect that will be considered is the cooling function, initially defined in Equation (7.12). The fundamental reasoning for introducing the cooling function into BEETS is an attempt to maximize the efficiency of the algorithm by minimizing the number of iterations required to obtain the optimized solution. If a relatively large evolution ratio, Equation (8.5), is selected without the implementation of a cooling function, the optimization may rapidly converge toward a solution, but may struggle to refine the structure as a direct function of the large ER value. On the other hand if a small ER value is selected, the number of iterations required to obtain the fundamental topology of the solution may be significantly increased reducing the efficiency of the software. Returning to the "original" cooling Equation (7.12), it can be seen that this is a function of the "temperature" at iteration 0 (I0) and the number of iterations passed since I0. Implementing the cooling in such a way is likely to have a detrimental effect upon the "widened search space" enabled by the introduction of Tabu Search as previously presented. The reason for this is that a direct implementation of (7.12) would lead to a continuous decrease in the relative fractional volume change per iteration, significantly narrowing the search area. Instead, it is therefore decided to introduce an amended Tabu Search cooling function that only becomes active toward the end of the optimization as the volume approaches the final target volume V_{end}, Equation (8.4). The activation of the SA cooling function will be determined by the use of Expression (8.16)

$$TV_{j+1} \le V_{\text{end}} \cdot (1 \pm COOLIT) \le TV_{j+1} \tag{8.16}$$

Equation (8.16) simply states that if the updated target volume for the next iteration, TV_{j+1}, lies within a given interval of the final target volume V_{end}, determined by the evolution ratio (ER), the defined cooling function will be activated

$$ER_{\text{cool}} = ER_{\text{cool}} - ER_{\text{cool}} \cdot SF_{\text{cool}} \tag{8.17}$$

In (8.17), ER_{cool} is the updated evolution ratio, which may be reset if Equation (8.16) becomes untrue at any given stage. At the first iteration where Equation (8.16) activates Equation (8.17), ER will be used to replace ER_{cool} on the right-hand side of Equation (8.17). SF_{cool} is a user-defined scale factor dictating the rate of the cooling process. SF_{cool} must adhere to Equation (8.18)

$$0.01 \le SF_{\text{cool}} \le 0.99 \tag{8.18}$$

Once an ER_{cool} value has been determined, this will replace ER in Equation (8.5) to calculate an updated TV_{j+1} value. Equations (8.16), (8.17), and (8.18) thus represent the methodology in which the cooling function principle of simulated annealing is introduced into the BEETS algorithm and software.

With the above changes and the "original" BESO algorithm illustrated in Figure 5.3, the flowchart for the overall BEETS algorithm can now be defined as illustrated in Figure 8.6.

The flowchart in Figure 8.6 is the backbone upon which the BEETS software has been programmed. The next section will explain the overall details of the software programming and algorithm implementation process, including initial verification of the algorithm and software.

8.2 ALGORITHM IMPLEMENTATION AND SOFTWARE DEVELOPMENT

The purpose of this section is to explain the overall software development process, including the implementation of the BEETS algorithm and some of the intricate differences between the flowchart of Figure 8.6 and the actual software. The first task of any software development process is to select a suitable programming language/software platform.

8.2.1 Selection of programming language and software platform

Many different options are available for any new software development, the majority of which can be used in conjunction with each other. Programming languages such as Fortran, C, and C + + are commonly used as a base for commercial finite element software for example LS-DYNA and Altair HyperWorks. There are many reasons for this; one of the primary reasons is likely the fact that these programming languages:

a. Generally have extensive mathematical libraries and advanced functions;
b. Can be coupled with a wide array of other languages, such as Java, Html, etc., to, for example, create advanced graphic user interfaces (GUIs);

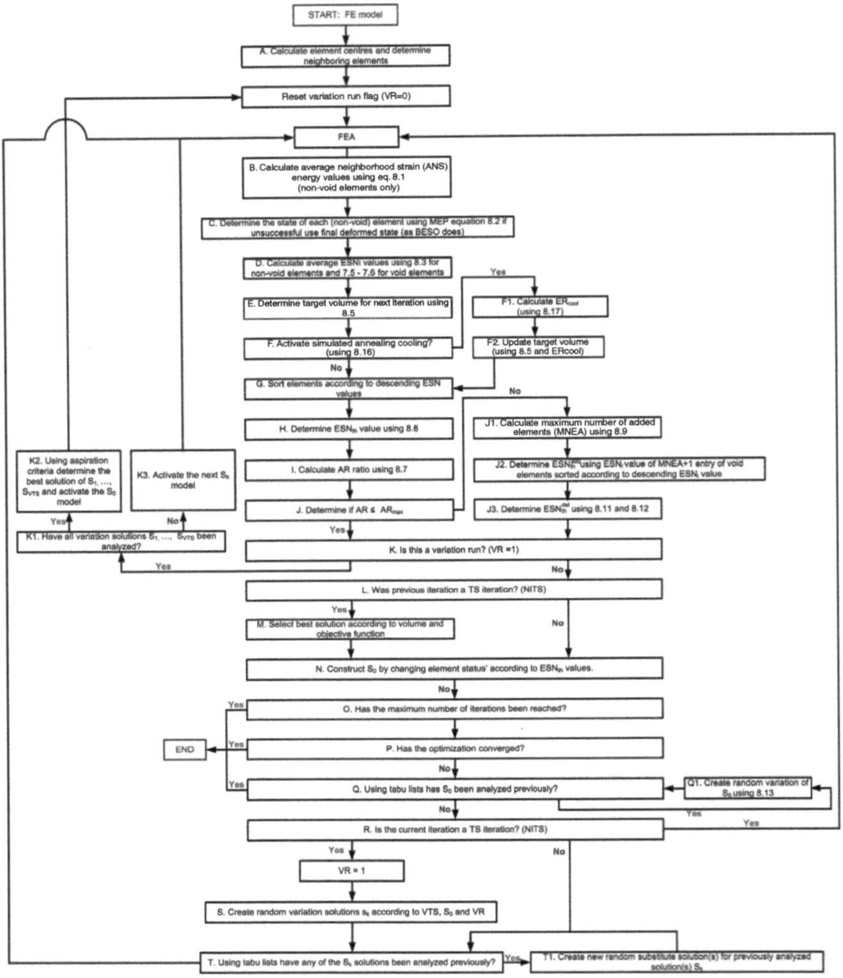

FIGURE 8.6 BEETS optimization algorithm flowchart.

c. Are very versatile and can run across multiple software platforms (e.g. Linux, Unix, and Microsoft Windows) with only minor changes to the software;

d. Are public, which means that in general no royalties or software license fees are applicable to any software created using any of these languages;

e. Can be very specifically tailored to maximize CPU and memory capacities, especially when handling large amounts of data.

The above facts imply that choosing a language such as C++ could indeed be advantageous for versatility, efficient hardware utilization, and high-performance computing, and must therefore be serious contenders for a longer term strategic development of any new optimization algorithm such as BEETS, if the potential of the algorithm can be justified. The advantages of such programming languages do, however, come at a cost, as they can be somewhat cumbersome to work in, and generally require specialist knowledge to fully utilize the potential in terms of, for example, efficient hardware utilization. Remembering that the purpose of the software and algorithm activities of this chapter is to demonstrate an example of algorithm development and investigate the feasibility and potential of the BEETS algorithm; for large-deformation nonlinear topology optimization it is decided to utilize "validated" commercial software wherever possible in an attempt to ensure that the "true" effects of the BEETS algorithm are clear. This will most likely correspond to the majority industrial engineering scenarios where it is unlikely that a new "dedicated" FE solver will be programmed (and validated) from scratch. In other words, it is decided to utilize commercial FE software for the FE analysis and to reduce the programming to predominately focus on the actual BEETS algorithm example. At this point, is should be noted that this choice does not rule out the usage of the programming languages used above, such as Fortran. With this change, the software is, however, not required to actively solve any FE model, but merely send an FE model to the FE solver, extract/read the relevant data to inform the BEETS optimization algorithm before updating the FE model and subsequently sending the updated model back to the FE solver. The core of the tasks ahead can, therefore, be solved using matrices and arrays, suggesting that commercial mathematics software such as MatLab or Maple may also be a suitable choice.

There is, however, one significant issue with utilizing such software; namely the fact that it is not possible for, for example, MatLab to adapt an FE file, which in general can be considered to be in text format. Hence, it is deemed that a scripting language is most suitable for the task ahead. For simplicity the example will utilize a Windows-based platform and Microsoft PowerShell is ultimately chosen for the task, as it enables the user to easily access and execute programs, conduct mathematical operations, as well as reading, writing, and adapting files of any given format. The BEETS software is, therefore, programmed in Microsoft PowerShell and requires an active LS-DYNA license for FEA; although this can be relatively easily changed to any other solver, FE-based or otherwise. LS-DYNA was chosen for the solver in this example; as it is a commercial FE code aimed specifically at nonlinear and dynamic analysis problems, is widely used within the automotive sector for, for example, crash analysis, and can be controlled via a command prompt (directly from within PowerShell). LS-DYNA is set up as an explicit FE solver; hence, all subsequent analysis/optimization will utilize explicit FEA unless otherwise stated. It should be noted that any FE solver can be used for this purpose, including freeware.

8.2.2 Overview of BEETS PowerShell software

As previously stated, the BEETS software is based upon the overall flowchart in Figure 8.6. The software has been compiled using a modular approach, meaning that a series of scripts, subroutines, and functions are called from the main script. There are two main advantages to this approach:

a. Obtaining an overview of the intricate details of the software is eased as individual steps (Figure 8.6) are simplified. This may also reduce the complexity of any debugging activities;

b. The software becomes modular, which means that replacing, upgrading, adding, or even removing aspects/parts of the software is easily achievable. This is an important attribute for future research and development.

The BEETS optimization software consists of a total of 12 PowerShell files (*.ps1). The file names, a narrative description, a link to the flowchart (figure 8.6) steps covered, as well as the equations used within each *.ps1 file are listed in Table 8.2.

There are a few requirements for the FE/LS-DYNA input file:

a. The optimization will only be conducted on *SHELL elements.

b. The shell elements must be defined in the global XY-plane only.

c. The file must contain a *SET_SHELL set containing all *SHELL elements within the model.

d. The *DATABASE_ASCII card must have the ELOUT option enabled.

The main reason for point d above is to ensure that the strain values are available for the BEETS optimization algorithm. Previously in this chapter, it was suggested that the individual element strain energy be the measurable quantity for the optimization algorithm. Extracting the element strain energy from an LS-DYNA FEA is, however, not a straightforward task. The only direct method of extracting the element strain energy in LS-DYNA is to model every single element as an individual part leading to a very extensive "organizational" task, significantly (and unnecessarily) reducing the versatility of the software. It is possible to extract the element strain values by postprocessing via LS-PrePost, a dedicated LS-DYNA pre- and postprocessor. The output format of this is, however, not a "conventional" text format, making it impossible to read the data into PowerShell for the purpose of optimization. Hence, it is decided to change the input data into the BEETS software from strain energy to actual strain values, as these are readily available and can also be considered a reasonable representation of the strain energy. This could potentially have an effect upon the accuracy of the results; however, the strain should serve as a valid means for relative comparison between individual elements as intended. Due to the fact that large deformation, and thereby large strains, are expected, it is decided to use the total (sum) of strain values ε_{tot} to represent the individual element as defined in (8.19)

$$\varepsilon_{\text{tot}} = \varepsilon_{11} + \varepsilon_{22} + \varepsilon_{33} + \varepsilon_{12} + \varepsilon_{13} + \varepsilon_{23} \tag{8.19}$$

As implied above, changing the measurable quantity from strain energy to strain values may have an effect upon the accuracy of the algorithm, and warrants further in-depth analysis. For the remainder of this chapter, the total strain value will however be used as defined by Equation (8.19).

TABLE 8.2 Overview of PowerShell Files

File Name	Description	Steps Covered (Figure 8.6)	Equations Used
MAIN.ps1	This is the main file that prompts the user for inputs; it contains the optimization iteration loops and calls all functions (subroutines) from separate *.ps1 files.	Start, end, K–T	
Element_Sensitivity_ Numbers.ps1	This file calculates the element sensitivity numbers (ESN)	D	(8.3)
Update_FE_file.ps1	This file updates the LS-DYNA keyword file for subsequent analysis of the updated design		
Determine_ESNth.ps1	This file calculates the target volume for the next iteration, determines threshold sensitivity values, including checking the admissions ratio and controlling the cooling function.	F–J	(8.5) (8.6) (8.7) (8.8) (8.9) (8.10) (8.11) (8.12) (8.16) (8.17) (8.18)
MEP.ps1	This file contains the maximum entropy principle methodology to determine the likely state of each individual element	C	(8.2)
Element_Strain_ Values.ps1	This file assembles the raw element strain data extracted by Read_elout_ strain.ps1. The file also calculates the ANS_i values as well as calling the MEP.ps1 file	B	(8.1)
Read_elout_strain.ps1	This file extracts the raw element strain values from the elout file created from the LS-DYNA solver.		
Determine_ neighboring_ elements.ps1	This file determines the neighboring elements of any given element as a function of the center coordinates of the elements, and a user defined distance.	A, B	
Read_FE_data.ps1	This file reads the vital FE data from the *.key / *.k LS-DYNA keyword file, such as number of elements, nodal coordinates, etc.	A	

(Continued)

TABLE 8.2 Overview of PowerShell Files *(cont.)*

File Name	Description	Steps Covered (Figure 8.6)	Equations Used
Remove_lines.ps1	This file is a "service file"; it contains a simple script to remove blank lines from the *.key / *.k files to ensure no errors occur when data is read using other *.ps1 files		
Read_file.ps1	This file is a "service file"; it contains a simple script which opens a file browser enabling the user to identify the input *.key / *.k file		

The BEETS software will prompt the user for the inputs listed in Table 8.3.

During the optimization process the software creates a new folder per optimization iteration within the folder that contains the base *.key / *.k file (Figure 8.7). The *key / *.k file is copied into the individual folders and altered according to the optimization criteria.

The FEA is subsequently solved from the individual iteration folders, meaning that output files are contained within these folders (Figure 8.8a). For Tabu Search iterations, variation runs are created within subfolders inside the iteration folder (Figure 8.8b).

In addition to the actual updated *.key / *.k files, the software also outputs general data about the overall optimization procedure, the input values, and specific data about the individual iterations. This data is stored in the Main_INFO.txt file placed in the main folder, that is, the folder that contains the original *.key / *.k file selected by the user. An example of the output is illustrated in Figure 8.9.

In addition to the general output data from the MAIN_INFO.txt file, the total volume and total strain values (per iteration) are exported to two separate *.csv files that can subsequently be imported to, for example, Excel.

With the above, the main features of the BEETS software have now been presented, including the algorithm development steps, overall flowchart, and significant variable names. The following section will examine the feasibility of the software for structures exposed to large (nonlinear) deformations.

8.3 LINEAR TOPOLOGY OPTIMIZATION CASE STUDIES

This chapter will contain a series of case studies using the BEETS software, and will aim to compare BEETS to other previously presented algorithms and software. The first case study will be used to verify the fundamental algorithm embedded within the BEETS software.

TABLE 8.3 BEETS User-Defined Inputs

Input Prompt	PowerShell Variable Name	Notes
Please enter the maximum number of iterations using an integer	Max_iter	Max_iter defines the maximum number of optimization iteration loops.
Please enter the frequency of Tabu Search iterations using an integer	TSI	TSI defines how often a TS iteration (Figure 8.5) is completed. For example, a value of 2 will dictate that iterations 2, 4, 6, etc., are Tabu iterations.
Please enter the number of variation runs per TS iteration using an integer	VTS	VTS defines how many variation runs (Figure 8.5) are completed per Tabu search iteration.
Please enter the Tabu Search fraction per variation run	TSF	TSF defines how many elements randomly are set to void for any given variation run. For example, if a model contains 100 elements and TSF is set to 0.1 10 elements will be randomly removed. It should be noted that this does not take varying element sizes (volumes), boundary conditions, strain levels, etc., into account.
Please enter an iteration number from which the Tabu search iterations should start	TABU_offset	TABU_offset provides an opportunity to offset the start of the TS iterations; this may; for example; be used to stabilize a given optimization run by avoiding random removal of elements at the initial iterations.
Use Tabu list (Y/N)	TABU_LIST	If Yes is selected, a Tabu list will be created to avoid repetition of running previous configurations/models.
Please enter the fraction of elements to be included in the Tabu list	TLF	This number dictates how many element statuses will be monitored per iteration. For example, if a given model (at iteration 0) contains 100 elements and TLF is set to 0.1 the status of 10 elements (selected at random) will be stored for each optimization iteration.
Please enter the maximum admissions ratio as a fraction	ARMAX	ARMAX defines the maximum fraction of elements which can be added to a model for any given iteration. For example, if a model contains 1000 (nonvoid) elements in a given iteration and ARMAX is set to 0.03, a maximum of 30 elements can be changed from void to-non-void for the next optimization iteration. It is advised that ARMAX is not set less than 0.03 to encourage the growth of additional elements where required. (Huang & Xie, 2008).
Please enter the filtering radius	RADIUS	RADIUS defines the maximum distance between two element centers to be used with the Von Neumann neighborhood calculations (Figure 8.3).

(Continued)

TABLE 8.3 BEETS User-Defined Inputs *(cont.)*

Input Prompt	PowerShell Variable Name	Notes
Please enter the target volume fraction	TVF	TVF defines the final target volume based upon the model volume at iteration 0. Using Read_FE_data.ps1, Table 8.2, the model volume is calculated by summation of individual shell volumes (areas). This volume is then multiplied by TVF to obtain the target volume.
Please enter the evolution ratio	ER	ER defines the fractional change of volume between two consecutive iterations as defined in Equation (8.5).
Please enter the optimization convergence tolerance	TERMTOL	TERMTOL determines if the optimization terminates before MAX_iter is reached, provided that the average change over the previous three iterations is less than or equal to TERMTOL.
Use Cooling function (Y/N)	COOLUSE	COOLUSE is a Boolean operator which dictates whether or not the Cooling Function is included in the optimization.
Please enter the Cooling interval	COOLIT	COOLIT is used to determine the interval in which the cooling function, that is, Equation (8.17) becomes active. For example if COOLIT = 0.1 and TVF equals 0.5 the cooling function will be active if the target volume of any given iteration is between 0.4 and 0.6.
Please enter the cooling scale factor (between 0.01 and 0.99)	SFCOOL	SFCOOL is only activated if COOLUSE is set to 1 (on); SFCOOL dictates the rate of cooling using Equation (8.17).
Use entropy (Y/N)	MEPUSE	MEPUSE is a Boolean operator, which dictates whether or not the maximum entropy principle is employed. If not, the optimization software will follow the BESO algorithm, potentially including the Tabu Search iterations.

The purpose of this section is to verify that the base algorithm of BEETS, which in essence is a BESO algorithm and follows the principles of BESO and obtains results similar to that of BESO optimization. For the first example, a simple bracket structure will be analyzed. The general FE model is illustrated in Figure 8.10.

The model illustrated in Figure 8.10 will be fully constrained at points A and B and exposed to two single point loads in the negative Y-direction as indicated by F. The model in Figure 8.10 contains 625 elements and utilizes the MAT_001:Elastic LS-DYNA keyword 971 material model. This is an isotropic

Name	Type
Iteration_0	File folder
Iteration_1	File folder
Iteration_2	File folder
Iteration_3	File folder
Iteration_4	File folder
Iteration_5	File folder
Iteration_6	File folder
Iteration_7	File folder
Iteration_8	File folder
Iteration_9	File folder
Iteration_10	File folder
Iteration_11	File folder
Iteration_12	File folder
Iteration_13	File folder
Iteration_14	File folder
Iteration_15	File folder
Iteration_16	File folder
Iteration_17	File folder
ADD_ELEMENTS.txt	TXT File
Main_INFO.txt	TXT File
plate_crush.key	LS-DYNA Keyword File
Strain_sum_history.csv	Microsoft Excel Comma Separated Values File
Volume_history.csv	Microsoft Excel Comma Separated Values File

FIGURE 8.7 Main folder contents.

Name	Type
elout	File
glstat	File
matsum	File
messag	File
nodout	File
Plate8.dpf01	DPF01 File
Plate8.key	LS-DYNA Keyword File
Plate8.otf	OpenType font file
Plate8.ptf	LS-DYNA Binary Plot File
Plate8.ptf01	PTF01 File
Plate8.ptf02	PTF02 File
Plate8ESV_it_1.txt	TXT File
Plate8SHELL_ELEMENT_FILE.txt	TXT File
Plate8SHELL_SET_FILE.txt	TXT File
spooles.res	RES File
status.out	OUT File

(a)

Name	Type
Variation_Run_0	File folder
Variation_Run_1	File folder
Variation_Run_2	File folder
Variation_Run_3	File folder
Plate8.key	LS-DYNA Keyword File

(b)

FIGURE 8.8 Example folder contents (a) standard iteration, (b) Tabu search iteration.

```
Main_INFO.txt - Notepad
File Edit Format View Help
Output from BEETS Structural Optimiser
Run initiated on:
Script location: D:\Entropy\MatLab\Multi_constraint_example
Input file: C:\Users\Case_Study_Example\crush.key
Maximum number of Iterations: 25
Frequency of Tabu Search Iterations: 26
Variation Runs per Tabu Search Iteration:
Tabu Search Fraction per Variation Run:
Maximum Admissions Ratio : 0.05
Filtering Radius: 0.0001
Target volume fraction: 0.3
Evolutionary ratio: 0.1
Optimisation termination tolerance: 0.03
Coooling interval:
Coooling scalefactor:

Data from iteration 0:
Current volume: 2500
Target volume: 2250
ESNth: 0.1303349
ESNthA:
ESNthD:
New volume: 2252
Iteration 0 contained: 624 elements
Iteration 0 was completed on XX/XX/XXXX 15:29:50

Data from iteration 1:
Current volume: 2252
Target volume: 2026.8
ESNth: 0.12283200956881
ESNthA:
ESNthD:
New volume: 2028
Iteration 1 contained: 562 elements
Iteration 1 was completed on XX/XX/XXXX 15:32:19

Data from iteration 2:
Current volume: 2028
Target volume: 1825.2
ESNth: 0.0910336936358362
ESNthA:
ESNthD:
New volume: 1828
Iteration 2 contained: 506 elements
Iteration 2 was completed on XX/XX/XXXX 15:34:32
```

FIGURE 8.9 Example of Main_INFO.txt file content.

linear elastic material model. The specific material parameters used are listed in Table 8.4.

The loading F will be applied in a quasi-static manner; however, the FE analysis will utilize explicit computation, and hence a time-history definition of the load is required. This has been defined as illustrated in Figure 8.11. It should be noted that the termination time for the analysis will be 1.1 s and that the load value is merely an indication of no load (0) and full load (1). The actual load magnitude will be defined by a scale factor, which initially will be set to 200 N.

8.3.1 BEETS case study 1

Using the above-described model as an input file to the BEETS optimization software, the model was run using the optimization parameters defined in Table 8.5.

With the above parameters set, the BEETS software uses the "standard" BESO algorithm as previously presented. The optimization completed after iteration 23 as the average change in volume over the final three iterations was 2.99%. Figure 8.12 illustrates the total strain as a function of the iteration number.

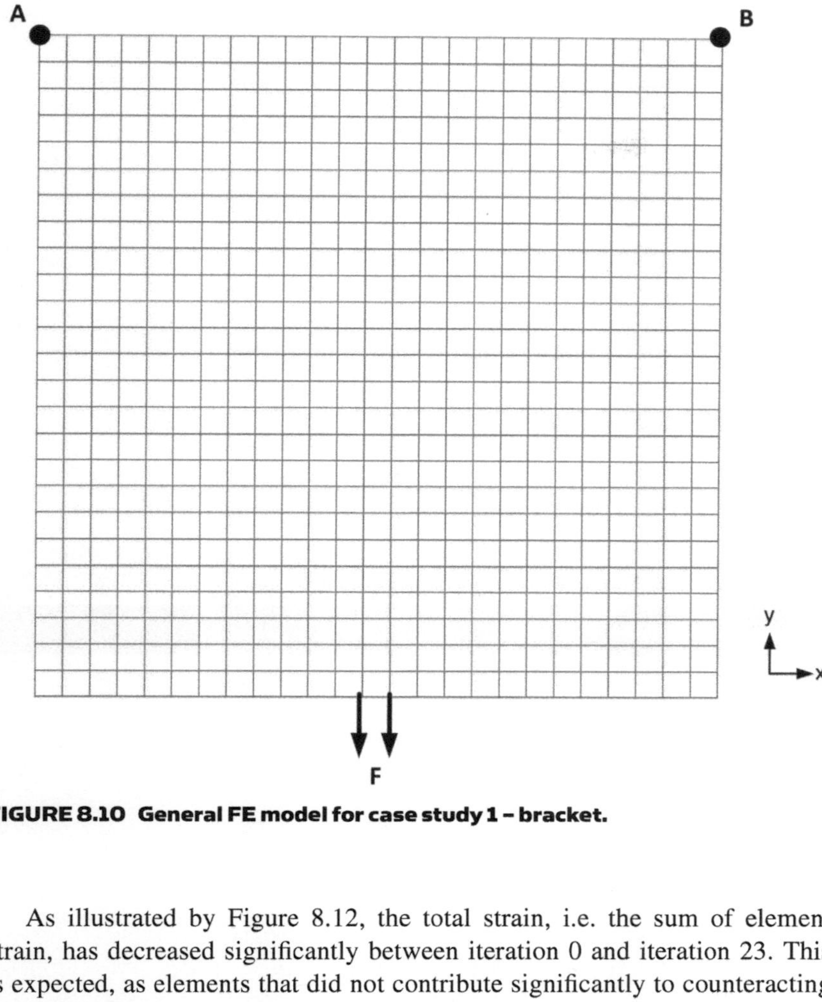

FIGURE 8.10 General FE model for case study 1 – bracket.

As illustrated by Figure 8.12, the total strain, i.e. the sum of element strain, has decreased significantly between iteration 0 and iteration 23. This is expected, as elements that did not contribute significantly to counteracting the applied load have been removed from the optimized structure. The lowest strain value is, however, not found in iteration 23, but in iteration 18. At this stage, it is convenient to review the volume as a function of iteration number; this graph can be found in Figure 8.13.

TABLE 8.4 Material Parameters for Case Study 1 – Bracket

Parameter	Value	Unit
Young's modulus	210,000	MPa
Poisson's ratio	0.3	

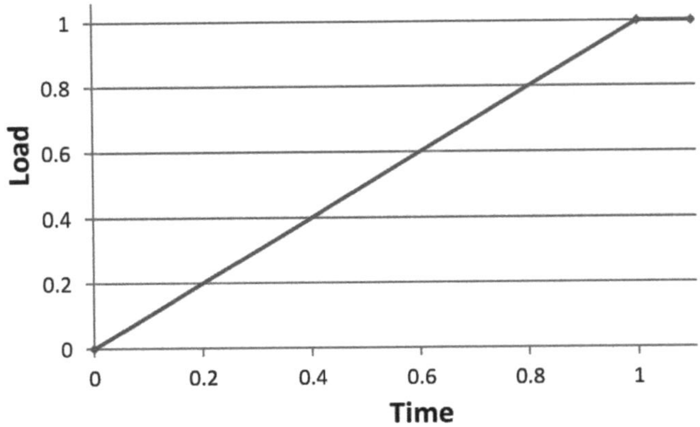

FIGURE 8.11 Load vs. time.

TABLE 8.5 BEETS Input Parameters for Case Study 1

PowerShell Variable	Value
Max_iter	25
TSI	26 (no Tabu Search included)
VTS	N/A
TSF	N/A
TABU_offset	N/A
TABU_LIST	N/A
TLF	N/A
ARMAX	0.03
RADIUS	0.0 (no neighborhood influence)
TVF	0.1
ER	0.1
TERMTOL	0.05
MEPUSE	No
COOLUSE	No
COOLIT	N/A
SFCOOL	N/A

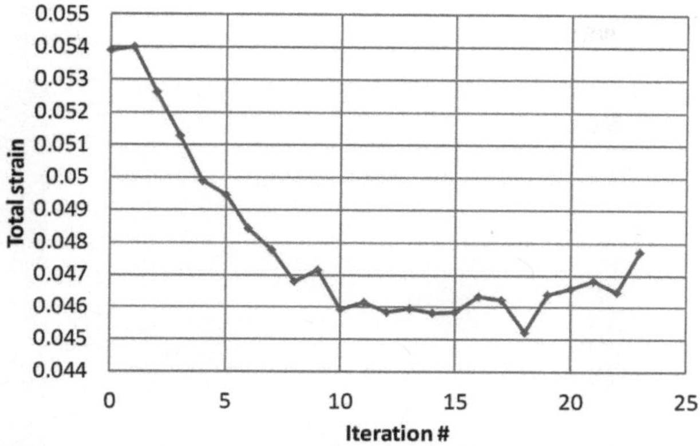

FIGURE 8.12 Sum of element strain per iteration case study 1.

Figure 8.13 clearly indicates a steady convergence tendency with a continuous reduction of volume or elements; this is substantiated when reviewing the MAIN_INFO.txt file that confirms that no elements were added at any point during the optimization. The tendency in Figure 8.12 is therefore not repeated in Figure 8.13. The tendency of Figure 8.12 could perhaps be partially attributed to numerical (rounding off) error. It is, however, more likely that the tendency observed in Figure 8.12 can be explained by reviewing the strain plots of selected iterations. Figure 8.14 illustrates the Von Mises strain values (which are representative of the total strain values in this case) for iteration 0 at the termination time, $t = 1.1$. Figure 8.14 clearly shows that there are a significant number of elements with very low strain values, indicating that these can be removed during the subsequent optimization process.

Figure 8.15 is the "corresponding" stage of iteration 18 to that of iteration 0 (Figure 8.14). As illustrated in Figure 8.15, a significant number of elements have been removed from the element; it can also be seen that the maximum

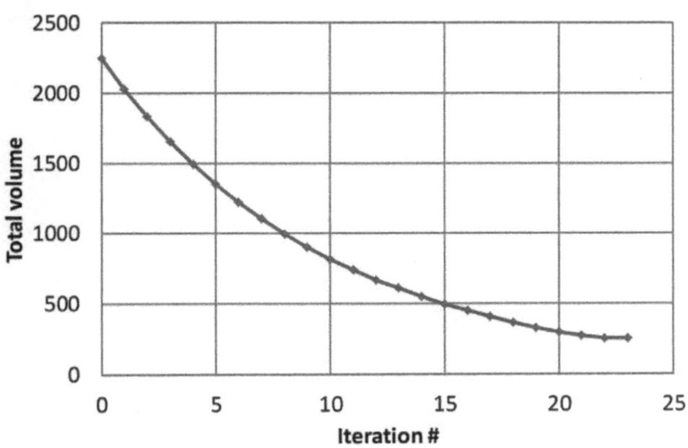

FIGURE 8.13 Total volume per iteration case study 1.

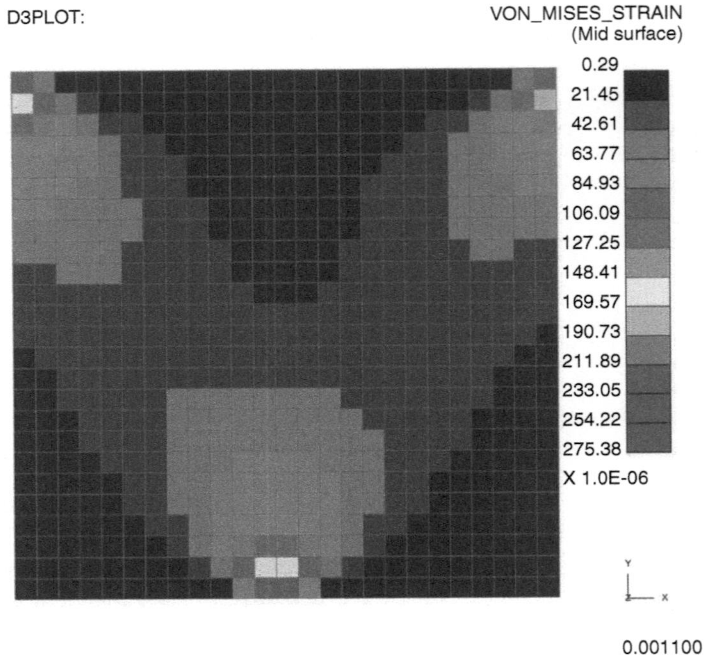

FIGURE 8.14 Total strain (Von Mises strain) case study 1, iteration 0, $t = 1.1$.

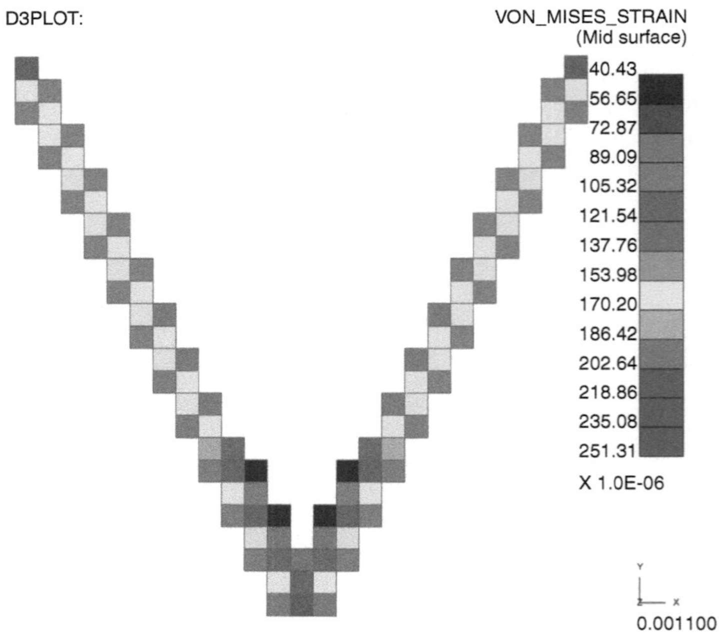

FIGURE 8.15 Total strain (Von Mises strain) case study 1, iteration 18, $t = 1.1$.

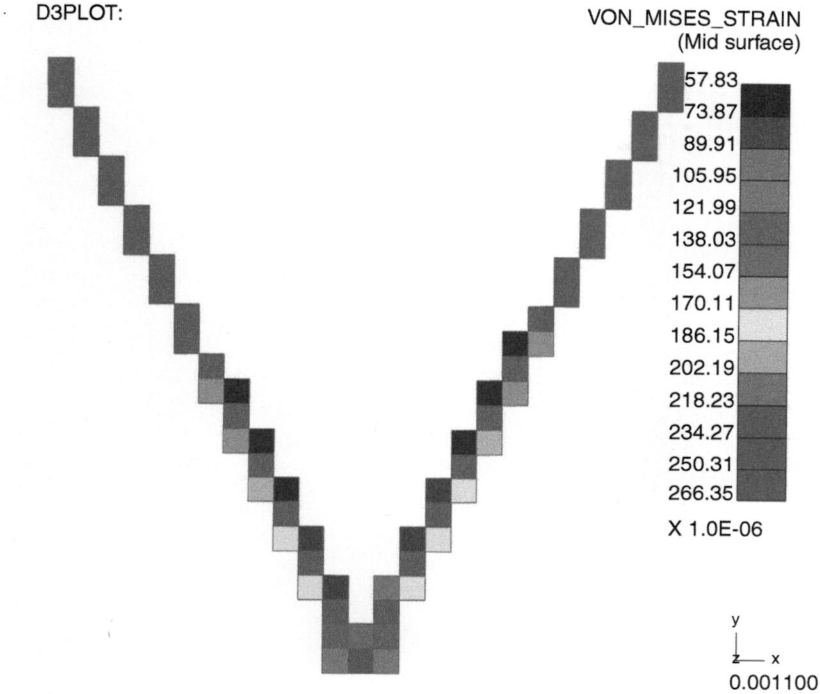

FIGURE 8.16 Total strain (Von Mises strain) case study 1, iteration 23, *t* = 1.1.

strain level has decreased from iteration 0 to iteration 18, both of which would contribute to the drop seen in Figure 8.12.

Figure 8.16 is the corresponding stage of iteration 23 to those of iteration 0 (Figure 8.14) and 18 (Figure 8.15). Comparing the three figures, the reason for the trend seen in Figure 8.12 can be identified. Although the number of elements has decreased between iterations 15 and 23, it is clear that the maximum strain value has increased; Figure 8.16 also indicates that the average strain value per element has increased as expected.

Despite the above discussions, the optimized structure illustrated by Figure 8.16 is indeed an optimized structure given the input model and associated data. It should be noted that the BEETS algorithm contains no symmetry constraints; the above bracket is symmetric purely because the model including boundary conditions is symmetric. The stress, strain, and displacement values confirm that the model has remained in the linear elastic region of the material curve. Figure 8.13 also confirms that the TVF value of 0.1 (or 10%) has been obtained.

8.3.2 BESO case study 1

A free (2D) downloadable version of the BESO algorithm is available from RMIT (2014), the authors of which are Huang and Xie; this software is

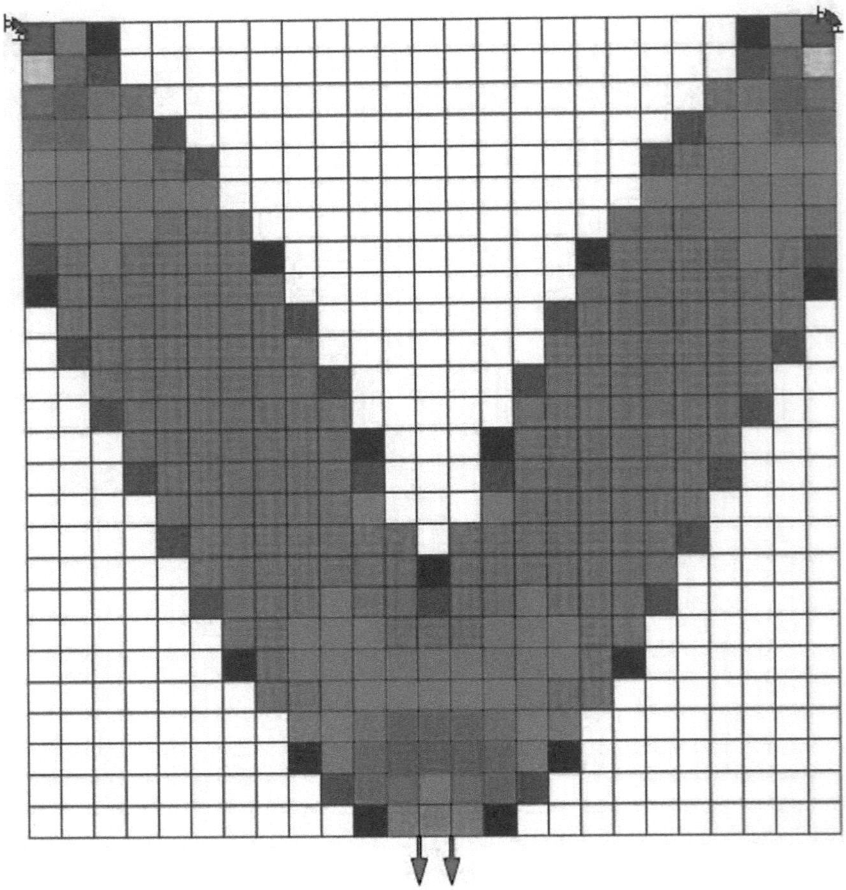

FIGURE 8.17 Optimized bracket from BESO software.

therefore a reasonable representation of the BESO algorithm capability and functionality, and can serve as a good means of comparison to the fundamental BESO algorithm of BEETS. Using the BESO software, it is possible to repeat the bracket optimization study presented above, using identical parameters and general model setup. The only aspect it is not possible to repeat is the filtering radius, as the BESO software does not allow for a filtering radius less than the element size. The results obtained from the BESO software can be seen in Figure 8.17.

Figure 8.17 reveals that the general shape of the bracket is identical between the BEETS and BESO algorithms; the major difference is most likely caused by the filtering radius, as explained above. Reviewing the output data from the BESO software, including Figure 8.17, it is clear that the algorithm has not met its target volume fraction objective of 10%. Nevertheless, there is seemingly a good correlation between BEETS and BESO in terms of the proposed bracket topology for this simple example.

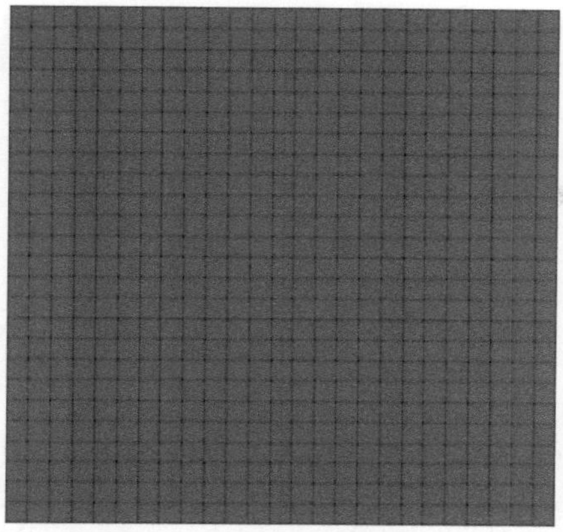

FIGURE 8.18 Optimized bracket using VDM-SIMP unconstrained optimization.

8.3.3 VDM-SIMP case study 1

In this subsection, the bracket of case study 1 will be optimized using the VDM-SIMP approach, using a linear static FE solver as presented in Chapter 4. All other attributes of the model are identical to those of the LS-DYNA model previously presented. The objective of the optimization was set to minimize volume with no constraints applied in accordance with the BEETS and BESO models above. The outcome from the VDM-SIMP optimization is illustrated in Figure 8.18.

As anticipated, the outcome of the VDM-SIMP optimization is not a feasible engineering design; the reason for this infeasible result is, as previously discussed, the fact that the VDM-SIMP optimization in general will not "function correctly" using unconstrained optimization, as is the case with BEETS and BESO as presented above. The VDM-SIMP optimization model was, therefore, repeated this time including a maximum compliance constraint (of 10 times the compliance of the initial model) imposed. The outcome of this optimization is illustrated in Figure 8.19.

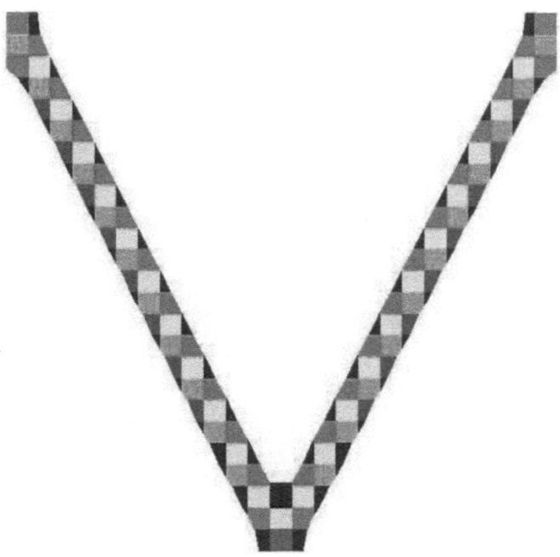

FIGURE 8.19 Optimized bracket using compliance constrained VDM-SIMP optimization.

As illustrated by Figure 8.19, it can be seen that the resulting bracket is indeed identical to the BEETS result, and very similar to that of BESO.

With the above in mind, it is clear that the bracket design problem above is a fairly simplistic optimization problem; a more advanced (linear static) case study will now be considered.

8.3.4 BEETS case study 2

For case study 2 a Warren bridge structure will be analyzed, as illustrated in Figure 8.20.

The Warren Bridge is fully constrained at point A and simply constrained in the Y-direction at point B in Figure 8.20. The model contains 3500 shell elements of unit thickness and is subjected to a single point load indicated by F in Figure 8.20. The model utilizes the material parameters listed in Table 8.4, as well as the general loading curve defined in Figure 8.11; the total load magnitude applied was 100 N in the negative Y-direction (Figure 8.20).

With the above parameters set, the BEETS software uses the "standard" BESO algorithm, as was the case for case study 1. The optimization of case study 2 completed after iteration 75 when Max_iter, Table 8.6, was reached. Figure 8.21 illustrates the total element strain as a function of the iteration number for case study 2.

Figure 8.21 indicates that a tendency to oscillate exists, as the total element strain value after iteration 12 varies between the two "extremes" of approximately 0.274 and 0.263 for all successive iterations. This tendency is initially confirmed by viewing the total volume per iteration from the case study (Figure 8.22).

Reviewing the MAIN_INFO.txt file from case study, it was confirmed that the target volume was indeed 2100, explaining the oscillation tendency. This thereby indicates that the optimization could have been terminated after approximately 12 iterations; however, this was not the case, as TERMTOL was set too low and thus not activated. Figure 8.23 illustrates the optimized structure after iteration 12 and the "final" structure after iteration 75.

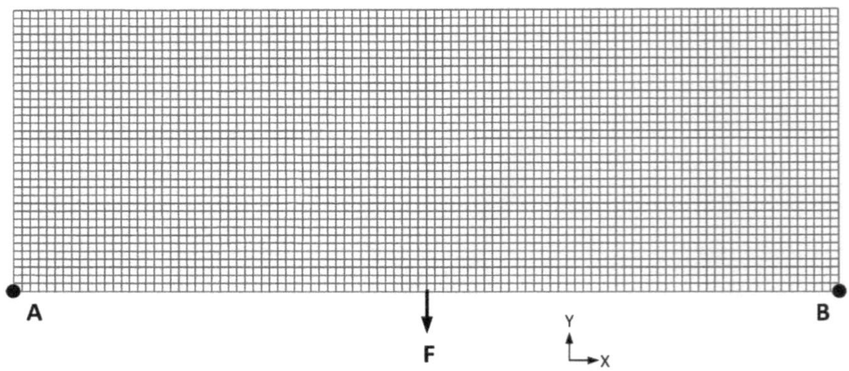

FIGURE 8.20 General FE model for case study 2 – Warren bridge.

TABLE 8.6 BEETS Input Parameters for Case Study 2 – Warren Bridge

PowerShell Variable	Value
Max_iter	75
TSI	76 (no Tabu Search included)
VTS	N/A
TSF	N/A
TABU_offset	N/A
TABU_LIST	N/A
TLF	N/A
ARMAX	0.03
RADIUS	0.0 (no neighborhood influence)
TVF	0.6
ER	0.04
TERMTOL	0.01
MEPUSE	No
COOLUSE	No
COOLIT	N/A
SFCOOL	N/A

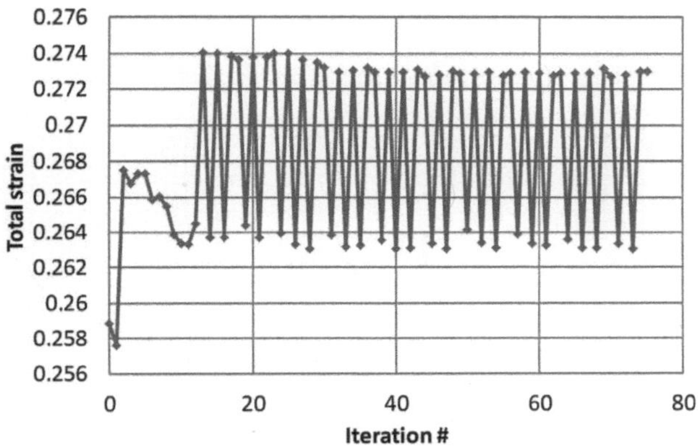

FIGURE 8.21 Sum of element strain per iteration case study 2.

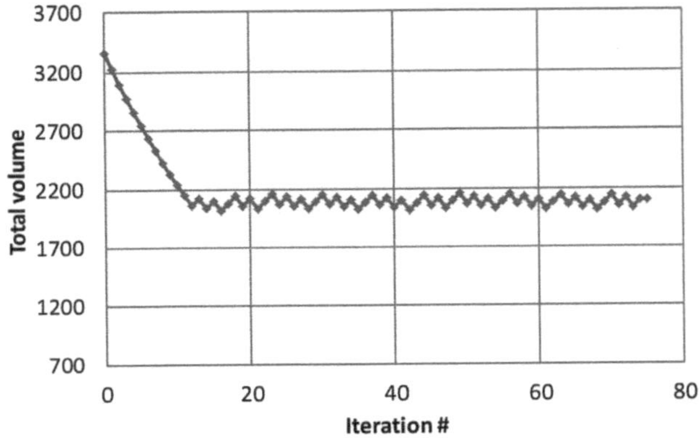

FIGURE 8.22 Sum of volume per iteration case study 2.

Figure 8.23 clearly illustrates that the difference between the two iterations is minimal. Due to the BEETS settings (Table 8.6), the algorithm used is, in essence, the BESO algorithm, as previously discussed. Figures 8.21–8.23 are thus good indicators of the previously discussed "oscillation tendency." One of the aims of the BEETS algorithm is, as previously defined, to reduce this tendency by using a cooling function and Tabu lists; however, neither were activated in this particular case. Despite the oscillation tendency, the results of the analysis seem logical given the boundary conditions, material parameters, and general optimization settings applied.

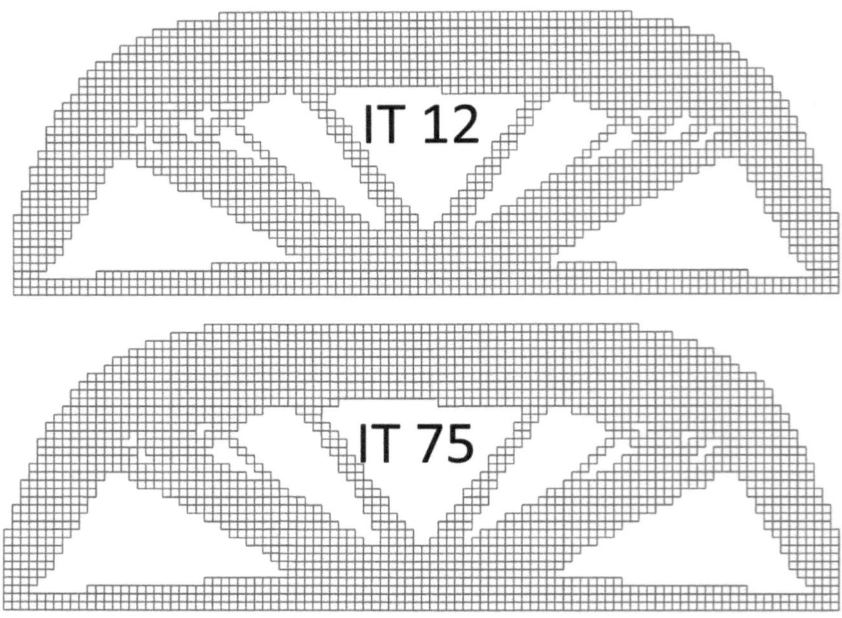

FIGURE 8.23 Case study 2 iteration 12 and iteration 75.

8.3.5 BESO case study 2

Using the BESO software (RMIT, 2014), the Warren bridge model was recreated and optimized using the relevant parameters from Table 8.6. Figure 8.24 illustrates the outcome from the BESO software.

Comparing Figures 8.24 to 8.23, some similarities do appear; there are, however, also some subtle differences, particularly around the "internal holes" in the structure. In an attempt to get a better overview of these differences, iteration 75 from Figure 8.23 has been overlaid onto Figure 8.24 in Figure 8.25.

The top image in Figure 8.25 illustrates Figure 8.23 on the top of Figure 8.24, whereas the bottom image in Figure 8.25 represents Figure 8.24 on the top of Figure 8.23. Areas A, B, and C in Figure 8.25 thus represent areas where the

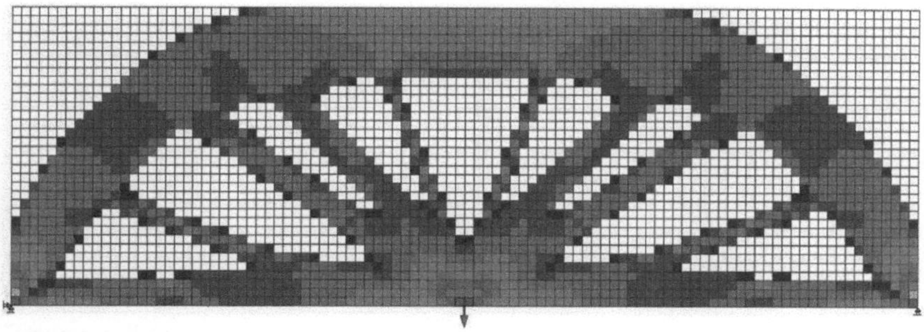

FIGURE 8.24 Optimized Warren bridge from BESO.

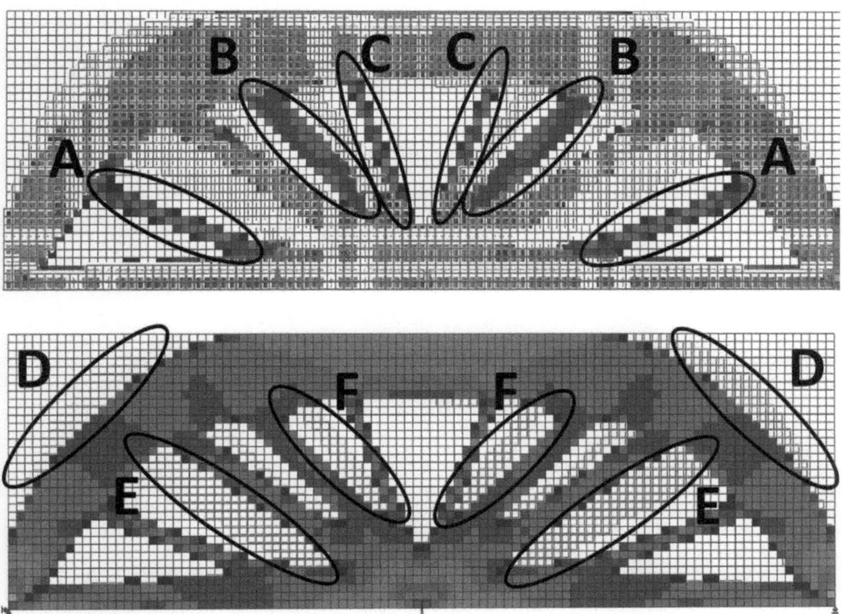

FIGURE 8.25 Overlay of BEETS and BESO results.

BESO software has retained elements while the BEETS software has not. Analog areas D, E, and F in Figure 8.25 indicate areas where the BEETS software has retained elements whereas the BESO software has not. The results thus indicate that the BEETS software has grown "wider" than the BESO software, areas D (Figure 8.25). Generally, it can also be noted that the "supports" of the BEETS results, i.e. E and F (Figure 8.25), lie in between those of the BESO results. In other words, Figure 8.25 reveals that E lies in between A and B; in addition, F lies in between B and C. It is not known exactly what has caused these differences; it may be due to the neighborhood influence, which as previously stated cannot be turned off in the BESO software. Other factors could relate to slight differences in the algorithm, the exact algorithm within the BESO software is not known; in addition, the BESO software does utilize linear static FEA. Hence, there may be a fundamental difference in the results guiding the optimization.

8.3.6 VDM-SIMP case study 2

The Warren bridge model was also solved using VDM-SIMP optimization. The model utilized the same optimization settings as used in case study 1, including a compliance constraint. The resulting topology from this optimization is illustrated in Figure 8.26.

The "external profile" of the optimized structure in Figure 8.26 is similar to those obtained using the BEETS and BESO software; the "internal supports" are, however, significantly different. The VDM-SIMP results utilize two main supports, whereas the BEETS results (Figure 8.23) indicate four supports, and finally the BESO software (Figure 8.24) indicates eight supports. The volumes of the three results are almost identical: 2100 for BEETS, 2096 for BESO, and 2070 for VDM-SIMP, despite the fact that the volume was not specified for the latter, as opposed to BEETS and BESO that aimed for the specific target volume of 2100. The above results thus highlight the potential differences between the three software/algorithms that may be profound for relatively simplistic problems such as the two illustrated above. In an attempt to determine whether or

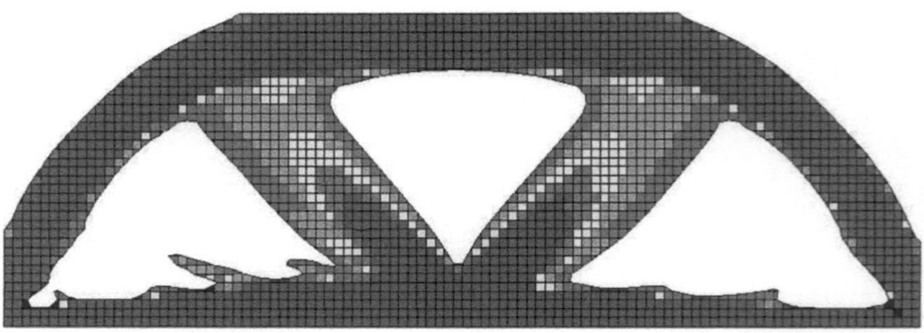

FIGURE 8.26 Optimized Warren bridge using compliance constrained VDM-SIMP optimization.

not the above differences may be affected by the different types of FEA, an additional case study is defined.

8.3.7 BEETS case study 3

The setup of case study 3 is, in essence, identical to case study 1, including material characteristics element size and the external load curve; the only difference is the number of boundary conditions and the specific locations where they are applied. These differences are illustrated in Figure 8.27.

As indicated in Figure 8.27, the model of case study 3 is exposed to compressive loading exerted by four single point loads. The loads all follow the load versus time curve defined in Figure 8.11 and each has a magnitude of 100 N. All DOF are constrained on all nodes along the A–A line in Figure 8.27. This optimization problem should have two distinctly different solutions when approached from a static and a dynamic FE viewpoint. The static solution should theoretically be a column, whereas the dynamic solution should resemble and upside down V. The BEETS specific optimization input parameters for case study 3 are defined in Table 8.8.

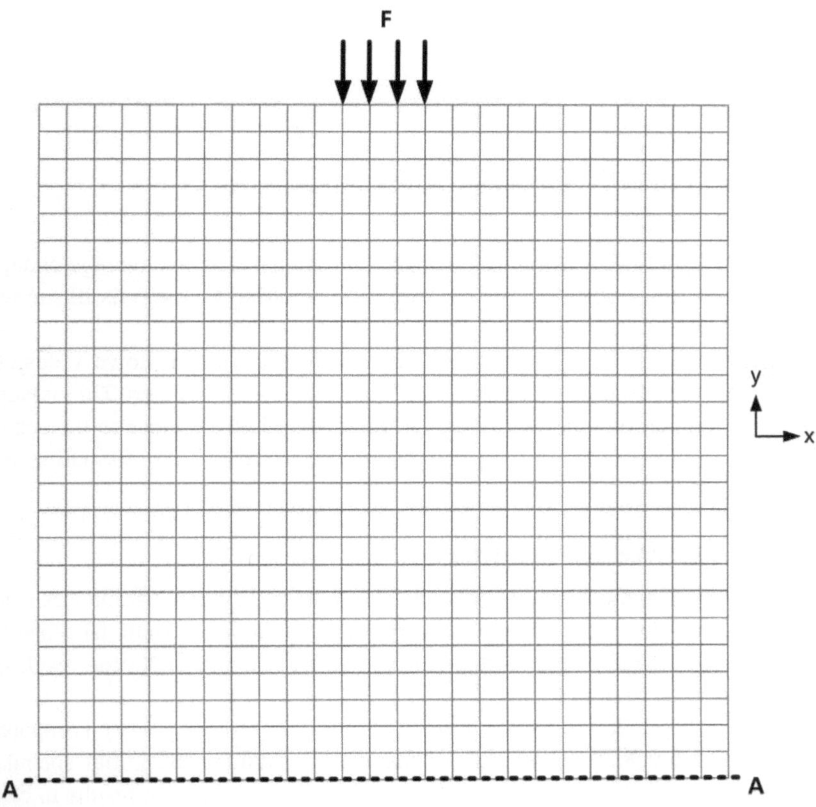

FIGURE 8.27 FE model for case study 3.

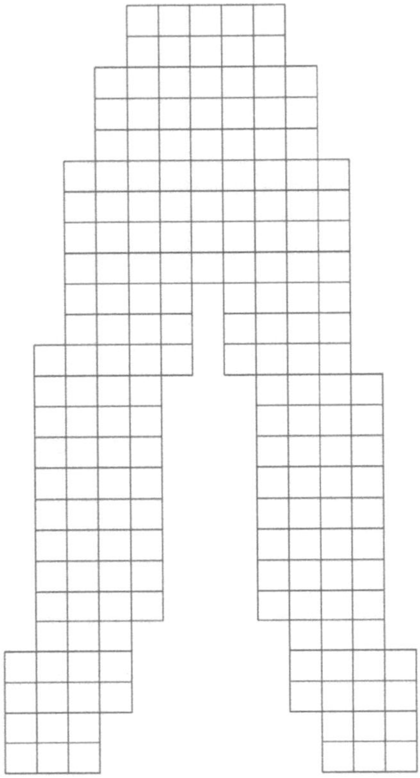

FIGURE 8.28 BEETS optimization result from case study 3.

The BEETS optimization terminated at iteration 13, as the average change over the final three iterations was 2.46%. The resulting topology is illustrated in Figure 8.28.

As anticipated, the resulting structure resembles and upside down V despite the fact that the loading is applied in a quasi-static-like manner. This is supported by the fact that the two "legs" of the V are relatively close to the center line of the structure indicating a relatively low/moderate rate of deformation and load application.

8.3.8 BESO case study 3

Using an identical model setup including relevant optimization parameters from Table 8.7, the above case study was analyzed using the linear static BESO software. The optimization converged and terminated at iteration 20; the resulting topology is illustrated in Figure 8.29.

Figure 8.29 clearly depicts a more "column-like" structure when compared to that of Figure 8.28. These results thereby substantiate the initial speculation that the difference between BESO and BEETS optimization results in case study 2, which may be influenced by the difference in FEA approach.

TABLE 8.7 BEETS Input Parameters for Case Study 3

PowerShell Variable	Value
Max_iter	25
TSI	26 (no Tabu Search included)
VTS	N/A
TSF	N/A
TABU_offset	N/A
TABU_LIST	N/A
TLF	N/A
ARMAX	0.03
RADIUS	0.0 (no neighborhood influence)
TVF	0.3
ER	0.1
TERMTOL	0.05
MEPUSE	No
COOLUSE	No
COOLIT	N/A
SFCOOL	N/A

8.3.9 VDM-SIMP case study 3

Adapting the above model setup into VDM-SIMP and including a compliance, constraint the optimization study was repeated. The VDM-SIMP optimization study converged at iteration 19; the resulting topology is illustrated in Figure 8.30.

Figure 8.30 thereby supports the statement that the "static" solution to the optimization problem of case study 3 is a column, whereas Figure 8.28 supports the statement that the "dynamic" solution is an upside down V. Case study 3 thereby supports the notion that the difference in results of case study 2 is influenced by the difference in FE solution methodology. The difference between the two "columns" of Figures 8.29 and 8.30 may be caused by the neighborhood influence included in the BESO software. For case study 3, it is also interesting to note that the BEETS algorithm terminated at iteration 13, whereas the BESO software converged at iteration 20 and VDM-SIMP at iteration 19.

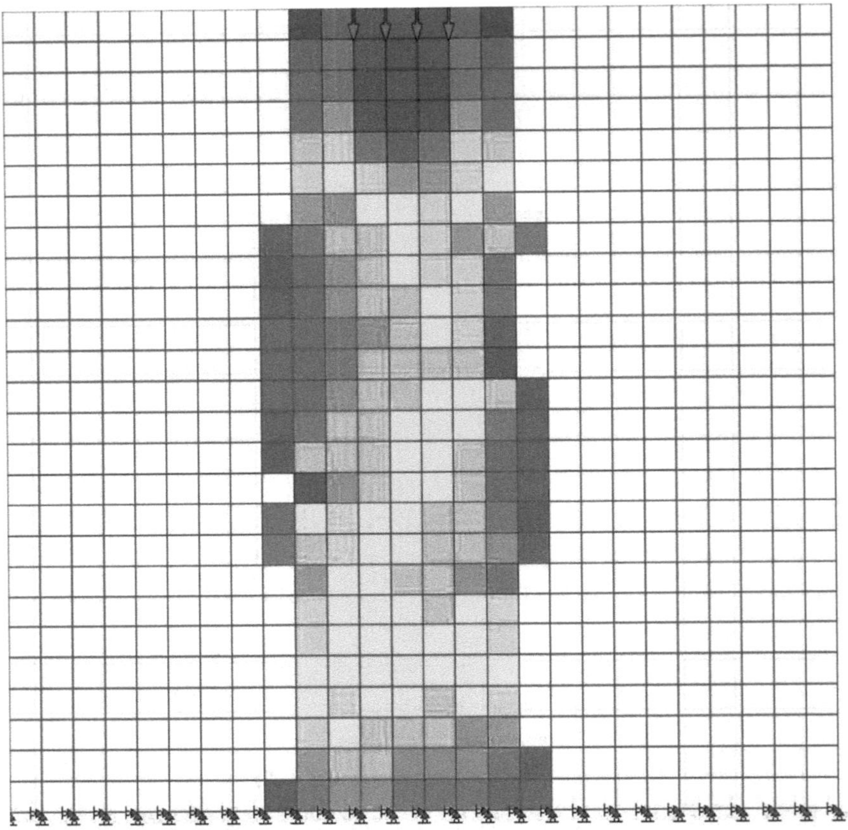

FIGURE 8.29 BESO optimization result from case study 3.

Based on the three case studies above, it can reasonably be argued that the "basic" results from the BEETS software are sensible, thus initially verifying that the software provides "correct" and "relevant" results. From this point onward, the focus of attention will now be turned to the overall area of interest: nonlinear topology optimization.

8.4 NONLINEAR TOPOLOGY OPTIMIZATION CASE STUDIES

This section will cover a series of case studies aimed at determining the suitability of the BEETS algorithm for nonlinear topology optimization. The initial case studies will, in essence, consist of utilizing the "basic" BESO algorithm of the BEETS software. The starting point will contain material as well as geometrical nonlinearity. Due to various factors, including current limitations of the programming of the BEETS software in addition to considerations in relation to stability of the FEA, the fundamental setup of the FE model has been adapted. Instead of applying the external load directly onto individual nodes, the external

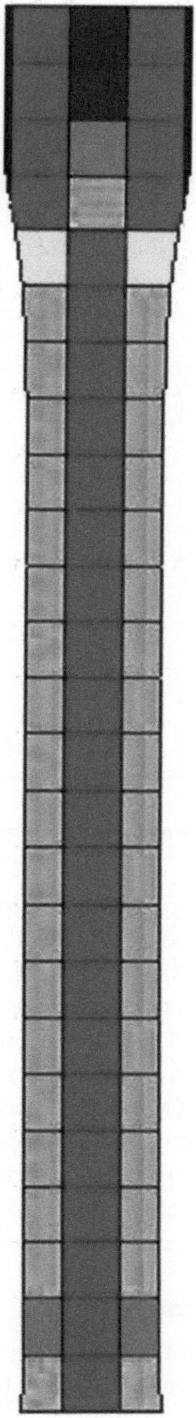

FIGURE 8.30 VDM-SIMP optimization result from case study 3.

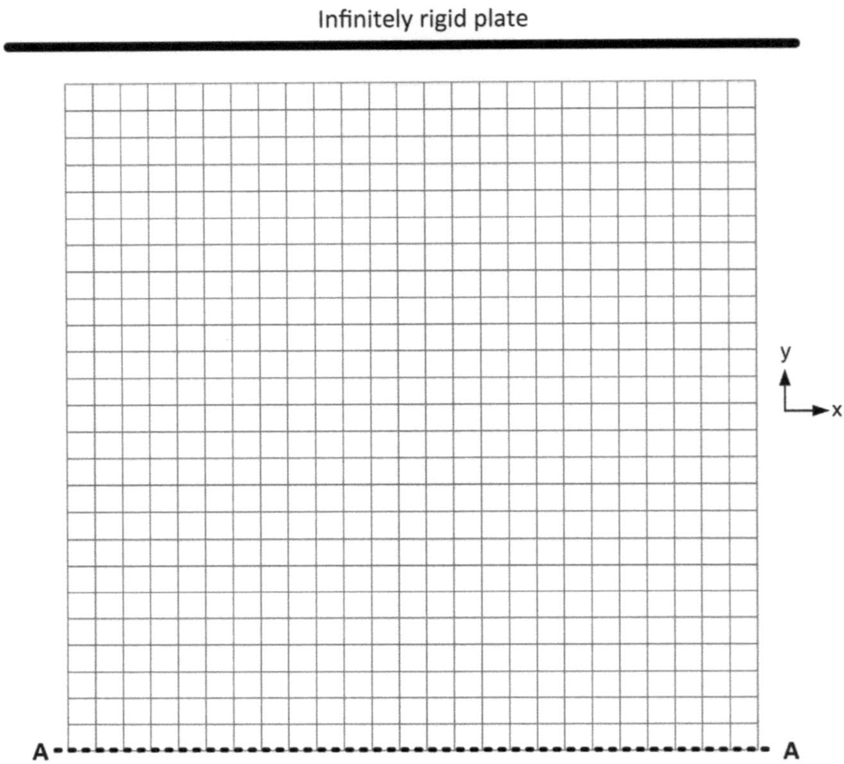

FIGURE 8.31 Fundamental FE model for nonlinear case studies.

load will now be applied by defining an infinitely rigid plate moving at a constant velocity. The general setup of the "revised model" is illustrated in Figure 8.31.

The infinitely rigid plate (IRP) in Figure 8.31 will move in the negative y-direction at a predefined constant velocity. The model will be fully constrained at all nodes along the A–A line. The IRP will have a predefined mass; hence, the external load is a function of the mass and the velocity of the IRP.

The material data used for the nonlinear case studies will be based on data from actual physical test of Boron steel (Bastien, 2014). The LS-Dyna material model used will be *MAT_024:Piecewise_linear_plasticity. This is an isotropic linear-elastic piecewise plastic material model with strain rate sensitivity. The general material parameters apart from the actual test curves are listed in Table 8.8.

8.4.1 BEETS case study 4

For this case study, the mass of the IRP will be 100 kg, and the initial velocity in the negative y-direction (Figure 8.31) will be 5 m/s, or 18 km/h, which in comparison to a vehicle crash scenario is a relatively modest impact velocity. The BEETS optimization parameters for this particular case study will remain unchanged to those from case study 3, that is, as specified in Table 8.7.

TABLE 8.8 Material Properties for Nonlinear Case Studies

Parameter	Value	Unit
Young's modulus	210,000	MPa
Poisson's ratio	0.3	
Yield stress	374.9	MPa

A better understanding of the fundamental differences between case study 4 and the preceding linear optimization case studies can be obtained by reviewing Figure 8.32 that represents the final deformed shape of the structure at iteration 0 before any optimization has occurred.

Figure 8.32 clearly illustrates a significant level of deformation and can therefore be argued to represent a structure exposed to large deformations. The maximum displacement resultant in the final deformation shape of Figure 8.32

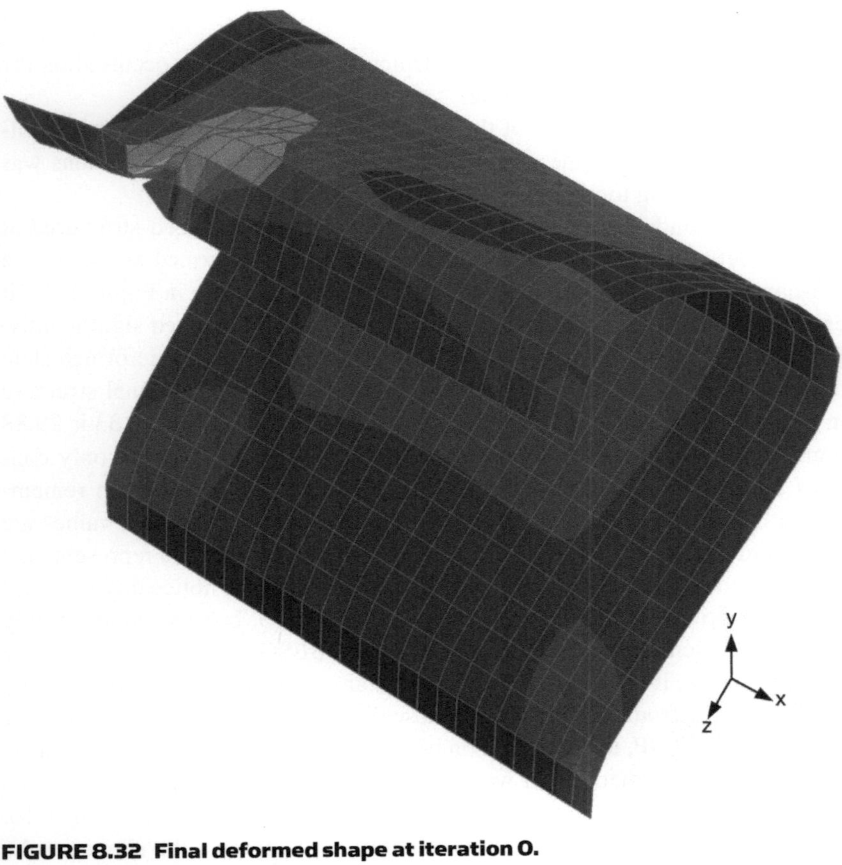

FIGURE 8.32 Final deformed shape at iteration 0.

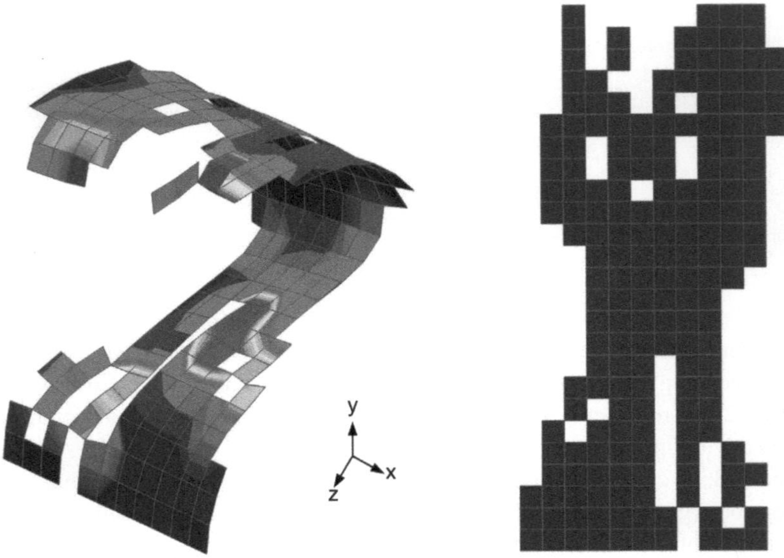

FIGURE 8.33 BEETS optimization results case study 4.

is 29.50 mm. Figure 8.32 also clearly depicts that a failure has occurred on the "top left-hand side" of the structure.

The BEETS optimization of the updated FE model presented above terminated at iteration 13, as the average change over the final three iterations was 3.3%. The resulting topology is illustrated in Figure 8.33.

The left-hand side of Figure 8.33 illustrates the optimized structured at its final deformed shape. When comparing the final deformed shape on the left-hand side of Figure 8.33 to that of the original structure in Figure 8.32, it can be seen that the overall deformation shape has not changed significantly. Furthermore, the maximum resultant displacement at the final deformed state has only changed significantly as well. The value from the original structure in Figure 8.32 is 29.5 mm; the corresponding value from Figure 8.33 is 29.88 mm. At this point, it should be noted that the optimized structure only contains 30% of the volume of the original structure; it is also worth remembering that no stipulations with respect to maximum displacement values are applied to the optimization. The right-hand side of Figure 8.33 represents the undeformed optimized structure, which unsurprisingly is noticeably different from the optimized BEETS structure of case study 3, but does interestingly bare some resemblance to the results from the BESO software of case study 3. At this stage, it does, however, not make any sense to directly compare the BEETS results from case study 4 to those obtained via the BESO software or indeed VDM-SIMP, the main reason being the significant levels of nonlinearity and large deformations. It was also decided not to conduct any comparison studies to the ESLM, as these will be further investigated in Chapter 10. The only relevant remaining option is therefore to compare the above results

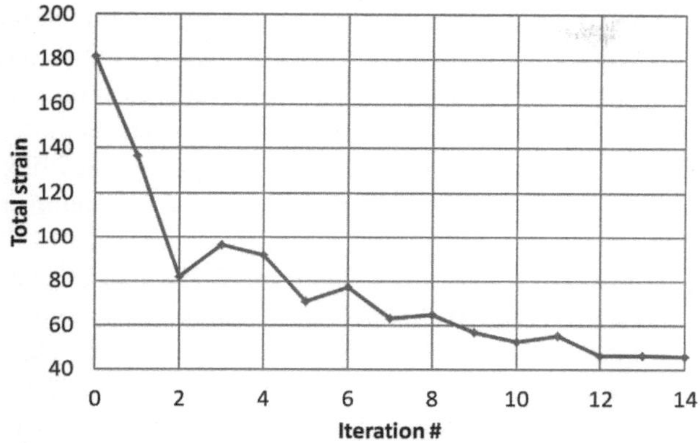

FIGURE 8.34 Sum of element strain per iteration case study 4.

to the HCA algorithm of LS-TaSc initially presented in Section 5.2; the purpose of these case studies is, however, to evaluate the change of results from the fundamental BESO algorithm, which resides inside the BEETS software. Once these are understood the results from the BEETS algorithm could be compared and evaluated against those of LS-TaSc.

Figure 8.35 and the volume history (Figure 8.35) represent the strain energy history from case study 4.

Figure 8.34 clearly indicates that the sum of element strain has substantially decreased from iteration 0 to iteration 14 as a significant decrease has occurred in the total strain contained within the structure, despite the fact that the structure is still withstanding an identical amount of external loading. The information in Figure 8.34 coupled with the marginal increase in maximum displacement is a positive indicator of an optimized structure, given the fact that the volume, or mass, of the structure has also decreased by 70%.

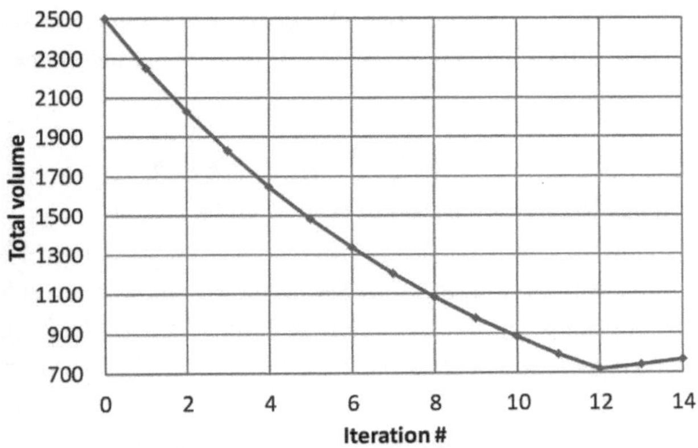

FIGURE 8.35 Sum of volume per iteration case study 4.

It should be noted that given the "initial" (symmetric) setup of the FE model illustrated in Figure 8.31 it is highly unlikely that the optimized topology is asymmetric as indicated in Figure 8.33. However, based on the initial deformed structure in Figure 8.32, which in essence provides the initial information to the BEETS optimizer the results of Figure 8.33 are credible; assuming the FEA results are credible – as will be the case for the remaining case studies of this chapter.

The volume history of case study 4 (Figure 8.35) also indicates that the optimization has converged. At this point, it is however worth remembering that the evolutionary ratio (ER) dictates a consistent rate of change in elements per iteration. This, coupled with the maximum allowed admissions ratio (ARMAX) and the termination tolerance (TERMTOL), dictates when the optimization will stop, provided the cooling function is not activated. This means that it can generally be accurately predicted at which iteration the optimization will terminate. It could be argued that a good rule of thumb is that TERMTOL should be less than both the ER and the ARMAX values; it naturally follows that ER must be larger than TERMTOL. This can be mathematically formulated as Expression (8.20)

$$TERMTOL < ARMAX < ER \tag{8.20}$$

Expression (8.20) will be used as a rule of thumb for all subsequent BEETS analyses.

The focus will now shift entirely onto the BEETS algorithm; the remaining case studies will primarily focus on five aspects of the BEETS algorithm/software:

1. Effects of ER and ARMAX magnitudes
2. Cooling function influence
3. Neighborhood influence
4. Tabu search influence
5. Effects of entropy

All the five points above will be the individual focus of attention in the subsequent sections; for all cases, the fundamental FE model of Figure 8.31 and the material parameters in Table 8.8 will be utilized unless otherwise specified. This also means that BEETS case study 4 will be used as the "benchmark" for the subsequent studies.

8.4.2 BEETS case study 5: effects of ER and ARMAX magnitudes

The aim of this case study is to initially investigate the effects of ER and ARMAX magnitudes. Expression (8.20) of Section 8.4.1 suggested a rule of thumb for the relationship between the ER, ARMAX, and TERMTOL values. Using this rule of thumb, is it possible to increase the ER and ARMAX magnitudes to reduce the overall computation time of the optimization problem? In order to get an overview of this, the optimization model from BEETS case study 4 is repeated using the input parameters listed in Table 8.9.

TABLE 8.9 BEETS Input Parameters for Case Study 5

PowerShell Variable	Value
Max_iter	25
TSI	26 (no Tabu Search included)
VTS	N/A
TSF	N/A
TABU_offset	N/A
TABU_LIST	N/A
TLF	N/A
ARMAX	0.1
RADIUS	0.0 (no neighborhood influence)
TVF	0.3
ER	0.35
TERMTOL	0.05
MEPUSE	No
COOLUSE	No
COOLIT	N/A
SFCOOL	N/A

As defined by Table 8.9, the only changes between BEETS case study 4 and case study 5 are, therefore, a significant increase in ER and ARMAX values. This should mean that the optimization will be able to reach its target volume fraction (TVF) of 0.3 in approximately three iterations; it is subsequently possible to significantly increase the number of elements, as ARMAX has been increased to 0.1.

The optimization study concluded due to the maximum number of 25 iterations being reached. At this point, it is therefore very clear that the desired effect of reduced overall CPU time was not obtained. Reviewing the change of the structure during the initial three iterations provides a very direct and unsurprising clue as to why this is (Figure 8.36).

As indicated in Figure 8.36, the optimization algorithm has simply become "too aggressive" as too much material or elements are removed in the initial iterations, causing the (optimization) model to become unstable. The explicit FE model does, however, remain stable allowing the analysis and optimization to continue. Figure 8.37 illustrates the final structure obtained from case study 5. This bears some parallels to the ESLM case studies of Chapter 10, where not

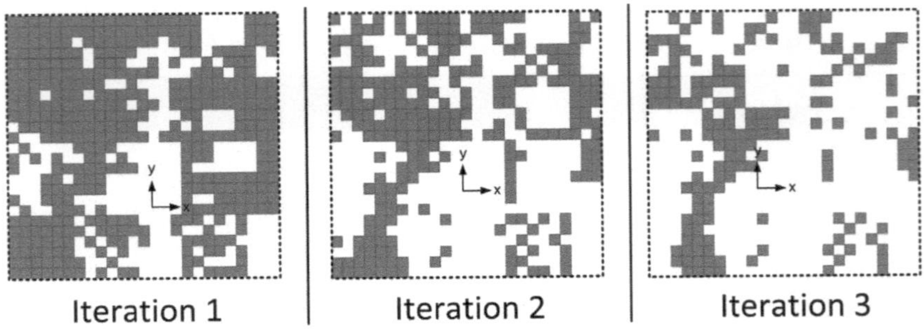

FIGURE 8.36 First three iterations BEETS case study 5.

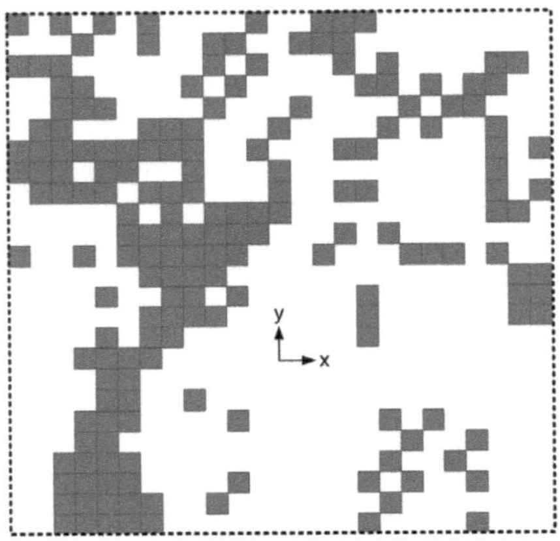

FIGURE 8.37 Final structure from BEETS case study 5.

only the optimization but also the FE analysis became unstable, causing the optimization to error terminate.

Figure 8.38 illustrates the volume history for BEETS case study 5. Observing the trends in the figure, it can be seen that the volume goes below the target value of 750 (corresponding to a TVF value of 0.3) after approximately three iterations.

From iteration 3 to iteration 4 the volume increases by 9.5%, being very close to the maximum allowable value, meaning that the volume exceeds the target value of 750. The optimization, therefore, reduces the value by 35% according to the ER value, meaning that the volume is significantly less than the target value at iteration 5. From iteration 5 to iteration 10, the volume steadily increases by 10% for each iteration, as dictated by the ARMAX value. At iteration 10, the volume has once again exceeded the target volume and is decreased by 35% according to the ER value. From this point onward, this cyclic increase and decrease is repeated until the maximum number of iterations is reached, as shown in Figure 8.38. Part of this trend identified in Figure 8.38 is an inherent flaw within the BESO algorithm, simply because the final target volume is not necessarily specified as a target volume for the individual iteration. This will naturally lead to the cyclic tendencies seen in Figure 8.38. Although case

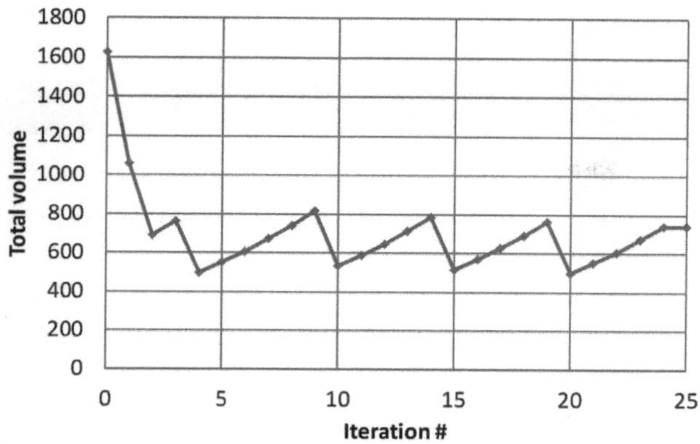

FIGURE 8.38 Sum of volume per iteration case study 5.

study 5 utilizes "extreme" ER and ARMAX values, it nevertheless highlights a flaw that may be solved by the introduction of a cooling function which will be the focus of attention in the following section.

8.4.3 BEETS case study 6: cooling function influence

The introduction of a cooling function may help to incrementally decrease or increase the "aggressiveness" of the optimization algorithm to better control and guide the optimization toward the target volume, thus reducing or completely eliminating the cyclic behavior illustrated in Figure 8.38. BEETS case study 6 is therefore a "replica" of case study 5 with the cooling function activated. The BEETS input parameters for case study 6 are listed in Table 8.10.

As defined in Table 8.10, the cooling function has been activated and will become active within an interval of ±10% of the final target volume. Within this range the ER value will be updated according to the SFCOOL value of 0.5. Using the cooling function, the optimization terminated at iteration 6, as the average change over the last three iterations was 3.04%. The resulting topology remains unfeasible, as it is very similar to the topologies of Figures 8.36 and 8.37. Figure 8.39 represents the volume as a function of iteration for case study 6.

As indicated in Figure 8.39, the introduction of the cooling function has completely removed the cyclic behavior of the BESO algorithm indicated in Figure 8.38. The particular cooling parameters COOLIT and SFCOOL do, of course, warrant further analysis and must be carefully chosen for each individual analysis; at this point the investigation into the cooling function will, however, be temporarily abandoned to focus upon the next adaptation of BESO in the BEETS software, namely the neighborhood influence.

8.4.4 BEETS case study 7: neighborhood influence

The purpose of this section is to initially investigate the influence of the neighboring elements. For this particular purpose the BEETS input parameters for case study 7 are listed in Table 8.11.

TABLE 8.10 BEETS Input Parameters for Case Study 6

PowerShell Variable	Value
Max_iter	25
TSI	26 (no Tabu Search included)
VTS	N/A
TSF	N/A
TABU_offset	N/A
TABU_LIST	N/A
TLF	N/A
ARMAX	0.1
RADIUS	0.0 (no neighborhood influence)
TVF	0.3
ER	0.35
TERMTOL	0.05
MEPUSE	No
COOLUSE	Yes
COOLIT	0.1
SFCOOL	0.5

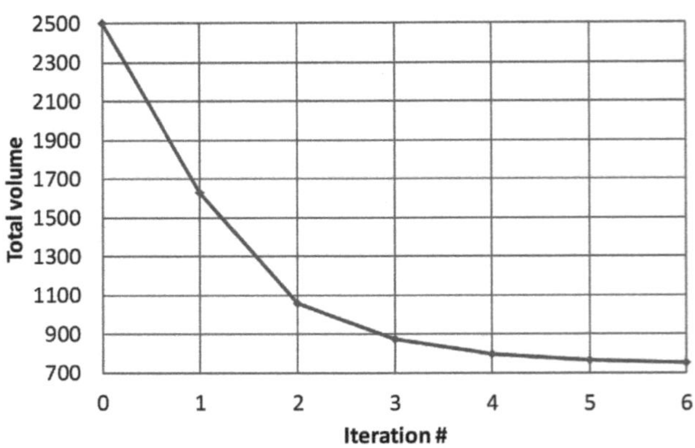

FIGURE 8.39 Sum of volume per iteration case study 6.

TABLE 8.11 BEETS Input Parameters for Case Study 7

PowerShell Variable	Value
Max_iter	25
TSI	26 (no Tabu Search included)
VTS	N/A
TSF	N/A
TABU_offset	N/A
TABU_LIST	N/A
TLF	N/A
ARMAX	0.05
RADIUS	0 (M1); 2.1 (M2); 4.1(M3); 6.1(M4)
TVF	0.3
ER	0.1
TERMTOL	0.03
MEPUSE	No
COOLUSE	No
COOLIT	N/A
SFCOOL	N/A

The only change between BEETS case study 4 and 7 is, thereby, the RADIUS parameter. As indicated in Table 8.11, case study 7 will actually contain four models: M1, M2, M3, and M4. M1 will not contain any neighborhood influence but has been included, as the ARMAX and TERMTOL parameters are not identical to those of BEETS case study 4. The average element size of the model is 2 mm, which means that the distance between two adjacent element centers is 2 mm. M2 should, therefore, contain the immediately adjacent elements only, whereas M3 and M4 will contain an increasing amount of neighborhood elements/influence, as illustrated in Figure 8.3.

All four models of case study 7 terminated due to convergence prior to the maximum number of iterations. M1, M2, and M3 all terminated at iteration 24 with an average percentage difference over the final three iterations of 2.51%. M4 terminated at iteration 17 with an average volume change over the final three iterations of 2.93%. Figure 8.40 contains the total volume per iteration for all four models.

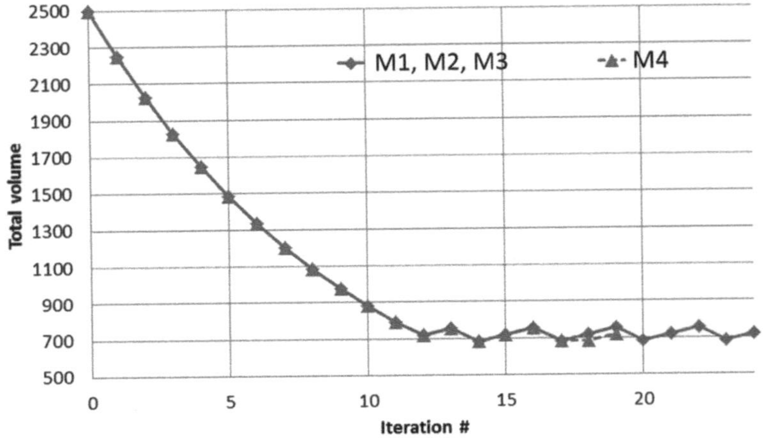

FIGURE 8.40 Sum of volume per iteration case study 7.

As expected, Figure 8.40 reveals that all four models follow the same pattern for volume reduction because this is dictated by the ER and ARMAX parameters, remembering that no cooling function is activated in case study 7. Figure 8.40 also reveals a slight change between M1–M3 and M4 from iteration 17 onward, causing the optimization of M4 to terminate premature to the other three models. This is an interesting and rather unexpected fact, which indicates that the neighborhood function may also have an influence on the termination and indeed help reduce the cyclic tendencies that may occur if the cooling function is not activated as discussed in the previous sections. Figure 8.40 thereby also indicates "locking" of element sensitivity numbers, demonstrating that difficulties with "singling out" elements have been introduced, preventing the optimization from reaching the iteration specific target volume. The existence of this tendency may be further substantiated by observing the final optimized structures as depicted in Figure 8.41.

The dotted boxes in Figure 8.41 denote the design volume or the outline of the "original" FE model. Figure 8.41 thereby reveals that the load paths of the optimized structure are situated toward the left-hand side of the design volume. This may initially seem odd as the model setup, including constraints and applied loading, is completely symmetrical. The reason for this fact is, however, very simply that a single element fails (due to a very high strain level) on the left-hand side of the model during the initial FE analysis at iteration 0, which can also be seen in Figure 8.32.

Based on the results of all four models as illustrated in Figure 8.41, some interesting initial conclusions may be drawn:

1. The inclusion of the neighboring elements undoubtedly has a significant effect upon the resulting topology. This is immediately obvious by comparing M1 to M2. There is also a significant difference between M2–M3 and M2–M4. Although there is detailed difference between M3–M4, there is also a general similarity in the location of the "primary load path." This is not the case when comparing M2–M3 and M2–M4.

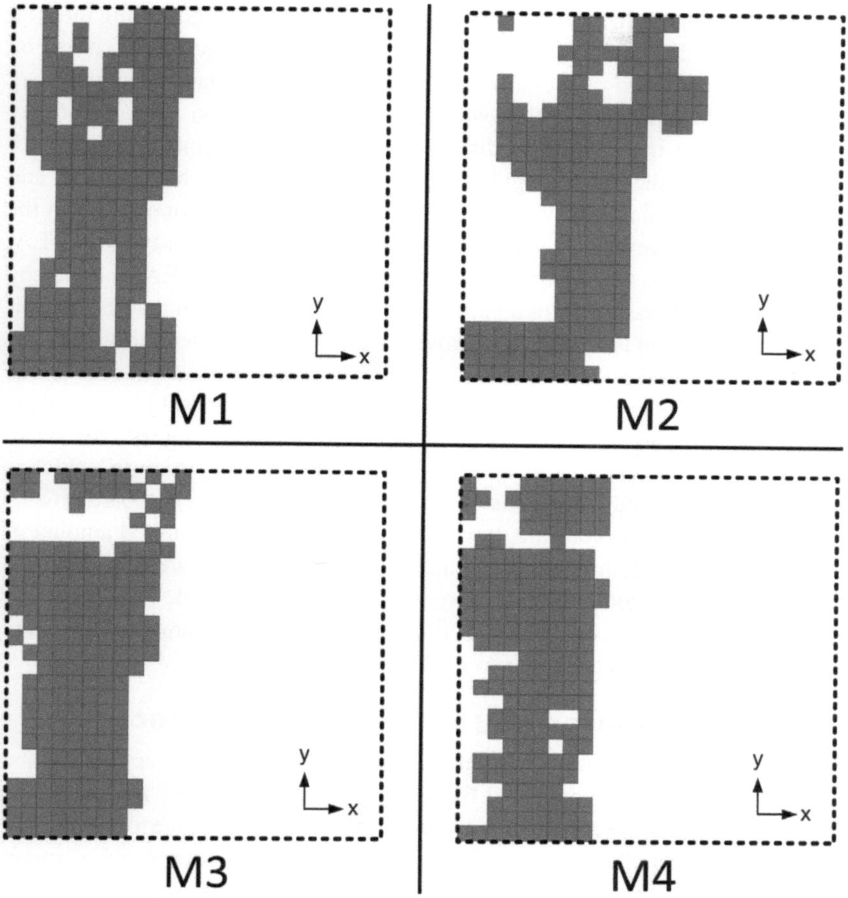

FIGURE 8.41 Optimized shapes case study 7.

2. Including only the immediately adjacent neighboring elements seemingly creates a structure prone to "bending." This statement is purely based on intuition; it is, however, clear that there is not a "straight" load path between the points of applied loading and the constraints (Figure 8.31). In fact, by observing the results from M2 in Figure 8.41, it can be argued that a significant moment is introduced based on this "misalignment." This conclusion is nevertheless not straightforward as this may be obvious in the "undeformed" structures (Figure 8.41); significant deformation and "self-contact" of the structure during deformation is, however, not easily taken into account.
3. Increasing the neighborhood influence via the RADIUS parameter seemingly leads to "instable structures" and "isolated elements." As previously mentioned, the introduction of the neighborhood influence may prevent the optimization from "singling out elements" for deletion. This is, of course, one of the main reasons for introducing the neighborhood influence in the first place so that the checkerboard effect may be reduced, making the

neighborhood influence a positive attribute. However, by observing the "top left-hand side" of the results of M2, M3, and M4 in Figure 8.41, it can be seen that "isolated element(s)" not connected to the "main structure" may also start to appear. In other words, there are element(s) that are directly connected to the applied loading but not to the constraints. This is more prominent in the "high neighborhood influence" models, M3 and M4, and is of course not a positive attribute that can be partially contributed to the neighborhood influence. These elements may lead to instability of the FEA; however, due to the "inherent stability" of explicit FE, the FEA, and thereby the optimization, is able to continue. It does, however, seem that the optimization algorithm is not able to remove these elements, as their individual element sensitivity number will remain relatively high. Furthermore, it can be seen that the high neighborhood influence models M3 and M4 have an inherent instability as they have only a single element connecting the "top structure" where the external loading is applied to the "lower structure" which is directly connected to the constraints, in other words a checkerboard effect.

The above observations are, however, based upon a very limited amount of information, and additional models and scenarios must be carefully explored before any general conclusions and trends can be substantiated. The above discussions are, therefore, not meant as "final conclusions" but merely as indicators for additional analysis and areas of exploration for future work.

8.4.5 BEETS case study 8: Tabu search influence

The purpose of this section is to initially explore the effects of applying Tabu Search (TS) onto the fundamental BESO algorithm. The primary reason behind the application of TS is an attempt to avoid premature convergence toward a local optimum point; in other words, to expand the search within the solution space. For BEETS case study 8, the optimization parameters are defined in Table 8.12.

The optimization of case study 8 terminated after iteration 25 due to the maximum allowable number of iterations being reached. Figure 8.42 represents the volume as a function of iteration number for case study 8.

As indicated in Figure 8.42 case study 8, also exhibited the cyclic behavior of the fundamental BESO algorithm in the absence of a cooling function. Figure 8.42 also indicates that the introduction of TS has not had any effect upon the change in volume per iteration. This is, however, not unexpected, as the target volume for TS runs within a given iteration number is identical to that of the "normal" iteration. In other words, the introduction of TS will not lead to any drastic changes of volume; instead, it may lead to a different solution/base structure for subsequent iterations. If the base structure has changed as a function of TS, the information will be stored in the Main_INFO.txt file output from the BEETS software. Reviewing this file for case study 8, it seems that in this case study no change in base structure took place as a function of introducing TS. There may be several reasons for this, including but not limited to:

1. The optimum structure for the individual iterations is in fact the structure obtained using the "fundamental BESO" algorithm. This could be further

TABLE 8.12 BEETS Input Parameters for Case Study 8.

PowerShell Variable	Value
Max_iter	25
TSI	3
VTS	5
TSF	0.2
TABU_offset	5
TABU_LIST	Y
TLF	0.5
ARMAX	0.05
RADIUS	0.0 (no neighborhood influence)
TVF	0.3
ER	0.1
TERMTOL	0.03
MEPUSE	No
COOLUSE	No
COOLIT	N/A
SFCOOL	N/A

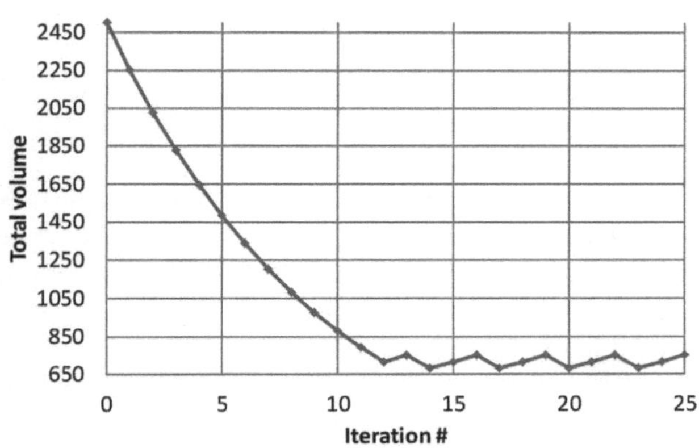

FIGURE 8.42 Sum of volume per iteration case study 8.

explored by exploring other FE models, preferably where "known" optimum solutions are available.
2. The design volume is "too coarse" for the effects of TS to become noteworthy. This could be further explored by increasing the element count.
3. The evaluation expression for "measuring" the effectiveness of TS is incorrect and/or too simplistic. At present, this is simply measured by dividing the sum of strain (for the entire model) by the volume of that model.
4. The constraint of utilizing an identical volume for the TS may be too restrictive. Allowing the individual TS iterations to reduce the volume simultaneously with "randomly" removing/adding elements may enhance the influence of the TS iterations.

As was the instance with case study 7, the above statements are based on a very limited amount of information and cannot be justified to be generally applicable. The statements should merely be considered as pointers for additional research, analysis, and scrutiny.

8.4.6 BEETS case study 9: effects of MEP

The purpose of this section is to initially explore the effects of introducing the maximum entropy principle (MEP) into the fundamental BESO algorithm. This case study will contain a total of four models. M1 and M2, which will be the first areas of attention, will simply include the FE model originally introduced at the start of Section 8.4, as illustrated in Figure 8.31 and utilizing the material parameters in Table 8.8, with the entropy parameter deactivated (M1) and activated (M2). The input parameters for the BEETS optimizations of case study 9 are defined in Table 8.13.

M1 and M2 of BEETS case study 9 terminated at iteration 25 due to the convergence criteria being met, as the average change of volume over the final three iterations was 2.51%. Figure 8.43 represents the volume as a function of iteration number for case study 9 M1 and M2.

Comparing the two graphs in Figure 8.43 to each other, it is clear that there is no difference at all between the graphs indicating that the introduction of MEP has had no effect upon the volume as a function of iteration; this is, however, as expected.

Figure 8.44 represents the total element strain per iteration, which indicates a significant difference between the two graphs from M1 and M2 suggesting a significant influence of introducing MEP.

Figure 8.44 reveals that the model without MEP (M1) generally detects a lower total strain value as opposed to M2 (with MEP) for the first three iterations. Thereafter, the total strain of M2 is generally lower than M1, and consistently lower from iteration 15 onward. Due to the methods in which MEP has been implemented in the BEETS software, as described in Section 8.2, it was anticipated that the MEP model would indicate higher total strain values (due to the method of summation); the fact that the MEP model consistently displays lower total strain values from iteration 15 onward is a strong indicator that this model (M2) is a more "optimized" model than M1, as the volume of the two

TABLE 8.13 BEETS Input Parameters for Case Study 9.

PowerShell Variable	Value
Max_iter	25
TSI	26 (no Tabu Search included)
VTS	N/A
TSF	N/A
TABU_offset	N/A
TABU_LIST	N/A
TLF	N/A
ARMAX	0.05
RADIUS	0.0 (no neighborhood influence)
TVF	0.3
ER	0.1
TERMTOL	0.03
MEPUSE	No (M1, M3)/Yes (M2, M4)
COOLUSE	No
COOLIT	N/A
SFCOOL	N/A

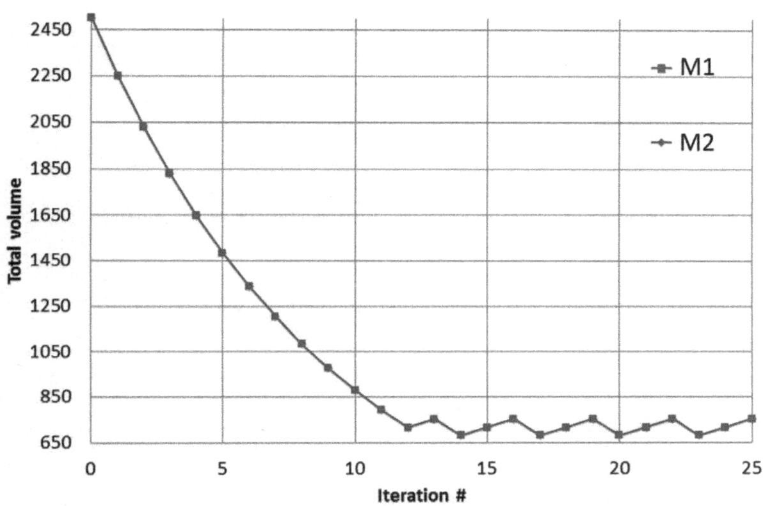

FIGURE 8.43 Sum of volume per iteration case study 9 M1 and M2.

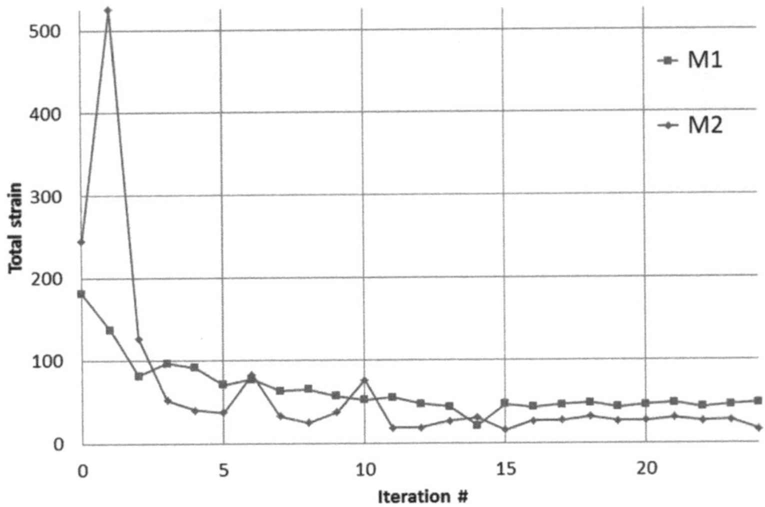

FIGURE 8.44 Sum of element strain per iteration case study 9 M1 and M2.

remains identical, as evidenced by Figure 8.44. The identical volume is confirmed when reviewing the resulting FE models from M1 and M2, which both contain 188 elements after the optimization. The resulting FE models/topology from the two models are illustrated in Figure 8.45.

Based on Figure 8.45, there can be no doubt that the introduction of MEP has a significant effect upon the results of the optimization. On one hand, the results from M2 demonstrate a very distinct load path on the far left of the design volume but on the other hand also seems to "suffer" from checkerboard effect. The effects of introduction MEP are therefore very interesting and most certainly warrant further analysis. In order to further analyze this, an additional two models are introduced: M3 and M4. The fundamental setup of these two models

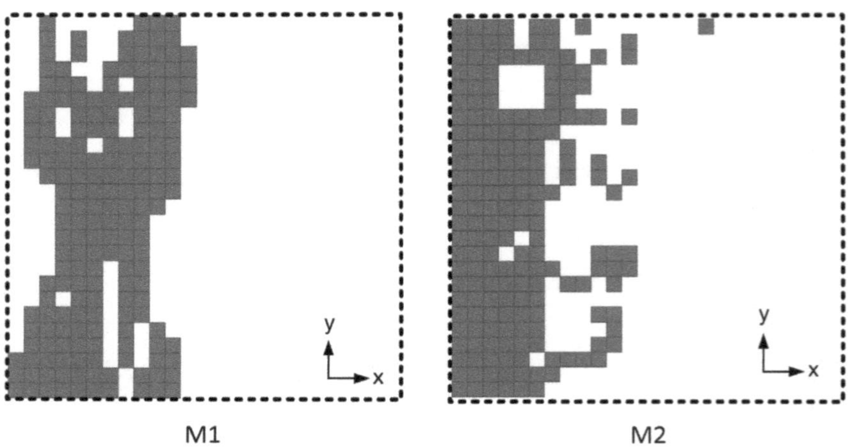

FIGURE 8.45 Resulting topologies case study 9 M1 and M2.

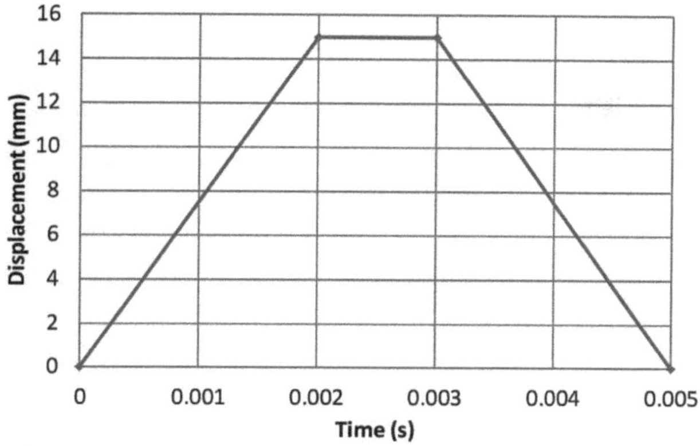

FIGURE 8.46 Load vs. time M3 and M4 case study 9.

is identical to those of M1 and M2; the method in which the loading is applied upon the infinitely rigid plate is, however, different. Observing a "typical" automotive crash scenario, it is clear that the loading does not necessarily remain consistent throughout the impact; in fact, there is generally a point where the two impacting vehicles move away from each other, which is generally referred to as "rebound." In the context of an FE-based crash analysis this means that the maximum deflection does not necessarily occur at the final timestep, which is very likely to have an effect upon optimization using BESO. But what are the effects using BEETS? In order to initially investigate this, the rigid plates of M3 and M4 will be subjected to a prescribed motion, not an initial velocity following the values listed in Figure 8.46.

In Figure 8.46, the maximum displacement value of 15 mm has been found by observing the maximum displacement value of the rigid plate from case study 4. In case study 4, this value was, however, obtained at 0.005 s; as defined in Figure 8.46, the maximum displacement will initially be obtained at 0.002 s, indicating a significant increase in strain rate. The two models M3 and M4 will thereby represent not only a difference in when the maximum deflection occurs but also a significant increase in deformation velocity.

Both M3 and M4 terminated at iteration 25 due to the convergence criteria being met; the average change of volume over the final three iterations was 2.51%. The volume versus iteration graphs for M3 and M4 are identical to those of M1 and M2 illustrated in Figure 8.43 and will therefore not be illustrated. The total element strain as a function of iteration number for both M3 and M4 is illustrated in Figure 8.47.

Analog to the trends identified in Figure 8.44, Figure 8.47 also indicates that the MEP enabled model (M4) initially displays a higher total strain value than the model without MEP (M3). From iteration 6 onward, the total strain is consistently less for the MEP-enabled model; furthermore, the total strain value is seemingly very consistent, almost constant from iteration 10 onward.

The resulting topologies from M3 and M4 are illustrated in Figure 8.48.

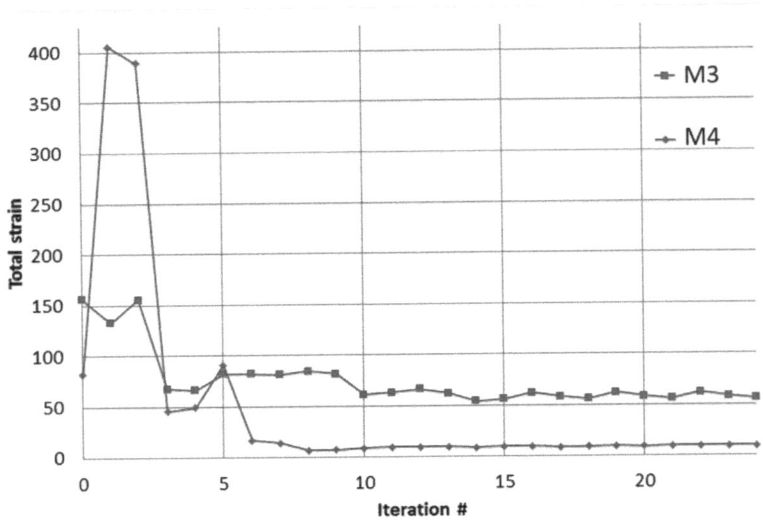

FIGURE 8.47 Sum of element strain per iteration case study 9 M3 and M4.

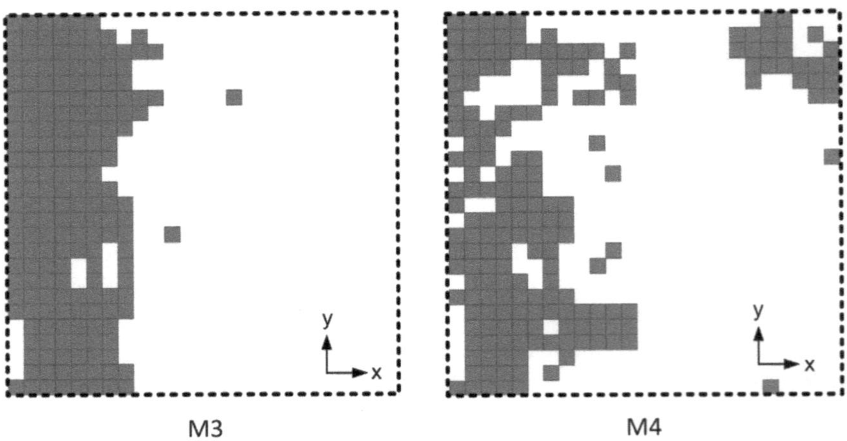

FIGURE 8.48 Resulting topologies case study 9 M3 and M4.

Figure 8.48 reveals a relatively simple and well-defined structure for the MEP-disabled model (M3), whereas the MEP-enabled structure, M4, is less well defined and seemingly subjected to checkerboard effect. The overall trends from M3 and M4 are thus very similar to those of M1 and M2; in general:

1. The MEP-enabled models seemingly produce more optimized structures. This statement is purely based upon the total element strain-to-volume ratio.
2. The MEP-enabled models are generally exposed to the checkerboard effect. This is very visible in both Figures 8.45 and 8.48.

Although case study 9 contains four models, the above statements remain based on a very limited amount of information and cannot be justified to be generally applicable. The statements should, therefore, merely be considered as pointers for additional research, analysis, and scrutiny.

At this point the "Individual" variations of BESO in the BEETS algorithm have been separately investigated; the combined effect/influence has, however, not. This will be the focus of attention in the next section.

8.4.7 BEETS case study 10: combined effect of BEETS

The purpose of this section is to briefly investigate the combined effects of the principles and adaptations implemented into the BEETS algorithm/software versus the "original" BESO algorithm. For this purpose the "original" FE model presented in Section 8.4 will be utilized. The input parameters for case study 10 M1 of the BEETS optimization study are defined in Table 8.14.

The "benchmark model" for case study 10 in general is M1 from case study 7, where ER, ARMAX, and TERMTOL are identical to those defined

TABLE 8.14 BEETS Input Parameters for Case Study 10 M1

PowerShell Variable	Value
Max_iter	25
TSI	3
VTS	5
TSF	0.3
TABU_offset	1
TABU_LIST	Yes
TLF	0.5
ARMAX	0.05
RADIUS	2.1
TVF	0.3
ER	0.1
TERMTOL	0.03
MEPUSE	Yes
COOLUSE	Yes
COOLIT	0.1
SFCOOL	0.5

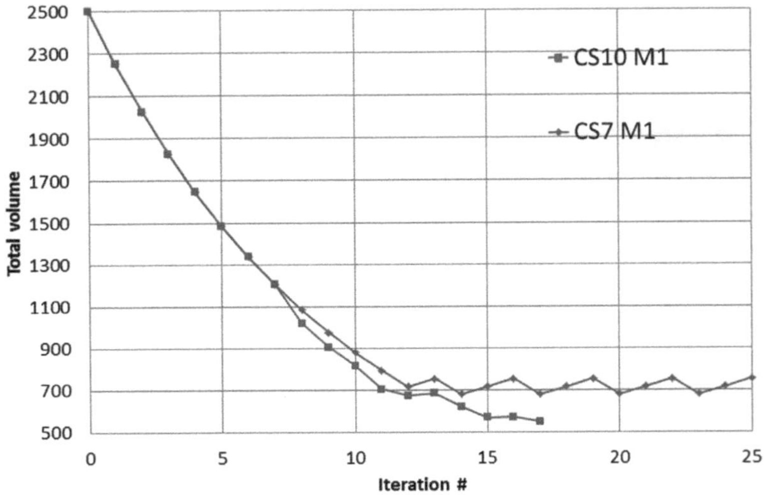

FIGURE 8.49 Sum of volume per iteration case study 10 M1 and case study 7 M1.

in Table 8.14; hence, M1 from case study 7 will be used for comparison. Figure 8.49 illustrates the sum of volume per iteration of case study 10 M1 and also includes case study 7 M1 for comparison.

As indicated in Figure 8.49, the BEETS algorithm terminates prior to the BESO algorithm, avoiding the cyclic behavior of the BESO algorithm. Figure 8.49 does, however, also reveal that the BEETS algorithm undershoots the target volume of 750. The primary reason for this can be found by reviewing the Main_INFO.txt file. After iteration 9 the volume is in excess of the final target volume of 750; however, as the volume has entered the cooling interval the ER value has been decreased. During iteration 10, the software determines that a number of elements need to be added to the model; however, due to the reduced ER ratio and the limiting ARMAX value, the software is unable to add as many elements as desired, leading to an undershoot in actual volume after iteration 10. This is simply because the ARMAX value is not cross-linked with the reduction in ER leading to an inconsistency in iteration target volume (which was 775 for iteration 10) and the actual volume of the model after the iteration (which was 704 for iteration 10). This trend continues until the optimization criterion has been met, leading to a significant undershoot in target volume. This problem needs to be resolved by reviewing the interaction between ER, ARMAX, and the cooling function.

The Main_INFO.txt file also reveals that the TS has an effect upon the overall optimization as "alternative" more efficient solutions were chosen based upon the TS at iterations 3, 6, 9, and 15. Figure 8.50 illustrates the strain per iteration for case study 10 M1 and case study 7 M1.

As indicated in Figure 8.50, the MEP-enabled model (M1 of case study 10) initially displays a higher total element strain value, which then drops below

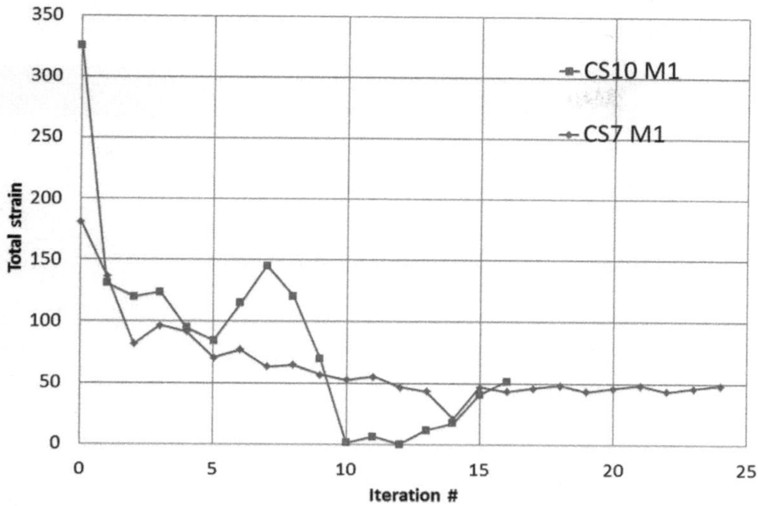

FIGURE 8.50 Sum of element strain per iteration case study 10 M1 and case study 7 M1.

that of the MEP disabled model (M1 of case study 7), despite a significantly lower volume. The resulting topologies from both models are illustrated in Figure 8.51.

As can be seen in Figure 8.51, the MEP-enabled model is subjected to the checkerboard effect, however not necessarily to a large extent. The main issue with case study 10 M1 is that the model is "disconnected," as there is no direct connection between the external loading applied at the top and the constraint applied at the bottom. This could, however, be a function of the undershoot phenomenon discussed above and to a greater extent a function of the limited neighborhood influence. In an attempt to determine whether the latter is true or not an

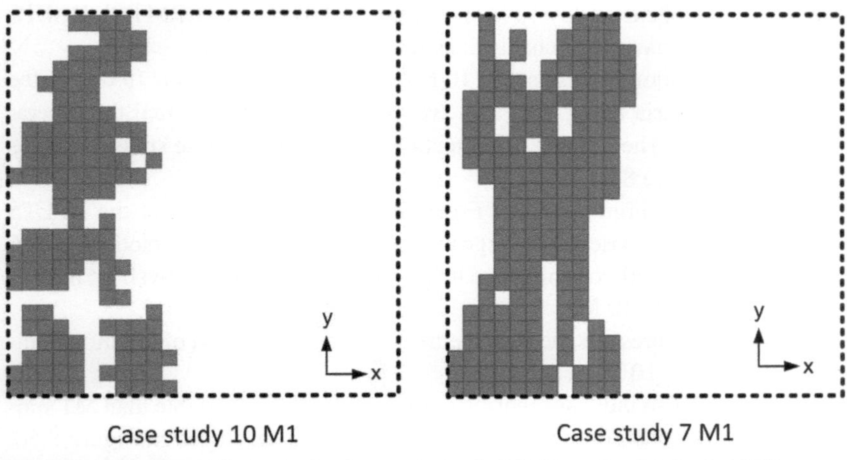

FIGURE 8.51 Resulting topologies case study 10 M1 and case study 7 M1.

TABLE 8.15 BEETS Input Parameters Case Study 10 M2

PowerShell Variable	Value
Max_iter	25
TSI	3
VTS	5
TSF	0.3
TABU_offset	1
TABU_LIST	Yes
TLF	0.5
ARMAX	0.05
RADIUS	6.1
TVF	0.3
ER	0.1
TERMTOL	0.01
MEPUSE	Yes
COOLUSE	Yes
COOLIT	0.1
SFCOOL	0.5

additional model is introduced. The additional FE model, case study 10 M2, is identical to that of case study 10 M1; only the RADIUS and TERMTOL BEETS input parameters have been changed, as specified in Table 8.15.

The optimization of case study 10 M2 terminated at iteration 15 due to the convergence criteria being met. The average change over the final three iterations was 0.87%. The volume as a function of iteration for case study 10 M2 is illustrated in Figure 8.52.

By comparing Figure 8.52 to Figure 8.49, it can be seen that the BEETS optimization more closely converges toward the final volume fraction of 750 and that the neighborhood influence helps control this, as no undershoot has occurred in case study 10 M2.

Figure 8.53 represents the total element strain as a function of iteration number for case study 10 M2.

Figure 8.53 also indicates that case study 10 M2 is more stable than M1 simply because the total strain is less fluctuant than that displayed in Figure 8.50. Figure 8.54 represents the resulting topology from case study 10 M2.

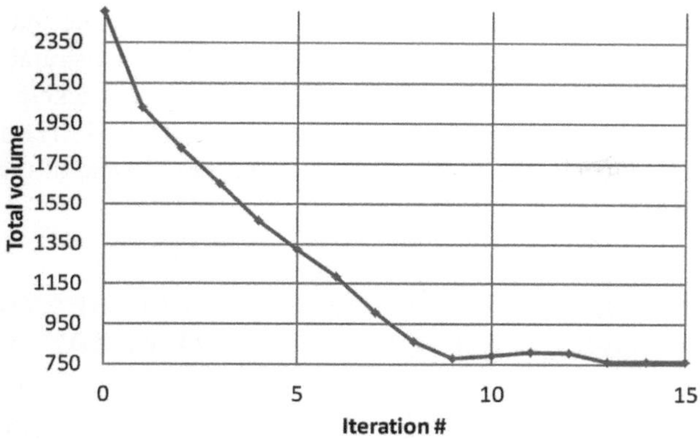

FIGURE 8.52 Sum of volume per iteration case study 10 M2.

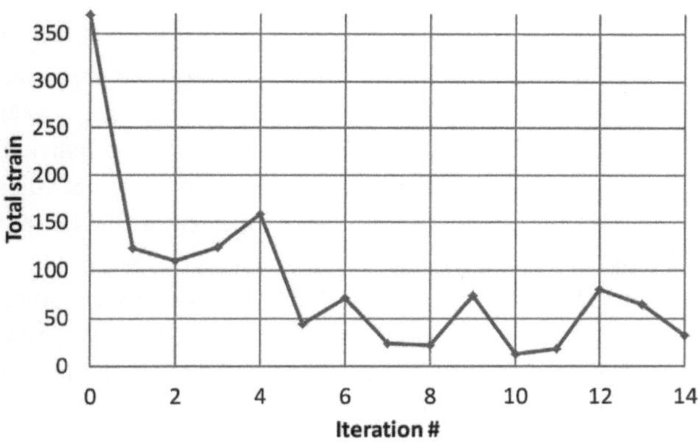

FIGURE 8.53 Sum of element strain per iteration case study 10 M2.

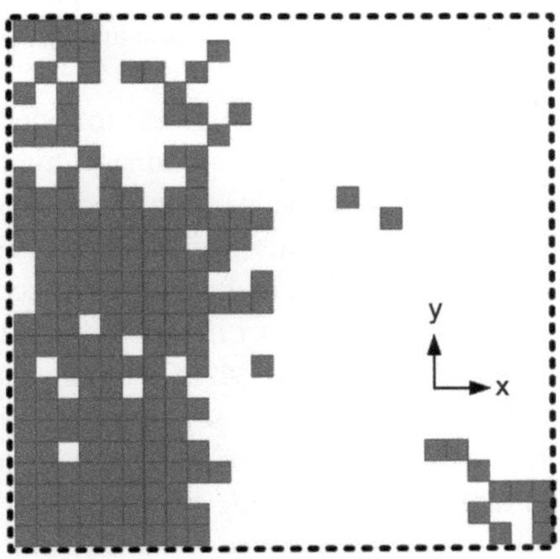

FIGURE 8.54 Resulting topology case study 10 M2.

Comparing Figure 8.54 to Figure 8.51, it can be seen that the former retains the direct link between the external loading and the constraints. There is, however, no denying that Figure 8.54 is also prone to the checkerboard effect, perhaps even more so than case study 10 M1 of Figure 8.51. The increase in neighborhood influence, therefore, seems to have eased some of the problems from case study 10 M1 but introduced or increased others.

8.5 CONCLUSION OF THE POTENTIAL OF BEETS FOR NONLINEAR TOPOLOGY OPTIMIZATION

Based on the case studies documented in Sections 8.3 and 8.4, it is possible to draw some general preliminary conclusions of the suitability of the BEETS algorithm for nonlinear topology optimization. Firstly, it should be emphasized that the conclusions are not final and that they are based upon relatively limited amount of information in essence based on a single case study with variable parameters; hence, little can be concluded about the overall versatility and robustness of the algorithm. Ignoring these limitations for the time being, it is clear that the BEETS algorithm has introduced some changes with respect to the BESO algorithm, and attempted to reduce and in some cases even eliminate the limitations of the BESO algorithm. The disadvantages of particular interest were initially listed in Table 8.1. Table 8.16 lists these "original" disadvantages and summarizes the findings from the BEETS algorithm based upon the case studies of this chapter.

Based on the previous case studies and the outcomes summarized in Table 8.16 above, it can be concluded that further refinements and adaptations of the BEETS software would be advantageous before an extended case study, including robustness, sensitivity, and versatility, is conducted. From the preceding case studies four main areas of interest for future work are suggested:

1. Objective function definition and monitoring. Although it has been suggested that the strain energy is the best measure for the objective function, this has not been analyzed. Furthermore, the current implementation of monitoring and assessing the total strain is rather crude and simplistic, as previously discussed.
2. MEP effects and implementation. Although the introduction of MEP seemingly has had a positive effect upon the volume-to-total strain ratio, it also seems to have introduced higher levels of the checkerboard effect. Furthermore, the actual influence of the intermediate strain values and the "built-in" neighborhood influence need to be further addressed. Finally, the MEP influence upon the total strain value per iteration must also be advanced.
3. Direct link and influence between ARMAX, cooling function, and ER. As discussed throughout Section 8.4, there seems to be an inherent flaw in the BEETS algorithm which causes the optimization to undershoot or even overshoot the final target volume simply because the ARMAX ratio is not affected by the cooling function and vice versa. The link between ARMAX, the cooling function, and the ER must therefore be reviewed.

TABLE 8.16 Findings in Response to "Original" BESO Disadvantages

"Original" BESO Advantage from Table 7.1	Importance of Resolving Disadvantage	Preliminary Conclusions of BEETS Based Upon Case Studies
Quasistatic loading assumption.	Medium	The quasistatic load assumption plays a lesser role with the introduction of MEP. A number of case studies have shown that a structure may be optimized even when exposed to very high strain rates. The limitations of the quasistatic loading assumption are still those of explicit FEA, and thereby dependent upon the timestep size.
Nonconvergent and potential oscillation between "equal" design solutions.	High	This problem has been addressed by the introduction of the cooling function which clearly reduced or even eliminated the oscillation phenomenon in case study 6.
No direct consideration of "intermediate" deformation modes/levels.	High	This has been addressed by the introduction of the MEP; case study 9 revealed the suggested benefits of this supposedly leading to a more optimized solution; the feasibility of these solutions could, however, be questioned.
Relatively high computational cost.	Low	This aspect has, in essence, not been analyzed. The primary reason for this is simply that this is highly dependent upon the actual programming; handling and storage of large data sets. This falls outside the scope of the current work meaning that no relevant and factual comparison can be completed at this stage.
Significant issues with high levels of nonlinear material behavior.	Medium	The BEETS model does not seem to have experienced any noteworthy problems with the higher levels of nonlinear behavior or even higher strain rates. However, the limited scope of these initial case studies must be taken into consideration in this context.
Low dispersion of search within potential solution space, high probability of "converging" toward local optimum points.	High	It is evident that the potential solution space has been significantly increased (on a variable scale) due to the adaptation of the TS, which in certain cases actively changed the structure in given iterations.

4. Limitations of current BEETS version. At present, the BEETS software only caters for structures in a 2D geometrical space. This needs to be expanded to 3D to review more relevant case studies, such as, for example, an automotive crush can or B-pillar.

Once the above issues have been addressed and a revised version of the BEETS software created, more advanced case studies should be solved and compared to results from, for example, LS-TaSc, as well as conducting physical experiments to verify the performance, validity, and indeed manufacturing feasibility of the results obtained.

The purpose of this chapter was to highlight some of the general issues a nonlinear topology optimization algorithm must resolve. In addition to this the purpose was to demonstrate how algorithms may be developed, adapted, refined and customized for specific or general applications. The ideology behind this is to demonstrate that it is possible to develop bespoke algorithms as well as inspire the reader to actively contribute to the continued development of optimization principles, methods, algorithms and software.

REFERENCES

Bastien, C. (2014). Confidential LS-Dyna data from research project. 28.11.14.

Chuang, C. H., & Yang, R. J. (2012). Benchmark of topology optimization methods for crashworthiness design. Available from: <http://www.dynalook.com/international-conf-2012/optimization-metal-forming18-a.pdf> Accessed 22.09.14.

Chuang, C. H., & Yang, R. J. (2012a). Benchmark of topology optimization methods for crashworthiness design. Available from: <http://www.dynalook.com/international-conf-2012/optimization-metal-forming18-a.pdf> Accessed 22.09.14.

Huang, X., & Xie, Y. M. (2008). Topology optimization of nonlinear structures under displacement loading. Available from: <http://www.sciencedirect.com/science/article/pii/S0141029608000205> Accessed 02.05.14.

RMIT. (2014). Software – RMIT University. Available from: <http://www.rmit.edu.au/research/cism/software> Accessed 08.08.14.

Singh, S. (2014). *Re-design of car seat rails and its height elevation system*. Unpublished MEng Thesis. Coventry, UK: Coventry University.

Chapter | nine

Applications of Concept Nonlinear Optimization

CHAPTER OUTLINE

9.1 INTRODUCTION

In the field of vehicle safety, crashworthiness highlights a high level of nonlinearity which can be caused by numerous physical phenomena, such as the unpredictability of buckling mode of the vehicle structure (case of the frontal crash), the interaction of the engine and other ancillaries in the engine bay, the high strain rate effect of the materials, etc... A vehicle contains few hundred components (including body panels), each having some influence on the vehicle impact response in crash dynamics scenarios.

This chapter will propose a vehicle architecture BIW structural optimization which will focus on panel sizing and grade selection. From this study, it will be evidenced that safety sub-systems, should not be included in this phase of the optimisation, but computed separately, as discussed in chapter 10. Safety subsystems refer, for example, to occupant restraint systems and the whiplash system design. A better strategy to optimize these safety systems would be to use a sled approach where the crash pulses could be an input to the occupants' cabin. Using a sled-based model would allow a faster and more reliable optimization of each subsystem.

Another level of nonlinearity can come from the fact that the vehicle is optimized across different disciplines and in some cases (not covered in this chapter) across different physics (structure and fluid coupling, for example). In this chapter, the vehicle will also needs to combine a high performance in durability, handling, and crashworthiness, subsequently the optimization has to consider its solution across various design group boundaries.

Nonlinear Optimization of Vehicle Safety Structures. http://dx.doi.org/10.1016/B978-0-12-804424-7.00009-9

This study will look at optimizing a vehicle structure across two disciplines: vehicle safety, and ride and handling. The safety assessment will consider frontal, pole, side impacts, while the ride and handling loadcase will be characterized mainly by a vehicle torsional stiffness requirement. There are obviously more ride and handling parameters to consider; however, in order to illustrate the vehicle optimization process, only these four simultaneous loadcases will be considered.

A step-by-step approach will be proposed and will include some opportunities to reduce the number of variables for the optimization by using design variable sensitivities. Overall, a vehicle structural optimization is, in an industrial setting, a balance between finding a reasonable solution and the computation time and cost.

9.2 BACKGROUND OF THE VEHICLE OPTIMIZATION STUDY

It is believed that the future of automobiles will be driven by lightweight structures and highly efficient powertrains. The TARF-LCV EPSRC-funded project (toward affordable, closed-loop recyclable future-low carbon vehicle structures) (TARF, 2011) aims to provide a strong scientific and technological underpinning to future LCV development in areas of advanced materials, low carbon manufacturing technologies, holistic mass-optimized vehicle structure design, and closed-loop recycling of end of life vehicles.

The initial work undertaken computed the ideal loadpath of the vehicle with the help of linear topology optimization, as described in detail in Chapter 5. Following various research, it was documented that the most realistic boundary condition to apply in the case of modeling crash events was inertia relief, in favor of standard boundary conditions, by balancing the external loading with inertial loads and accelerations within the structure itself. Inertia relief has the benefit to prevent any structural loadpath definition skewing, as in crash events the vehicle structure is in motion and not physically restrained to its wheels.

The optimization method utilized was based on an SIMP algorithm, which penalized the vehicle structure design envelope Young's modulus, part of the stiffness matrix, leading to a vehicle mass reduction. It was documented that a SIMP penalty factor p of 1 was adequate to obtain useful structural results.

The initial optimization included a total of six individual loading scenarios, including pole impact, side barrier impact, roof crush, low-speed centerd rear impact, high-speed rear impact, and front impact (offset deformable barrier (ODB)). Linear topology results allowed some parts of the structure to be defined with credibility, like the vehicle safety cell, while some other sections needed extra considerations, especially in the areas prone to buckling. As discussed in the previous chapter, linear solvers are not suitable to compute bifurcation; hence, the vehicle front end and rear end were re-engineered to meet controlled collapse mode, while the natural load obtained in the linear topology optimization was used to design body panels (LCVT, 2011).

This re-engineering was based on the fact that the front crash structure should not exceed 300kN, as this load was calculated to be the initiation buckling load for the A pillar (Prem, 2013). More automatic shape and size of body panels routines have been developed to convert the beam model cross section properties (second moment of area) into closed sheet metal, constraining each body in white (BIW) geometry to meet the vehicle exterior styling (Christensen, 2013). The proposed model is the completion of the TARF vehicle architecture based on linear optimization. The optimization methodology investigated in this chapter will address the very local structural changes that cannot be computed by a global topology and localized linear sizing. The study focuses on a multidisciplinary optimization that will solely include size and material grades, as the vehicle package is very well-defined and no geometry modifications can be performed.

9.3 INITIAL VEHICLE CRASH PERFORMANCE

The TARF CAE vehicle model included simultaneous safety analysis requirements containing frontal ODB, side impact barrier, and pole impact, as well as a torsional rigidity requirement. In the original topology work, the torsional rigidity load case was not included because the torsional loads are very small compared with the magnitude of the other impact scenarios.

The TARF CAE model is a monocoque structure, which was discretized with an average mesh size of 5 mm for the frontal structural loadpath, while the door open panels (doors, boot, bonnet, etc.) were discretized to 10 mm. A variety of element types were also used, as listed in Table 9.1.

The total mass of the model is 1000 kg (TARF project target mass), which is made of 163 kg of the BIW structure and 372 kg for the ancillaries. Another 465 kg of nodal masses was distributed to the model to include the masses of all other components, such as seat, carpet, trim, harness, etc., such that the whole structure attains its target mass. The crash model has been solved using the explicit solver with a timestep of 0.56 μs, as explained in Chapter 2, leading to virtually no percentage added mass during the whole duration of the computation (Figure 9.1).

The frontal and side impact models used the finite element crash barriers that are commercially available (LSTC, 2015). These models contain detailed honeycomb barrier properties which will deform locally under impact, providing

TABLE 9.1 Element Type and Count in the TARF LS-Dyna Model

Element Types	Beams	Discretes	Shells	Solids
Number of elements	2926	11	597,621	84,147

FIGURE 9.1 TARF model: overlay of frontal and side impact setup including LSTC barriers.

local structural force responses. Consequently, it is not possible to simplify them to obtain a faster runtime and at the same time obtain believable results representative of a EuroNCAP frontal and side impact tests.

In the case of the pole impact, the pole during the test does not deform due to its construction; hence, it can be modeled as rigid. As a consequence, there are two options to model the pole:

- Meshing the pole as a finite element mesh and applying a RIGID (non-deformable) material property
- Use a RIGIDWALL command

The former will require adding more elements in the computation, as well as setting up some extra contacts between the vehicle structure and the meshed pole component. This method, in a computation sense, will work; however, there will be a computational cost to this, as the contact search algorithm will have to track more nodes and then use a penalty-based contact to repel the panels from the meshed pole component. When a penalty-based contact is present, there is a risk of instability and this needs to be considered.

The RIGIDWALL option, on the other hand, can be used for simple geometries. The RIGIDWALL, compared with the penalty-based contact, works more like a kinematic contact. When a node of the vehicle contacts the pole, it stops at the wall boundary due to its position relative to the wall; a force is then applied. This is a very efficient method for such loadcase, as it is computationally stable and faster than a penalty-based contact approach. In the case of the TARF pole impact computer model, the RIGIDWALL option was favored (Figure 9.2).

The impact velocities have been set as per Table 9.2.

The impact velocity for the side impact is slightly lower due to the type of barrier being used and weighed 1306 kg instead of 950 kg, based on the Euro NCAP Side impact test Protocol v6.0 (EuroNCAP, 2012). The side impact

FIGURE 9.2 TARF vehicle setup against a rigid pole.

TABLE 9.2 Impact Velocities for Each Loadcase

	Impact Speed (m/s)
Front crash	17.9
Side impact	11.7
Pole impact	8.3

velocity was adjusted so that the same impact kinetic energy is observed as per the EuroNCAP protocol.

The energy computation validation has been performed, showing that the models are stable (Figure 9.3). The total energy equates to the sum of the kinetic energy and the internal energy (energy of deformation), hence describing a conservative system. All the energy transfer is accounted for. This step is fundamental prior to performing any optimization, in order to ensure that the computer outputs can be interpreted in a scientific manner later on in the optimization stage. It can be noted that the front crash internal energy slightly dips down from time 0.03 s. It is accounted by the sliding energy taken from the friction between the vehicle and the RIGIDWALL (too small to plot in Figure 9.3). This energy loss is negligible, hence it has insignificant impact on the result interpretation.

In the case of the frontal and pole impact, the vehicle is in motion and the barriers are static, while in the side impact test it is the reverse, that is, a

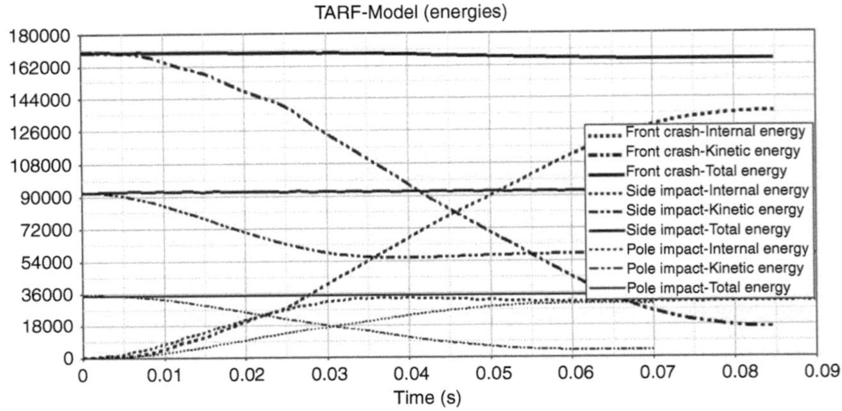

FIGURE 9.3 Energy balance validation of the crash models (front, side, and pole impacts).

950 kg barrier impacts the vehicle. Considering these facts, it can be noticed in Figure 9.3, as a check, that the total energy of the front crash is higher than that of the side impact, which is then higher than that of the pole impact.

The fourth loadcase, that is, torsional rigidity has been computed using an implicit solver as the structure does not need to plastically deform. Running this model in explicit would have also caused some issues with inertial effects, as to compute in a reasonable time, the torsional force would have needed to be applied in 0.1 s. This would have caused structural oscillations, which would have masked the structural displacement needed to accurately read the torsional rigidity value. Also, as the loading has to be applied during a longer duration, explicit computing is not the best method to use, as the solver can only compute one timestep at a time, which would be totally impractical. In the torsional rigidity case, there is no plasticity involved, nor large displacements; consequently, it is possible to use an implicit linear solver, while still conserving the spotweld tied contacts activated to keep the panels attached together. The torsional rigidity model is fundamentally the same as the safety models, except that the computation instruction was switched from explicit to implicit to change the solving method.

The same mesh is used for all the models, except for the boundary conditions. The vehicle is clamped at the rear turrets and a displacement is applied at the wheels center via a stiff spring, causing torsion. This displacement is generating a force via the springs. The torsional rigidity is measured by dividing the cross section moment in the *x*-axis against the wheel angle change (Figure 9.4).

The structural performance of the TARF vehicle is assessed against the four loadcases. The study only considered the structural integrity of the vehicle and did not consider the occupants' injuries, the main priority being to minimize structural intrusions within the cabin. In order to monitor the intrusions, springs of very low stiffness (0.1 N/mm) are included in the computer model. The intrusions are computed by monitoring the springs' length, as illustrated in Figure 9.5.

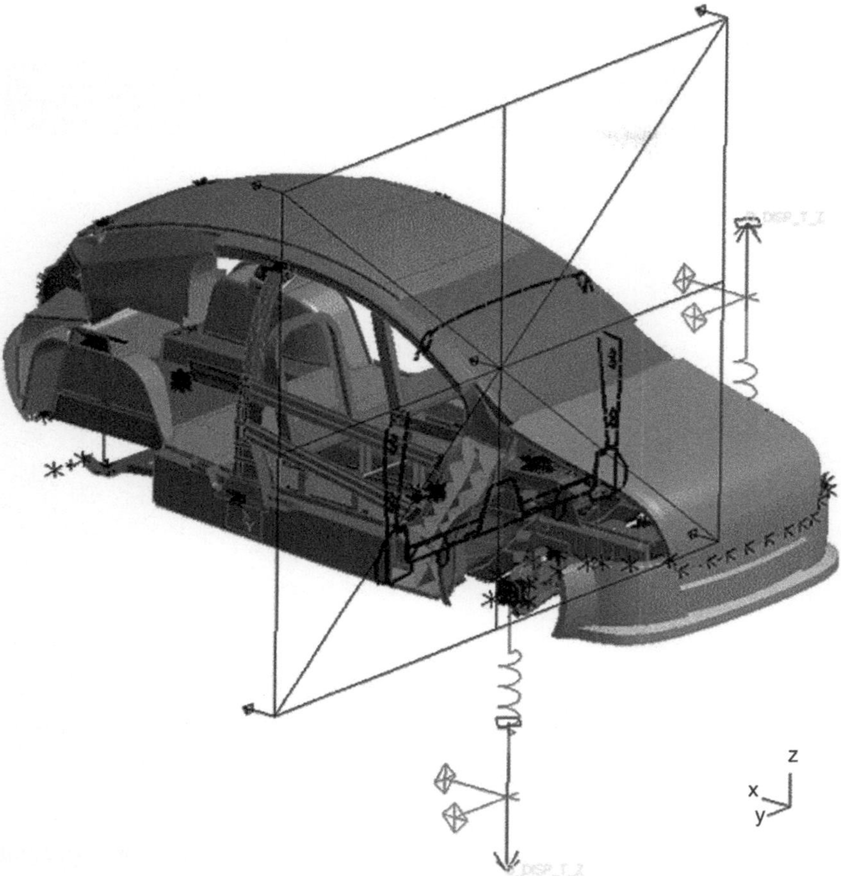

FIGURE 9.4 Torsional rigidity model setup (the plane represents a cross section force output).

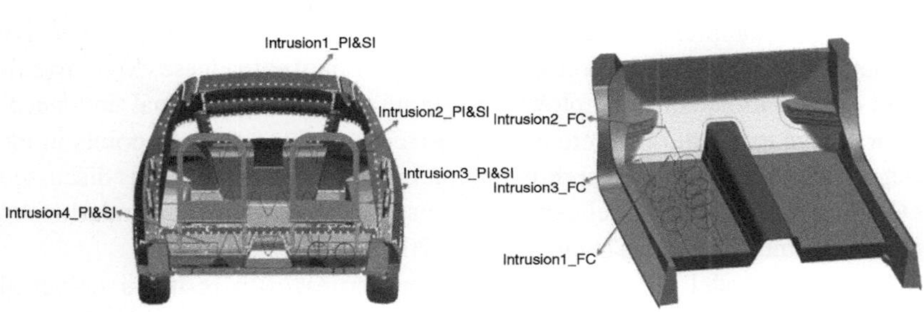

FIGURE 9.5 Location of sensors in TARF vehicle (left: side and pole, right: frontal measurements).

TABLE 9.3 Initial Response from TARF Model

Crash Mode	Sensor Variable Name	Location	Value	Units
Frontal crash	FC1_Intrusion	Footwell middle low	13.47	mm
	FC2_Intrusion	Footwell middle high	26.37	mm
	FC3_Intrusion	Footwell Outboard low	5.39	mm
	FC_acceleration	Chassis	445377.00	mm/s^2
Pole impact	PI1_Intrusion	B pillar top location	47.76	mm
	PI2_Intrusion	B pillar seatbelt location	245.52	mm
	PI3_Intrusion	B pillar pelvis location	275.52	mm
	PI4_Intrusion	B pillar sill location	214.05	mm
Side impact	SI1_Intrusion	B pillar top location	1.37	mm
	SI2_Intrusion	B pillar seatbelt location	52.27	mm
	SI3_Intrusion	B pillar pelvis location	165.00	mm
	SI4_Intrusion	B pillar sill location	76.42	mm
Torsion	Torsional_stiffness	N/A	9,483,970	Nmm/deg

The springs across the occupant the H-point location monitor the structural deformation of the vehicle in side and pole impact. The longitudinal springs capture the foot-well deformation in frontal crash, as listed in Table 9.3.

9.4 CHOOSING SAMPLING METHODS AND METAMODELS

The design of complex structures requires numerous computationally expensive simulations to understand the responses for a particular loadcase. Metamodel-based optimization can be employed to minimize the computational time needed for design exploration where design surfaces are fitted through points in the design space to construct an approximation to the design response, as discussed in Chapter 4. The metamodel can then be used instead of actual simulations to find the optimum variables (Stander et al., 2013).

Each metamodel has its own unique properties; as a result, a universal metamodel that performs best for all possible problems does not exist. It is important to understand the problem at hand and try to choose the best possible method. Another aspect to consider is whether the need of global approximation which covers the entire design space is needed or if a local approximation model is more appropriate. The complexity of the response which should be captured by the metamodel also influences its selection. Noise would be another aspect to take into consideration; an approximation model may perform better with the presence of noise, whereas an interpolating model may not be the right choice. However, the chances that an interpolation model would provide better predictions compared with another interpolation model in unknown points are not guaranteed, even in the absence of noise (Ryberg et al., 2012).

Metamodel comparative studies for multiple modeling criteria have been previously conducted and according to one study (Jin et al., 2001) which compared polynomial regression (PR), Kriging, multivariate adaptive regression splines (MARS), and radial basis functions (RBF) models, concluded that RBF metamodel performs best overall and are the best models when small and scarce sample sets are used, when both average accuracy and robustness are taken into consideration. For very difficult problems of large scale and of high-order nonlinearity, the MARS model was found to be most accurate, provided that large sample sets were used. The performance would, however, deteriorate if the sample size reduces, in which case RBF performs best. It was also found that Kriging is very sensitive to noise since it interpolates the sample data. It is recommended that PR be used for low-order nonlinearity problems due to its high accuracy and ease of use. PR is, however, not accurate for highly nonlinear problems. It has been proposed that PR should be implemented first to investigate whether a reasonable fit can be obtained (Clarke et al., 2005); however, studies have shown that support vector regression (SVR) had the best performance overall when accuracy and robustness need to be taken into consideration (Clarke et al., 2005). More recent studies (Li et al., 2010) also indicate that SVR performed best in terms of prediction accuracy and robustness for stochastic problems; the next best model was found to be Kriging. The models that were compared include artificial neural network, RBF, Kriging, SVR, and MARS. Although there have been contradicting studies (Kim et al., 2009) that concluded that Kriging and moving least squares (MLS) provided accurate metamodels when compared with SVR and RBF models. It is not possible to come to a clear conclusion on which metamodel performs best overall as there are a number of parameters that can be tuned while building an approximation. The results can also vary considerably based on how well these parameters have been tuned; the software used also influences the results.

The main concept would be to use several metamodels to perform optimization continuously which would lead to the best design possible (Ryberg et al., 2012). Multidisciplinary design optimization is a process in which multiple disciplines such as crash, NVH, or torsional rigidity are included within the optimization. This allows for variable optimization process to consider different loadcases and to find an optimum that meets all of the constraints established while adhering to the optimization objective.

Two different types of the response surfaces (metamodels) have been proposed to fit the sample data and used polynomial response surface and radial basis function methodologies.

9.4.1 Polynomial response surface methodology

Polynomial metamodels are also known as response surface models and are used in response surface methodology (RSM). The RSM are defined as a set of statistical and mathematical methods for developing, improving, and optimizing processes and products (Ryberg et al., 2012). These models are developed by a process of fitting a regression model to a dataset of variable settings and corresponding responses. Polynomial terms are recommended to be used for the screening process of the variables and for investigating the variable cross-influence when responses are being determined (Stander et al., 2013). When the sequential response surface method is used, which is an iterative process of finding the optimum variables, polynomial response surface methodology is commonly utilized, mainly due to computational cost (Stander et al., 2003).

9.4.2 Radial basis functions (RBF)

Radial basis function methods are the means to approximate the multivariate function (Martin et al., 2003). It is one of the primary tools for interpolating multidimensional scattered data. The method's ability to handle arbitrary scattered data to easily generalize to several space dimensions and to provide spectral accuracy has made it particularly popular in several different types of applications, including cartography, neural networks, medical imaging, and more particularly in our scenario, the numerical solution of partial differentia equations (Wright, 2003).

The method uses linear combinations of a radially symmetric function based on Euclidean distance or other such metric to approximate response functions (Jin et al., 2001). The radial basis neural network consists of a three-layer topology (input, hidden, and output). The input layer is linear and the hidden layer is made up of nonlinear radial units, each of which responds to only a local region of input space. The output layer creates an approximation over the entire design space through input–output mapping after performing a biased weighted sum of these units using the Hardy's multiquadrics functions or Gaussian function (Stander and Goel, 2008). This method is more suited to capture nonlinear response surface definitions compared with its polynomial count part method.

9.5 DESIGN OF EXPERIMENT (DOE) AND GLOBAL SENSITIVITIES

It is always possible to optimize a vehicle structure directly, bypassing the DOE stage; nevertheless, this step is very important as it allows the analyst and engineer to understand how the structure behaves in each loadcase, as well as capturing the key design factors and contributors to the vehicle performance. This stage also provides the engineer and analyst with design directions and a list of potential compromises that may be conducted on the vehicle architecture without considerable losses in vehicle performance.

Consequently, performing a DOE phase analysis is always very useful. In the case of the TARF-LCV vehicle, the influence of the panel thickness and materials on the vehicle responses to various crash scenarios had not been studied extensively. DOE is therefore the perfect tool to understand what efficient design possibilities are available to the designer.

- Which components provides the most contribution to the vehicle structural performance? (Need to be mainly acted upon.)
- Which components make the least contribution to the vehicle structural performance? (Can be ignored or can be compromised for other design requirements.)

Most of the aspects of the model have been estimated to best fit current market leading vehicles. Thanks to DOE techniques, it is possible to explore the vehicle design and how the panels influence the responses to individual load-cases. Performing a variance-based sensitivity analysis using Sobol's indices, it is possible to determine the significance of each engineering design variables, hence eliminating redundant variables from the optimization stage.

The baseline front crash structure used in this study is based on the sizing performed in Chapter 5 (Method 2: detailed frontend design); hence, it has been assumed that the panels in the frontend structure, that is, the crush cans and the longitudinal, were already optimized. The emphasis of the proposed DOE study aims to capture the importance of the role of other strategic panels, assuming the main frontend architecture is set. This analysis is just an example and it is, of course, also acceptable to add the front end as a variable, however at a computational cost.

Consequently, considering Figure 9.6, it can be noted that the longitudinal, crush cans and bumper beams are not selected, but the other main structure is.

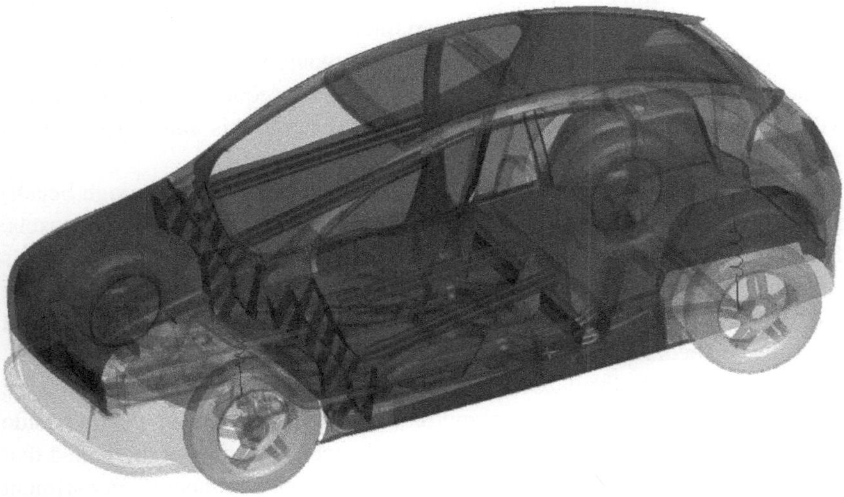

FIGURE 9.6 Dark color indicates variable locations on the model.

It can be noted that the bonnet has also been selected, which may come as a surprise. The main reason for this variable selection was that the bonnet was fully bonded to the structure and that in the frontal loadcase, its deformation energy was significant. Consequently, it has also been selected as part of the DOE experiment.

Each computer model contains 24 parameters. Some parameters can be made constants and some can be activated as design variables. As mentioned earlier, the TARF-LCV model has been set up for three crash scenarios and one torsional loadcase. These loadcases will fully share all variables among them, and it was determined that 17 variables were necessary as part of the study. These 17 variables consist of 14 panel thickness variables and 3 discrete material variables that would represent a material grade changes (800, 1019, and 1143MPa). It can be noticed that as the vehicle's architecture is symmetrical, each variable will affect both sides of the vehicle structure.

The variables chosen for the study are listed in Table 9.4.

The side impact beams have been separated as inner and outer sections, as depicted in Figure 9.7.

Each analysis will have a set of responses, for which a polynomial metamodel of linear approximation order will be associated. This approximation will be based on a space-filling sampling method A space-filling algorithm which minimizes the maximum distance between the experimental design points for a given number of points (Stander et al., 2013). The default set of simulation points for 17 variables is 28; however, 38 simulation points were considered for the analysis to provide more data for the polynomial fit.

The 13 responses monitored are the vehicle mass, intrusion, and acceleration for the 4 loadcases.

It is then possible to assert which of the 38 runs across the four models meet the design requirements, that is, the design constraints (Table 9.5), as well as defining the influence of each design variable.

It has to be noted that the vehicle total mass would vary during the study due to the attributed change in panel thickness. As a consequence, the total energy will change accordingly as it is directly proportional to this mass change (for a constant velocity).

The constraints were obtained by taking the average value through benchmarking studies conducted on existing vehicles for the side and pole impacts. The constraints for the front passenger compartment footwell intrusion levels are considerably lower than the limit established by Euro NCAP frontal offset test. There has been considerable research over the years suggesting footwell intrusion and pedal movements resulting in significant injuries to the occupants' lower legs. A study from 2003 by NHTSA (Eigen and Glassbrenner, 2003), based on crash result data from approximately 58,000 non-vehicle nearside occupant injuries and 62,000 vehicle nearside occupant injuries, indicated that many moderate and maximum injuries take place with occupant compartment intrusions being less than 150 mm. Injury severities AIS 2+ and AIS 3+ (kill or

TABLE 9.4 Variables Used for the DOE Study

Variable Number	Variable Name	Part Description	Continuous Variables Nominal (mm)	Min (mm)	Max (mm)
1	B_Pillar	Door: B-Pillar L/R	0.8	0.6	0.85
2	Bonnet	Bonnet	0.5	0.4	0.6
3	Door_B1	Door beams inner section (L/R)	1.5	0.6	1.8
4	Door_B2	Door beams outer section (L/R)	0.6	1.0	1.8
5	Door_In	Door: Inner L/R	1.5	0.6	1.8
6	Floor_Sa	BIW: Floor-reinforcement L/R	2.5	1.5	2.6
7	Floor_Tu	BIW: Floor-Tunnel	1.5	1.0	1.8
8	Lower_A	BIW: Lower-A-Pillar-Reinforcement L/R	2.0	1.0	2.0
9	Roof_Pa	BIW: Roof-Panel	1.5	1.0	1.6
10	Seat_Pa	BIW: Rear-Seat-Panel	1.5	1.0	2.0
11	Side_Pa	BIW: Side-Panel L/R	1.5	08	1.5
12	Sill_In	BIW: Sill-Inner L/R	2.0	1.0	2.1
13	Sill_Out	BIW: Sill-Outer L/R	2.0	1.0	2.1
14	Wheel_Pa	BIW: Wheel-Arch-Panel L/R	1.5	1.0	1.5
			Discrete Variables (MPa)		
15	mat_B1	Door beams inner grade (800, 1019, 1143)	800	1019	1143
16	mat_B2	Door beams outer grade (800, 1019, 1143)	800	1019	1143
17	mat_BP	Door: B-Pillar L/R grade (800, 1019, 1143)	800	1019	1143

seriously injured) were also reported for intrusion levels greater than or equal to 30 mm. It must also be noted that the fleet in consideration was non-uniform in nature and factors such as age and seat belt use were not taken into account.

The variable influence on the individual loadcases is depicted in the global sensitivity plots in Figures 9.8–9.11.

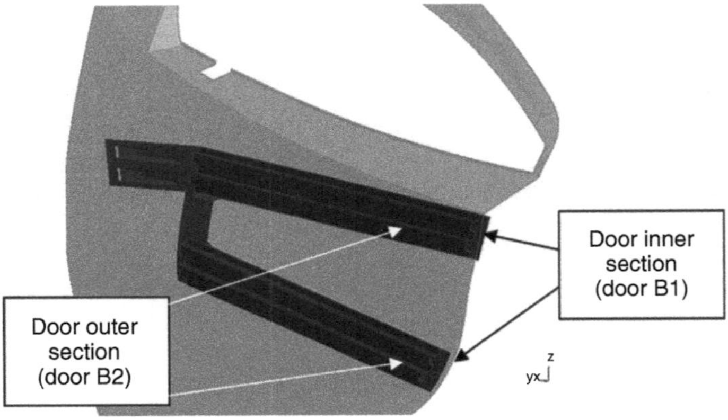

FIGURE 9.7 TARF door with focus on side impact beams (outer, corrugated, Door_B2 and inner, flat, Door_B1).

TABLE 9.5 Constraints for the DOE of Crash Models

Constraint Number	Constraint Name	Value	Upper Bound	Lower Bound
1	FC1_Intrusion	20 mm	x	
2	FC2_Intrusion	30 mm	x	
3	FC3_Intrusion	20 mm	x	
4	FC_acceleration	480,000 mm/s^2 or 48.9 g	x	
5	PI1_Intrusion	151 mm	x	
6	PI2_Intrusion	320 mm	x	
7	PI3_Intrusion	349 mm	x	
8	PI4_Intrusion	318 mm	x	
9	SI1_Intrusion	20 mm	x	
10	SI2_Intrusion	145 mm	x	
11	SI3_Intrusion	232 mm	x	
12	SI4_Intrusion	150 mm	x	
13	Torsional_stiffness	8.8 kNm/deg		x

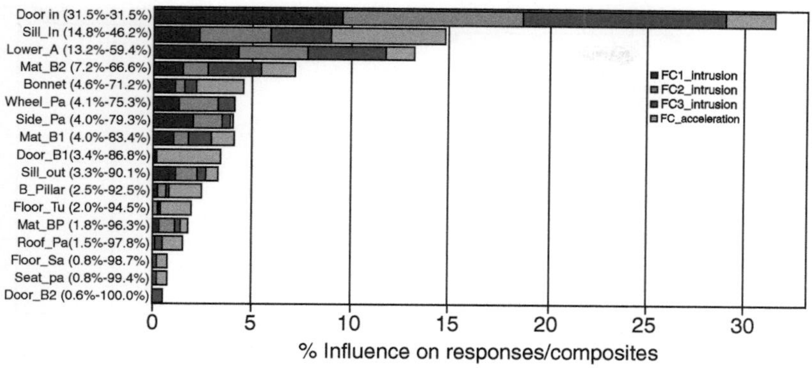

FIGURE 9.8 Global sensitivity plot for frontal ODB test.

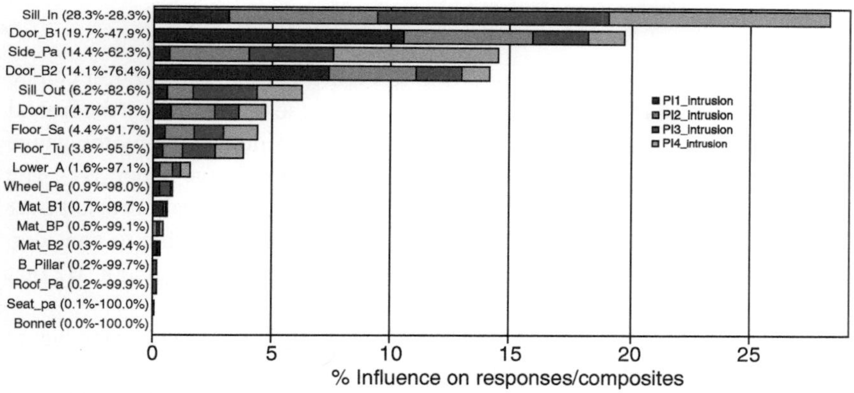

FIGURE 9.9 Global sensitivity plot for pole impact test.

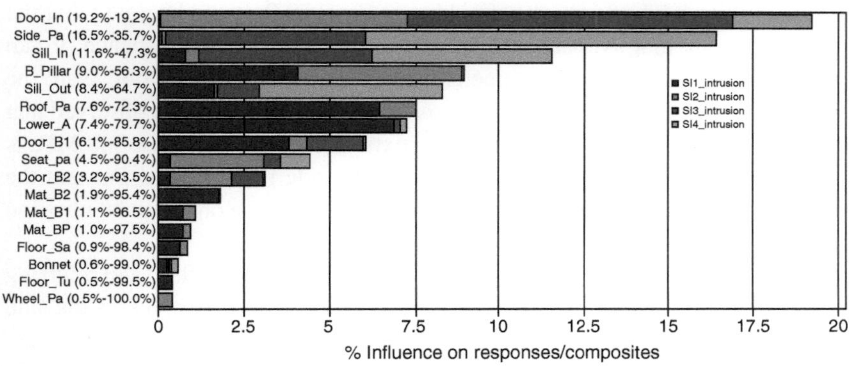

FIGURE 9.10 Global sensitivity plot for side impact test.

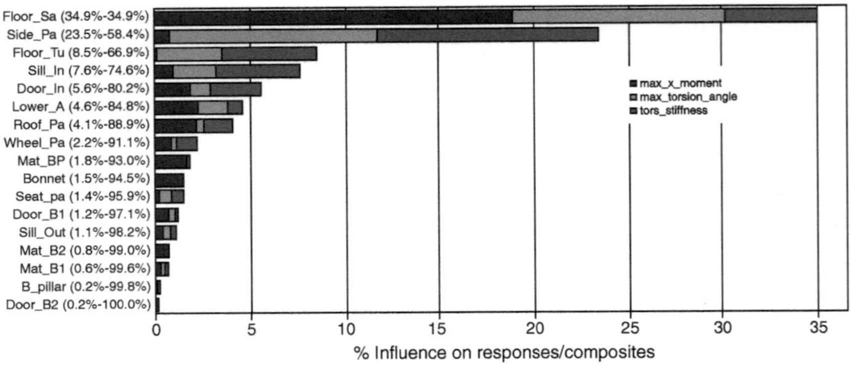

FIGURE 9.11 Global sensitivity plot for torsion test.

Considering Figure 9.8 and referring to Figure 9.5 and Table 9.4, it can be noted that the door inner panel gauge (Door_in) has the most effect on the frontal intrusions, except on the frontal acceleration which is dominated by the gauge change of the sill inner (Sill_in) and the door beam inner gauge (Door_B1). The lower A pillar (Lower_A) is also influential on the ODB frontal intrusion over all the rest of the variables and looks to be an important variable to consider. Mat_B2 seems to have some relevance; however, it ties up with Door_B2 that has the least effect on the ODB event.

In the case of the pole impact, looking at Figure 9.9 and referring to Figure 9.5 and Table 9.4, the sill inner gauge (Sill_in) has the most influence on the middle to lower pole intrusions, while the door beams (Door_B1 and Door_B2) have the most influence on the side intrusion located at the highest Z lateral displacement output. The door beams include a thickness change, as well as a material grade change. Looking at Figure 9.9, it can be noted that the material grades mat_B1 and mat_B2 have very little influence; hence, it can be concluded that the main parameters affecting Door_B1 and Door_2 are the thickness and not the grade change. Looking at Table 9.4, it can be observed that the grade remains very high and that any extra elasticity given past 800 MPa is not judged significant in the current study. The side panels (Side_Pa) and the sill outer panel gauges will have some important influence on the pole intrusion around the occupant's hip location.

In the case of the side impact loadcase, depicted in Figure 9.10 and referring to Figure 9.5 and Table 9.4, the door inner gauge (Door_in) as well as the side panel gauge (Side_Pa) and the sill inner (Sill_in) have some considerable effect on side impact intrusion.

It can be seen that the side panel and the sill inner both have the most influence on the side impact intrusion located at the base of the vehicle structure, while the door inner gauge has more influence on the intrusions located at the middle on the structure.

The roof panel, door beam, and B pillar gauge (B_Pillar) have the benefit of affecting the intrusion in the upper area of the vehicle.

In the torsional rigidity loadcase depicted in Figure 9.11 and referring to Figure 9.5 and Table 9.4, it can be noticed that the gauge of the side panels (Side_Sa) has the most effect on the torsional stiffness of the vehicle, followed by the sill inner (Sill_in), the floor tunnel (Foor_Tu), and the floor reinforcement (Floor_Sa) gauges. From the GSA/Sobol's indices output based from Figure 9.8 to Figure 9.11, it is also possible to capture the importance of the 17 variables pertaining to individual loadcases.

Looking at Table 9.6, it can be observed that adding a gauge variable for the bonnet, the roof panel, the rear seat panel, the sill outer, and the wheel arch panels is not a good use of computation time, as their effect in all four loadcases is secondary.

TABLE 9.6 Important Variables Classification Based on Sobol's Indices

Variable Number	Variable Name	Part Description	Important (Y/N)	Loadcases
1	B_Pillar	Door: B-Pillar L/R	Y	SI
2	Bonnet	Bonnet	N	
3	Door_B1	Door beams inner section L/R	Y	PI
4	Door_B2	Door beams outer section L/R	Y	PI
5	Door_In	Door: Inner L/R	Y	FC, SI
6	Floor_Sa	BIW: floor-reinforcement L/R	Y	T
7	Floor_Tu	BIW: floor-Tunnel	Y	T
8	Lower_A	BIW: lower-A-pillar-reinforcement L/R	Y	FC
9	Roof_Pa	BIW: roof-panel	N	
10	Seat_Pa	BIW: rear-seat-panel	N	
11	Side_Pa	BIW: side-panel L/R	Y	PI, SI, T
12	Sill_In	BIW: sill-inner L/R	Y	FC, PI, SI, T
13	Sill_Out	BIW: sill-outer L/R	N	
14	Wheel_Pa	BIW: Wheel-Arch-Panel L/R	N	
15	mat_B1	Door: Beam1 L/R (800, 1019, 1143)	N	
16	mat_B2	Door: Beam2 L/R (800, 1019, 1143)	N	
17	mat_BP	Door: B-Pillar L/R (800, 1019, 1143)	N	

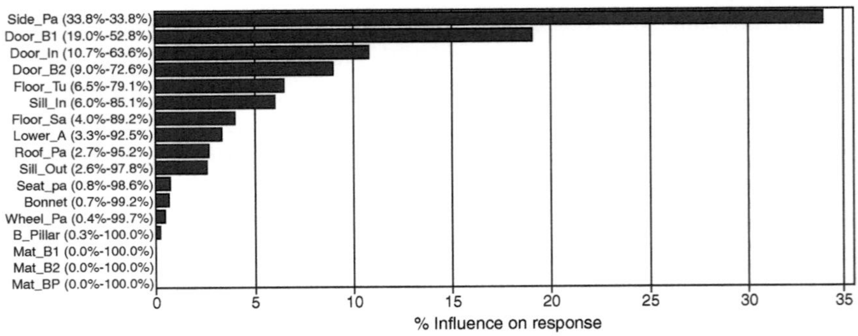

FIGURE 9.12 Global sensitivity plot for the objective function: mass.

Using global sensitivity plots, based on Sobol's indices, it is also possible to capture which variables have the most influence on the TARF vehicle mass. In this case, among the variables selected, changing the gauge of a side panel (Figure 9.12) has greatest influence on the vehicle mass, simply because it is a larger panel.

Considering the responses computed from the DOE, 7 feasible design solutions were obtained which meet all of the constraints set in Table 9.5. From the DOE study, it is possible to identify the potential mass savings for the TARF vehicle structure (Figure 9.13). The percentage reduction, compared with the

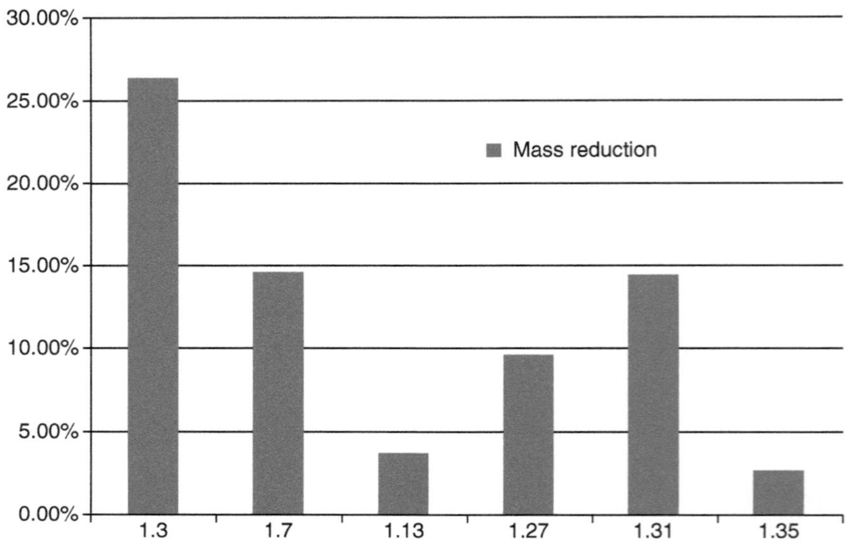

FIGURE 9.13 Structural mass reduction potential from DOE simulation points based on the 14 selected panels.

nominal panel gauge setup, is directly associated with the sum of the chosen variables' component mass reduction due to the change in panel thickness. Along with the baseline run, simulation points 1.3, 1.7, 1.13, 1.27, 1.31, and 1.35 were the only feasible options available.

The mass reductions, displayed in Figure 9.13, are not necessarily the most ideal solutions to our optimization problem. Consequently, selecting the results from model 1.3 as the final design solution, which represents the lightest structure, may be intuitively tempting. It has to be observed, however, that choosing model 1.3 is no guarantee of obtaining the lightest and most stable solution. An optimization stage is therefore necessary.

As a starting point of this optimization, one of these models can be taken as references, based on the results of Figure 9.13. It is not a good practice to take either the lightest solution or the heaviest solution, but a model configuration for which the solution will be found in the heavier and lighter design space around about this model. The choice would then be models 1.7 and 1.31.

It was decided that certain variable values from run 1.31 would be considered for the optimization study as starting variables. Although this model does not provide the optimum values, it will propose a better starting point than the original model considered.

9.6 STRUCTURAL OPTIMIZATION

9.6.1 Optimization setup

In order to optimize the TARF vehicle, metamodels are required to capture complex responses. RBF using Hardy's Multi-Quadrics transfer function was considered the main metamodel for this study, along with space filling as the sampling method. Space filling was again taken into account because the space-filling algorithm does the positioning of all the points in a single cycle (Stander et al., 2013). Since a global optimization is being undertaken, various options are available, such as genetic algorithm (GA), adaptive simulated annealing (ASA), and a combination of global optimizers with local gradient-based optimization method known as Hybrid algorithm. The Hybrid algorithms are better suited for global optimization as they are both efficient and effective. ASA with Leapfrog Optimizer for Constrained minimization (LFOPC) was chosen as the algorithm for the optimization stage.

From the DOE study, dominant variables were established for individual loadcases and their significance on the mass of the vehicle structure. Variables 1, 3, 4, 5, 8, 11, and 12 from Table 9.6 were considered for the optimization stage, comprising seven panel thickness variables. The influential variables (6 and 7) unique to the torsion models listed in Table 9.6 were kept constant (Floor_Sa and Floor_Tu) and were assigned the baseline values and not the values from run 1.31.

This decision is made to further reduce the number of variables in the optimization process, hence reducing the computation time. This approach reduces the variables from 9 down to 7.

All the other variables with insignificant influence on individual responses were not considered for the optimization phase. Metamodels have been generated for each loadcase and are depicted in Figures 9.14, 9.17, 9.20 and 9.23.

Once the metamodels are generated, it is important to investigate the difference between the computed value by the solver and the extrapolated value through the response surface. The residual allows this to be observed by depicting small vertical bars linking the computed point to the extrapolated point in the response surfaces in Figures 9.14, 9.17, 9.20, and 9.23. Visually, the response surfaces generated through the computed points look adequate for all the loadcases.

It is also possible to observe the accuracy of the metamodels by visually looking at the quality of the fit using the R^2 (squared) value. R^2 is a statistical measure of how close the data is to the fitted regression line. Linear regression calculates an equation that minimizes the distance between the fitted line and all of the data points. Technically, ordinary least squares regression minimizes the sum of the squared residuals, or the percentage of the response variable variation, that is, explained as the ratio variation/total variation.

The R^2 value is always between 0 and 1.0.

- A low R^2 value indicates that the region of interest is either too large or too small and that the gradients are not trustworthy.
- A value of 1.0 for R^2 indicates a perfect fit. However, the value will not warn against an over-fitted model with poor prediction capabilities.

It can be observed the R^2 value is computed from a line; hence, least square regression on its own cannot determine whether the prediction is biased, or, in our case, highly nonlinear. It is, therefore, important to use the least squares regression, as well as the residual plots to generate a realistic assessment. Consequently, both metrics are important to certify whether the response surfaces are adequate.

9.6.2 Frontal impact loadcase

In all the cases of the frontal loadcase, the response surfaces are RBF functions that follow the sample points closely. In the case of the frontal loadcase (Figures 9.14, 9.15, and 9.16), it can be noted that the residual is small and that the least square regression fit value is above 0.7, which is indicating a fair level of correlation. It can be noted that in the case of the X-Acceleration response that the middle on the response surface is very convex, which is suggesting a high level of nonlinearity. It can be seen that the surface is stretched when away from the sampling points, which is suggesting that maybe more points may be needed to investigate whether the curvature at the center of this response surface is genuine or is due to insufficient number of data points. Three points suggest unfeasible design with a high residual on the X-Acceleration response surface. One of the reasons may be that, in the acceleration response, the event is recorded by an accelerometer. Accelerometers are very sensitive to the computer model timestep as they are the first entity to be computed as part of the explicit loop, as discussed in Chapter 2. Adding to this, the acceleration is filtered with a CFC60, which

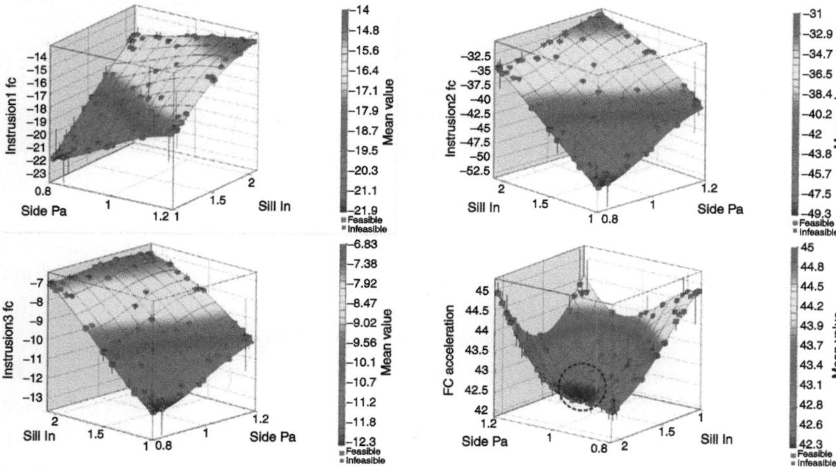

FIGURE 9.14 Response surfaces for frontal loadcase with residuals.

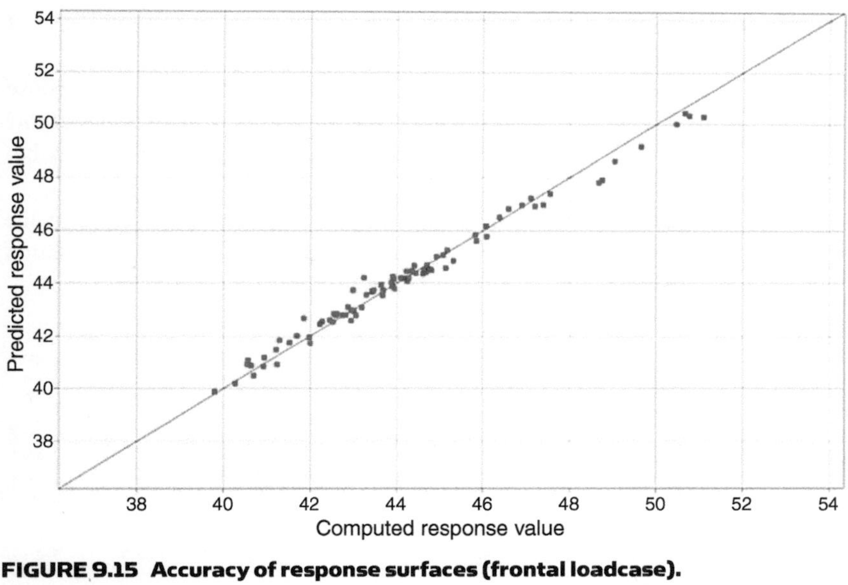

FIGURE 9.15 Accuracy of response surfaces (frontal loadcase).

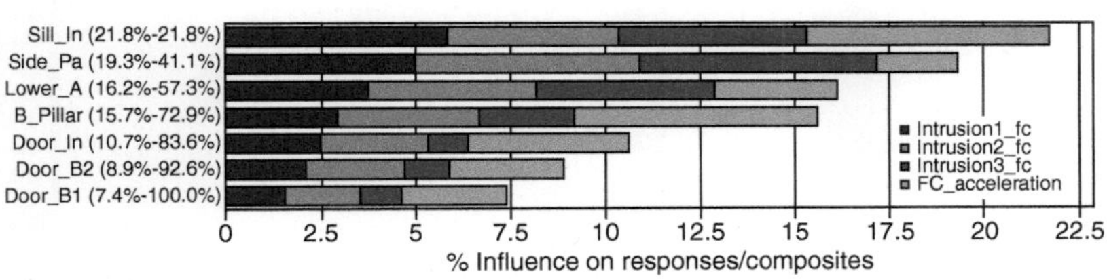

FIGURE 9.16 Sensitivities (frontal impact).

removes any unnecessary noise; however, due to the complex interaction between engine bay components and the vehicle structure, the response may be sensitive.

The displacement springs where the variables are recorded are connected to the vehicle footwell. There are numerous interactions inside the engine bay and the vehicle structure because of panels and components contacting each other. As a consequence, the overall responses recorded in a frontal crash scenario are highly nonlinear.

Looking at Figure 9.15, the R^2 value is 0.7 (minimum), which is suggesting a fair correlation between the computer results and the respective estimations from the response surfaces. It can be observed that, as explained before, the curvature in the X-Acceleration is not captured with this method.

The panel sensitivities solely for the frontal impact are depicted in Figure 9.16. It can be observed that the sill inner (Sill_In) has the most influence for all the variables studied, while the variable Door_B1 has the least influence on the frontal crash loadcase.

9.6.3 Pole impact loadcase

In the case of the pole impact, it can be observed that the response surfaces are more planar because there are fewer structural interactions than in the case of the frontal crash (Figures 9.17, 9.18, and 9.19).

There does not seem to be any distinctiveness in the pole impact response surfaces, as they visually pass through the sample points without any extreme change of curvature. The residual plots are showing minimal differences between computer and extrapolated samples (Figure 9.17).

Considering the accuracy plot (Figure 9.18), it can be observed that the value of R^2 is, in general, in the order of 0.9, which suggests an excellent correlation that visually looks tighter than the front crash loadcase (Figure 9.15).

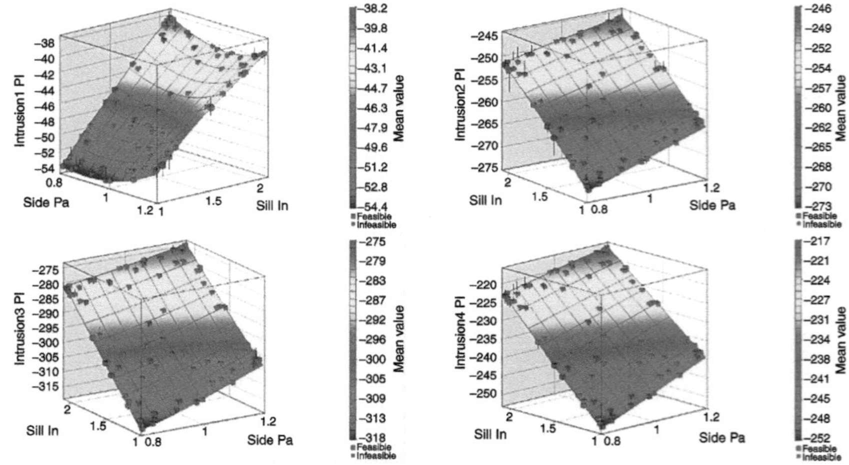

FIGURE 9.17 Response surfaces for pole impact loadcase with residuals.

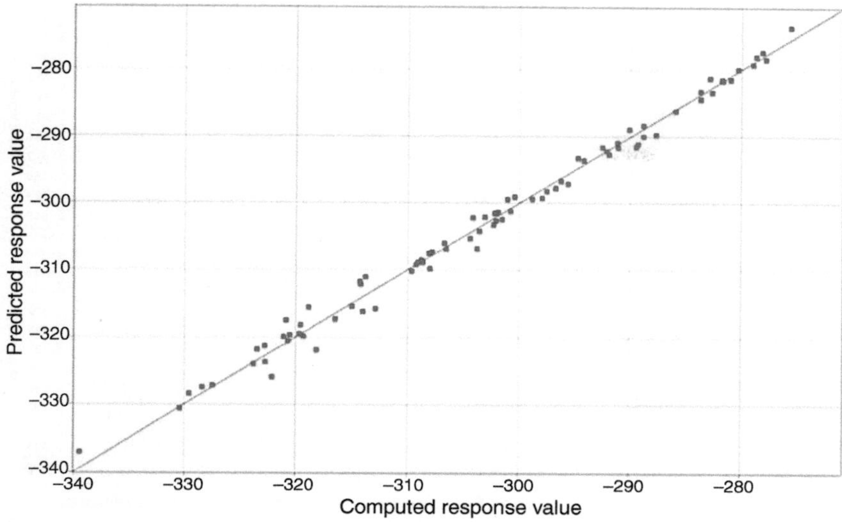

FIGURE 9.18 Accuracy of response surfaces (pole impact loadcase).

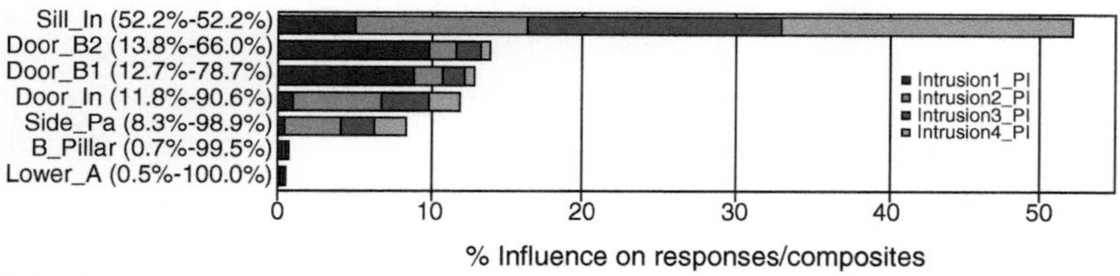

FIGURE 9.19 Sensitivities (pole impact).

The pole impact sensitivity analysis is suggesting that the sill inner (Sill_In) has the most influence in the lower-to-middle/upper side intrusions, while door reinforcements Door_B1 and Door_B2 have the most influence on the top lateral intrusion (Intrusion1_PI).

9.6.4 Side impact loadcase

In the case of the side impact, the same comments can be made as in the case of the pole impact (Figures 9.20, 9.21 and 9.22). In all the residual plots, residuals are small and there are no localized changes of curvature (Figure 9.20), unlike in the X-Acceleration response surface of the frontal crash (Figure 9.14).

The accuracy plots for the side impact loadcase (Figure 9.21) are comparable with the pole impact loadcase, where the R^2 value is in the order of 0.9.

Considering the sensitivity of the side impact loadcase (Figure 9.22) the door inner, sill inner, and B pillar have the most influence of the four variables studied.

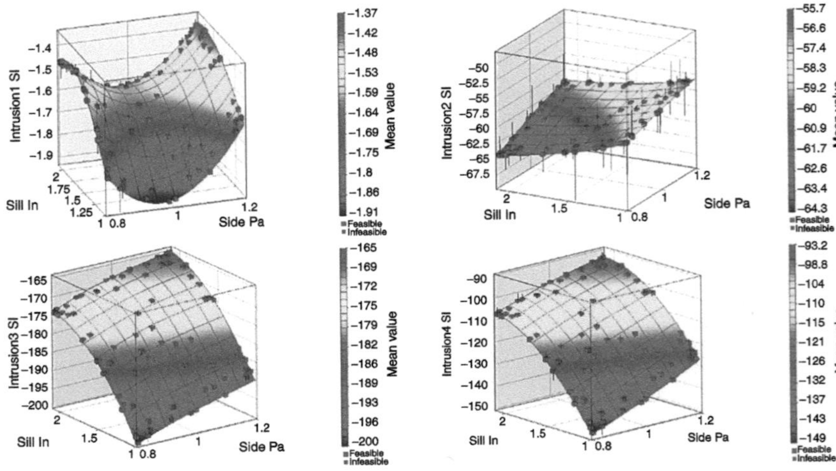

FIGURE 9.20 Response surfaces for side impact loadcase with residuals.

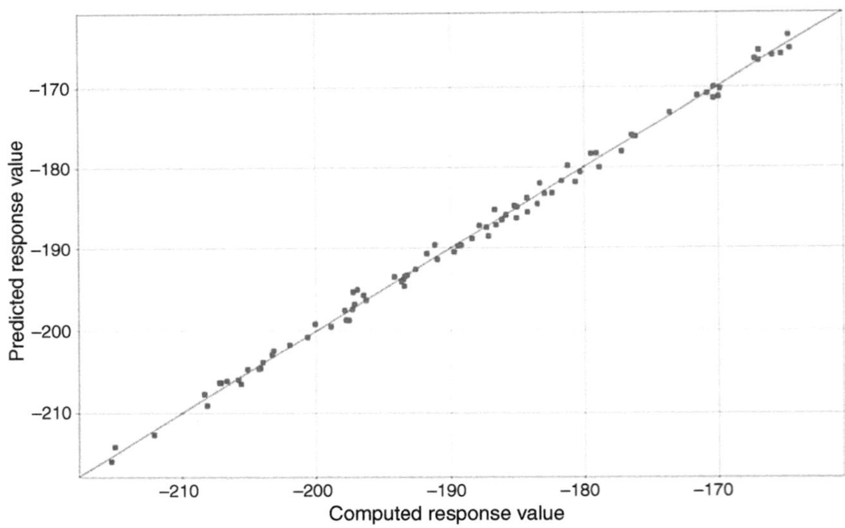

FIGURE 9.21 Accuracy of response surfaces (side impact loadcase).

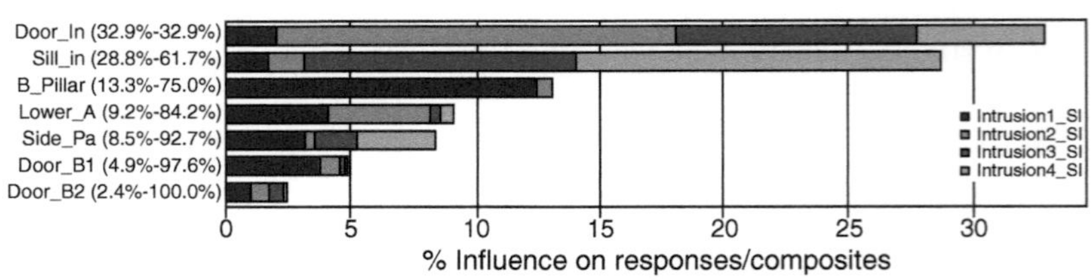

FIGURE 9.22 Sensitivities (side impact).

9.6.5 Torsional rigidity loadcase

Considering the torsional rigidity loadcase, it can be noticed that, in general, the curvature of the response surfaces is large; however, some large residuals can be observed at the boundaries of the design space, as illustrated in Figure 9.23. This is also captured in the correlation graph in Figure 9.24, where the distances

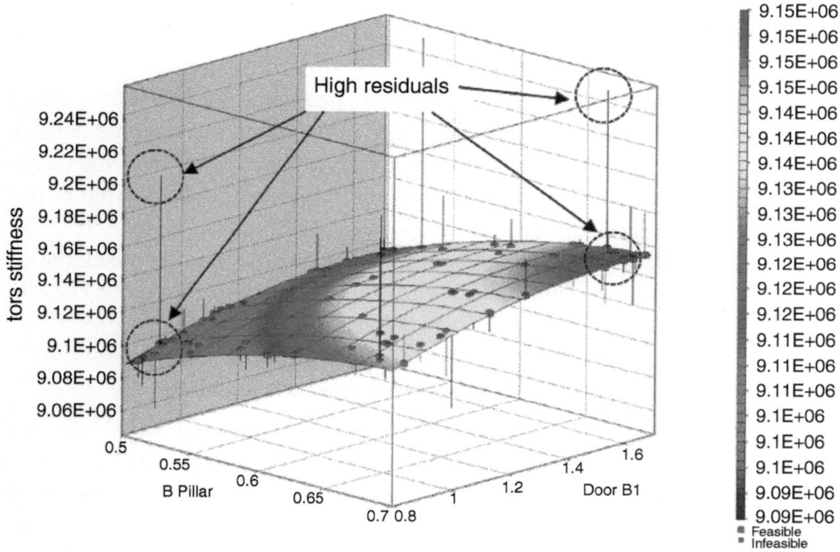

FIGURE 9.23 Response surfaces for torsional stiffness loadcase with residuals (discrepancies in circles).

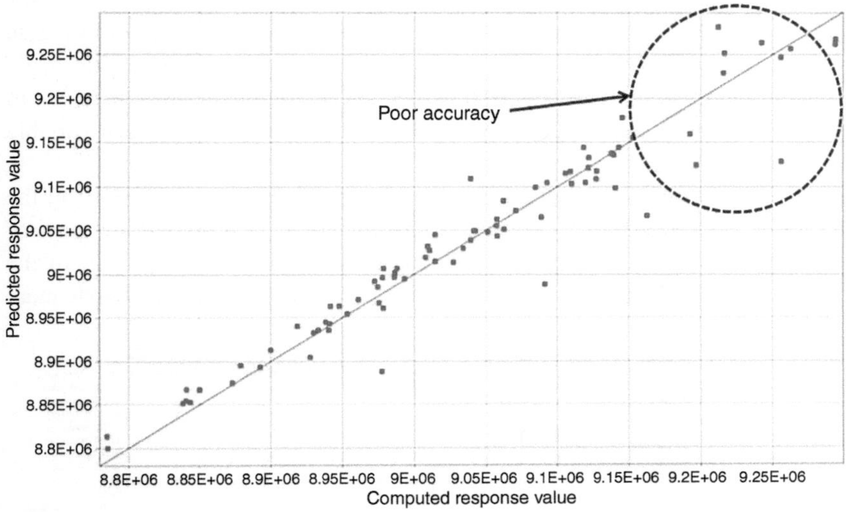

FIGURE 9.24 Accuracy of response surfaces (torsional stiffness loadcase) – (discrepancies in circles).

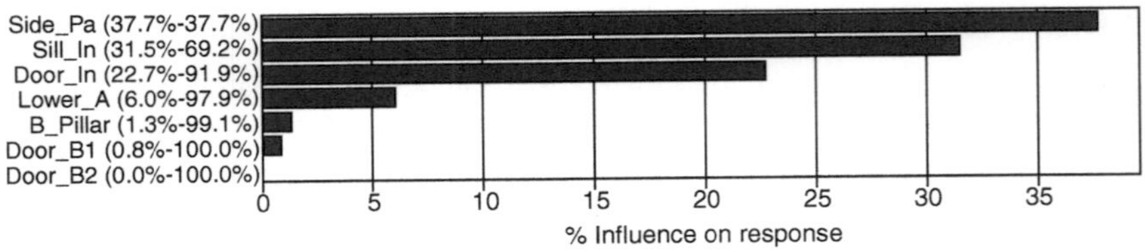

FIGURE 9.25 Sensitivities (torsional stiffness).

between the computed points and their extrapolated value are noticeable. This is suggesting that when the optimization is carried out along the response surface, should the global solution require a torsional rigidity response near the design boundary, then there may be some results discrepancies between what is predicted and the validation model.

The torsional stiffness seems to be mainly affected by the side panel, the sill inner, and the door inner (Figure 9.25) and the vehicle mass (Figure 9.27) by the vehicle side panel, door inner and doors B1 and B2.

9.6.6 General model responses and objective

The previous sections allowed investigating the influence of the seven variables over each of the loadcases independently. This can be very useful, as each sensitivity model can provide the designer and analyst with some guidance on which area to address to modify the vehicle response for a specific loadcase. Obviously, when these sensitivities are considered independently, then a manual change can affect other loadcase performance, which may not be desirable. Consequently, should a manual optimization using solely on engineering expertise take place at this stage, it is important to be aware of this and also consider the other sensitivity plots in order to find a compromise solution.

Combining all the sensitivities together can be complicated for someone to interpret; nevertheless, Figure 9.26 highlights the importance of each variable for all the loadcases.

The optimization study used mainly the same vehicle model across all the loadcases. As the variables were fully shared across each model, the mass of each model is the same. This is an important point to consider as part of the optimization; if one wishes to reduce the mass, it is easier to minimize one mass

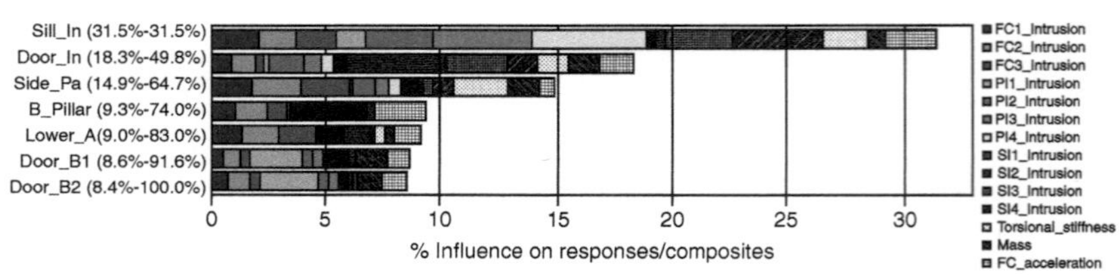

FIGURE 9.26 Overall model sensitivity (all loadcases and responses).

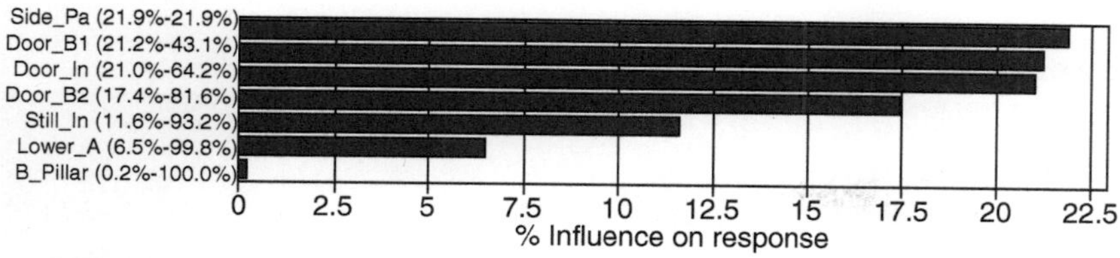

FIGURE 9.27 Sensitivities (mass).

objective. Keeping a shared mass variable across all the models has reduced the study to a single objective optimization and not a multi-objective optimization. Previous studies have indicated that partially sharing variables based on elimination of variables using sensitivity plots results in a faster convergence than other optimization methods, such as sequential (Craig et al., 2002).

The mass sensitivity responses are depicted in Figure 9.27 and are suggesting that the vehicle side panel, Door B1, Door inner, and Door B2 have the most effect on the mass.

9.6.7 Optimization results

An optimum solution was found while performing a single stage optimization, however it was noticed that one variable just missed the upper bound design constraints, as illustrated in Table 9.7.

It can be observed that variable "FC2_intrusion" is violating the design constraint by 1.45 mm. It can also be noticed that reducing the vehicle mass, in this case, has reduced slightly the impact deceleration severity, from 45.40 to 42.64 g and the side impact intrusion (highest location) from 1.37 to 1.14 mm. All the other variable gauges have been decreased in order to "detune" the vehicle structure, that is, make it more compliant by removing mass, while respecting the design boundaries (except for the "FC2_intrusion"). In all the cases, the gauges of the optimized solution are smaller than or equal to the baseline (Table 9.8).

In order to find a solution for the "FC2_intrusion" constraint violation, one has to compute more sample points to refine the response surfaces. This is certainly possible; however, there is a cost/time balance that also needs to be considered against the possible gain of achieving 1.45 mm of structure stiffness improvements. It would be proposed, for example, with the support of the frontal impact sensitivity (Figure 9.16) to change the local shape of the key panels which have influence on the variable "FC2_intrusion." A local change of second moment of area could achieve this while improving further all the other loadcases.

Considering the difference of masses of the components optimized, it can be concluded that the vehicle mass, based on the early optimized model, has been reduced by 27.53 kg. The baseline model weighed 330.15 kg, was reduced to 315.75 after DOE, and to 302.62 kg after full optimization. Consequently, the final optimized vehicle mass is 972.47 kg (from 1000 kg).

TABLE 9.7 Optimization Outcome

Scenarios	Response	Original without Optimization	Optimum	Upper Bound	Pass/ Fail	Units
Frontal crash	FC1_Intrusion	13.47	19.57	20	P	mm
	FC2_Intrusion	26.37	41.45	40	F	mm
	FC3_Intrusion	5.39	11.26	20	P	mm
	FC_acceleration	45.40	42.64	48	P	g
Pole impact	PI1_Intrusion	47.76	88.96	151	P	mm
	PI2_Intrusion	245.51	293.60	320	P	mm
	PI3_Intrusion	275.52	329.67	349	P	mm
	PI4_Intrusion	214.05	255.10	318	P	mm
Side impact	SI1_Intrusion	1.37	1.14	20	P	mm
	SI2_Intrusion	52.27	54.58	145	P	mm
	SI3_Intrusion	164.996	211.701	232	P	mm
	SI4_Intrusion	76.4197	141.316	150	P	mm
Torsion	Torsional_ stiffness	9483970	8772560	*8500000 (Lower Bound)*	P	Nmm/deg

TABLE 9.8 Optimization Results – Gauge Changes from Original Design

Variables	Baseline (mm)	DOE Run 1.31 (mm)	Optimum (mm)
B_PILLAR	0.80	0.60	0.60
DOOR_B1	1.50	1.00	0.80
DOOR_B2	1.00	1.80	0.60
DOOR_IN	1.50	1.70	1.00
LOWER_A	2.00	1.20	1.00
SIDE_PA	1.50	1.00	0.80
SILL_IN	2.00	1.00	1.50
FLOOR_SA	2.50	1.50	2.50
FLOOR_TU	1.50	1.20	1.50

The response of the TARF-LCV structure for the optimized variables meets virtually all except one of the constraints put forth for the optimization phase. Although the design constraints were met, the final solution did highlight an issue with the upper A-pillar "T" junction in the frontal crash scenario (Figure 9.28) and illustrated in Figure 9.29. Due to a lack of reinforcement in this region and predominantly due to the reduction in the gauge of the lower A-pillar reinforcement, the passenger compartment is being compromised slightly. The force transferred through the hydroformed upper A-pillar post is 16.6 kN which is well below its critical buckling load, which has been estimated at 50 kN using a linear buckling criteria. The optimized vehicle A-pillar structure is, therefore, stable.

Maybe as a future optimization it would be possible to output the compressive force (buckling load) traveling through the A-pillar and to use it as

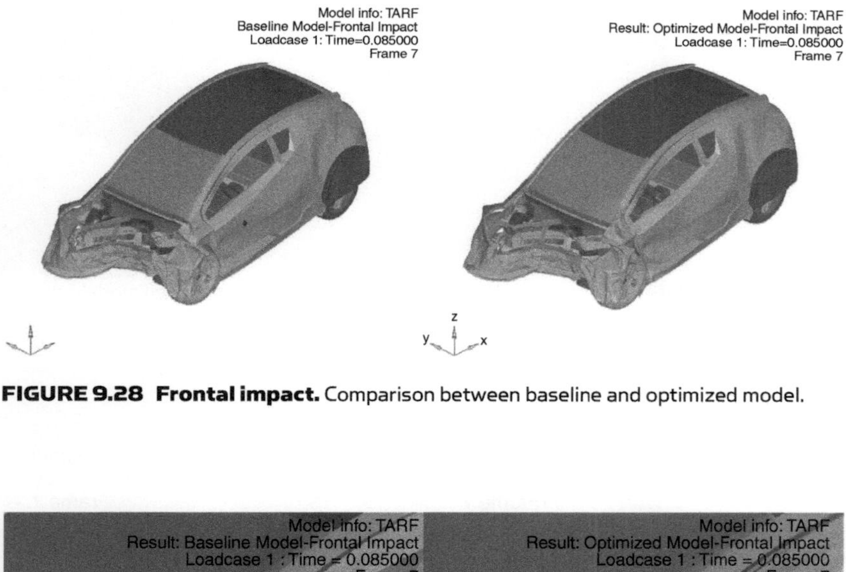

FIGURE 9.28 Frontal impact. Comparison between baseline and optimized model.

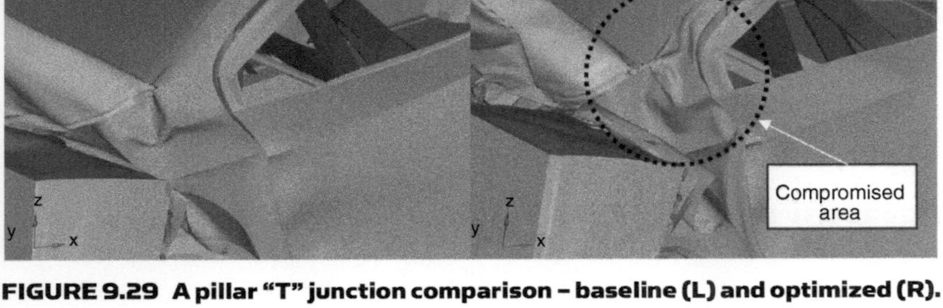

FIGURE 9.29 A pillar "T" junction comparison – baseline (L) and optimized (R).

structural frontal crash structural stability criteria. It can, therefore, be concluded that there are many approaches to structural optimization.

Considering the pole and side impact scenarios, it can be noted that the side intrusions are reasonable and that the optimization process has indeed allowed the structure to be more compliant. In both cases, the deformations of the optimized structure are more pronounced, as illustrated in Figures 9.30 and 9.31.

The same comment can be made for the torsional stiffness in Figure 9.32, where the chassis rotates more for the same amount of twisting load applied to the shock towers.

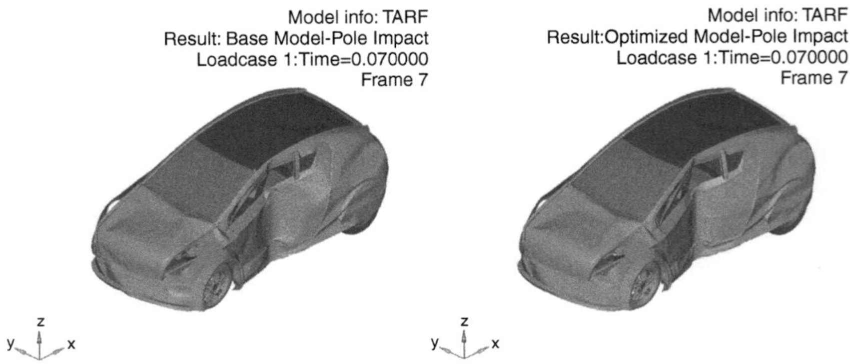

FIGURE 9.30 Pole impact. Comparison between baseline and optimized model.

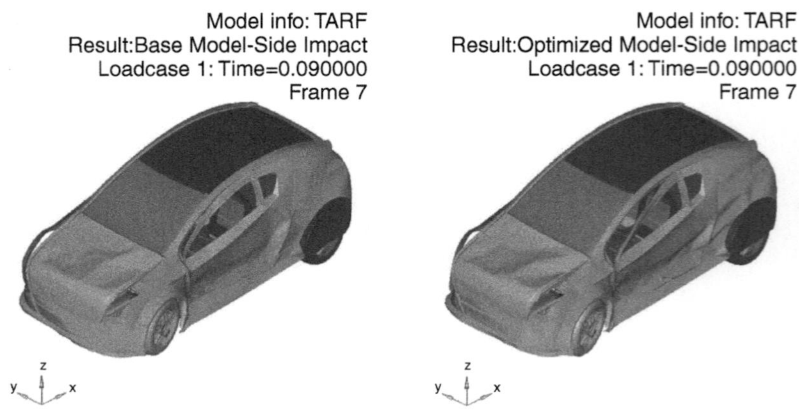

FIGURE 9.31 Side impact. Comparison between baseline and optimized model.

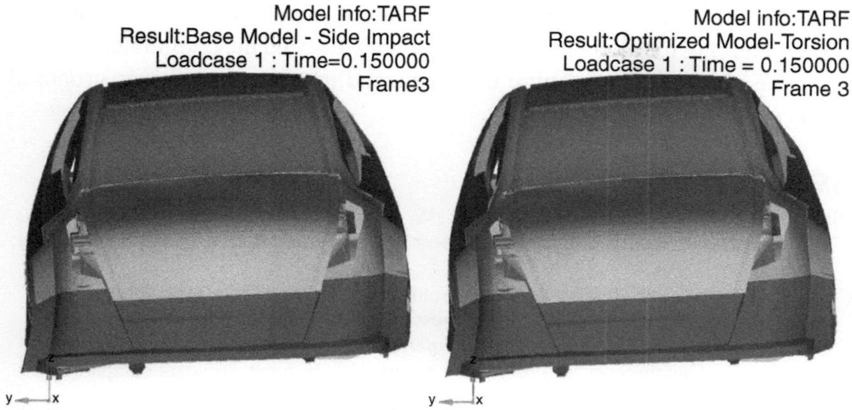

FIGURE 9.32 Torsional stiffness. Comparison between baseline and optimized model (displacement magnified ×30).

9.7 CONCLUSIONS

This chapter has shown that multidisciplinary design optimization is a very useful tool for identifying the global behavior of a vehicle structure. The sensitivity analysis conducted through the DOE study helped to identify the importance of the design variables with respect to the three crash safety scenarios and torsional loadcases. These variables are related to panel gauge and material grade stiffness variations and could have included some global panel shape changes. The geometry change was not conducted because of the package restriction available and the maturity of the vehicle design. As panel section alteration, changing the second moment of area would be a very efficient technique to manage stiffness and impact energy (bending and buckling). It is recommended that shape variables should be included in the design optimization as early as possible in the early concept stages to obtain improved performance over simple gauge changes.

The use of polynomial metamodel for the DOE study resulted in a lower number of simulation points being used. The sensitivity analysis before the optimization allowed for redundant variable elimination from the optimization stage. The polynomial metamodel ease of use, implementation and fairly accurate predictions through the response surface allowed for the design exploration, which yielded in an optimum value with significant mass reduction compared with the initial TARF model. The single stage optimization method is considered ideal for a limited simulation budget and is suited for global design exploration. RBF metamodel is suited to capture nonlinear responses and provides a good approximation of the optimum result for a small set of simulation points. The study facilitated a significant reduction in the mass of the vehicle. The best design would need to be investigated by using several metamodels and an even higher sampling for each metamodel to numerically find the optimum and investigate the cost benefit to acquire these data and improve the solution proposed in this study.

REFERENCES

Clarke, S. M., Griebsch, J. H., & Simpson, T. W. (2005). Analysis of support vector regression for approximation of complex engineering analysis. *Journal of Mechanical Design*, *127*(6), 1077–1087.

Craig, K. J., Stander, N., Dooge, D. A., & Varadappa, S. (2002). Multidisciplinary design optimization of automotive crashworthiness and NVH using LS-OPT. In *Proceedings from 7th international Ls-dyna user's conference*.

Eigen, A.M., & Glassbrenner, D. (2003). The relationship between occupant compartment deformation and occupant injury. NHTSA Technical Report, 37.

European New Car Assessment Programme: Euro NCAP (2012) Side Impact Testing Protocol V6.0 Available from <www.euroncap.com/en/for-engineers/protocols/adult-occupant-protection/>Accessed 01.03.15.

Jin, R., Chen, W., & Simpson, T. (2001). Comparative studies of metamodelling techniques under multiple modelling criteria. *Structural and Multidisciplinary Optimization*, *23*(1), 1–13.

Kim, B. -S., Lee, Y. -B., & Choi, D. -H. (2009). Comparison study on the accuracy of metamodeling technique for non-convex functions. *Journal of Mechanical Science and Technology*, *23*(4), 1175–1181.

LCVTP. (2011). Low Carbon Vehicle Technology Project. Re: BG/AM1011, partly funded by ERDF and AWM. Project based on Tata 'Beacon' vehicle layout. Participation of Coventry University for vehicle aerodynamics, HVAC and Vehicle Body Structure.

Li, Y., Ng, S., Xie, M., & Goh, T. (2010). A systematic comparison of metamodeling techniques for simulation optimization in decision support systems. *Applied Soft Computing*, *10*(4), 1257–1273.

LSTC. (2015). LSTC Barrier Models. Available from <http://www.lstc.com/products/models/barriers> Accessed 10.06.15.

Martin, D. B., Ciarlet, P. G., Iserles, A., Kohn, R. V., & Wright, M. H. (2003). *Radial basis functions theory and implementations*. Cambridge: Cambridge University Press, p. 271.

Ryberg, A. B., Bäckryd, R. D., & Nilsson, L. (2012). *Meamodel-based multidisciplinary design optimization for automotive applications*. Linköping: Linköping University, Technical Report.

Stander, N., Roux, W. J., Giger, M., Redhe, M., Fedorova, N., & Haarhoff, J. (2003). Crashworthiness optimization in LS-OPT: case studies in metamodeling and random search techniques. In *Proceedings from 4th European LS-DYNA conference*. Ulm, Germany.

Stander, N., & Goel, T. (2008). Metamodel sensitivity to sequential adaptive sampling in crashworthiness design. In *10th International LS-DYNA user's conference*.

Stander, N., Roux, W., Basudhar, A., Effleston, T., Goel, T., & Craig, K. (April 2013). *LS-OPT user's manual, Version 5. 0*. California, USA: Livermore Software Technology Corporation.

TARF. Towards affordable closed-loop recyclable future-low carbon vehicle structures (2011) EPRSC funded project. <http://gow.epsrc.ac.uk/NGBOViewGrant.aspx?GrantRef=EP/I038616/1> Accessed 01.03.14.

Wright, B. (2003). Radial basis function interpolation: numerical analytical developments. PhD Thesis, submitted at the Faculty of the Graduate School of the University of Colorado. Available from <https://amath.colorado.edu/faculty/fornberg/Docs/GradyWrightThesis.pdf> Accessed 10.07.14.

Chapter | ten

Optimization for Refinement of Vehicle Safety Structures

CHAPTER OUTLINE

10.1 INTRODUCTION

In the previous chapters, various optimization techniques were discussed, addressing the development of vehicle safety structures.

The following chapter will highlight two important vehicle safety cases relating to occupant and pedestrian protection.

It is evident that such requirements could be directly included within a holistic framework of vehicle optimization; however this chapter will analyze each of them as a unique loadcase Consequently, this approach will reduce the number of optimization variables and highlight that such analyses are on their own very complex and highly nonlinear. In a crash event, once the intrusions are minimized, the restraint system is needed to reduce the occupant's deceleration levels, as well as limiting head and chest contact forces against the steering wheel. The pedestrian protection loadcase, on the other hand, can have an important influence in the vehicle front end. Such legislation can affect the vehicle styling, as well as the vehicle structure compliance that is, bumper stiffness, top of the vehicle grille, as well as the bonnet. This chapter will present 2 typical optimization problems one based on a full frontal restraint system design and the other one on the implementation of the lift strategy for a deployable pedestrian bonnet system.

Nonlinear Optimization of Vehicle Safety Structures. http://dx.doi.org/10.1016/B978-0-12-804424-7.00010-5

10.2 OCCUPANT PROTECTION ANALYSIS CASE STUDY

10.2.1 Introduction

The case study will assess the implementation of a restraint system for a vehicle impacting a rigid wall at 35 mph. The model used in this research is based on a multibody model coded in Madymo, in which selected variables were parameterized to illustrate the optimization of a restraint system.

In this example, it is assumed that the crash pulse signature is recorded as depicted in Figure 10.1. This crash pulse contains typical phases:

- Time 0 to 0.02 s, usually represents the collapse mode of the vehicle bumper crush can structure and the flex of the bumper system.
- Time 0.02 to 0.05 s, usually represents the collapse mode of the vehicle front longitudinal structure, including bonnet and shotguns;
- Time 0.06 s, usually depicts the peak of deceleration with components in the engine bay contacting each other and creating large resistance to the vehicle motion.

Not wearing the seatbelt, in general, can be tragic, as recorded in the Global Status Report On Road Safety 2013 report (WHO, 2013). Wearing a seatbelt reduces the risk of a fatal injury by 40–50% for drivers and front seat occupants. On Nebraska roadways, for example, 565 unbelted vehicle occupant fatalities were recorded between 2006 and 2010, which is an average of 113 fatalities per year. This accounts for 49% of all traffic fatalities during the five-year period and approximately 51% of all vehicle occupant fatalities (Nebraska Strategic Highway Safety Plan, 2013). In the UK, based on an AA Streetwise survey, 1 of every 20 drivers (5%) does not wear his seatbelt (BBC News, 2011).

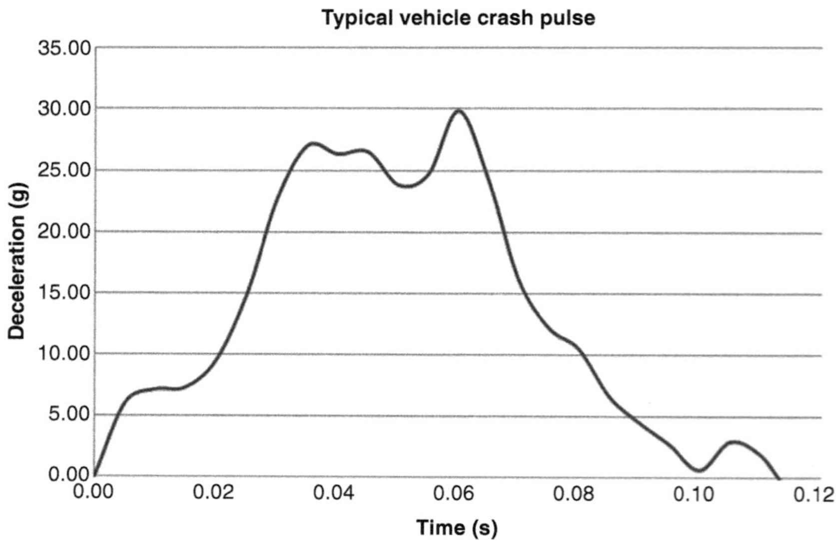

FIGURE 10.1 Typical vehicle crash pulse.

TABLE 10.1 Unbelted Cases from NASS Databases

	Number of Samples	Percentage	Percentage
Killed (AIS 6)	138	11.6%	41.0%
Seriously injured (AIS 3-5)	349	29.4%	
Moderately injured (AIS 1-2)	596	50.2%	
No injuries	105	8.8%	
All	1188	100.0%	

It has also been well-documented (NHTSA, 2007) that the use of the seatbelt at the time of the crash makes a difference in the need for hospitalization, as people not wearing their seatbelts during the accident were more likely to be hospitalized compared with those wearing it (32% vs. 19%).

Considering the National Automotive Sampling System Database (NASS) (1997–2003) (NHTSA, 2013c), it can also be suggested that, for unbelted cases, killed and seriously injuries represent 41% of all unbelted cases, for the front left row (driver side) as displayed in Table 10.1.

When this crash pulse is exerted on to an unbelted occupant, then, based on the Madymo computer model used in this study, the following injury outcomes are observed (Table 10.2) and concurs with the statements above.

The values in Table 10.2 recorded are statistically above any survival values (NHTSA, 2015), especially HIC, chest acceleration, chest deflection, and neck injuries.

The occupant kinematics is shown in Figure 10.2 against time.

At time 0, the occupant is in position on his seat. The first 0.02 s, when the vehicle crush cans are collapsing, the occupant moves forward relative to the vehicle cabin interior and gains inertia. One has to remember that the crash test dummy is passive and cannot brace.

At time 0.04 s, the arms are leaving the steering wheel and the pelvis has moved forward further horizontally, as well as vertically; however, because of

TABLE 10.2 Injuries Inflicted on Unbelted Occupant

Maximum Injuries (No Restraint System)									
HIC	Chest Acc. (g)	Chest Deflection (mm)	NCE	NCF	NTE	NTF	Left Femur (N)	Right Femur (N)	Head Acc. (g)
70,505	106.5	82.8	1.9	1.9	1.3	0.78	6423.5	6005.3	1400

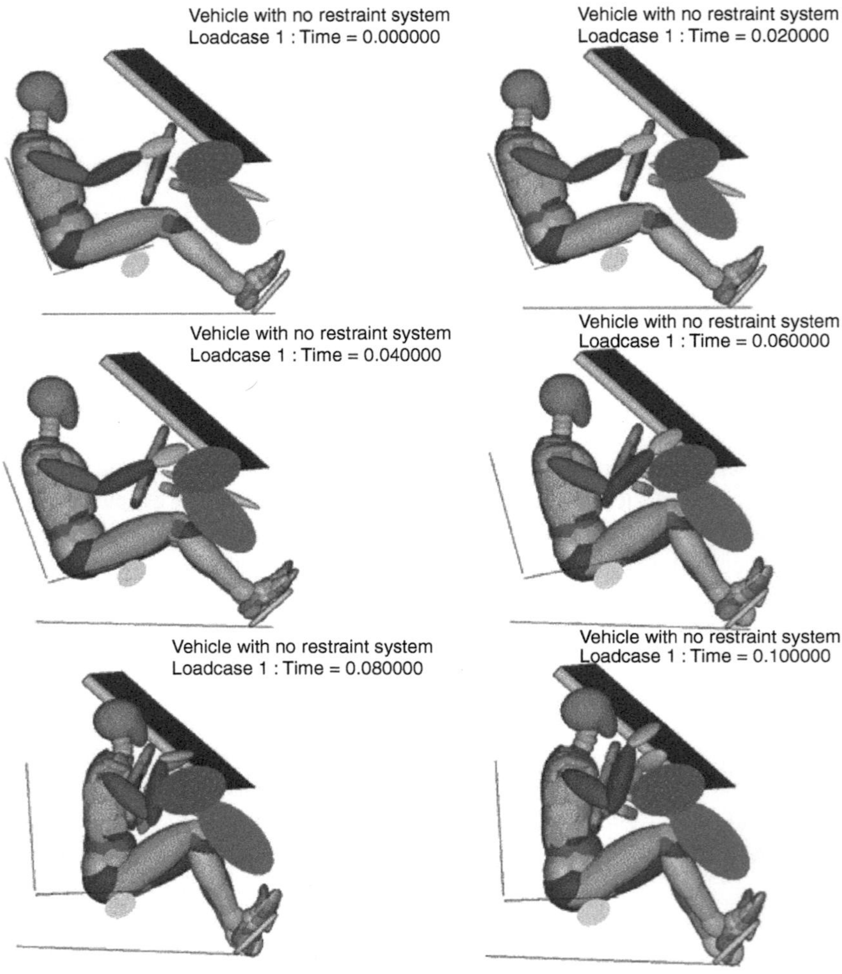

FIGURE 10.2 Unbelted occupant kinematics.

the inertia force, the thighs are deforming the seat cushion and initiate contact with the anti-submarining pan, represented by a circle under the seat pan. At this point, the knees contact the knee bolster. A rise in pelvis forces can be observed in Figure 10.3. The thorax's slight rise in deceleration is created by the contacts generated by the head and the knees, as they both are a loadpath to the thorax. Same comment is given for the head.

At time 0.06 s, the knees and pelvis do not move anymore as they are fully in contact with the knee bolster and the anti-submarining pan. The chest contacts the steering column (maximum thorax acceleration) and the chest forces then start to rise exponentially.

At time 0.078 s, the thorax force is highest and the torso rotates about the pelvis (locked between the anti-submarining pan and the knee bolster), which

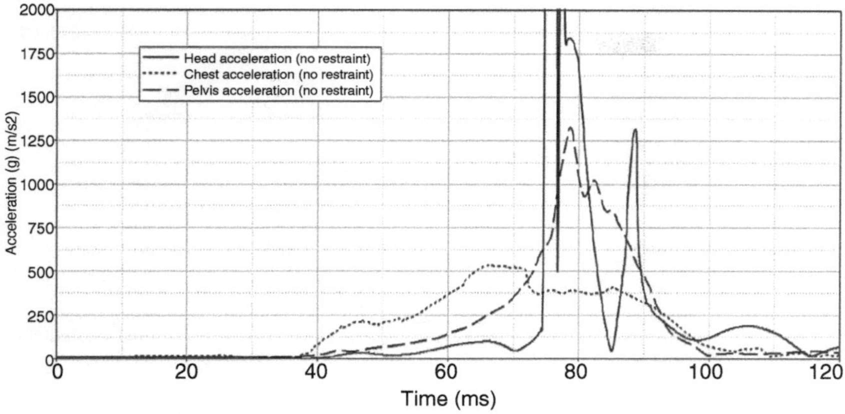

FIGURE 10.3 Occupant acceleration response (no restraint system). Head acceleration cropped at 200 g for clarity (max 1400 g).

leads to the head impacting the windscreen. This contact can be observed in Figure 10.3, as the deceleration is then extreme.

This scenario can be alleviated if the proper restrain system is fitted to the vehicle.

Restraint systems differ in technologies, but usually contain the following hardware features:

- An airbag, which inflates and creates a cushion between the head, thorax, and the steering wheel/steering column;
- A seatbelt, which keeps the occupant coupled to the seat and prevents ejection;
- A pretensioner, which removes the seatbelt slack and couples the occupant to the seat as early as possible;
- A load-limiter, which releases the seatbelt load to at a set value to reduce chest forces and deflections.

It is proposed to carry over a pre-existing restraint system comprising of a seatbelt assembly and an airbag module. Following this implementation, the occupant's injuries are listed in Table 10.3 with the kinematics depicted in Figure 10.4.

Looking at Figure 10.4, the pelvis acceleration curves increase sharply at time 0.016 s, which is caused by the firing of the seatbelt pretensioner which

TABLE 10.3 Injuries Inflicted on a Restrained Occupant

Maximum Injuries (With Restraint System)									
HIC	Chest Acc. (g)	Chest Deflection (mm)	NCE	NCF	NTE	NTF	Left Femur (N)	Right Femur (N)	Head Acc. (g)
347	38.6	34.8	0.00	0.10	0.12	0.16	73	2936	60

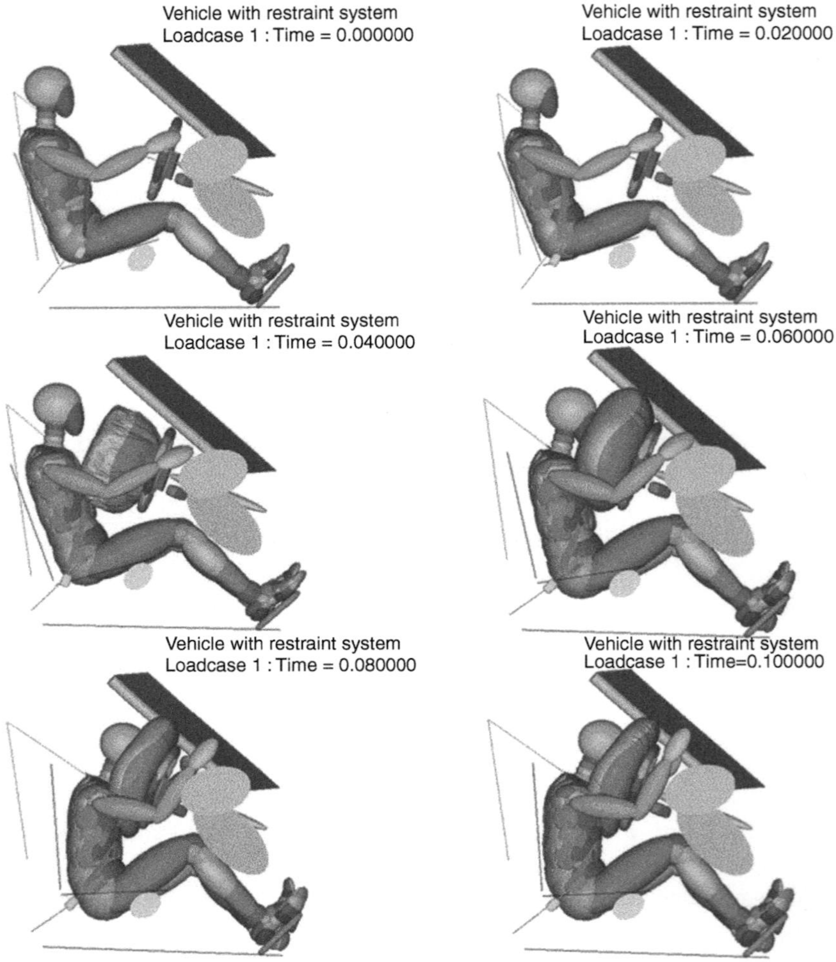

FIGURE 10.4 Belted occupant kinematics.

aims to remove all of the slack in the belt system. The pretensioner is needed to improve the coupling of the occupant to the seat.

At time 0.058 s, the head acceleration increases because of the contact between the head and the airbag (Figure 10.5). At this point in time, it can be wondered why an airbag should be used. Following the captions in Figure 10.4, it can be noted that the occupant's head is not contacting the windscreen and the steering wheel anymore. A cross section through the occupant and the restraint system in Figure 10.6, is proving that no head contact is visible with the steering wheel during the whole duration of the crash event. Consequently, the airbag is used to manage the occupant's inertial forces and, when well engineered, can prevent hard head contacts against the steering wheel.

When the two scenarios are overlaid, as depicted in Figure 10.7, it can be noted that the decelerations in the restrained scenario in the first 0.06 s are higher, which means that energy is removed from the occupant, while in the unbelted

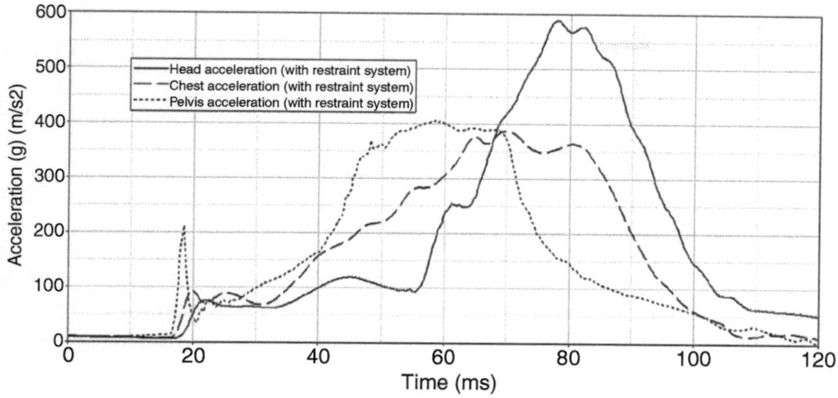

FIGURE 10.5 Occupant acceleration responses (with restraint system).

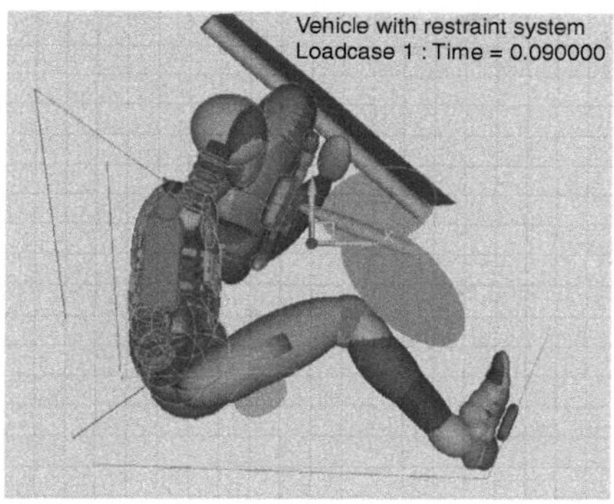

FIGURE 10.6 Occupant head missing windscreen and steering column thanks to restraint system (time 0.09 s).

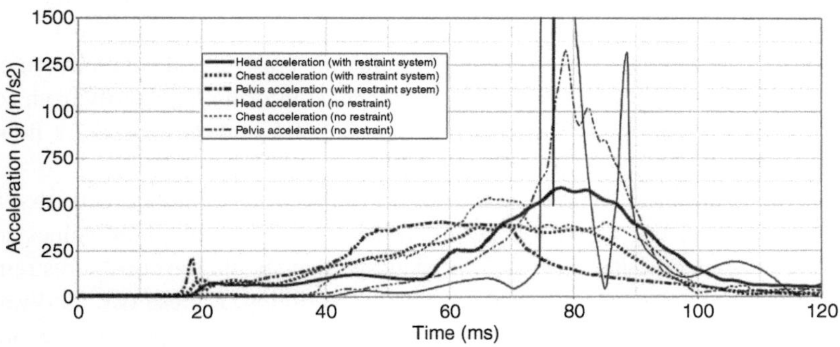

FIGURE 10.7 Occupant acceleration comparison between unbelted and restrained scenarios.

one, before 0.040–0.060 s, very little energy is dissipated. Consequently, the occupant will impact the cabin structure with more momentum, which will lead to higher injuries.

This example clearly illustrates the need of restraint systems in vehicle design as it correlates with the NHTSA findings listed in Table 10.1.

10.2.2 Optimization parameters

As the restraint system has been carried over from another motor vehicle; it is proposed to investigate whether there are any opportunities to optimize it further.

Optimizing a restraint system can be very complex, as there are many design variables that can be tuned to achieve better safety performance. These parameters can include, among others:

- Inflation pressure
- Airbag volume
- Airbag geometry
- Airbag folding pattern
- Airbag weave stiffness (warp/weft)
- Airbag cloth permeability
- Airbag venting
- Airbag firing time
- Seatbelt pretensioner firing time
- Seatbelt load limiter
- Steering column collapsible collapse force characteristic
- Anti-submarining pan collapse force characteristic
- "D" loop ring position
- "D" loop geometry and friction
- Etc.

It is proposed in this study to carry over a restraint system for another vehicle and to change simply the pyrotechnic airbag charge, the pretensioner firing time, and the airbag firing time.

In all restraint systems, one of the first actions is to couple the occupant to the seat before the airbag is deployment. Consequently, the timing of activation of the seatbelt pretensioner is assumed to be earlier or equal to the firing time of the airbag.

In the study undertaken, it is supposed that the crash signal is confirmed by the electronic control unit (ECU) that has received information mainly from accelerometers (usually positioned on the engine block and the lower seat floor, located at the base of the "B" pillar).

Looking at Figure 10.8, the pretensioner in the initial vehicle is activated at 16 ms and the airbag deploys at 20 ms. It has to be noted that these values are vehicle dependent; consequently, it is proposed that should the current restraint system be implemented on this new vehicle, the timing of these two hardware features may be changed between 14 and 18 ms for the seatbelt pretensioner and 18 to 24 ms for the airbag firing times.

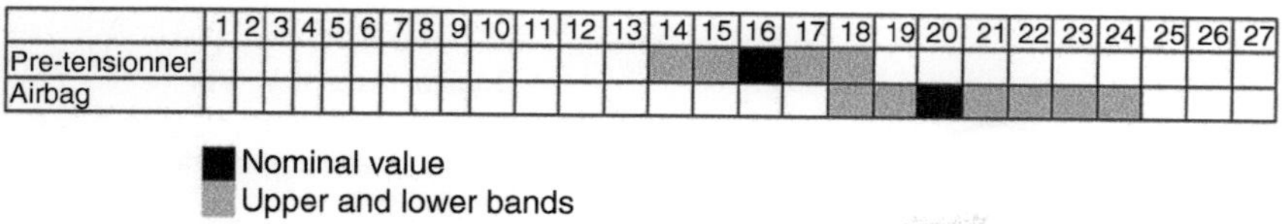

FIGURE 10.8 Firing sequence range of seatbelt pretensionner and airbag deployment.

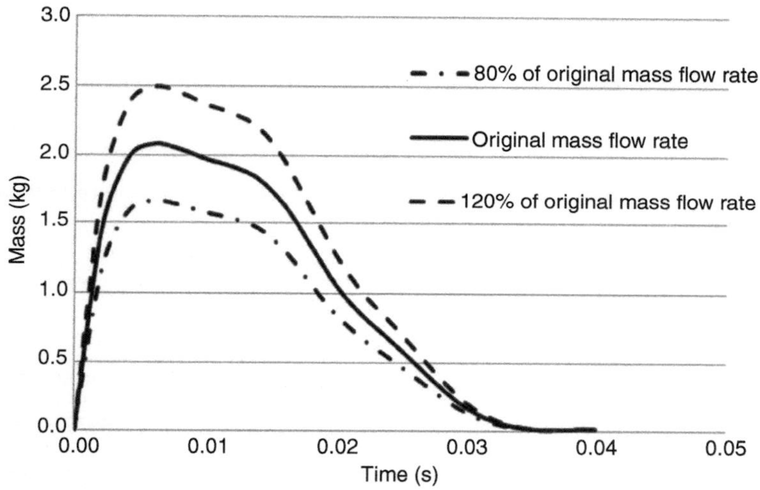

FIGURE 10.9 Airbag inflator characteristics variable spread (single-stage deployment).

It is also envisaged in this study that the airbag inflator may be considered a design variable and hence vary arbitrarily from 80 to 120% of the original inflator flow rate, as depicted in Figure 10.9. In reality, a different airbag inflator may have a different response than a straight scalar; however this proposed method of investigation would provide a design direction on the strength of the inflator to use. It has to be noted, however that this simple scaling method would not be suitable to replace a single stage inflator by a dual stage inflator, as illustrated in Figure 10.10, as the flow rate shape is very different, that is, less in magnitude but remains constant, in general, longer than a single-stage device.

10.2.3 Optimization setup

The optimization method used in this study is based on a gradient-based solution. The methodology is explained in the flowchart illustrated in Figure 10.11.

10.2.4 Step 1: definition of variables and responses

Step 1, the initial step, sets up the initial computer run with the nominal airbag system design variables, that is, seatbelt pretensioner, airbag firing time, and

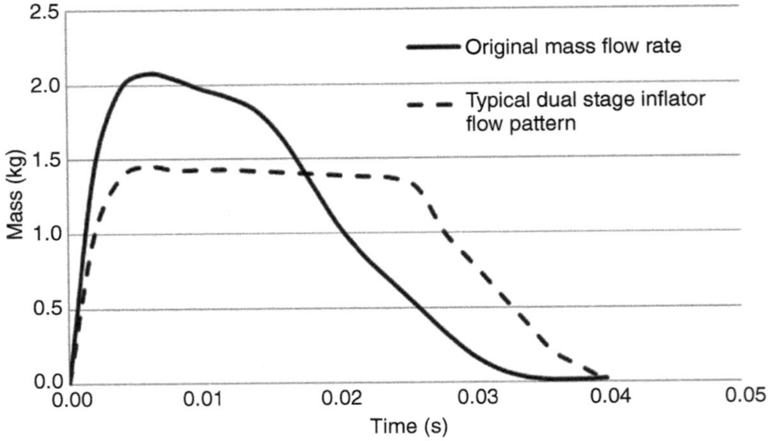

FIGURE 10.10 Typical difference of inflation between single-stage and double-stage inflation.

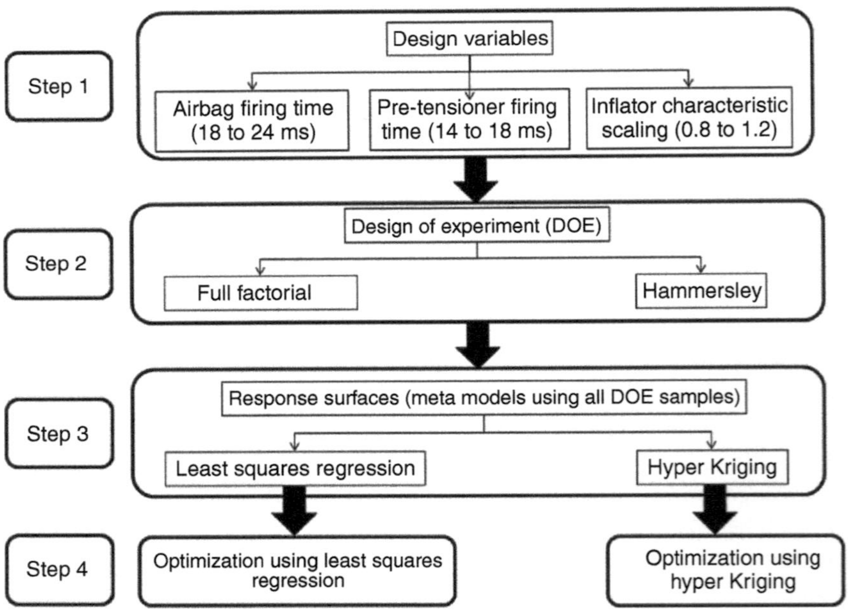

FIGURE 10.11 Occupant safety optimization flow chart.

inflator characteristic scalar to 16ms, 20ms and 1.0, respectively. This step also defined the output which will need to be extracted in each computer run, and feed into Step 2.

The outputs proposed for the study are:

- Head injury criteria (HIC)
- Chest acceleration (g)
- Chest deflection (mm)

- Neck injury (compression–extension) – NCE (no unit)
- Neck injury (compression–flexion) – NCF (no unit)
- Neck injury (tension–extension) – NTE (no unit)
- Neck injury (tension–flexion) – NTF (no unit)
- Femur load left (N)
- Femur load right (N)

Step 1 also provides the variable ranges that will be taken into account in Step 2.

10.2.5 Step 2: design of experiment (DOE)

Step 2 creates design permutations using two methods, one being full factorial and the other Hammersley.

Full factorial creates all the combinations possible from the set variable and variable levels available. Considering the problem, there are three variables (seatbelt pretensioner, airbag firing time, and inflator characteristic scalar), each having two levels.

The full factorial permutations are, therefore automatically computed using Equation (10.1) (number of permutations) and illustrated in Figure 10.12:

$$(\text{number of variables})^{\text{number of levels}} = 2^3 = 8 \text{ permutations} \qquad (10.1)$$

Looking at the sampling point location, it can be noted that the response at the corners of the design space will be very well-captured, while the accuracy at the center of the design space will be lacking as no sampling points are used. Consequently, using full factorial sampling on its own is not good practice, hence the need to sample the void between the points located at the extremities of the design space and consequently the use of the Hammersley method.

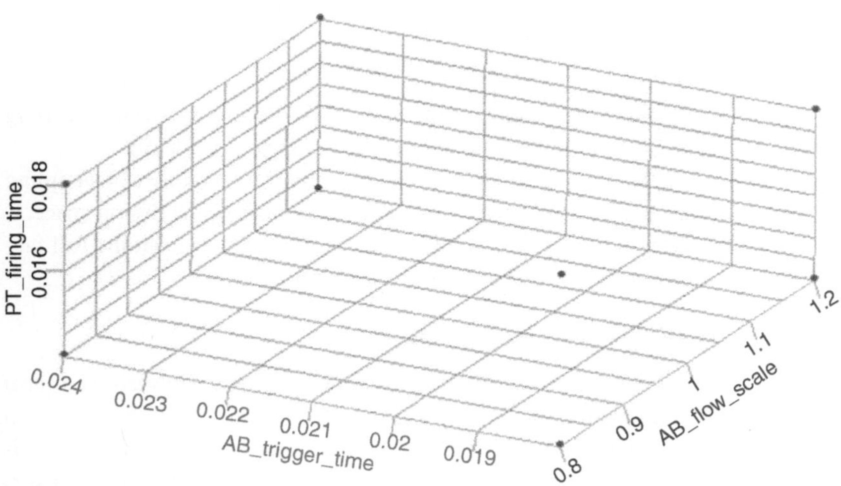

FIGURE 10.12 Full factorial sample selection.

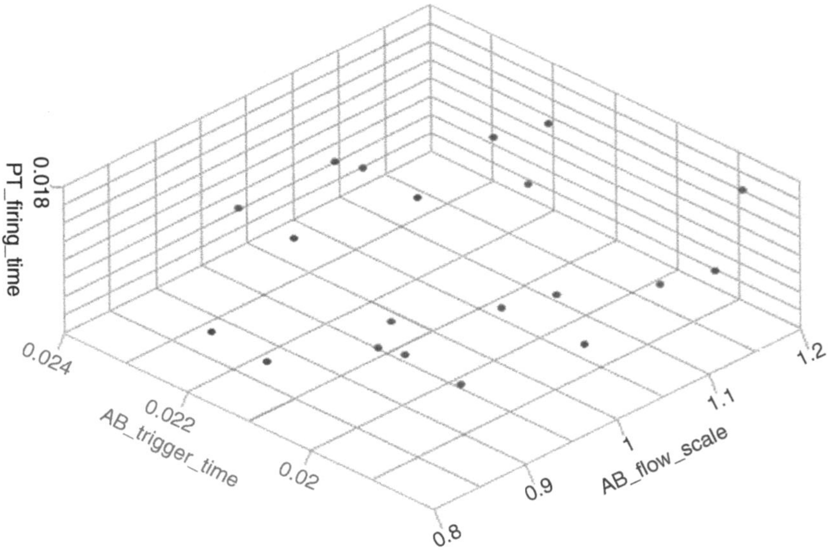

FIGURE 10.13 Hammersley sample selection (20 samples).

Hammersley sampling belongs to the category of quasi-Monte Carlo methods. This technique uses a quasi-random number generator, based on the Hammersley points, to uniformly sample a unit hypercube. Hammersley sampling is an efficient sampling technique that provides reliable estimates of output statistics using reasonably fewer samples (Viana, 2013).

The minimum number of Hammersley permutations is, therefore computed using Equation (10.2) and illustrated in Figure 10.13:

$$\text{Minimum number of Hammesley sample points} = \frac{(N+1)(N+2)}{2} = \frac{(3+1)(3+2)}{2} = 10 \text{ permutations} \tag{10.2}$$

It is possible, therefore to increase the number of Hammersley points as long as they are above the minimum number required in Equation (10.2).

In this study, it has been decided to double the minimum number of sampling points to 20 Hammersley points to achieve a better capture of the design space.

The full DOE will include the combination of the samples from the full factory and the samples from the Hammersley selection.

10.2.6 Step 3: response surfaces

This stage creates a meta-model of the Madymo multibody model. The term "meta-model" is understood as a simplified mathematical model that acts as a replacement from the multi-body element model in order to compute the physical responses of the restraint system. This meta-model is in the shape of a surface, hence the terms "Response Surfaces."

Considering the sample population computed by the DOE and illustrated in Figures 10.12 and 10.13, there can be, in general, two major types of response surfaces:

1. The response surface aims to pass through a plane that represents the average responses (least square regression (LSR)). This plane passes through the center of gravity of all the responses and the slope is calculated as being the best compromise (smallest distance) between responses and the plane, using a least square distance of each point to that plane. The advantage of this method is that it is very fast to compute and usually converges to a solution. However, if the plane is far from some responses, in case of severe nonlinearity, this convergence will be incorrect. This method is efficient for small levels of nonlinearity.

It can be noticed in the ISO view in Figure 10.14 that the response surface is represented by a plane, within the design space set in this study, which suggests that the chest deflection increases the earlier airbag triggers and the higher inflation scalar.

It can be noticed in Figure 10.15 that the plane is not passing through some of the points; hence, some optimization inaccuracy will result in the chest deflection computed from this response surface.

It can also be observed that the later the airbag trigger time, the higher the chest accelerations for a constant airbag inflation scalar (Figure 10.16).

Again, the least square regression (LSR), depicted in Figure 10.17, fails to capture adequately the occupant's chest accelerations as a meta-model; however, it is apparent that the chest acceleration trend is going downward like the sample

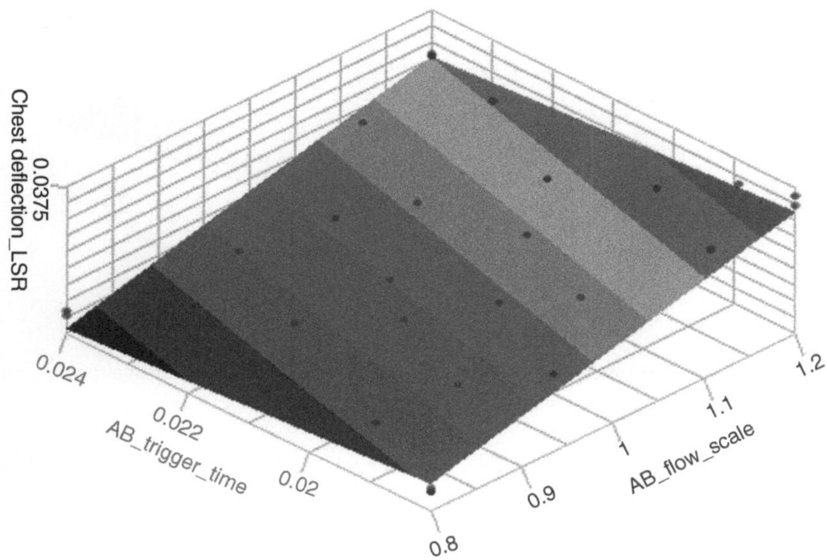

FIGURE 10.14 LSR response surface. Chest deflection as a function of airbag trigger time and airbag flow scaling value (ISO view).

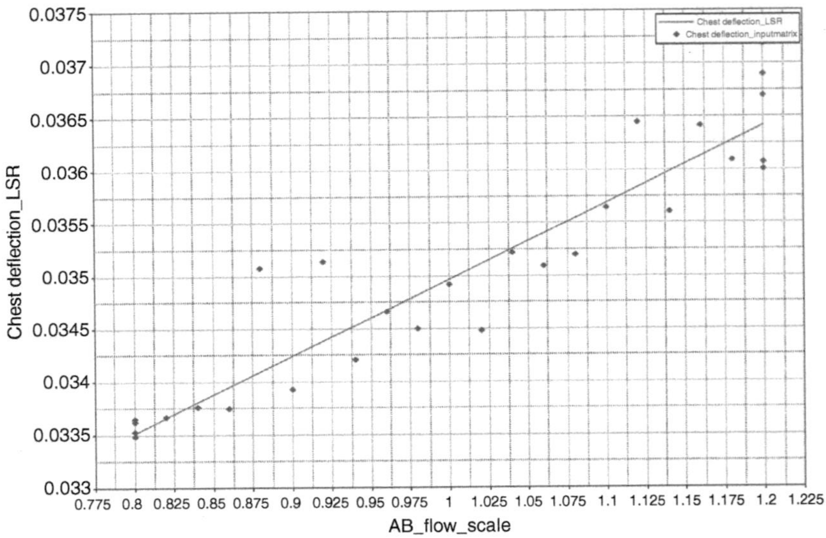

FIGURE 10.15 LSR response surface. Chest deflection as a function of airbag trigger time and airbag flow scaling value.

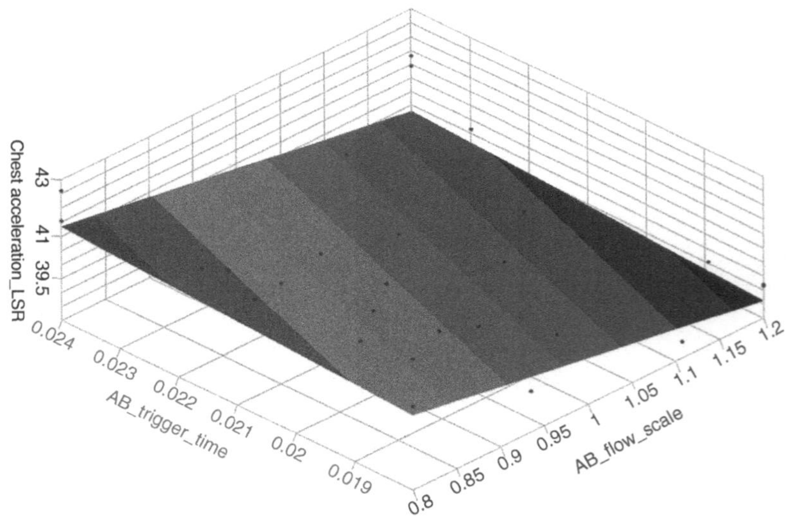

FIGURE 10.16 LSR response surface. Chest acceleration as a function of airbag trigger time and airbag flow scaling value (ISO view).

responses. Consequently, it can be said that such a meta-model would provide a qualitative design direction, but not a quantitative one.

2. The response surface aims to pass through every single response; this method is called Hyper-Kriging (HK). The advantage of such a method is that the response surface embraces more response at the sampling points,

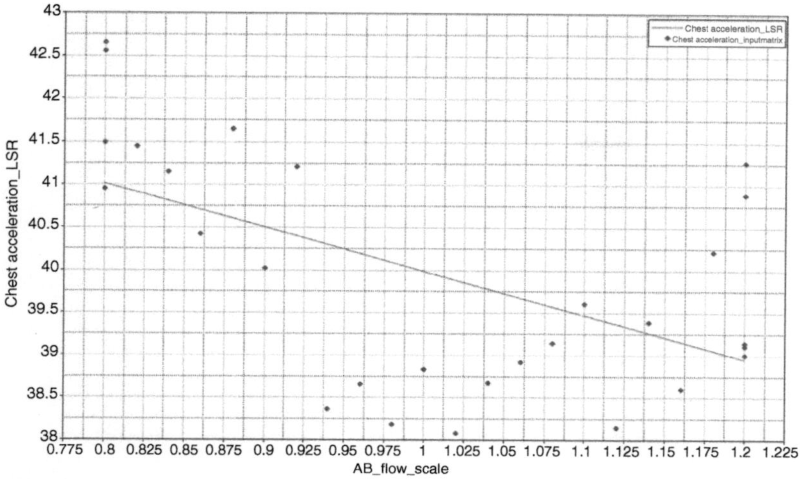

FIGURE 10.17 LSR response surface. Chest acceleration as a function of airbag flow scaling value.

which should yield to a higher definition of the response surfaces, hence theoretically a more accurate meta-model. The drawback of such meta-models is that for very sharp changes of peaks and troughs in the response surface definition, convergence will be difficult to achieve.

Looking at Figures 10.18–10.21, it can be noted that the response surfaces have more definition.

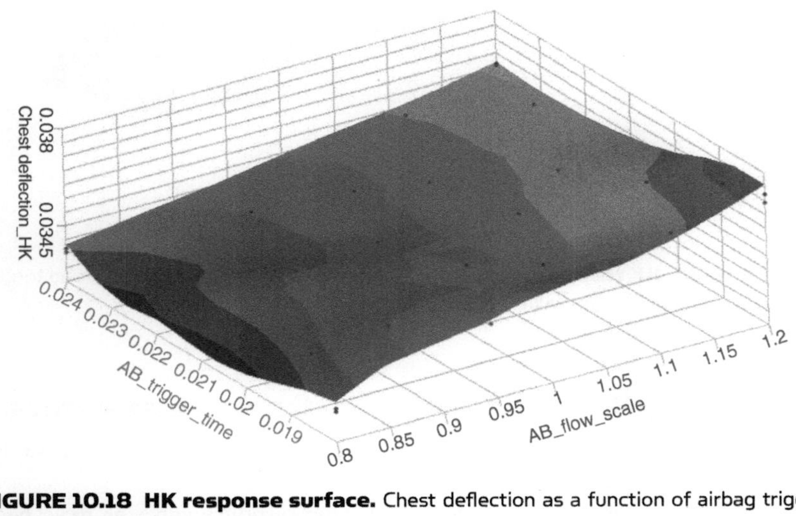

FIGURE 10.18 HK response surface. Chest deflection as a function of airbag trigger time and airbag flow scaling value (ISO view).

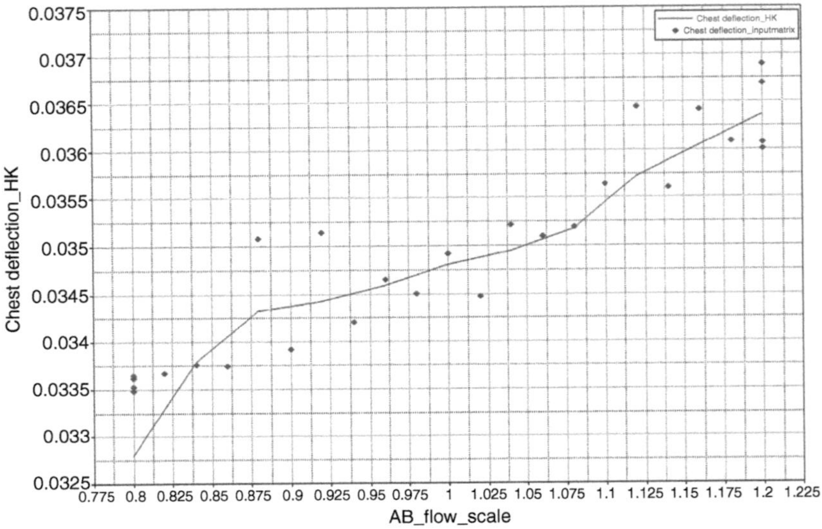

FIGURE 10.19 HK response surface. Chest deflection as a function of airbag flow scaling value.

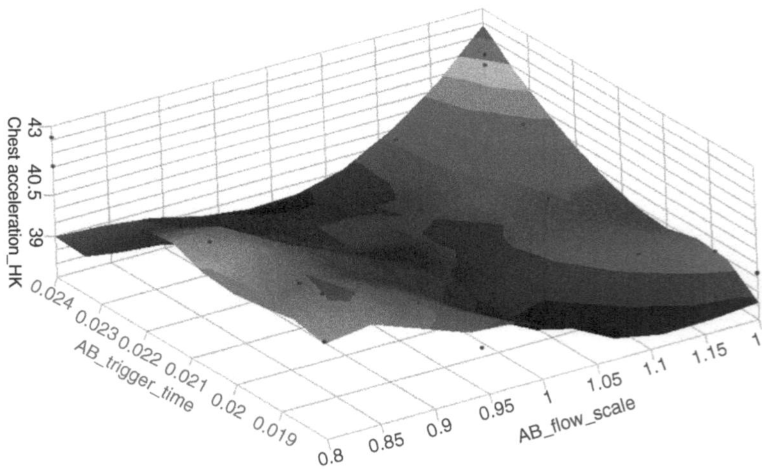

FIGURE 10.20 HK response surface. Chest acceleration as a function of airbag trigger time and airbag flow scaling value (ISO view).

10.2.7 Step 4: optimization using the response surfaces

The aim of the optimization is to investigate whether the carry-over restraint system could be further improved.

Now that the design of experiment and the response surfaces have been computed, an optimum injury output level can be evaluated, based on the

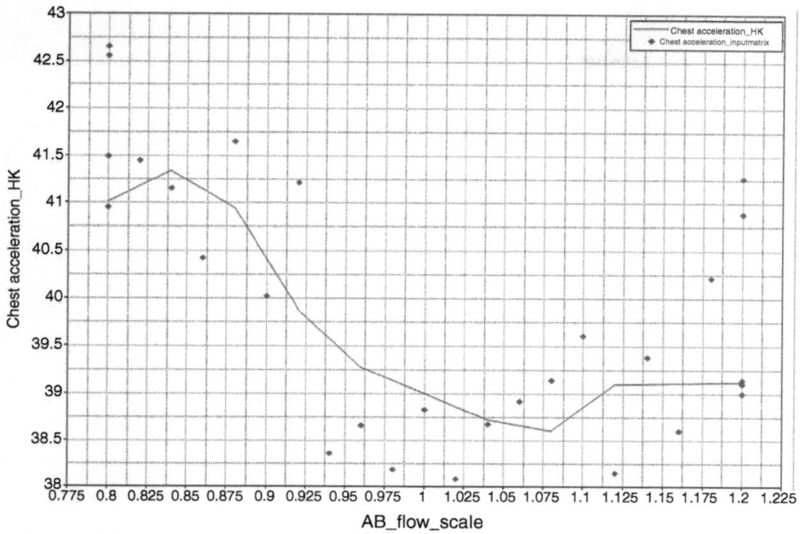

FIGURE 10.21 HK response surface. Chest acceleration as a function of airbag flow scaling value.

response surface method. One of the clear advantages of using response surfaces is that an optimum solution can be computed within seconds versus hours (and in some cases days), should the optimizationn be 'direct' and use the multi-body model instead.

The list of safety outputs are listed in Table 10.4 and contain the bounds of the injuries for the optimization. The optimization will aim to minimize the injuries (Table 10.5), while making sure that no injury is traded over another one.

The values of airbag flow scaling, airbag firing time, and pretensioner firing time can be computed, based on the constraints and objectives from Tables 10.4 and 10.5 and are listed in Table 10.6.

It can be observed that the optimum values computed between the two methods are vastly different, proposing different restraint system strategies.

The least square solution would propose a more aggressive airbag system than the Kriging solution with a later activation of the seatbelt pretensioner. It can be noted that the airbag trigger time is virtually similar.

Following these optimization results, it is essential to recompute the actual injury values based on the optimization parameter obtained. The reason for this is that the optimization are based on response surfaces for which the fit may not be perfect, as can be observed in Figures 10.19 and 10.21. These comparisons can be visualized in Figures 10.22–10.27.

It can be noted from Figure 10.22 that the optimization suggests a better HIC outcome than the initial restraint system; however recomputing the optimized values contradicts this proposal, and suggests a HIC value above the constraint value set within the optimization.

TABLE 10.4 List of Constraints in the Restraint System Optimization

Injury Type	Constraint Bounds	Target Value	Initial Value	Units
HIC	≤	350	347	–
Chest deflection	≤	0.035	0.034	m
Chest acceleration	≤	40	38.6	g
Left femur	≤	3000	73	N
Right femur	≤	3000	2936	N
NCE	≤	1.0	0.00	–
NCF	≤	1.0	0.12	–
NTE	≤	1.0	0.16	–
NTF	≤	1.0	0.16	–

TABLE 10.5 List of Objectives in the Restraint System Optimization

Injury Type	Objective Type
HIC	Minimize
Chest deflection	Minimize
Chest acceleration	Minimize
Left femur	Minimize
Right femur	Minimize
NCE	Minimize
NCF	Minimize
NTE	Minimize
NTF	Minimize

TABLE 10.6 Optimization Parameter Solutions from Both Response Surfaces

	Airbag Flow Scaling	Airbag Trigger Time	Pretensioner Trigger Time
Initial restrain System setup	1.000	0.020	0.016
Least square Regression solution	1.200	0.021	0.018
Kriging solution	1.056	0.020	0.014

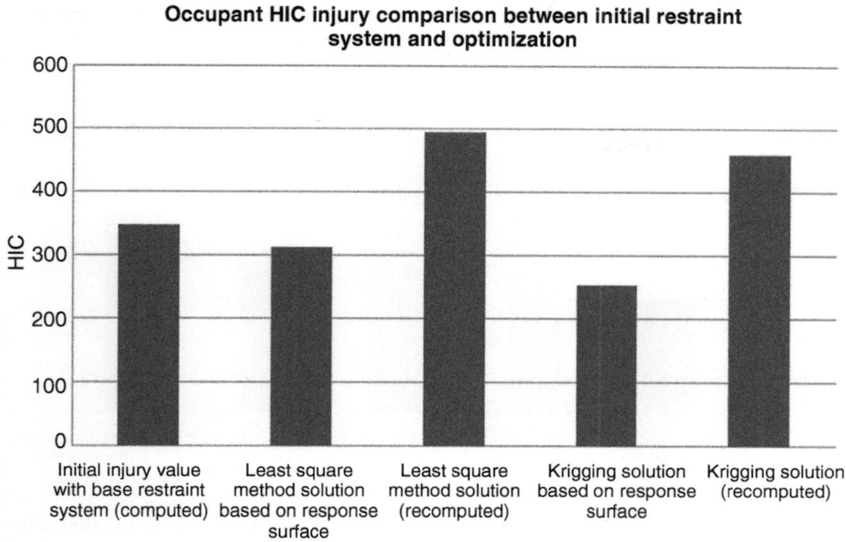

FIGURE 10.22 HIC injury comparison between original restraint system and optimization proposals.

In Figure 10.23 the optimization suggests a better acceleration outcome than the initial restraint system in the case of the Kriging-based optimization. Recomputing the optimized values confirms that the chest acceleration is higher (39.25 g vs. 38.5 g); however it is still under the constraint set at 40 g.

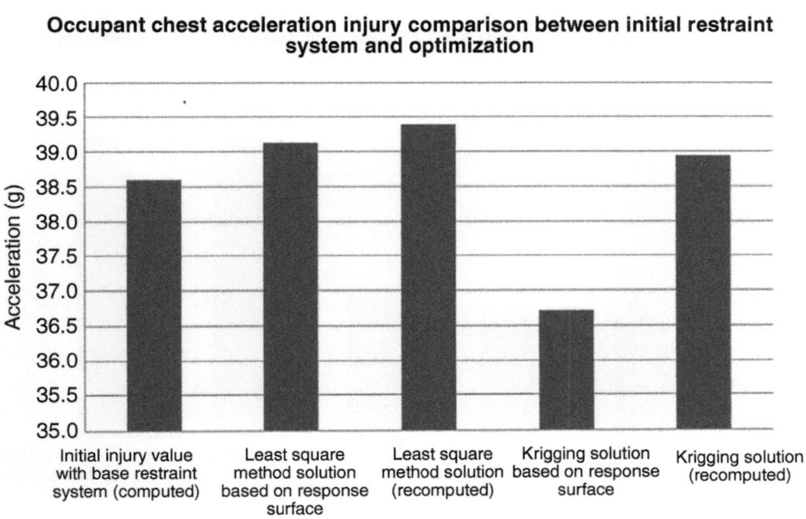

FIGURE 10.23 Chest acceleration injury comparison between original restraint system and optimization proposals.

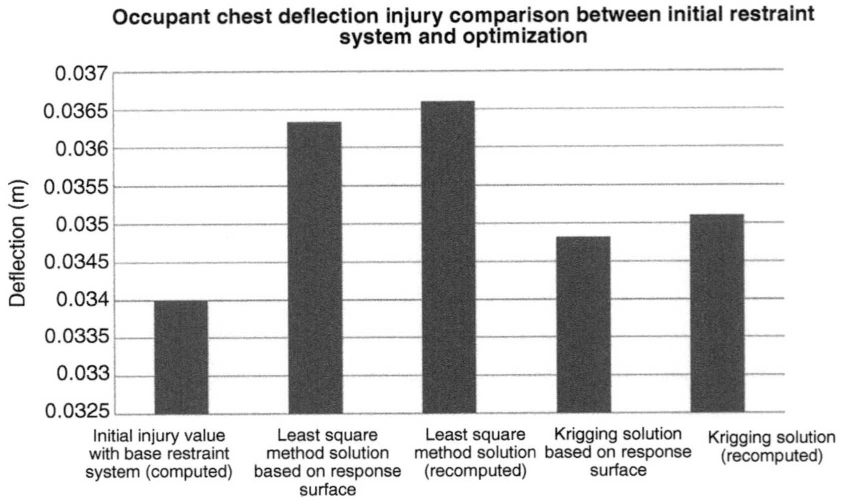

FIGURE 10.24 Chest deflection injury comparison between original restraint system and optimization proposals.

Considering the occupant chest deflection injuries (Figure 10.24), the predicted deflections based on the Kriging interpolation suggest a better solution than the original restraint system configuration, but again the recomputation proves otherwise, as well as border line to the optimization constraint set at 0.035 m.

The occupant right femur load is expected to perform better with the new restraint system parameters (Figure 10.25), including the recomputed model. The Kriging interpolation does not suggest any possible improvements.

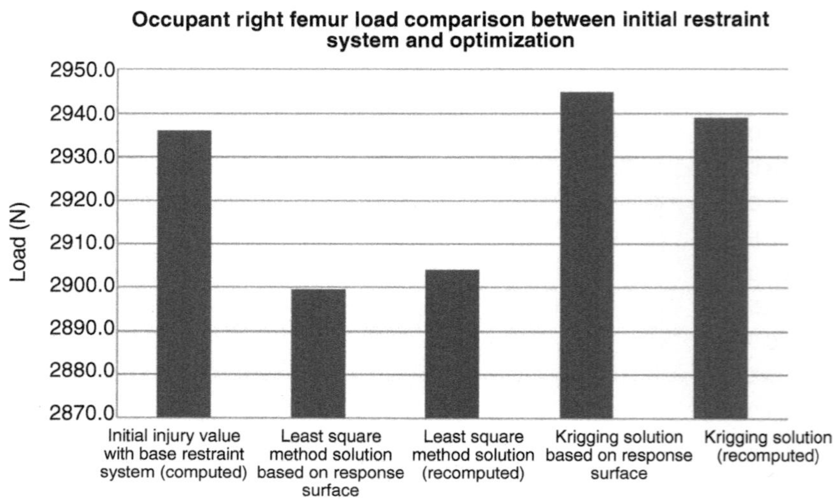

FIGURE 10.25 Right femur load injury comparison between original restraint system and optimization proposals.

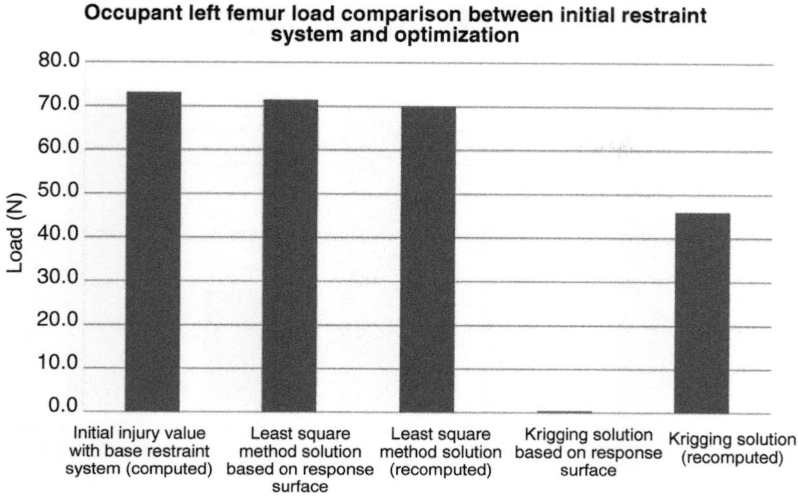

FIGURE 10.26 Left femur load injury comparison between original restraint system and optimization proposals.

Considering Figure 10.26, it can be observed that the femur loads are under the 3000 N load requirements set as part of the constraints. The LSR regression suggests a modest improvement, while the Kriging-based optimization advocates a major reduction in left femur injury loads.

In order to understand the improvement suggested by the Kriging-based optimization, the shape of the response surface needs to be considered (Figure 10.27).

The shape of the response surface is concave and returns a low left femur force for an airbag flow rate of 1.056 and an airbag trigger time of 0.020 s.

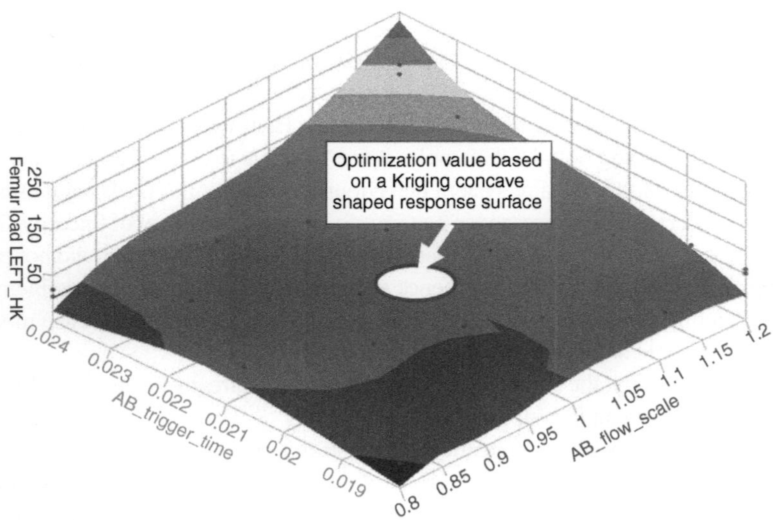

FIGURE 10.27 Left femur load Kriging response surface.

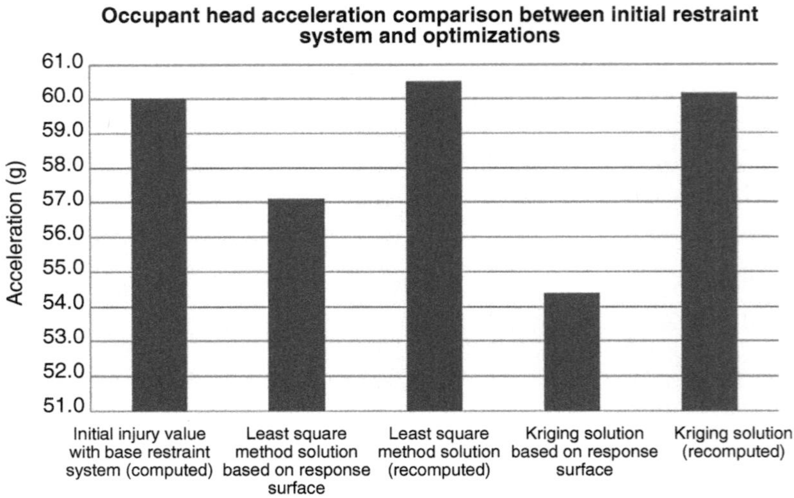

FIGURE 10.28 Head acceleration injury comparison between original restraint system and optimization proposals.

Consequently, the efficiency of the optimization will greatly depend on the build of the response surface.

The occupant head acceleration value theoretically is improved, based on both response surfaces; however the recomputation suggests otherwise (Figure 10.28).

10.2.8 Conclusion of the occupant study

The occupant study has shown that the original model was believable, as the findings confirm that using an airbag reduces occupant injuries. All the injury values computed in the original model are reasonable and suggest a safe system which has potential of improvement using optimization.

A design of experiment was conducted to capture suitable permutations within the design space, i.e. parameters based on their maximum values using the full factorial method, as well as intermediate ones using the Hammersley sampling methodology. By combining both DOE samples it was possible to create a comprehensive understanding of the restraint system behaviour. Looking at the DOE responses, it could be observed that the response patterns do not show any level of linearity. The response surfaces computed using linear (LSR) and "curved" (Kriging) did not pass through most of the response data, hence suggesting a highly nonlinear problem.

When this eventuality is present, it is essential to recompute the base model with the variable solutions obtained by the meta-models.

It the case study, the carry-over model happened to already be very suitable for the vehicle system analyzed.

If it was not possible to optimize this restraint system, within the set of parameters chosen and within their spread of design values. This study reinforces the fact that optimizing the full BIW with the full restraint system together is not a sound methodology, It has been observed that the response of a restraint system, including seatbelt and airbag was non-linear and that it would remain non-linear should more engineering parameters be used. Combining full vehicle model and restraint system together in one holistic computer model will generate would a lot of noise during the optimization and it is doubtful whether any meaningful solution can be extracted.

10.3 PEDESTRIAN PROTECTION ANALYSIS CASE STUDY

The aim of this study is to improve the head impact performance of a concept steel pedestrian bonnet. The computer model developed suggests that this vehicle does meet the GTR9 pedestrian head impact requirements (HIC of 1917 computed); consequently, this case study will investigate whether deploying the bonnet would improve a child head-bonnet impact against the top of the vehicle battery box. It is well-known that such deployable technologies create a bigger gap between the bonnet inner panel and the hard contact areas of the engine bay.

How can the rear lifters deployment functions be optimized to minimize the head injuries of a 6-year-old child (Figure 10.29)?

The study will be performed using LS-Dyna which is an explicit solver using the finite element method.

The first question that may be asked is where is the child head relative to the bonnet before the deployment sequence can start?

A 50th percentile 6-year-old (Serre et al., 2009; CDC 2012) is 1196 mm tall. Assuming the vehicle bumper height to be 609 mm from the ground, it is

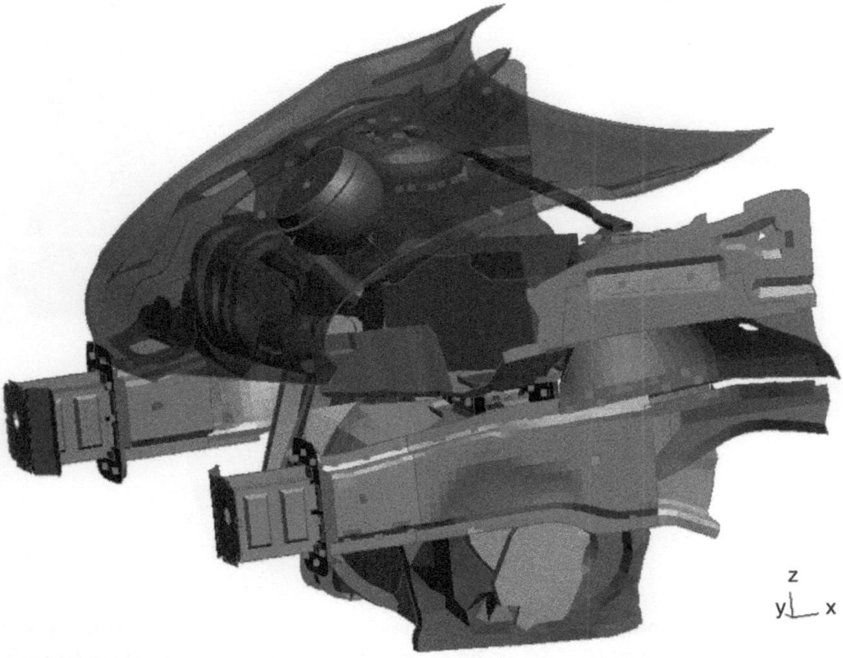

FIGURE 10.29 Head impact scenario: 6-year-old impacting battery box. Bonnet in a closed position.

suggested that the length of body that will swing onto the front grille and bonnet is the difference between these two values, that is, 587 mm.

Consequently, as the GTR9 test velocity is 35 km/h (or 9722 mm/s), it would take 0.060 s for the 6-year-old child's torso and head to "wrap around" the bonnet before head contact takes place.

In order for the bonnet to deploy, a sensor must provide a signal to the main vehicle CPU, which will then make a decision on whether a pedestrian bonnet deployment should occur or not. There can be many eventualities that could trigger unwanted deployments, such as a shopping trolley impact in a car park or impacting a pheasant in a countryside road; hence, a deployment decision time of 0.03 s would sound a good estimate.

Consequently, the head will be 0.03 s away from hitting the closed bonnet before deployment starts, or 295 mm (at 35 km/h), as illustrated in Figure 10.30.

FIGURE 10.30 Comparison between head impactor start position: Near contacting the bonnet (GTR9) vs. head impactor in flight 0.03 s before bonnet starts to deploy. Bonnet shown here in a closed position.

10.3.1 Parameterization of the problem

The rear of the bonnet will deploy/lift following a prescribed displacement motion curve that may have the pattern illustrated in Figure 10.31.

The lifting loadcurve will always start at time 0 s (point "A" in Figure 10.31, representing the deployment time, when all the delays have been added to the computation) and always end at time 0.1 s (points "C" in Figure 10.31). The runtime for this deployable bonnet computation is 0.08 s; hence, the loadcurve definitions must be greater than the end time of the run to avoid any unwanted curve interpolations.

The lifting will be defined by the end time when the bonnet's final upward position is desired (horizontal axis of points "B" in Figure 10.31) and its magnitude (ordinate of points 'B' in Figure 10.31).

The input required as design variables is:

- The end time of the bonnet lift $x_{(B)}$
- The maximum bonnet lift $y_{(B)}$

It will be assumed that $y_{(C)} = y_{(B)}$ once the maximum bonnet height is reached. It is assumed that $x_{(C)}$ is always 0.1 s, which is more than the main head impact runtime needed for HIC computation.

The front of the bonnet latch is only allowed to rotate about the transverse Y direction to create a rotation while keeping the number of contacts to a minimum and maximize runtime efficiency.

Consequently, the variables are defined in Table 10.7.

The nominal run is performed, using a lift height of 80 mm which is completed within a 0.030 s deployment duration. The head injury criterion (HIC) is 587.5 and the maximum head impact acceleration is 123.8 g, as depicted in Figure 10.32.

It can be noticed from Figure 10.32 that the head impact pattern is presenting two distinct peaks.

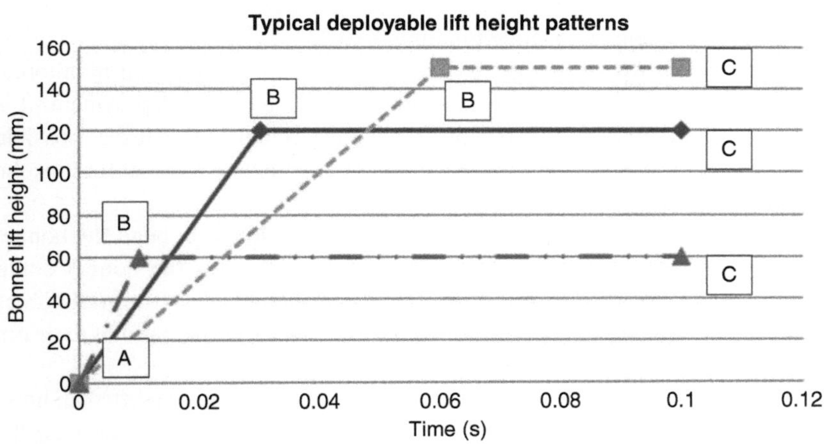

FIGURE 10.31 Deployment rear latch motion pattern.

TABLE 10.7 Variables for Deployment Bonnet Study

	Lower Value	Nominal Value	Maximum Value
End of lift time (s)	0.001	0.030	0.060
Lift height (mm)	0.0	80.0	150.0

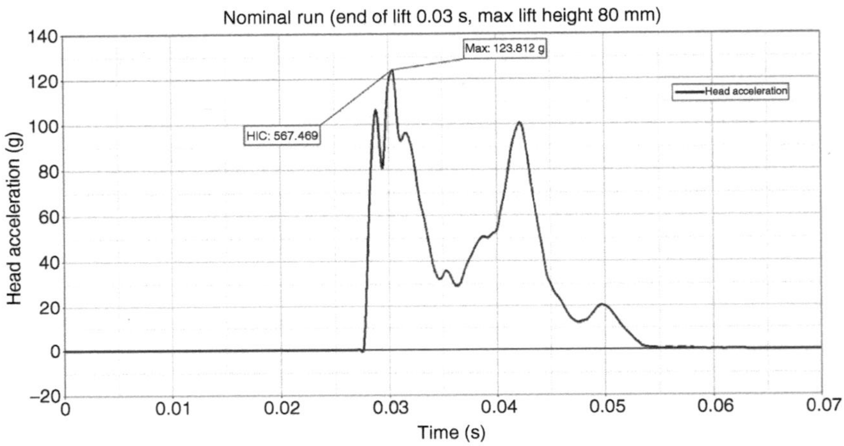

FIGURE 10.32 Head impact response vs. time graph (nominal run). HIC and maximum deceleration.

The first peak, which is located around 0.028–0.032 s, represents the first contact time, when the head is decelerated by the bonnet mass and stiffness.

After 0.032 s, the pulse is reducing, meaning that the structure is collapsing and moving downward, but at time 0.040 s, the head is contacting against the battery box, which is causing the acceleration pulse to rise sharply.

The impact event is illustrated in Figure 10.33.

In the analysis performed, the bonnet lift speed has also been monitored, as this can be also a very influencing parameter in the bonnet deployment. It is instinctively very important to position the bonnet as high and as fast as possible to provide the best package head clearance. However, there is a cost for this: the structural integrity of the bonnet.

If the bonnet is lifted too quickly, the forces could either bend the bonnet at the lifters' attachments, or, if the bonnet is too heavy, the bonnet could be permanently bent after the maximum lift is achieved, as inertial forces would to continue its vertical travel while the corners are secured at their top position.

There are no fixed rules on what speed the bonnet needs to be lifted at, however when a solution is computed, one would like to choose the one with the lowest deployment speed.

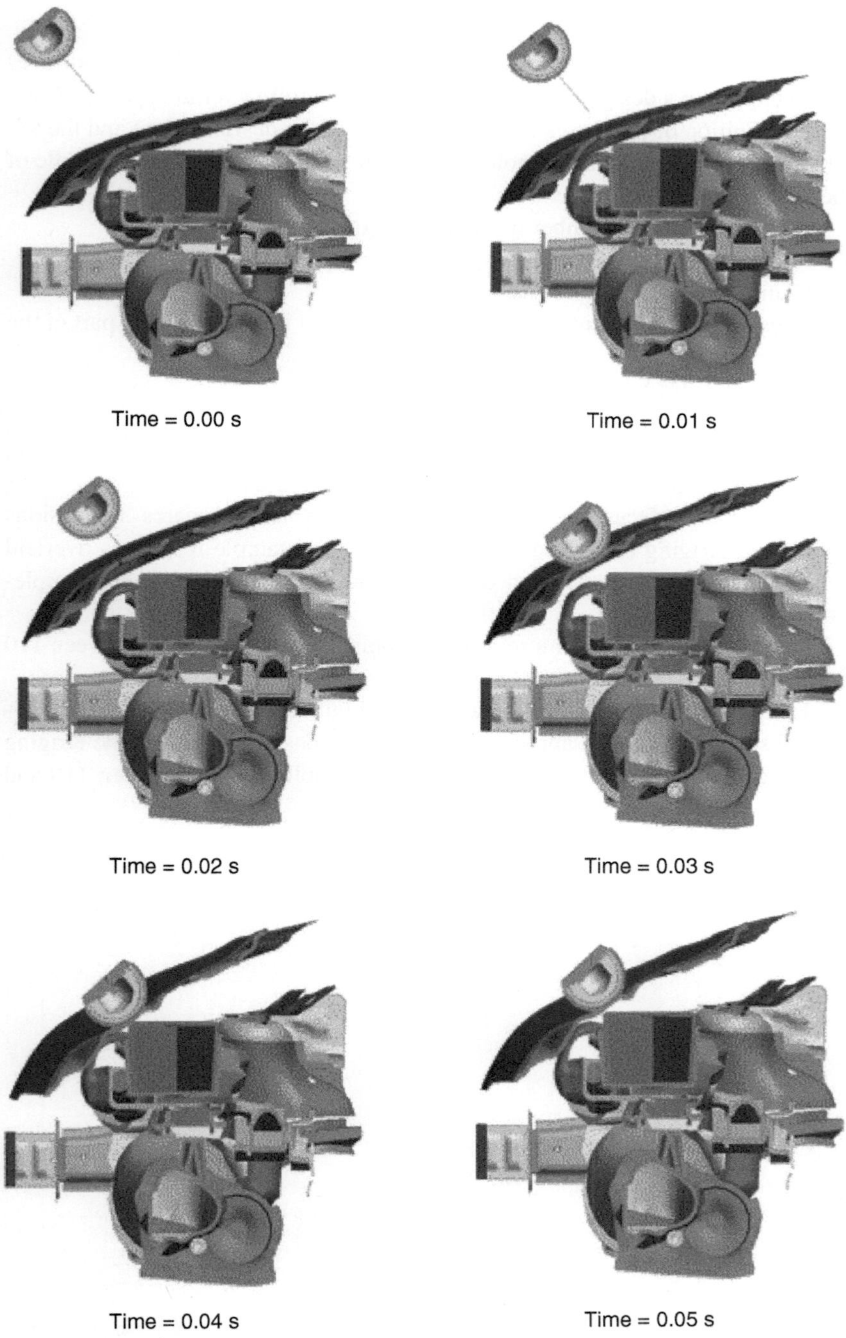

FIGURE 10.33 Head impact kinematics as function of time.

10.3.2 Optimization method

The method used is also the same as Figure 10.11, whereby response surfaces are created from a design of experiment based on a full factorial approach (used in the definition of responses in the boundaries of the design space) and the addition of Hammersley sample points to include more definition in the middle of the design space.

Consequently, the number of samples used is:

- Full factorial: four samples
- Hammersley: 12 samples

It was observed that more definition was needed in the following part of the design space:

- 105 mm $\leq$ lift height $\leq$ 150 mm
- 0.004 s $\leq$ end of lift $\leq$ 0.016 s

Consequently, another Hammersley DOE was undertaken in order to add more sampling points and obtain more defined response surfaces.

Response surfaces are computed using the least squares regressions (LSR) and Kriging (HK) and their accuracy to the sample responses overlaid (Figures 10.34–10.37, extracted with a constant value of deployment completion time of 0.030 s).

It can be observed that more sampling points have been added between 110 and 150 mm lift height.

In both head acceleration meta-models (Figures 10.34 and 10.35), the response surface fits are an approximate to the response points; even the Kriging response surface is poor in spite of the extra sampling point between 110 and 150 mm.

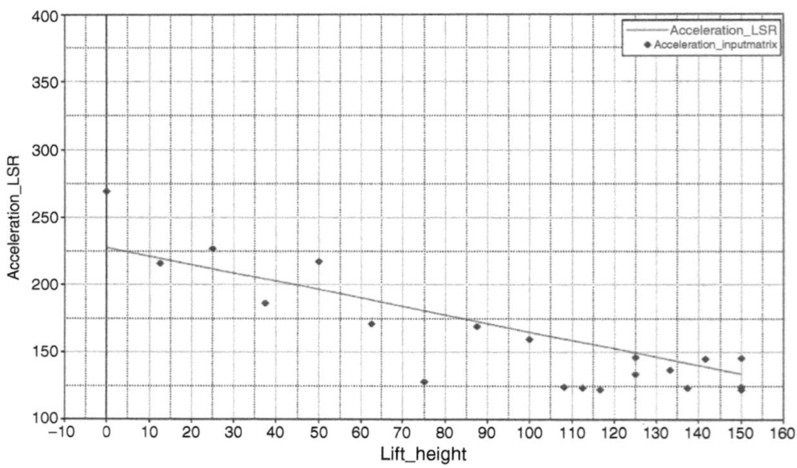

FIGURE 10.34 Discrepancies between LSR acceleration response surface and response sample points (for a bonnet lift completed in 0.030 s).

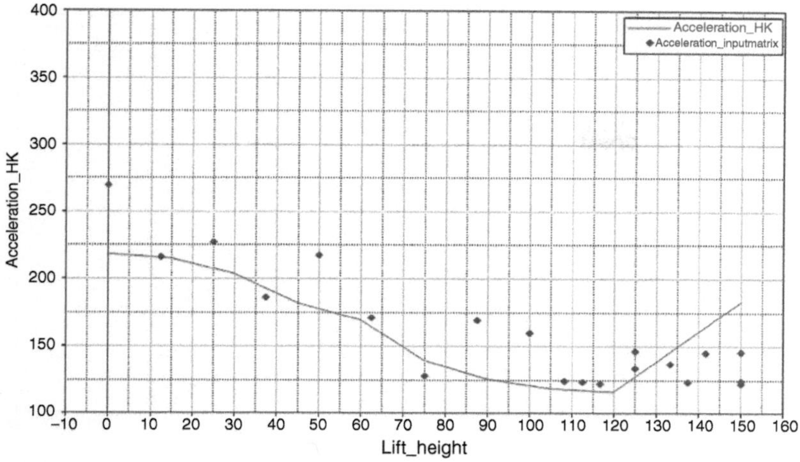

FIGURE 10.35 Discrepancies between HK acceleration response surface and response sample points (for a bonnet lift completed in 0.030 s).

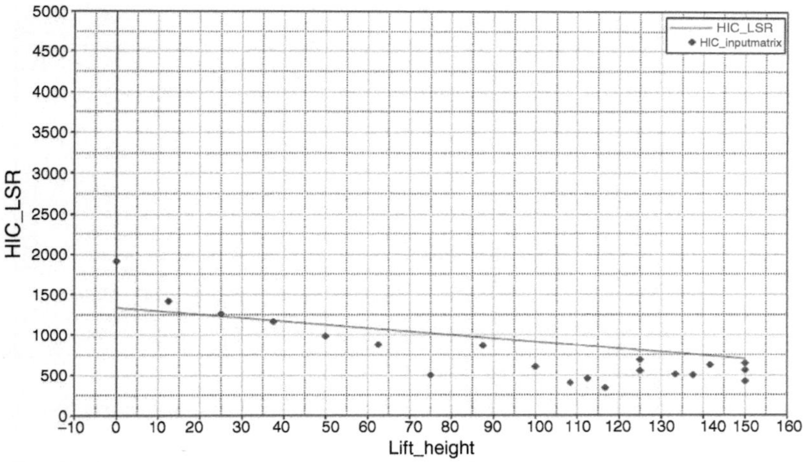

FIGURE 10.36 Discrepancies between LSR HIC response surface and response sample points (for a bonnet lift completed in 0.030 s).

On the other hand, it can be observed that the response surface fits better the sample HIC responses, as illustrated in Figure 10.36 and Figure 10.37.

The bonnet lift speed has been computed by dividing the bonnet lift height by the lift duration. It can be noted that the LSR interpolation in Figure 10.38 does not represent a physical speed, as some values are negative, suggesting a downward deployment, which is not what the bonnet is doing! The response surfaces are flat; hence, it will be possible to find a stable minimum. However, the real value of the deployment speed will need validation because of this

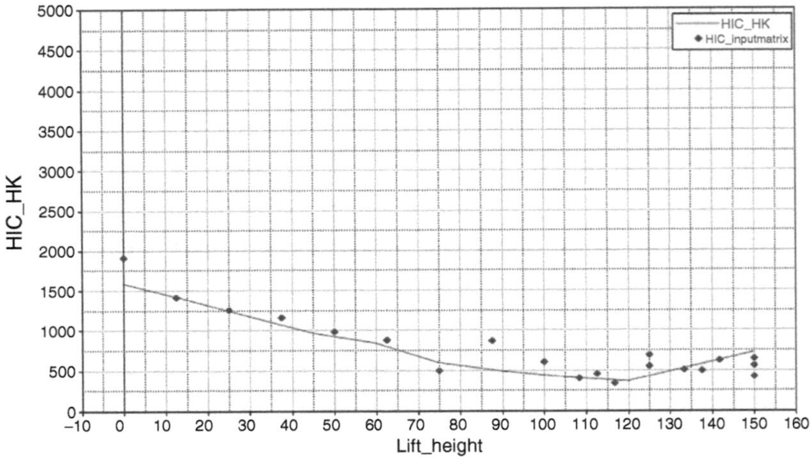

FIGURE 10.37 Discrepancies between HK HIC response surface and response sample points (for a bonnet lift completed in 0.030 s).

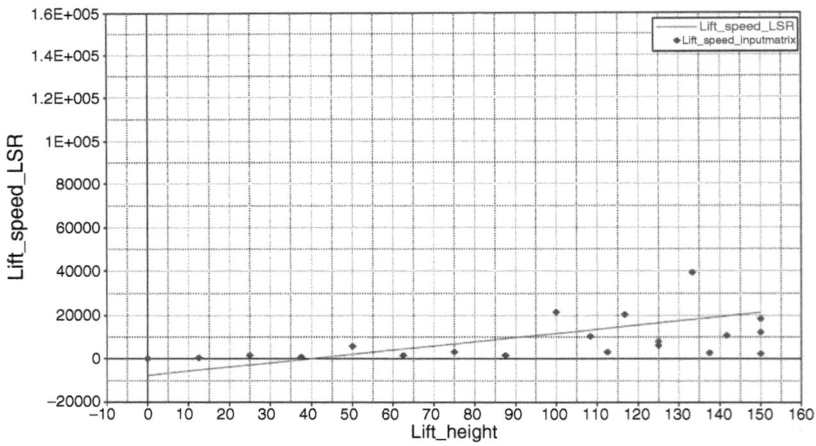

FIGURE 10.38 Discrepancies between LSR deployment speed response surface and response sample points (for a bonnet lift completed in 0.030 s).

negative deployment direction. In the HK response surface, as illustrated in Figure 10.39, the bonnet deployment speed is logical, as it does not contain any negative values; however after 100 mm deployment height, the response surface and the response samples diverge.

The optimization constraints and objectives are listed in Table 10.8.

10.3.3 Optimization solution

The results from the optimization are listed in Table 10.9.

It can be observed that the lift speed value computed from the LSR response surface is too low (25 mm/s while the true value must be 2356.5 mm/s).

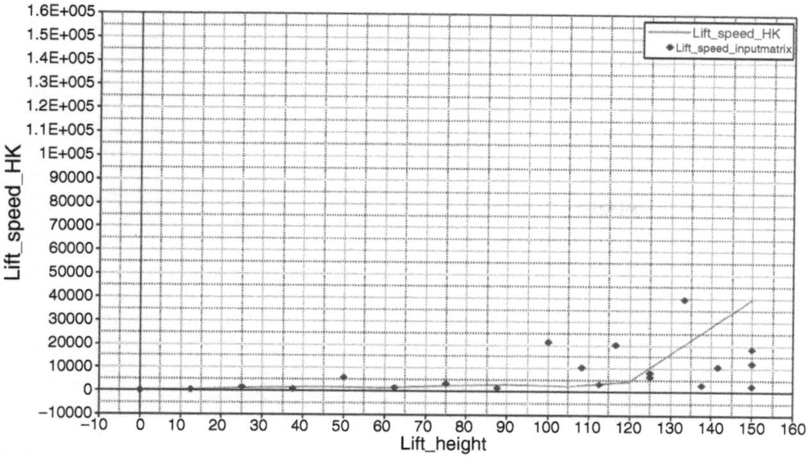

FIGURE 10.39 Discrepancies between HK deployment speed response surface and response sample points (for a bonnet lift completed in 0.030 s).

TABLE 10.8 Pedestrian Deployable Bonnet Constraints and Objectives

	Constraints		
	Bounds	Value	Units
Acceleration	≤	120	g
HIC	≤	600	–
	Objectives		
HIC		Minimize	
Lift speed		Minimize	

TABLE 10.9 Optimization Solutions (Computed vs. Validated)

Optimization Response Surface Utilized		Lift Height (mm)	End Lift (s)	HIC	Acceleration (g)	Lift Speed (mm/s)
LSR	Optimization computed from response surface	141.39	0.060	579	119	25.6
	Validation using LS-Dyna	141.39	0.060	679	122	2356.5
HK	Optimization computed from response surface	105.98	0.029	323	109	1319.0
	Validation using LS-Dyna	105.98	0.029	490	125	3654.5

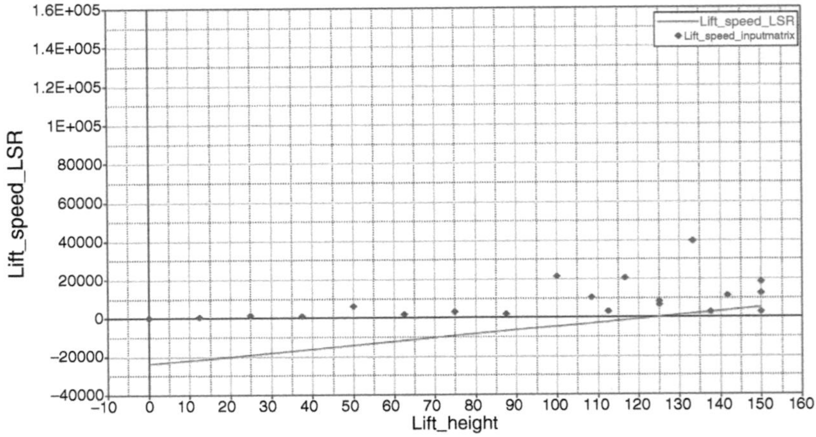

FIGURE 10.40 Discrepancies between LSR deployment speed response surface and response sample points (for a bonnet lift completed in 0.060 s).

This is due to the fact that, as previously explained, the response surface is inadequate and nonphysical for lift heights of 140 mm, as the lift speed is negative (Figure 10.40). Consequently, the validation of the lift speed is necessary. The same is observed with the Kriging fit, where an expected lift speed was optimized to 1319.0 mm/s, while the true lift based on the lift height and duration was certified to be at 3654.5 mm/s.

Regardless of these lift speed values, the response surfaces are relatively planar; hence, the minimum has been achieved. It is then possible to verify the damage sustained by the bonnet during the deployment before the head impact occurs (Figure 10.41).

It can be observed at time 0.030 s that the maximum plastic strain at the lift areas is in the order of 4.7% during the deployment phase and can reach a maximum value during the impact up to 9.1%. In the bonnet inner window area (Figure 10.41), some minor plastic strain can be observed. The magnitude of this plastic strain is not enough to bend the bonnet permanently, as it is not located in the main bonnet structural members that run alongside the bonnet armature.

Consequently, the two proposed solutions suggest a structurally safe lifting speed.

It may be argued that during the optimization, more constraints, such as plastic strain, could have been added. It takes an experienced CAE analyst to judge whether the solution is structurally sound, as the loading in the rear section of the bonnet could have induced "artificial" localized plastic strains that could have skewed the optimization. This structural validation can only be done "*a posteriori*"; consequently, the plastic strain constraint has been ignored.

It can be noted that each optimization is proposing a different strategy: the LSR proposal is suggesting a slow lift at 2356.5 mm/s that lifts the bonnet up to 141.39 mm, while the HK advocates a faster lift (3654.5 mm/s) but less high (105.98 mm). Also, both optimizations have met the constraints set in Table 10.8.

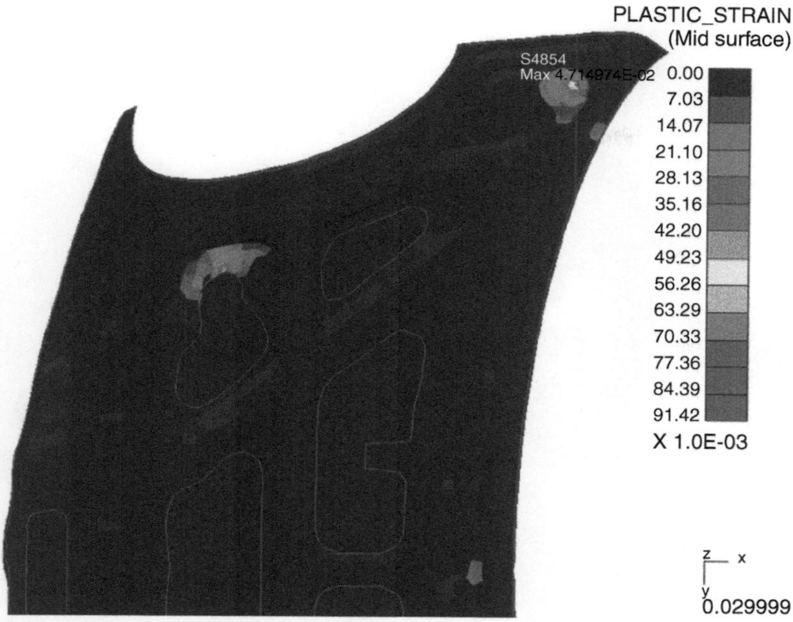

FIGURE 10.41 Maximum strain values during the deployment phase prior to head impact (HK solution, which contains the highest lift speed).

These two solutions are proposing two different head impact responses that are depicted in Figure 10.42.

Each solution contains a double peak; when the deceleration increases, this means that a force is applied to the head, while when it reduces it means that the applied force is reducing in intensity. This first peak, in both cases, is due to the inertia of the bonnet the head needs to overcome, as well as the bonnet's

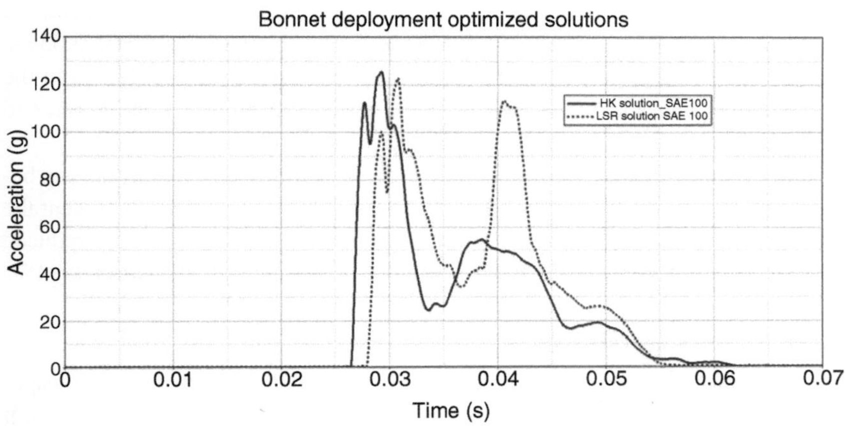

FIGURE 10.42 HK and LSR head impact solutions: acceleration responses.

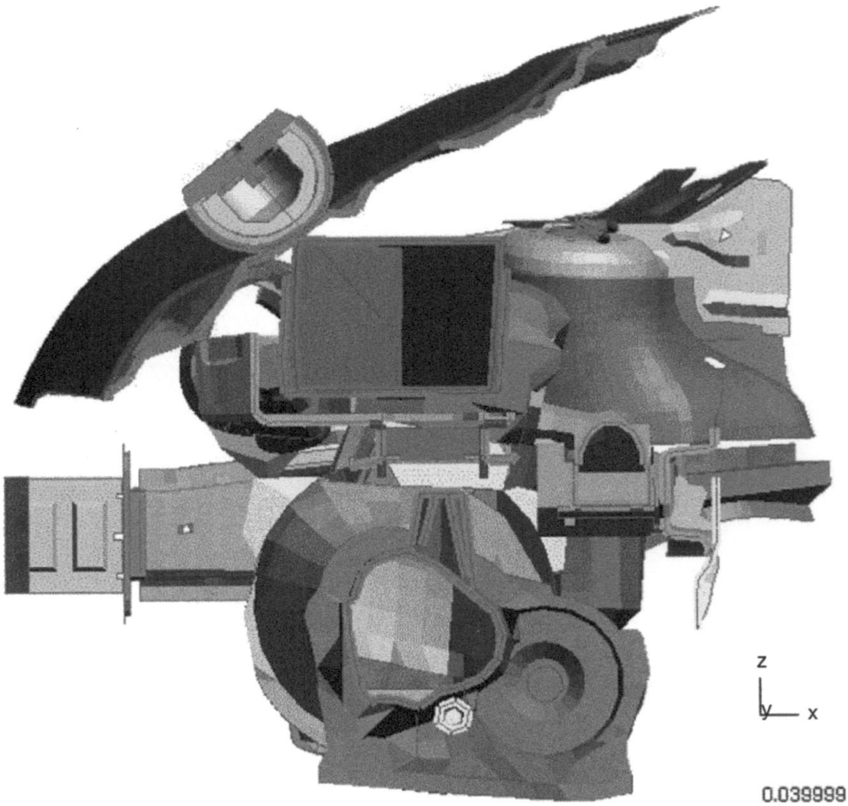

FIGURE 10.43 LSR solution suggests a hard contact against the battery box.

lifting momentum and stiffness. It can be noted that the first peak contains three bumps, hence relating to the same physics in both scenarios.

The second peak in each scenario occurs for different reasons. In the case of the Kriging, the second peak is lowest. After the first part of the impact, the bonnet moves downward, hence reducing its force on the head. At time 0.035 s, the curvature of the bonnet outer is changing, pulling the bonnet skin from the bonnet edges, and, like the stretched skin of a drum, causing a resistive force to the head motion.

For the LSR solution, the bonnet lift height is 36 mm lower; hence, the gap between the impacting headform and the battery box is not sufficient to prevent the contact from occurring (Figure 10.43). Nevertheless, lifting the bonnet reduced the kinetic energy from 164.0 to 9.5 J, which is a considerable reduction of 94%.

10.4 CONCLUSIONS

Nonlinear optimization is a complex matter in the field of vehicle safety, especially when occupants and pedestrian impactors are used to compute injuries. It has been observed that when complex systems were involved in the optimization, it was challenging to converge to a reliable solution, that is, a solution that would return the same results when recomputed by the original solver.

In pedestrian protection, the nonlinearity of the materials, as well as the computation of the injuries, is a real problem. In general, it is possible to achieve a reasonable correlation for accelerations. As accelerations are the first component computed from the explicit solver, they are some of the most reliable components to validate against, although some filtering may be needed and can add to the complication. As the problem is usually solving Newton's second law, acceleration is always the preferred one. Correlating on velocities and displacements is more complicated, as these tensors are computed from the acceleration by differentiation, and hence will create a smoother signal that will be extremely sensitive to any small changes in the acceleration response. As the acceleration response is a function of the element size and the timestep, this is a losing battle from the start.

In the case of occupant safety, the sheer number of injury criteria to meet makes the nonlinear problem even more complicated to solve.

The strain rate of the material models always causes some nonlinearity in the force/acceleration responses, as the strain rate computed at each timestep dictates the change of material stiffness characteristics from the material model.

One of the most difficult injuries to capture is the HIC, as it is based on the acceleration value and an injury window, which is very sensitive to the shape of the acceleration curve. Consequently, the optimization of HIC can diverge greatly between the response surface and the recomputed value.

Always investigate the shape of the response surfaces to estimate a level of confidence in the optimization solution. Optimization findings must always be recomputed to validate the solution.

REFERENCES

BBC News. (2011). One in 20 drivers not belting up, survey suggests. Article available from <www.bbc.co.uk/news/uk-13775695> [1.07.13] BBC. (2012). UN: Six billion mobile phone subscriptions in the World. Article available from <www.bbc.co.uk/news/technology-19925506> [1.07.13].

CDC. (2012). Anthropometric Reference Data for Children and Adults: United States, 2007–2010. Technical document. Available from <www.cdc.gov/nchs/data/series/sr_11/sr11_252.pdf> Accessed 14.02.15.

Nebraska Strategic Highway Safety Plan. (2013). Fatal crashed involving unbelted vehicle occupants. Technical Report. Available from <www.transportation.nebraska.gov/safe/tysummit/2012/presentations/1-Fatal-Crashes-Unbelted.pdf> Accessed 8.07.13.

NHTSA. (2007). 2007 motor vehicle occupant safety survey. Technical Report (Volume 4 *Crash* Injury and Emergency Medical Services Report) Available from < www.nhtsa.gov/Driving + Safety/Research + & + Evaluation/2007 +Motor + Vehicle + Occupant + Safety + Survey> Accessed 8.07.13.

NHTSA. (2013c). National Automotive Sampling System (NASS). Technical Database Available from <www.nhtsa.gov/NASS> Accessed 8.07.13.

NHTSA. (2015). Injury risk curves and protection reference values. Technical report. Available from <www.nhtsa.gov/cars/rules/rulings/80g/80gii.html> Accessed 11.01.15.

Serre, T., Lalys, L., & Lecoz, S. (2009). Anthropometric data of the 3 and 6 year-old child regarding the position of the safety belt. Technical report from CASPER FP7. Available from < media.univ-lyon1.fr/iea-dhm2011/abstracts/2272.pdf> Accessed 14.02.15.

Viana, P. (2013). Things you wanted to know about the Latin hypercube design and were afraid to ask. Conference paper. Available from <www2.mae.ufl.edu/mdo/Papers/5176.pdf>. In: *10th World Congress on structural and multidisciplinary optimization*, May 19–24, 2013, Orlando, FL.

WHO. (2013). Global status report on road safety 2013. Technical report (World Health Organisation). Available from <www.who.int/violence_injury_prevention/road_safety_status/2013/report/en/>. Accessed 1.06.13.

Chapter | eleven

The Future of Structural Optimization and Vehicle Safety

CHAPTER OUTLINE

The purpose of this chapter is firstly to briefly summarize the highlights of the previous chapters in this book, as well as to provide an insight into the future of structural optimization, including the daunting challenges ahead.

11.1 VEHICLE ARCHITECTURES, STRUCTURES, AND SAFETY REQUIREMENTS

Vehicle safety has greatly improved over the last 40 years thanks to the introduction of crash test standards that investigate the injury criteria tolerance withstood by occupants and pedestrians. Despite crash test protocols being different between the US, the European Union, and the rest of the world, all have the same main objective: to reduce injuries and fatalities taking the loadcases that are relevant to each region into account. If roof crush and the unbelted loadcases are tested in the US, these tests are judged not relevant to the European market, as all occupants must wear their seatbelts by law and the European accident pattern is not skewed toward rollover crashes. The crash modes are also very different in frontal, where a vehicle must withstand a 50 km/h rigid wall impact in its full overlap, compared with a 40% overlap in Europe against a deformable honeycomb barrier. The outcome leads to very different vehicle architectures that are

Nonlinear Optimization of Vehicle Safety Structures. http://dx.doi.org/10.1016/B978-0-12-804424-7.00011-7

on their own an engineering prowess to overcome, and even more challenging should the vehicle be sold in more than one continent. In general, if the US legislation dictates testing for more crash diversity compared with the rest of the world, European legislation tends to consider more injury criteria, adding to the technical challenges. One of the trends observed is that there is an effort to harmonize the safety tests in particular in connection with pedestrian protection tests, thanks to the GTR9, which has harmonized the impactor shapes, impact zones, and impact speeds for most markets. Vehicle safety is, however, still evolving and new loadcases are being introduced to reflect accidents observed in different geographical regions. In the US, for example, a new 20% frontal overlap crash test is being introduced, which causes a new technical challenge, as the rigid barrier usually misses the bumper system and the crush can, leading to the crash load being carried through the wheel and along the A-pillar. In this connection, the new technical challenges include the prediction of the wheel motion, as its rotation inboard or outboard would change the force pattern transferred to the safety cage. It is believed that this loadcase would induce cabin rotation and some technical challenges for occupant motion prediction and restraint system deployment trigger and airbag shape designs, due to the event variability. In the case of new loadcases, one also needs to consider which crash test dummy or computer tool would be adequate to assess the occupant injuries, as a cabin rotation would create a scenario that may not be suited for the Hybrid III frontal impact dummy, indicating that a new tool may be required. New computer tools are constantly being developed and, maybe, in the future the tool of choice to improve vehicle safety even further will be human computer models. Some studies have shown that in the case of a pedestrian accident reconstruction, the THUMS human model shows similarities with the victim postmortem report (Wen and Bastien, 2015). This study has looked into pedestrian spleen, heart, liver, and lungs to assess the damage. The work presented was a major step in addressing the computation of detailed internal organ injuries; however, it still falls short of addressing the nuances of injuries per organ at the subtle levels of abbreviated injury scales (AIS), which are subjectively provided by the surgeon during the pedestrian autopsy. The immediate breakthrough urgently needed in vehicle safety occupant and pedestrian assessment will be achieved when AIS can be computed from a human model. As the human model is omni-directional, it may be a universal tool of choice for vehicle engineering. The next level of improvement will then be reached when gender, age, and pre-existing medical conditions can be considered within the vehicle safety design; only then can true vehicle passive safety design be achieved.

From 2016 onward, no vehicle can be awarded 5 stars in the EuroNCAP test if automatic emergency braking (AEB) is not fitted to the vehicle (EuroNCAP, 2015). This new feature holds great safety benefits to reduce vehicle-to-vehicle and vehicle-to-pedestrian collisions, and it is believed that it will change the accident landscape, as vehicles will impact with lower speeds, and thereby lower kinetic energy. There will be a period of transition where vehicles with and without AEB will coexist and it is foreseen that most vehicles in the US and

EU will have AEB by 2025 (Ashmore, 2015). It is suggested that future collision modes will be significantly different from the ones currently tested as part of EuroNCAP and other legislative authorities; hence, it is expected that future safety tests will evolve to reflect these forthcoming changes (Khan, 2015). The introduction of autonomous vehicles is also likely to lead to different modes of accidents. Research has found that the occupant stance prior to pre-braking is important, as this dictates the occupants' position relative to the airbag system (Bastien, 2014). When vehicles are fully autonomous, it is not currently understood how the vehicle cabin will be utilized by the occupants, nor what postures the occupants will adopt.

11.2 NUMERICAL TECHNIQUES FOR STRUCTURAL ASSESSMENT OF VEHICLE ARCHITECTURES

Vehicle safety computation mostly utilizes explicit computational techniques, as almost all modes of impacts are of short duration (less than 200 ms), and involve inertia transfer, large rigid body rotation, and bifurcation. To date, this has been the preferred method for computing high-speed impact and low-speed damageability vehicle integrity as well as occupant and pedestrian injuries, due to the fact that injuries relate to a transient load phenomenon. In general, the only loadcase known to require implicit computation and using explicit computation is the FMVSS216 roof crush planum crush, due to contact instability and high level of nonlinearity caused by plasticity and large deformations. Reviewing the development of implicit computing algorithms, it has been observed that implicit solvers are becoming more resilient and it is envisaged that such solvers could be utilized for the numerical assessment of roof crush safety loadcases in the near future.

The explicit method uses an incremental process that computes the stress wave traveling through the structure and the occupant. The computation of the motion of this stress wave is very sensitive to the computer model setup, especially the mesh size, shape, and the material properties. The mesh size and shape will rule the "depth" that the stress wave travels through the structure at every single timestep. A small timestep will provide refined and more accurate stress wave propagation. However, because of the nature of explicit computation, there is a high computational cost to pay, as each computation step is mandatory and cannot be skipped, unlike implicit finite element analysis methods, which can converge more rapidly but are not currently suited for crash applications, as justified above. A coarse mesh discretization enabling a larger timestep will generate unrealistic contact forces; a fine mesh is always recommended for model validation. In most FE-based computer models, including optimization, convergence studies, structural deformation energy is a good parameter to use for convergence check. Failing to analyze a multitude of models observing the smoothness of the deformed shape can be a subjective and useful estimation of convergence. For a typical full vehicle crash test model utilizing steel or aluminum material properties, discretizing with a mesh size of 5 mm is a reasonable starting point.

11.3 OPTIMIZATION TECHNIQUES FOR NONLINEAR STRUCTURAL OPTIMIZATION

Throughout this book the reader has been introduced to a number of optimization techniques, from the very basic definitions of optimization through to highly complex case studies and examples of vehicle structures and occupant protection. The aim has been to develop the readers' understanding of the fundamental principles, such as line search algorithms and gradient-based methods, through more "practical" iterative methods, such as sequential linear programming (SLP), optimality conditions, robust optimization, and meta-modeling. The purpose of this was to provide the reader with fundamental insights into the complexity of optimization in general, with the added complexity of numerical analysis through FE. With this background information, the focus of attention turned to the specific challenges of structural optimization in connection with vehicle architectures. Although the majority of optimization principles and techniques presented and discussed in this context are generally applicable, there has been a bias toward topology optimization. The main reason for this is that topology optimization in general may be perceived to be the most challenging type of structural optimization due to the very large number of design variables generally considered; however, other optimization disciplines such as shape or size are not without their challenges. In this context, the most widespread method of topology optimization, the VDM-SIMP approach, was presented in greater detail. Following the practical examples relating to concept vehicle safety structures, the focus was turned to an adaptation of the VDM-SIMP approach intended specifically for nonlinear topology optimization. Although the emphasis in this part of the book was on ESLM, the underlying aspiration was to highlight some of the general challenges and difficulties associated with nonlinear topology optimization through the use of a specific method and a relevant structural problem. The conclusions of this part of the book question the feasibility of this particular approach to nonlinear optimization, given the inherent linear static nature upon which the VDM-SIMP approach is based.

Following these chapters the emphasis was again altered, this time with the ambition of broadening the readers' horizon of the types of algorithms currently available, and also of commenting on what may be expected from an optimization algorithm. How much should the future of engineering relies upon optimization? In this context, a wide variety of optimization algorithms and principles from a number of disciplines were presented, compared, and evaluated in the context of nonlinear structural optimization. Following this, a list of "requirements" for a nonlinear topology optimization algorithm was proposed and an algorithm denoted as BEETS was subsequently defined, programmed, and evaluated for future potential. The aim of this specific part of the book was not necessarily to revolutionize the world of optimization but to inspire readers and engineers to contribute to the future of nonlinear optimization by asking questions, pointing out general areas of concern, demonstrating techniques, providing tools and examples, and certainly not least of all to stimulate ideas.

11.4 APPLICATION OF NONLINEAR OPTIMIZATION

Optimizing vehicle structures is a complex endeavor due to the delicate balance between legislative requirements: "nice to haves" and brand "DNA". Everything starts with the styling envelope of the vehicle at the beginning of a project and from this design envelope the cabin packaging requirements for the occupant, as well as the vehicle architecture, needs to be defined. The vehicle designer can be "conventional" and utilize existing vehicle architectures or optimize the loadpaths of the forces traveling through the vehicle safety structure by using linear static topology optimization. When used in the concept stages of the vehicle design, topology optimization can be a powerful tool that can propose the path of the forces traveling through the structure, as well as negate the use of other redundant structures. It has been observed that the use of inertia relief was favorable compared with single point constraints to prevent artificial structural definitions that would be skewed because of unrealistic boundary conditions. Using linear topology optimization can be very useful; this method does, however, have one major limitation, as it is not capable of accurately computing a loadpath for buckling loadcases due to the unstable bifurcation mode. Consequently, the area of the vehicle which would require the most caution is the frontal and rear crash structures, as they can collapse and buckle "flat." The B pillar and side structures, on the other hand, work mostly in bending; hence, the output suggested from the linear topology optimization can be exploited with some confidence. Once the path is computed, the next step is to convert the topology into a sheet panel design. This step is necessary, as the results of a topology optimization based on the vehicle design envelope in this case uses a 3D mesh, with no detailed sheet metal profile definitions required for manufacturing. It is then necessary to recompute the structure using arbitrary or detailed section shapes against the same loadcases to ensure that the deformations and the Eulerian linear buckling limits are not violated. All these steps that can be performed using a linear FE solver should provide a sound vehicle architecture base for subsequent detailed local optimization that cannot be extracted at the present stage, as plasticity and detailed buckling modes which have not yet been captured. As vehicle safety scenarios deal with time-dependent events (crash deceleration pulse), an explicit solver is required to compute and define such details as cross-sectional shapes, material gauges, and grades.

In order to optimize the final details of the vehicle, certain aspects thereof need to be parameterized; consequently, different methods can be utilized; for example, direct and indirect optimization methods. Due to the size of the computer models and the number of loadcases to be analyzed, the direct computational method is not preferential, as opposed to the optimization of the vehicle structure using an indirect method often based on meta-models. Meta-models are mathematical models that are able to replicate a complex FE model of the vehicle, yet remain adequately simple to be used for efficient and relatively fast computation. These models are often based on 3D response surfaces built from sample points of the original computer model, with simple or complex curvatures depending on the degree of interpolation required and the scatter of the responses. These surface models endeavor to match all the finite element model responses; however, there is often a residual that represents a difference between the responses from the sample points to the surface fit. A residual represents a level error; consequently, optimizations using response surfaces generated with high residuals will most likely not be able to determine the actual optimum solution. Subsequently, it is important to carefully and considerably choose the sampling and the response surface methods, as both will greatly influence the outcome of the optimization.

Despite meta-models being fast to compute, the more that are used in a single optimization, the more complex and lengthy the optimization becomes. It is, therefore, necessary to extract the key variables in the vehicle structures that are most influential in the context of each loadcase response. Methods such as ANOVA or global sensitivity plots are very useful to understand the effects of each variable upon the vehicle architecture performance. Such tools can then either be used to redesign the vehicle directly by using engineering know-how, or alternatively perform an even more refined optimization by only selecting specific key variables. Structural shape changes by means of morphing dimensions and shape feature variables are also possible. Morphing shapes can, however, be a significant challenge to set up in a computer model. The calculation following this is, however, not different from computing standard gauge changes; hence, shape changes have not been specifically addressed in this book.

11.5 THE FUTURE OF STRUCTURAL OPTIMIZATION IN ENGINEERING

In a world where high-performance computation (HPC) is a matter of course, it is "natural" to develop increasingly complex optimization tools providing the analyst with more options but also generating a tremendous amount of data relative to the amount of data available at the turn of the millennium, the early dawn of cluster computing. State-of-the-art HPC computing usually generates such large quantities of data that a new level of mindfulness is necessary to capture the essence of the vehicle structure, ensuring that the engineering team retains control enabling logical, feasible, and efficient decisions to be made. All analysts/engineers need to acquire or adapt to a different type of mindset when utilizing state-of-the-art computing and optimization techniques that should include an even greater awareness of the requirements for engineering expertise, know-how, and general "common sense". Leaving the optimization routines, algorithms, and programs to fully determine all aspects of vehicle safety design is not a responsible course of action, unless future optimization procedures possess a human engineering mind as part of its routines to enable the implementation of logical engineering decisions in the process, which is highly unlikely to be imminent. It should also be noted that in the future structural optimization may not necessarily be FE based; advances in mesh-less methods may, for example, lead to drastic changes in the future of engineering and consequently structural optimization as for example through geometric shape optimization.

Optimization is undoubtedly a very powerful tool in the right hands and can take a design from concept to near completion faster than has ever been possible. It is difficult to envisage the future engineering of a vehicle, including all associated safety requirements, without the help of HPC computation and optimization tools. Assuming that it is theoretically possible to optimize every aspect to its full extent, it is necessary to understand the computational restrictions and the level of certainty of the results obtained. Most optimizations are based on computer models that have some clear limitations, as they are mathematically ruled by the timestep, the element size, the material model, the failure characterization, and the strain rate effects (to name just a few). In a vehicle safety simulation, one also has to include the manufacturing effects, such as thinning, plastic strains, tolerances, effects of tool wear and cost as part of a manufacturing phase. As there is an inherent physical variability in almost every aspect of engineering, such as the magnitude of Young's modulus for each material batch, and even location within a given component, it is difficult to rely on a truly exact optimization solution; robust optimization must be carefully understood.

Despite these limitations, optimization techniques remain valuable tools and it is envisaged that their use at all levels of the vehicle design including the concept phase will drastically increase in the near future.

There has never been a better or more appropriate time to ensure that analysts utilizing optimization techniques have acquired the necessary engineering principles to guide the design of the next generation of vehicles. The future of excellence in vehicle design will certainly rely on the use of HPC and optimization but will also continue to rely on the engineering expertise of the analyst, as well as the capabilities and synergies between all aspects of the design, engineering, and manufacturing disciplines.

11.6 ADDITIONAL ASPECTS OF OPTIMIZATION

Throughout this book the emphasis has very much been aimed at addressing the "classical" mechanical and automotive engineering challenges associated with the development and refinement of vehicle safety structures, including aspects such as crash absorption energy. These have been observed and addressed in isolation; there are, however, many other aspects that are becoming increasingly topical, and therefore increasingly interesting to consider from an optimization viewpoint. These include:

- **Multiobjective optimization.** The primary challenge of almost any component relating to structural engineering of, for example, aircraft and automotive vehicles is to balance component mass with structural performance; the ideal scenario is low mass combined with high structural performance. For example, the primary front crash structure of a vehicle should be as light as possible, but also be as energy-absorbent as at all possible during an impact. These two "desirable quantities" generally counteract each other and increase the mass to obtain increased energy absorption. This can, of course, be done by selecting one quantity as the objective and the other as a constraint. This will most likely not lead to a truly optimum structure, as one quantity in effect becomes passive. Instead, multiobjective optimization should be employed in order to obtain the truly optimum structure by actively considering both objectives. This is, however, not a simple task; there are plenty areas of research and development within the field of multiobjective optimization ripe for exploration.

- **Multidiscipline and multiphysics optimization.** This is perhaps one of the most topical areas of optimization which generally goes hand in hand with multiobjective optimization. The introduction of multiphysics and/or multidiscipline optimization enables a much more cohesive and "compact" design evolution simply because multiple aspects such as aerodynamic performance, structural integrity, noise vibration and harshness (NVH), fatigue, and durability may all be defined, developed, evaluated, and optimized "simultaneously" facilitating a more "harmonized" design cycle. However, a focus of expansive research and development of modern-day multidiscipline/multiphysics/multiobjective optimization is still in its infancy.
- **Holistic optimization.** With advances in multidiscipline/multiphysics/multiobjective optimization, the prospect of holistic optimization becomes increasingly enticing. All aspects of engineering are under continued scrutiny and there is "pressure" to reduce carbon foot-prints and introduce green initiatives with an overall view of reducing pollution and preserving natural resources. Holistic optimization principles offer an interesting approach to successfully meeting these environment goals. What if an optimization tool could not only solve the structural challenges, but also find the most "efficient" method of manufacturing, transport, assembly and disassembly, for example, from a CO_2-equivalent emissions viewpoint? Holistic optimization tools could be developed to cater for this specific purpose.
- **Nonisotropic optimization.** One of the potential keys to partly solving the above-mentioned mass vs. performance issue is the introduction of nonisotropic materials, such as composites, that are becoming increasingly widespread in their use within many different fields of engineering, including automotive and aerospace. The introduction of nonisotropic materials does, however, significantly add to the complexity of any optimization task, simply because there is additional data to be considered, including a potentially large increase in design variables. This is particularly evident in the context of topology optimization where many different strategies could be taken in terms of where, when, and how the introduction of a nonisotropic material should be considered in the design process. It is not impossible to "extend" current optimization technologies such as the VDM-SIMP approach to nonisotropic materials, but is it an efficient and robust method?
- **Nonoptimization approach.** The term optimization has almost become synonymous with advanced algorithms and HPC computing in engineering circles; however, there are other avenues to explore. Although this book has utilized computing to demonstrate examples, etc., it is not the only way to

perform optimization. Before the engineering team decides to devote numerous hours and resources to advanced optimization iterations, the wider range of options for meeting set targets should be considered. In automotive engineering, the most obvious example is that of the primary crash structures. Although optimization techniques are under rapid development to cater for large nonlinear deformations, nonisotropic material behavior, etc., is it really the "ideal" method of reducing the mass of the vehicle? Surely, the "optimum" mass of the crash structure is 0 kg? This is, of course, not achievable through structural optimization, but is it achievable through other means? If a specific type of accident, for example, frontal impact, could be eliminated as a potential crash scenario (statistically speaking), it would enable the front crash structure to be completely removed. This may be possible, for example, through driverless cars, connected vehicles that "talk to each other," and/or through active safety measures such as in car radar technology or automatic breaking. Although these specific solutions may be a (not too distant) vision of the future, they demonstrate a point of considering optimization in a wider context.

REFERENCES

Ashmore, R. (2015). Principal engineer, Vehicle Safety, Jaguar LandRover. Private discussion during CPD module M37MAE "Vehicle Safety", Held at Coventry University on June 1, 2015.

Bastien, C. (2014). The prediction of kinematics and injury criteria of unbelted occupants under autonomous emergency braking. PhD thesis, Coventry University, October 2014.

Khan, R. (2015). Current PhD thesis. Supervisors: C. Bastien, S. Kanarachos. Preliminary title: "Effect of AEB on future vehicle architectures", TARF (Towards the Affordable Recyclable Future – Low Carbon Vehicle), EPSRC funded project.

SARTRES. (2011). Safe road trains for the environment. European Project. Available from <www.sartre-project.eu/en/Sidor/default.aspx> Accessed 1.07.13.

SARTRES. (2012). Platooned traffic can be integrated with other road users on conventional highways. Article Available from <www.sartre-project.eu/en/press/Documents/SARTRE%20final%20partner%20release.pdf> Accessed 1,07.13.

Wen, L., Bastien, C. (2015). Stability and sensitivity of THUMS pedestrian model and its trauma response to a real life accident. In: LS-Dyna European conference, Held in, June 15–17, 2015, Wurzburg, Germany.

Index

A

B

F

L

M

N

O

S

T

U

V

W

X

Y

Kolibri 360 GmbH
Ferdinand-Jühlke-Straße 7
99095 Erfurt, Deutschland
produktsicherheit@kolibri360.de